Problems and Solutions in
Mathematical Olympiad

High School 2

Other Related Titles from World Scientific

Problems and Solutions in Mathematical Olympiad
Secondary 3
by Jun Ge
translated by Huan-Xin Xie
ISBN: 978-981-122-982-4
ISBN: 978-981-123-141-4 (pbk)

Problems and Solutions in Mathematical Olympiad
High School 1
by Bin Xiong and Zhi-Gang Feng
translated by Tian-You Zhou
ISBN: 978-981-122-985-5
ISBN: 978-981-123-142-1 (pbk)

Problems and Solutions in Mathematical Olympiad
High School 3
by Hong-Bing Yu
translated by Fang-Fang Lang and Yi-Chao Ye
ISBN: 978-981-122-991-6
ISBN: 978-981-123-144-5 (pbk)

Problems and Solutions in
Mathematical Olympiad

High School 2

Editors-in-Chief

Zun Shan *Nanjing Normal University, China*

Bin Xiong *East China Normal University, China*

Original Author

Shi-Xiong Liu *Zhongshan Affiliated School of South China Normal University, China*

English Translator

Jiu Ding *School of Mathematics and Natural Sciences, University of Southern Mississippi, USA*

Copy Editors

Ming Ni *East China Normal University Press, China*

Ling-Zhi Kong *East China Normal University Press, China*

Lei Rui *East China Normal University Press, China*

East China Normal University Press

World Scientific

Published by

East China Normal University Press
3663 North Zhongshan Road
Shanghai 200062
China

and

World Scientific Publishing Co. Pte. Ltd.
5 Toh Tuck Link, Singapore 596224
USA office: 27 Warren Street, Suite 401-402, Hackensack, NJ 07601
UK office: 57 Shelton Street, Covent Garden, London WC2H 9HE

British Library Cataloguing-in-Publication Data
A catalogue record for this book is available from the British Library.

PROBLEMS AND SOLUTIONS IN MATHEMATICAL OLYMPIAD
High School 2

ISBN 978-981-122-988-6 (hardcover)
ISBN 978-981-123-143-8 (paperback)
ISBN 978-981-122-989-3 (ebook for institutions)
ISBN 978-981-122-990-9 (ebook for individuals)

For any available supplementary material, please visit
https://www.worldscientific.com/worldscibooks/10.1142/12088#t=suppl

Desk Editor: Tan Rok Ting

Typeset by Stallion Press
Email: enquiries@stallionpress.com

Printed in Singapore

Editorial Board

Board Members

Preface

It is said that in many countries, especially the United States, children are afraid of mathematics and regard mathematics as an "unpopular subject." But in China, the situation is very different. Many children love mathematics, and their math scores are also very good. Indeed, mathematics is a subject that the Chinese are good at. If you see a few Chinese students in elementary and middle schools in the United States, then the top few in the class of mathematics are none other than them.

At the early stage of counting numbers, Chinese children already show their advantages.

Chinese people can express integers from 1 to 10 with one hand, whereas those in other countries would have to use two.

The Chinese have long had the concept of digits, and they use the most convenient decimal system (many countries still have the remnants of base 12 and base 60 systems).

Chinese characters are all single syllables, which are easy to recite. For example, the multiplication table can be quickly mastered by students, and even the "stupid" people know the concept of "three times seven equals twenty one." But for foreigners, as soon as they study multiplication, their heads get bigger. Believe it or not, you could try and memorise the multiplication table in English and then recite it, it is actually much harder to do so in English.

It takes the Chinese one or two minutes to memorize $\pi = 3.14159\cdots$ to the fifth decimal place. However, in order to recite these digits, the Russians wrote a poem. The first sentence contains three words and the second sentence contains one $\cdots$ To recite π, recite poetry first. In our opinion, this just simply asks for trouble, but they treat it as a magical way of memorization.

Application problems for the four arithmetic operations and their arithmetic solutions are also a major feature of Chinese mathematics. Since ancient times, the Chinese have compiled a lot of application questions, which has contact or close relations with reality and daily life. Their solutions are simple and elegant as well as smart and diverse, which helps increase students' interest in learning and enlighten students'. For example:

"There are one hundred monks and one hundred buns. One big monk eats three buns and three little monks eat one bun. How many big monks and how many little monks are there?"

Most foreigners can only solve equations, but Chinese have a variety of arithmetic solutions. As an example, one can turn each big monk into 9 little monks, and 100 buns indicate that there are 300 little monks, which contain 200 added little monks. As each big monk becomes a little monk 8 more little monks are created, so $200/8 = 25$ is the number of big monks, and naturally there are 75 little monks. Another way to solve the problem is to group a big monk and three little monks together, and so each person eats a bun on average, which is exactly equal to the overall average. Thus the big monks and the little monks are not more and less after being organized this way, that is, the number of the big monks is $100/(3 + 1) = 25$.

The Chinese are good at calculating, especially good at mental arithmetic. In ancient times, some people used their fingers to calculate (the so-called "counting by pinching fingers"). At the same time, China has long had computing devices such as counting chips and abaci. The latter can be said to be the prototype of computers.

In the introductory stage of mathematics – the study of arithmetic, our country has obvious advantages, so mathematics is often the subject that our smart children love.

Geometric reasoning was not well-developed in ancient China (but there were many books on the calculation of geometric figures in our country), and it was slightly inferior to the Greeks. However, the Chinese are good at learning from others. At present, the geometric level of middle school students in our country is far ahead of the rest of the world. Once a foreign education delegation came to a junior high school class in our country. They thought that the geometric content taught was too in-depth for students to comprehend, but after attending the class, they had to admit that the content was not only understood by Chinese students, but also well mastered.

The achievements of mathematics education in our country are remarkable. In international mathematics competitions, Chinese contestants have won numerous medals, which is the most powerful proof. Ever since our country officially sent a team to participate in the International Mathematical Olympiad in 1986, the Chinese team has won 14 team championships, which can be described as very impressive. Professor Shiing-Shen Chern, a famous contemporary mathematician, once admired this in particular. He said, "One thing to celebrate this year is that China won the first place in the international math competition $\cdots$ Last year it was also the first place." (Shiing-Shen Chern's speech, *How to Build China into a Mathematical Power*, at Cheng Kung University in Taiwan in October 1990)

Professor Chern also predicted: "China will become a mathematical power in the 21st century."

It is certainly not an easy task to become a mathematical power. It cannot be achieved overnight. It requires unremitting efforts. The purpose of this series of books is: (1) To further popularize the knowledge of mathematics, to make mathematics be loved by more young people, and to help them achieve good results; (2) To enable students who love mathematics to get better development and learn more knowledge and methods through the series of books.

"The important things in the world must be done in detail." We hope and believe that the publication of this series of books will play a role in making our country a mathematical power. This series was first published in 2000. According to the requirements of the curriculum reform, each volume is revised to different degrees.

Well-known mathematician, academician of the Chinese Academy of Sciences, and former chairman of the Chinese Mathematical Olympiad Professor Yuan Wang, served as a consultant to this series of books and wrote inscriptions for young math enthusiasts. We express our heartfelt thanks. We would also like to thank East China Normal University Press, and in particular Mr. Ming Ni and Mr. Lingzhi Kong. Without them, this series of books would not have been possible.

Zun Shan and Bin Xiong
May 2018

Contents

Chapter 1

Maximum and Minimum Values

1. Key Points of Knowledge and Basic Methods

A. *Use basic inequalities to find maximum (minimum) values*

(1) Use mean value inequalities to find maximum (minimum) values.

By using mean value inequalities, one can conveniently find the minimum value of the sum of two positive numbers and the maximum value of the product of two positive numbers, under some constraint. Let $x, y \in \mathbf{R}_+$, $S(x, y) = x + y$, and $P(x, y) = xy$.

If $P(x, y)$ has a fixed value P, that is $xy = P$, then

$$S(x, y)_{\min} = 2\sqrt{P}$$

obtained when $x = y$; if $S(x, y)$ has a fixed value S, that is $x + y = S$, then

$$P(x, y)_{\max} = \frac{1}{4}S^2$$

obtained when $x = y$.

The conclusion is completely similar in the case of three, four, and even n positive numbers in general.

(2) Use the Cauchy inequality and other inequalities to find maximum (minimum) values.

Use the Cauchy inequality and other inequalities to lead to a constant (the maximum or minimum value), and then make sure to find at least a set of values for the variables so that the value of the objective function equals the constant.

B. *Use monotonicity of functions to find maximum (minimum) values*

If $f(x)$ is an increasing function on $[a, b]$, then $f(x)_{\min} = f(a)$ and $f(x)_{\max} = f(b)$.

If $f(x)$ is a decreasing function on $[a, b]$, then $f(x)_{\min} = f(b)$ and $f(x)_{\max} = f(a)$.

2. Illustrative Examples

Example 1. Suppose that three positive real numbers a, b, and c satisfy $a \geq 3$ and $ab + bc + ca = 16$. Find the minimum value of $2a + b + c$.

Solution. From the given condition,

$$(a + b)(a + c) = a^2 + ab + bc + ca$$
$$= a^2 + 16$$
$$\geq 3^2 + 16 = 25,$$

so

$$2a + b + c = (a + b) + (a + c)$$
$$\geq 2\sqrt{(a + b)(a + c)}$$
$$\geq 2\sqrt{25} = 10,$$

in which the equality holds when $a = 3, a + b = a + c$, and $ab + bc + ca = 16$, that is $a = 3$ and $b = c = 2$. Thus, the minimum value of $2a + b + c$ is 10.   □

Remark. To find a maximum (minimum) value, one needs to use properties and theorems to give a series of deformations and "same direction inequalities" for the given conditions and the objective function, till a constant is found. The key point is that all the equalities must be satisfied at some moment.

Example 2. Suppose that three real numbers a, b, and c satisfy $2^a + 4^b = 2^c$ and $4^a + 2^b = 4^c$. Find the minimum value of c.

Analysis. We may find the minimum value of 2^c first. Note the relationship among the bases of the exponential expressions. Let 2^a, 4^b, and 2^c be x, y, and z, respectively. Then the problem for the exponentials can be reduced to that for the resulting algebraic expressions.

Solution. Denote $2^a, 4^b$, and 2^c by x, y, and z, respectively. Then $x, y, z > 0$.

From the given conditions, $x + y^2 = z$ and $x^2 + y = z^2$, so

$$z^2 - y = x^2 = (z - y^2)^2 = z^2 - 2y^2z + y^4.$$

Hence, by using the arithmetic mean-geometric mean inequality, we have

$$z = \frac{y^4 + y}{2y^2} = \frac{1}{4}\left(2y^2 + \frac{1}{y} + \frac{1}{y}\right)$$

$$\geq \frac{1}{4} \cdot 3\sqrt[3]{2y^2 \cdot \frac{1}{y} \cdot \frac{1}{y}} = \frac{3}{4}\sqrt[3]{2}.$$

When $2y^2 = \frac{1}{y}$, that is $y = \frac{1}{\sqrt[3]{2}}$, the minimum value of z is $\frac{3}{4}\sqrt[3]{2}$ (with the corresponding value $\frac{\sqrt[3]{2}}{4}$ for x, satisfying the requirement).

Since $c = \log_2 z$, the minimum value of c is $\log_2 \frac{3}{4}\sqrt[3]{2} = \log_2 3 - \frac{5}{3}$. □

Example 3. Suppose that u, v, and w are positive real numbers satisfying the condition

$$u\sqrt{vw} + v\sqrt{wu} + w\sqrt{uv} \geq 1.$$

Find the minimum value of $u + v + w$.

Analysis. The involved function and the expression of the given condition have a big "difference," so we need to figure out how to make them close enough to each other. Using the arithmetic mean-geometric mean inequality can change the condition to $uv + vw + wu \geq 1$, which is a quadratic inequality. We may consider squaring $u + v + w$ to use the condition.

Solution. By using the arithmetic mean-geometric mean inequality and the given condition, we have

$$u\frac{v + w}{2} + v\frac{w + u}{2} + w\frac{u + v}{2} \geq u\sqrt{vw} + v\sqrt{wu} + w\sqrt{uv} \geq 1,$$

thus $uv + vw + wu \geq 1$. Therefore,

$$(u + v + w)^2 = u^2 + v^2 + w^2 + 2uv + 2vw + 2wu$$

$$= \frac{u^2 + v^2}{2} + \frac{v^2 + w^2}{2} + \frac{w^2 + u^2}{2} + 2uv + 2vw + 2wu$$

$$\geq 3uv + 3vw + 3wu \geq 3,$$

that is, $u + v + w \geq \sqrt{3}$. On the other hand, choosing $u = v = w = \frac{\sqrt{3}}{3}$ satisfies the condition and the value of $u + v + w$ is $\sqrt{3}$.

To summarize, the minimum value of $u + v + w$ is $\sqrt{3}$.   □

Remark. From the above three examples, we can see that when looking for a maximum (minimum) value, we should be very careful in determining the condition that makes the equality valid.

Example 4. Suppose that a, b, and c are nonnegative real numbers. Find the minimum value of

$$\sqrt{\frac{a}{b+c}} + \sqrt{\frac{b}{c+a}} + \sqrt{\frac{c}{a+b}}.$$

Analysis. It is difficult to apply a basic inequality directly, so we deform the expression:

$$\sqrt{\frac{a}{b+c}} + \sqrt{\frac{b}{c+a}} + \sqrt{\frac{c}{a+b}} = \frac{a}{\sqrt{a(b+c)}} + \frac{b}{\sqrt{b(c+a)}} + \frac{c}{\sqrt{c(a+b)}}.$$

Then by applying the arithmetic mean-geometric mean inequality to the denominator of each term of the right hand side expression in the above equality, we obtain that

$$\sqrt{\frac{a}{b+c}} + \sqrt{\frac{b}{c+a}} + \sqrt{\frac{c}{a+b}} \geq 2.$$

But a careful checking finds that the equality does not hold in the above inequality. The reason is that the premise of the previous deformation is that a, b, and c are all nonzero, but exactly one of them is allowed to be zero in the original expression! Obviously, there is at most one of a, b, and c to be zero. If one of them is zero, then without loss of generality, we may assume that $c = 0$. Now, the original expression $= \sqrt{\frac{a}{b}} + \sqrt{\frac{b}{a}} \geq 2$, in which the equality holds if and only if $a = b \neq 0$. This indicates that 2 is indeed the minimum value of the original expression.

Solution. It is obvious that there is at most one zero among a, b, and c. Suppose that there is one zero among a, b, and c. Without loss of generality, let $c = 0$. Then

$$\text{the original expression} = \sqrt{\frac{a}{b}} + \sqrt{\frac{b}{a}} \geq 2,$$

in which the equality holds if and only if $a = b \neq 0$.

If a, b, and c are all nonzero, then the original expression

$$= \frac{a}{\sqrt{a(b+c)}} + \frac{b}{\sqrt{b(c+a)}} + \frac{c}{\sqrt{c(a+b)}}$$

$$\geq \frac{2a}{a+(b+c)} + \frac{2b}{b+(c+a)} + \frac{2c}{c+(a+b)} = 2.$$

It is easy to see that the equality can never be satisfied in the above inequality.

In summary, when $a = 0$ and $b = c \neq 0$, $b = 0$ and $c = a \neq 0$, or $c = 0$ and $a = b \neq 0$, the original expression achieves its minimum value 2. □

Example 5. Find the minimum value of

$$\frac{1}{\sin^2 A} + \frac{1}{\sin^2 B} + \frac{4}{1+\sin C}$$

among all $\triangle ABC$.

Solution. From the arithmetic mean-geometric mean inequality and some trigonometric identity,

$$\frac{1}{\sin^2 A} + \frac{1}{\sin^2 B} + \frac{4}{1+\sin C} \geq \frac{2}{\sin A \sin B} + \frac{4}{1+\sin C}$$

$$= \frac{4}{\cos(A-B) - \cos(A+B)} + \frac{4}{1+\sin C}$$

$$\geq \frac{4}{1+\cos C} + \frac{4}{1+\sin C},$$

and the equality is satisfied if and only if $A = B$.

By the Cauchy inequality,

$$\frac{1}{1+\cos C} + \frac{1}{1+\sin C} \geq \frac{4}{(1+\cos C)+(1+\sin C)}$$

$$= \frac{4}{2+(\cos C + \sin C)}$$

$$\geq \frac{4}{2+\sqrt{(1^2+1^2)(\cos^2 C + \sin^2 C)}}$$

$$= 2(2-\sqrt{2}).$$

The equality in the above holds if and only if $\cos C = \sin C$, that is, $C = \frac{\pi}{4}$.

Summarizing the above, it is easy to see that $\frac{1}{\sin^2 A} + \frac{1}{\sin^2 B} + \frac{4}{1+\sin C}$ has the minimum value $8(2-\sqrt{2})$ when $A = B$ and $C = \frac{\pi}{4}$. $\square$

Some maximum (minimum) value problems do not need much knowledge of inequalities for solutions, but one needs to give an appropriate deformation or deep analysis and discussion to the objective function or the given condition.

Example 6. Suppose that x_1, x_2, and x_3 are nonnegative real numbers satisfying $x_1 + x_2 + x_3 = 1$. Find the minimum value and the maximum value of

$$(x_1 + 3x_2 + 5x_3)\left(x_1 + \frac{x_2}{3} + \frac{x_3}{5}\right).$$

Solution. By the given condition and the Cauchy inequality,

$$(x_1 + 3x_2 + 5x_3)\left(x_1 + \frac{x_2}{3} + \frac{x_3}{5}\right)$$

$$\geq \left(\sqrt{x_1}\sqrt{x_1} + \sqrt{3x_2}\sqrt{\frac{x_2}{3}} + \sqrt{5x_3}\sqrt{\frac{x_3}{5}}\right)^2$$

$$= (x_1 + x_2 + x_3)^2 = 1.$$

The equality is satisfied when $x_1 = 1, x_2 = 0$, and $x_3 = 0$, so the required minimum value is 1.

Now, we have

$$(x_1 + 3x_2 + 5x_3)\left(x_1 + \frac{x_2}{3} + \frac{x_3}{5}\right)$$

$$= \frac{1}{5}(x_1 + 3x_2 + 5x_3)\left(5x_1 + \frac{5x_2}{3} + x_3\right)$$

$$\leq \frac{1}{5} \cdot \frac{1}{4}\left[(x_1 + 3x_2 + 5x_3) + \left(5x_1 + \frac{5x_2}{3} + x_3\right)\right]^2$$

$$= \frac{1}{20}\left(6x_1 + \frac{14x_2}{3} + 6x_3\right)^2$$

$$\leq \frac{1}{20}(6x_1 + 6x_2 + 6x_3)^2 = \frac{9}{5}$$

with the equality being valid when $x_1 = \frac{1}{2}, x_2 = 0$, and $x_3 = \frac{1}{2}$, so the sought maximum value is $\frac{9}{5}$. $\square$

Example 7. Suppose that $|x_i| < 1$ for $i = 1, 2, \ldots, n$ and

$$|x_1| + |x_2| + \cdots + |x_n| = 19 + |x_1 + x_2 + \cdots + x_n|.$$

Find the minimum value of the positive integers n.

Solution. By the given condition,

$$19 = \sum_{i=1}^{n} |x_i| - \left|\sum_{i=1}^{n} x_i\right| \leq \sum_{i=1}^{n} |x_i| < n.$$

So $n \geq 20$.

On the other hand, when $n = 20$, we choose

$$x_i = \begin{cases} \dfrac{19}{20}, & 1 \leq i \leq 10, \\[2mm] -\dfrac{19}{20}, & 11 \leq i \leq 20. \end{cases}$$

Then $x_1, x_2, \ldots, x_{20}$ satisfy all the conditions of the problem.

To summarize, the desired minimum positive integer n is 20. $\square$

Remark. There are numerous groups of the numbers $\{x_1, x_2, \ldots, x_{20}\}$ that satisfy the given condition, but it is enough for us to find one group.

Finally, let us look at an example that requires us to give some deep discussion.

Example 8. Suppose that real numbers $x, y,$ and z satisfy $x^2 + y^2 + z^2 = 1$. Find the maximum value of

$$(x^2 - yz)(y^2 - zx)(z^2 - xy).$$

Solution. Let $f(x, y, z) = (x^2 - yz)(y^2 - zx)(z^2 - xy)$. From $f(x, 0, z) = -x^3 z^3$ and the assumption, the maximum value of $f(x, y, z)$ must be positive.

Since $f(x, y, z)$ is symmetric with respect to $x, y,$ and z, and since

$$f(x, y, z) = f(-x, -y, -z),$$

we may assume that $x \geq y \geq z$ and $x + y + z \geq 0$. Then $x^2 - yz \geq 0$.

If $f(x, y, z)$ achieves the maximum value at (x, y, z), then

$$f(x, y, z) - f(x, -y, -z) = -2x(x^2 - yz)(y^3 + z^3) \geq 0,$$

so $y^3 + z^3 \leq 0$.

If $z = 0$, then $y = 0$, and so $f(x, y, z) = 0$. Hence, it must be that $z < 0$, thus $y^2 - zx > 0$.

Since $f(x, y, z) > 0$, we see that $z^2 - xy > 0$. By the arithmetic mean-geometric mean inequality,

$$f(x, y, z) \leq \left[\frac{(x^2 - yz) + (y^2 - zx) + (z^2 - xy)}{3} \right]^3$$

$$= \left[\frac{\frac{3}{2} - \frac{1}{2}(x + y + z)^2}{3} \right]^3 \leq \frac{1}{8}.$$

The equality is valid in the above when $x^2 - yz = y^2 - zx = z^2 - xy > 0$ and $x + y + z = 1$. In particular, $f(x, y, z) = \frac{1}{8}$ when $x = \frac{1}{2}, y = \frac{\sqrt{5}-1}{4}$, and $z = -\frac{\sqrt{5}+1}{4}$.

Therefore, the maximum value of $f(x, y, z)$ is $\frac{1}{8}$.   □

3. Exercises

Group A

I. Filling Problems

1. If $\sqrt{x} + \sqrt{y} \le a\sqrt{x+y}$ for all positive numbers x and y, then the minimum value for a is _____.

2. Assume that x, y, and z are positive numbers that satisfy $xyz(x+y+z) = 1$. Then the minimum value of $(x+y)(x+z)$ is _____.

3. If $a > 0, b > 0$, and $a+b = 1$, then the minimum value of $\left(a + \frac{1}{a}\right)\left(b + \frac{1}{b}\right)$ is _____.

4. Four real numbers x, y, z, and w satisfy $x+y+z+w = 1$. Then the maximum value of $M = xw+2yw+3xy+3zw+4xz+5yz$ is _____.

5. Suppose that x and y are real numbers. The maximum value of z satisfying $x + y + z = 5$ and $xy + yz + zx = 3$ is _____.

6. Given $x, y \in \mathbf{R}_+$ that satisfy $x - 3\sqrt{x+1} = 3\sqrt{y+2} - y$. Then the maximum value of $x + y$ is _____.

II. Calculation Problems

7. Suppose that $\lg a < 0, \lg b < 0, \lg c < 0$, and $\lg(a + b + c) = 0$. Find the maximum value of $\lg(a^2 + b^2 + c^2 + 18abc)$.

8. Let $x, y \in \mathbf{R}_+$ and $S = \min\left\{x, y + \frac{1}{x}, \frac{1}{y}\right\}$. Find the maximum value of S.

9. Let real numbers a and b satisfy $a = x_1 + x_2 + x_3 = x_1 x_2 x_3$ and $ab = x_1 x_2 + x_2 x_3 + x_3 x_1$, where $x_1, x_2, x_3 > 0$. Find the maximum value of $p = \frac{a^2 + 6b + 1}{a^2 + a}$.

Group B

10. Given $x, y, z \in \mathbf{R}_+$. Let

$$P = \frac{x}{x+y} + \frac{y}{y+z} + \frac{z}{z+x},$$

$$Q = \frac{y}{x+y} + \frac{z}{y+z} + \frac{x}{z+x},$$

$$R = \frac{z}{x+y} + \frac{x}{y+z} + \frac{y}{z+x}.$$

Denote $f(x, y, z) = \max\{P, Q, R\}$. Find $f(x, y, z)_{\min}$.

11. Suppose that numbers $x_1, x_2, \ldots, x_{1991}$ satisfy the condition

$$|x_1 - x_2| + |x_2 - x_3| + \cdots + |x_{1990} - x_{1991}| = 1991.$$

Denote $y_k = \frac{1}{k}(x_1 + x_2 + \cdots + x_k)$ for $k = 1, 2, \ldots, 1991$. Find the maximum possible value of

$$|y_1 - y_2| + |y_2 - y_3| + \cdots + |y_{1990} - y_{1991}|.$$

12. Let $a, b,$ and c be positive real numbers satisfying $a + b + c \geq 1$. Find the maximum value of

$$\frac{a - bc}{a + bc} + \frac{b - ca}{b + ca} + \frac{c - ab}{c + ab}.$$

Chapter 2

Usual Methods for Proving Inequalities

1. Key Points of Knowledge and Basic Methods

To prove inequalities, one needs not only to apply basic properties of inequalities, theorems, and proven inequalities, but sometimes also to use methods of proving inequalities cleverly and flexibly. The most often used methods are described in the following sections.

A. *Method of comparison*

It is similar to the comparison of two real numbers from Chapter 8 in the first volume of the book.

B. *Synthetic method*

Starting from known facts (given conditions, properties of inequalities, theorems, proven inequalities), deduce the desired inequality step by step.

C. *Analytic method*

Starting from the inequality to be proved, look for a sufficient condition that makes the inequality valid. If this sufficient condition exists, then it can be asserted that the original inequality must be true.

D. *Method of stretching*

If it is difficult to prove an inequality, say $A > B$, one may consider looking for an intermediate quantity C as a passage by proving that $A > C$ and $C > B$. Then $A > B$ is satisfied by the transitivity of inequalities.

11

E.   *Method of contradiction*

When it is difficult to prove an inequality (such as $A > B$) directly, one may proceed by assuming the conclusion to be false (assuming $A \leq B$), and then lead to a contradiction. Then the original inequality is true via the indirect proof.

F.   *Method of mathematical induction*

We shall cover this part in later chapters.

2.   Illustrative Examples

We first look at an example that uses the synthetic method to prove an inequality.

Example 1. Suppose that three real numbers a, b, and c are all not less than -1 and satisfy $a^3 + b^3 + c^3 = 1$. Prove:

$$a + b + c + a^2 + b^2 + c^2 \leq 4,$$

and give the condition for the equality to be true.

Analysis. Note that $4 = 1 + a^3 + 1 + b^3 + 1 + c^3$. It is enough to prove

$$a + a^2 \leq 1 + a^3, \ b + b^2 \leq 1 + b^3, \ c + c^2 \leq 1 + c^3.$$

Proof. Note that when $x \geq -1$,

$$1 - x - x^2 + x^3 = (1 - x)^2(1 + x) \geq 0, \tag{2.1}$$

and the equality in (2.1) holds if and only if $x = \pm 1$.

From (2.1), we know that $x + x^2 \leq 1 + x^3$ when $x \geq -1$.

Substituting $x = a, b$, and c into the above inequality in succession and adding the resulting expressions, we have

$$a + a^2 + b + b^2 + c + c^2 \leq 1 + a^3 + 1 + b^3 + 1 + c^3 = 4.$$

The equality holds in the above inequality if and only if (a, b, c) equals $(1, 1, -1)$ and all rotations of $(1, 1, -1)$.   □

Example 2. Let a, b, and c be the three sides of a triangle and let $m > 0$. Prove:

$$\frac{a}{a+m} + \frac{b}{b+m} > \frac{c}{c+m}.$$

Proof. We use the analytic method. Since a, b, and c are all positive numbers,

$$\frac{a}{a+m} + \frac{b}{b+m} > \frac{c}{c+m} \tag{2.2}$$

$$\Leftarrow \frac{a(b+m) + b(a+m)}{(a+m)(b+m)} > \frac{c}{c+m}$$

$$\Leftarrow \frac{2ab + m(a+b)}{ab + m(a+b) + m^2} > \frac{c}{c+m}$$

$$\Leftarrow 1 + \frac{m^2 - ab}{2ab + m(a+b)} < 1 + \frac{m}{c}$$

$$\Leftarrow m^2 c - abc < 2abm + m^2(a+b)$$

$$\Leftarrow m^2(c - a - b) < ab(2m + c). \tag{2.3}$$

Since a, b, and c are the three sides of a triangle, $a + b > c$, so

$$m^2(c - a - b) < 0 < ab(2m + c),$$

that is, the inequality (2.3) is satisfied. Hence, the inequality (2.2) is valid. $\square$

Remark. In the above proof process, the implication notation "$\Leftarrow$" can be changed to "$\Leftrightarrow$", but in the proof process with the analytic method, it is not necessary to require "$\Rightarrow$" to be satisfied at any step.

Example 3. Assume that a, b, and c are any real numbers. Prove:

$$(a^2 + 2)(b^2 + 2)(c^2 + 2) \geq 9(ab + bc + ca).$$

Proof. Note that

$$(a^2 + 2)(b^2 + 2) \geq 3\left[\frac{(a+b)^2}{2} + 1\right]$$

$$\Leftrightarrow a^2 b^2 + 2(a^2 + b^2) + 4 \geq \frac{3a^2 + 3b^2}{2} + 3ab + 3$$

$$\Leftrightarrow a^2 b^2 + 1 + \frac{a^2 + b^2}{2} \geq 3ab. \tag{2.4}$$

By the arithmetic mean-geometric mean inequality, the above inequality is true.

Hence, from (2.4) and the Cauchy inequality,

$$(a^2 + 2)(b^2 + 2)(c^2 + 2) \geq 3 \left[\frac{(a+b)^2}{2} + 1 \right] (c^2 + 2)$$

$$\geq 3(a + b + c)^2 \geq 9(ab + bc + ca). \qquad \square$$

Example 4. Suppose that $a > b > 0$. Prove: $\sqrt{2}a^3 + \frac{3}{ab-b^2} \geq 10$.

Proof. Since $a > b > 0$,

$$0 < ab - b^2 = b(a - b) \leq \frac{[b + (a-b)]^2}{4} = \frac{a^2}{4}.$$

Hence,

$$\sqrt{2}a^3 + \frac{3}{ab - b^2} \geq \sqrt{2}a^3 + \frac{3}{\frac{a^2}{4}} \geq \sqrt{2}a^3 + \frac{12}{a^2}$$

$$= \frac{\sqrt{2}}{2}a^3 + \frac{\sqrt{2}}{2}a^3 + \frac{4}{a^2} + \frac{4}{a^2} + \frac{4}{a^2}$$

$$\geq 5\sqrt[5]{\frac{\sqrt{2}}{2}a^3 \cdot \frac{\sqrt{2}}{2}a^3 \cdot \frac{4}{a^2} \cdot \frac{4}{a^2} \cdot \frac{4}{a^2}}$$

$$= 10. \qquad \square$$

Remark. In the above proof, we used a deformation of the basic inequality: If $x, y \in \mathbf{R}_+$, then $xy \leq \frac{(x+y)^2}{4}$, which is a very useful result.

In addition, in order to eliminate a, we rewrote $\sqrt{2}a^3$ as the sum of two terms and $\frac{12}{a^2}$ as the sum of three terms, and consequently their product is a constant. This is a trick used frequently when solving a problem by applying mean value inequalities.

Example 5. Prove:

$$16 < \sum_{k=1}^{80} \frac{1}{\sqrt{k}} < 17.$$

Proof. We make the real number $\sqrt{k}$ "larger" and "smaller" respectively in the following way:

$$\sqrt{k-1} < \sqrt{k} < \sqrt{k+1},$$

which implies that

$$\sqrt{k} + \sqrt{k-1} < 2\sqrt{k} < \sqrt{k} + \sqrt{k+1} \ (k \in \mathbf{N}_+).$$

Thus,

$$\frac{1}{\sqrt{k} + \sqrt{k+1}} > \frac{1}{2\sqrt{k}} > \frac{1}{\sqrt{k-1} + \sqrt{k}},$$

namely

$$2(\sqrt{k+1} - \sqrt{k}) < \frac{1}{\sqrt{k}} < 2(\sqrt{k} - \sqrt{k-1}).$$

Therefore,

$$2(\sqrt{81} - 1) < \sum_{k=1}^{80} \frac{1}{\sqrt{k}} = 1 + \sum_{k=2}^{80} \frac{1}{\sqrt{k}} < 1 + 2(\sqrt{80} - 1).$$

Now, $1 + 2(\sqrt{80} - 1) < 1 + 2(\sqrt{81} - 1) = 17$, so

$$16 < \sum_{k=1}^{80} \frac{1}{\sqrt{k}} < 17.$$

□

Remark. The degree of changing an expression to a new one should be appropriate when proving inequalities with the method of stretching by means of "making it larger" or "making it smaller". For example, in the above equation, after obtaining $\frac{1}{\sqrt{k}} < 2(\sqrt{k} - \sqrt{k-1})$, if we had summed up the resulting 80 inequalities for $k = 1, 2, \ldots, 80$, then we would have had

$$\sum_{k=1}^{80} \frac{1}{\sqrt{k}} < 2\sqrt{80}.$$

But now $2\sqrt{80} > 17$, so the above strategy of "making it larger" is too much! Here, rewriting $\sum_{k=1}^{80} \frac{1}{\sqrt{k}}$ as $1 + \sum_{k=2}^{80} \frac{1}{\sqrt{k}}$ was a key step.

Example 6. Suppose that $a, b, c > 0$ and $a + b + c \geq abc$. Prove: Among following three inequalities

$$\frac{6}{a} + \frac{3}{b} + \frac{2}{c} \geq 2, \quad \frac{6}{b} + \frac{3}{c} + \frac{2}{a} \geq 2, \quad \frac{6}{c} + \frac{3}{a} + \frac{2}{b} \geq 2,$$

at least two are valid.

Analysis. Here, there are three inequalities, and "at least two" contains two cases: Exactly two inequalities are satisfied, or the three inequalities are all satisfied. But a direct proof is difficult, so we choose the method of contradiction. The negation of "two inequalities are satisfied" has also two cases: None of the three inequalities are satisfied, or exactly one inequality is valid.

Proof. Suppose that at most one inequality is satisfied. We divide the situation into two cases for the discussion:

(1) None of the three inequalities are satisfied, in other words

$$\frac{6}{a} + \frac{3}{b} + \frac{2}{c} < 2, \quad \frac{6}{b} + \frac{3}{c} + \frac{2}{a} < 2, \quad \frac{6}{c} + \frac{3}{a} + \frac{2}{b} < 2.$$

Adding the above inequalities leads to

$$11 \left(\frac{1}{a} + \frac{1}{b} + \frac{1}{c} \right) < 6,$$

that is,

$$\frac{1}{a} + \frac{1}{b} + \frac{1}{c} < \frac{6}{11}. \tag{2.5}$$

But from $a + b + c \geq abc$, we see that

$$\left(\frac{1}{a} + \frac{1}{b} + \frac{1}{c} \right)^2 \geq 3 \left(\frac{1}{ab} + \frac{1}{bc} + \frac{1}{ca} \right) = 3 \frac{a+b+c}{abc} \geq 3,$$

namely

$$\frac{1}{a} + \frac{1}{b} + \frac{1}{c} \geq \sqrt{3}. \tag{2.6}$$

The two inequalities (2.5) and (2.6) are contradictory.

(2) Exactly one of the three inequalities is satisfied. Without loss of generality, assume that

$$\frac{6}{a} + \frac{3}{b} + \frac{2}{c} < 2, \tag{2.7}$$

$$\frac{6}{b} + \frac{3}{c} + \frac{2}{a} < 2, \tag{2.8}$$

$$\frac{6}{c} + \frac{3}{a} + \frac{2}{b} \geq 2. \tag{2.9}$$

(2.7) plus (2.8) multiplied by 7 and then minus (2.9) lead to

$$14 > \frac{17}{a} + \frac{43}{b} + \frac{17}{c} > 17 \left(\frac{1}{a} + \frac{1}{b} + \frac{1}{c} \right),$$

that is,

$$\frac{1}{a} + \frac{1}{b} + \frac{1}{c} < \frac{14}{17}. \tag{2.10}$$

The inequality (2.10) is contradictory to (2.6).

In summary, the hypothesis is not true, and consequently the original proposition is true. □

Example 7. Let $x_1, x_2, \ldots, x_n$ be real numbers. Prove:

$$\left(\sum_{1 \le i < j \le n} |x_i - x_j| \right)^2 \ge (n-1) \sum_{1 \le i < j \le n} |x_i - x_j|^2.$$

Proof. Note that for every $1 \le i < j \le n$, there holds

$$|x_i - x_j| \le |x_i - x_k| + |x_k - x_j|.$$

Fixing i and j, and letting k be all of $1, 2, \ldots, n$ except for i and j, we have

$$(n-2)|x_i - x_j| \le \sum_{k=1}^n (|x_i - x_k| + |x_k - x_j|) - 2|x_i - x_j|$$

$$\le \sum_{1 \le k < l \le n} |x_k - x_l| - |x_i - x_j|.$$

Thus, $(n-1)|x_i - x_j|^2 \le \sum_{1 \le k < l \le n} |x_k - x_l||x_i - x_j|$.

Summing up the above inequalities for all $1 \le i < j \le n$ gives

$$\sum_{1 \le i < j \le n} (n-1)|x_i - x_j|^2 \le \left(\sum_{1 \le i < j \le n} |x_i - x_j| \right)^2.$$
□

In the following, we consider a more difficult problem.

Example 8. Suppose that $x, y, z \in \mathbf{R}_+$ and $x + y + z = 1$. Prove:

$$\frac{1}{1+x^2} + \frac{1}{1+y^2} + \frac{1}{1+z^2} \le \frac{27}{10}. \tag{2.11}$$

Analysis. Obviously, $0 < x, y, z < 1$. Note that the equality holds in the inequality (2.11) when $x = y = z = \frac{1}{3}$. In order to use the condition $x + y + z = 1$, we give a bold conjecture: There exist constants A and B such that the local inequality $\frac{1}{1+x^2} \le Ax + B$ is satisfied for $x \in (0, 1)$.

Adding up the three similar local inequalities seems to be able to solve the problem. From the fact that the inequality $\frac{1}{1+x^2} \leq Ax + B$ will become an equality when $x = \frac{1}{3}$, we know that the equation

$$(1 + x^2)(Ax + B) = 1$$

has a multiple root $x = \frac{1}{3}$. Let

$$(3x - 1)^2(ax + b) = 0.$$

Comparing the coefficients reveals that $A = -\frac{27}{50}, B = \frac{27}{25}, a = -\frac{3}{50}$, and $b = \frac{2}{25}$.

Proof. Since $x, y, z \in \mathbf{R}_+$ and $x + y + z = 1$, so $0 < x, y, z < 1$. Hence,

$$(3x - 1)^2 \left(x - \frac{4}{3} \right) \leq 0,$$

that is,

$$x^3 - 2x^2 + x - \frac{4}{27} \leq 0,$$

or namely

$$(x - 2)(1 + x^2) + \frac{50}{27} \leq 0.$$

Thus,

$$(1 + x^2)(2 - x) \geq \frac{50}{27}.$$

It follows that

$$\frac{1}{1 + x^2} \leq \frac{27}{50}(2 - x).$$

Similarly,

$$\frac{1}{1 + y^2} \leq \frac{27}{50}(2 - y),$$

$$\frac{1}{1 + z^2} \leq \frac{27}{50}(2 - z).$$

Adding up the three inequalities, we obtain that

$$\frac{1}{1 + x^2} + \frac{1}{1 + y^2} + \frac{1}{1 + z^2} \leq \frac{27}{50}[6 - (x + y + z)]$$

$$= \frac{27}{50}(6 - 1) = \frac{27}{10}. \qquad \square$$

3. Exercises

Group A

I. Filling Problems

1. Given $a, b \in \mathbf{R}_+$ such that $a + b = 1$. Then the minimum value of $\left(a + \frac{1}{a}\right)^2 + \left(b + \frac{1}{b}\right)^2$ is _____.

2. Suppose that nonnegative real numbers x and y satisfy $x + y \leq 1$. Then the solutions of the equation $8xy = 5x(1 - x) + 5y(1 - y)$ are _____.

3. Let n be a given positive integer such that $n \geq 2$, and $a_1, a_2 \ldots, a_n \in (0, 1)$ with $a_{n+1} = a_1$. Then the maximum value of $\sum_{i=1}^{n} \sqrt[6]{a_i(1 - a_{i+1})}$ is _____.

II. Calculation Problems

4. Let $a, b, c > 0$ satisfy $a + b + c \geq \frac{1}{a} + \frac{1}{b} + \frac{1}{c}$. Prove:
$$a^3 + b^3 + c^3 \geq a + b + c.$$

5. Prove: For any $x > \sqrt{2}$ and $y > \sqrt{2}$, there is the following inequality:
$$x^4 - x^3 y + x^2 y^2 - xy^3 + y^4 > x^2 + y^2.$$

6. Suppose that $a, b,$ and c are positive numbers satisfying $a^2 + b^2 + c^2 = 3$. Prove:
$$\frac{1}{1 + 2ab} + \frac{1}{1 + 2bc} + \frac{1}{1 + 2ca} \geq 1.$$

7. Let $x > 0$ and $y > 0$, and let $n \in \mathbf{N}_+$. Prove: $\frac{x^n}{1+x^2} + \frac{y^n}{1+y^2} \leq \frac{x^n + y^n}{1+xy}$.

8. Prove: The inequality
$$|x| + |y| + |z| - |x + y| - |y + z| - |z + x| + |x + y + z| \geq 0$$
is satisfied for all real numbers $x, y,$ and z.

9. Let $a, b,$ and c be real numbers satisfying $abc = 1$. Prove: At most two numbers among $2a - \frac{1}{b}, 2b - \frac{1}{c},$ and $2c - \frac{1}{a}$ are greater than 1.

Group B

10. Suppose that $a_1, a_2, \ldots, a_n$ and $b_1, b_2, \ldots, b_n$ are all positive real numbers and satisfy $\sum_{k=1}^{n} a_k = \sum_{k=1}^{n} b_k$. Prove:
$$\sum_{k=1}^{n} \frac{a_k^2}{a_k + b_k} \geq \frac{1}{2} \sum_{k=1}^{n} a_k.$$

11. Let a, b, and c be the lengths of the three sides of a triangle. Prove:

$$\frac{\sqrt{b+c-a}}{\sqrt{b}+\sqrt{c}-\sqrt{a}} + \frac{\sqrt{c+a-b}}{\sqrt{c}+\sqrt{a}-\sqrt{b}} + \frac{\sqrt{a+b-c}}{\sqrt{a}+\sqrt{b}-\sqrt{c}} \le 3.$$

12. Assume that three real numbers x, y, and z satisfy $x^2 + y^2 + z^2 = 2$. Prove:

$$x + y + z \le xyz + 2.$$

Chapter 3

Common Techniques for Proving Inequalities

1. Key Points of Knowledge and Basic Methods

A. *Variable substitutions*

Variable substitution is a common technique in solving mathematical problems. For some inequalities with rather complicated structures involving many variables and unclear relations among them, one may introduce new variables to transform (or partially transform) the original ones, realizing the purpose of simplifying the structure and presenting the characteristics of the problem. The usual substitutions are linear substitutions, trigonometric substitutions, fractional substitutions, and incremental substitutions.

B. *"Without loss of generality, assume that"*

For the proof of some symmetric inequalities of multiple variables (for example three variables a, b, and c), one can consider "without loss of generality, assume that" an ordered relation is provided, such as $a < b < c$ or $a \geq \max\{b, c\}$, etc. This amounts to increasing one more condition, and so decreasing the difficulty of the proof.

For the proof of some other symmetric inequalities of multiple variables (for example three variables a, b, and c), one can consider "without loss of generality, assume that" the sum of all variables equals a constant, such as $a + b + c = 1$ or $abc = 1$, etc. This also has the role of decreasing the difficulty of the proof.

C. *Construction method*

Sometimes, according to the characteristics or conditions of the inequality, one may introduce an appropriate figure, equation, function, special example, and etc. to realize the transformation of the proposition, which results in realizing the purpose of proving the inequality.

2. Illustrative Examples

We first look at several examples of substitutions.

Example 1. Suppose that x, y, and z are positive numbers and

$$x + y + z = \frac{1}{xyz}.$$

Find the minimum value of $(x+y)(y+z)$.

Analysis. From $\sqrt{xyz(x+y+z)} = 1$, we can recall an area formula of a triangle, so we construct a triangle with $a = x+y, b = y+z$, and $c = z+a$ as its three sides. By symmetry and $a + b = x + z + 2y = c + 2y > c$, we know that this triangle is well-defined.

Proof. Construct $\triangle ABC$ with the three sides

$$a = x+y, \quad b = y+z, \quad c = z+a.$$

Then the half circumference $p = \frac{1}{2}(a+b+c) = x+y+z$ and the area

$$S = \sqrt{p(p-a)(p-b)(p-c)} = \sqrt{(x+y+z)xyz} = 1.$$

On the other hand,

$$(x+y)(y+z) = ab = \frac{2S}{\sin C} = \frac{2}{\sin C} \leq 2, \tag{3.1}$$

and the equality in (3.1) holds if and only if $\angle C = 90°$. In the latter case, $(x+y)^2 + (y+z)^2 = (x+z)^2$, namely

$$y(x+y+z) = xz. \tag{3.2}$$

The choice of the real numbers $x = z = 1$ and $y = \sqrt{2} - 1$ satisfies (3.2), from which the equality of (3.1) is satisfied.

Hence, $(x+y)(y+z)$ has the minimum value 2. $\square$

Remark. A substitution similar to that in Example 1 is called a linear substitution. By its geometric meaning, the substitution of Example 1 is also referred to as the "tangent length substitution": Let a, b, and c be the lengths of the three sides of a triangle, and x, y, and z be the lengths of the line segments from the three sides divided by the tangent points of the inscribed circle. Denote

$$a = x + y, \quad b = y + z, \quad c = z + x.$$

Here $x, y, z \in \mathbf{R}_+$, the selection of which satisfies implicitly the important property that "the sum of any two sides is greater than the third side." The inverse substitution of the above substitution is

$$x = \frac{a + c - b}{2},$$

$$y = \frac{a + b - c}{2},$$

$$z = \frac{b + c - a}{2},$$

where $x, y, z, a, b, c \in \mathbf{R}_+$.

For those inequalities that involve (or imply) the sides of a triangle, sometimes we can consider applying the "tangent length substitution" or its inverse substitution to solve the problem.

Example 2. Let x_1, x_2, and x_3 be positive numbers. Prove:

$$x_1 x_2 x_3 \geq (x_2 + x_3 - x_1)(x_1 + x_3 - x_2)(x_1 + x_2 - x_3).$$

Proof. Let $a = x_2 + x_3 - x_1, b = x_1 + x_3 - x_2$, and $c = x_1 + x_2 - x_3$. Then

$$x_1 = \frac{b + c}{2}, \quad x_2 = \frac{c + a}{2}, \quad x_3 = \frac{a + b}{2}.$$

At most one of a, b, and c is not greater than 0. Otherwise, without loss of generality assume that $a \leq 0$ and $b \leq 0$. Then $x_3 \leq 0$, contradictory to $x_3 > 0$.

When exactly one of a, b, and c is not greater than 0, the inequality is obviously satisfied.

When a, b, and c are all positive numbers, the original inequality is reduced to

$$(b + c)(c + a)(a + b) \geq 8abc.$$

Since $b + c \geq 2\sqrt{bc}, c + a \geq 2\sqrt{ca}$, and $a + b \geq 2\sqrt{ab}$, the above inequality is clearly satisfied. $\qquad \square$

Remark. In the above process, by using the "tangent length substitution," we immediately transformed an "unclear" inequality into that with a clear structural characteristic.

Example 3. Assume that n positive real numbers $a_1, a_2, \ldots, a_n$ $(n \geq 3)$ satisfy

$$\frac{1}{1+a_1^4} + \frac{1}{1+a_2^4} + \cdots + \frac{1}{1+a_n^4} = 1.$$

Prove: $a_1 a_2 \cdots a_n \geq (n-1)^{\frac{n}{4}}$.

Proof. Let $a_i^2 = \tan x_i$ with $x_i \in \left(0, \frac{\pi}{2}\right)$ $(i = 1, 2, \ldots, n)$. Then the given condition becomes

$$\sum_{i=1}^{n} \cos^2 x_i = 1.$$

By the arithmetic mean-geometric mean inequality,

$$\sin^2 x_i = 1 - \cos^2 x_i \geq (n-1) \left(\prod_{j=1, j \neq i}^{n} \cos x_j \right)^{\frac{2}{n-1}} \quad (i = 1, 2, \ldots, n).$$

Multiplying the above inequalities produces

$$\prod_{i=1}^{n} \sin^2 x_i \geq (n-1)^n \prod_{i=1}^{n} \cos^2 x_i,$$

namely

$$\prod_{i=1}^{n} \tan^2 x_i \geq (n-1)^n.$$

Hence, $\prod_{i=1}^{n} a_i = \left(\prod_{i=1}^{n} \tan^2 x_i \right)^{\frac{1}{2}} \geq (n-1)^{\frac{n}{4}}$. $\qquad \square$

Remark. Here we adopted a trigonometric substitution. From the analogy between algebraic (in-)equalities and trigonometric identities, we can figure out needed different substitutions of new formats. As an example, for such structures as $a^2 + b^2 \leq r^2$, one may choose the substitutions $a = t \cos \theta$ and $b = t \sin \theta$ with $0 \leq t \leq r$.

Proof. Make substitutions $\frac{1}{1+a_i^4} = x_i$ $(i = 1, 2, \ldots, n)$. Then

$$a_i^4 = \frac{1 - x_i}{x_i}$$

and

$$x_1 + x_2 + \cdots + x_n = 1.$$

Using the arithmetic mean-geometric mean inequality, we have that

$$(a_1 a_2 \cdots a_n)^4 = \frac{1 - x_1}{x_1} \frac{1 - x_2}{x_2} \cdots \frac{1 - x_n}{x_n}$$

$$= \frac{(x_2 + x_3 + \cdots + x_n)(x_1 + x_3 + \cdots + x_n) \cdots (x_1 + x_2 + \cdots + x_{n-1})}{x_1 x_2 \cdots x_n}$$

$$\geq \frac{(n-1) \sqrt[n]{x_2 x_3 \cdots x_n}(n-1) \sqrt[n]{x_1 x_3 \cdots x_n} \cdots (n-1) \sqrt[n]{x_1 x_2 \cdots x_{n-1}}}{x_1 x_2 \cdots x_n}$$

$$= (n-1)^n,$$

and so $\prod_{i=1}^n a_i \geq (n-1)^{\frac{n}{4}}$. □

Example 4. Suppose that $a, b,$ and c are positive real numbers satisfying $abc = 1$. Prove:

$$\frac{1}{1 + 2a} + \frac{1}{1 + 2b} + \frac{1}{1 + 2c} \geq 1.$$

Proof. Let $a = \frac{x}{y}, b = \frac{y}{z},$ and $c = \frac{z}{x}$. Then the original inequality is equivalent to

$$\frac{y}{2x + y} + \frac{z}{2y + z} + \frac{x}{2z + x} \geq 1. \tag{3.3}$$

The Cauchy inequality implies that

$$[x(x + 2z) + y(y + 2x) + z(z + 2y)] \left(\frac{x}{x + 2z} + \frac{y}{y + 2x} + \frac{z}{z + 2y} \right)$$
$$\geq (x + y + z)^2,$$

in other words,

$$(x + y + z)^2 \left(\frac{x}{x + 2z} + \frac{y}{y + 2x} + \frac{z}{z + 2y} \right) \geq (x + y + z)^2.$$

Thus, (3.3) is valid, and so the original inequality is satisfied. □

Remark. Because of $abc = 1$ and the symmetry, we made fractional substitutions $a = \frac{x}{y}, b = \frac{y}{z},$ and $c = \frac{z}{x}$, so creating a condition for the application of the Cauchy inequality.

Example 5. For an integer $n \geq 2$, let positive real numbers $a_1, a_2, \ldots, a_n$ satisfy

$$\sum_{i=1}^{k-1} a_i \leq a_k \ (2 \leq k \leq n).$$

Prove: $\sum_{i=1}^{n-1} \frac{a_i}{a_{i+1}} \leq \frac{n}{2}$, and find a condition for the equality to be satisfied.

Proof. Let $x_1 = a_1$ and $x_k = a_k - \sum_{i=1}^{k-1} a_i \ (2 \leq k \leq n)$. Then

$$x_{k+1} - x_k = a_{k+1} - 2a_k \ (1 \leq k \leq n-1).$$

Consequently,

$$2\sum_{i=1}^{n-1} \frac{a_i}{a_{i+1}} = \sum_{i=1}^{n-1} \left(1 - \frac{x_{i+1} - x_i}{a_{i+1}}\right)$$

$$= n - \frac{x_1}{a_1} - \sum_{i=1}^{n-1} \frac{x_{i+1} - x_i}{a_{i+1}}$$

$$= n - \frac{x_n}{a_n} - \sum_{i=1}^{n-1} x_i \left(\frac{1}{a_i} - \frac{1}{a_{i+1}}\right) \leq n. \tag{3.4}$$

The last step in the above is because $x_i \geq 0$ for $1 \leq i \leq n$ and $a_i \leq a_{i+1}$ for $1 \leq i \leq n-1$.

The equality in (3.4) holds if and only if $x_2 = x_3 = \cdots = x_n = 0$, namely

$$a_2 = a_1, \ a_3 = 2a_1, \ \ldots, \ a_n = 2^{n-2}a_1. \qquad \square$$

Example 6. Suppose that real numbers a, b, and c satisfy $a + b + c = 1$ and $abc > 0$. Prove:

$$ab + bc + ca < \frac{\sqrt{abc}}{2} + \frac{1}{4}.$$

Proof 1. If $ab + bc + ca \leq \frac{1}{4}$, then the proposition is true.

Suppose that $ab + bc + ca > \frac{1}{4}$. Without loss of generality, let $a = \max\{a, b, c\}$. Then $a \geq \frac{1}{3}$ from $a + b + c = 1$. Now,

$$ab + bc + ca - \frac{1}{4} \leq \frac{(a+b+c)^2}{3} - \frac{1}{4} = \frac{1}{12} \leq \frac{a}{4} \tag{3.5}$$

and

$$ab + bc + ca - \frac{1}{4} = a(b + c) - \frac{1}{4} + bc$$

$$= a(1 - a) - \frac{1}{4} + bc$$

$$\leq \frac{1}{4} - \frac{1}{4} + bc = bc, \tag{3.6}$$

in which the equality in (3.5) is satisfied when $a = \frac{1}{3}$ and the equality in (3.6) is satisfied when $a = \frac{1}{2}$, so the equalities in (3.5) and (3.6) cannot be satisfied at the same time.

Since $ab + bc + ca - \frac{1}{4} > 0$, multiplying (3.5) and (3.6) gives

$$\left(ab + bc + ca - \frac{1}{4}\right)^2 < \frac{abc}{4},$$

that is,

$$ab + bc + ca - \frac{1}{4} < \frac{\sqrt{abc}}{2},$$

so

$$ab + bc + ca < \frac{\sqrt{abc}}{2} + \frac{1}{4}. \qquad \square$$

Proof 2. Since $abc > 0$, so there are one positive number and two negative numbers among a, b, and c, or all three are positive numbers. For the former situation, without loss of generality assume that $a > 0$ and $b, c < 0$. Then

$$ab + bc + ca = b(a + c) + ca < b(a + c) = b(1 - b) < 0,$$

hence the conclusion is obviously true.

For the latter case, without loss of generality assume that $a \geq b \geq c > 0$. Then $a \geq \frac{1}{3}$ and $0 < c \leq \frac{1}{3}$. Now,

$$ab + bc + ca - \frac{\sqrt{abc}}{2} = c(a + b) + \sqrt{ab}\left(\sqrt{ab} - \frac{\sqrt{c}}{2}\right)$$

$$= c(1 - c) + \sqrt{ab}\left(\sqrt{ab} - \frac{\sqrt{c}}{2}\right).$$

Since $\sqrt{ab} \geq \frac{\sqrt{b}}{3} \geq \frac{\sqrt{c}}{3}$ and $\sqrt{ab} \leq \frac{a+b}{2} = \frac{1-c}{2}$, we see that

$$c(1 - c) + \sqrt{ab}\left(\sqrt{ab} - \frac{\sqrt{c}}{2}\right) \leq c(1 - c) + \frac{1 - c}{2}\left(\frac{1 - c}{2} - \frac{\sqrt{c}}{2}\right)$$

$$= \frac{1}{4} - \frac{3c^2}{4} + \frac{c\sqrt{c}}{4} + \frac{c}{2} - \frac{\sqrt{c}}{4}.$$

Thus, it is enough to prove $\frac{3c^2}{4} - \frac{c\sqrt{c}}{4} - \frac{c}{2} + \frac{\sqrt{c}}{4} > 0$, namely

$$3c\sqrt{c} - c - 2\sqrt{c} + 1 > 0. \tag{3.7}$$

Since $0 < c \leq \frac{1}{3}$,

$$\frac{1}{3} - c \geq 0. \tag{3.8}$$

From the arithmetic mean-geometric mean inequality,

$$3c\sqrt{c} + \frac{1}{3} + \frac{1}{3} \geq 3\left(3c\sqrt{c} \cdot \frac{1}{3} \cdot \frac{1}{3}\right)^{\frac{1}{3}} = \sqrt[3]{9}\sqrt{c} > 2\sqrt{c}. \tag{3.9}$$

Adding up (3.8) and (3.9) gives (3.7), so the original inequality is true. □

Example 7. Let $a, b,$ and c be positive numbers. Prove:

$$\sqrt{abc}(\sqrt{a} + \sqrt{b} + \sqrt{c}) + (a + b + c)^2 \geq 4\sqrt{3abc(a + b + c)}. \tag{3.10}$$

Analysis. This inequality seems complicated, but after noting that it has homogeneous symmetry, we may let $abc = 1$, so that the original inequality can be simplified.

Proof. Without loss of generality, let $abc = 1$. Then the original inequality (3.10) is transformed to

$$\sqrt{a} + \sqrt{b} + \sqrt{c} + (a + b + c)^2 \geq 4\sqrt{3} \cdot \sqrt{a + b + c}. \tag{3.11}$$

Note that $\sqrt{a} + \sqrt{b} + \sqrt{c} \geq 3\sqrt[3]{\sqrt{a} \cdot \sqrt{b} \cdot \sqrt{c}}$, so we let $\sqrt{a + b + c} = t$ (>0) and prove the following inequality

$$3 + t^4 \geq 4\sqrt{3}t \tag{3.12}$$

that is stronger than (3.11).

It is easy to see that $t = \sqrt{a + b + c} \geq \sqrt{3}$, hence

$$3 + t^4 = 3 + \frac{t^4}{3} + \frac{t^4}{3} + \frac{t^4}{3} \geq 4\sqrt[4]{3\left(\frac{t^4}{3}\right)^3}$$

$$= \frac{4}{\sqrt{3}}t^4 \geq \frac{4t}{\sqrt{3}}\sqrt{3}^2 = 4\sqrt{3}t.$$

Hence, the inequality (3.12) is valid, and thus the original inequality is true. □

Remark. Different techniques were used to transform the inequality (3.10) to the inequality (3.11), which was then replaced by the stronger inequality (3.12). That is exactly the key point of the proof. One may also simplify the inequality by letting $a + b + c = 1$ in the present example, and the interested reader can try this approach.

Finally, we look at an example that will be solved by constructing a figure.

Example 8. Let x, y, and z be real numbers such that $0 < x < y < z < \frac{\pi}{2}$. Prove:

$$\frac{\pi}{2} + 2 \sin x \cos y + 2 \sin y \cos z > \sin 2x + \sin 2y + \sin 2z.$$

Proof. The inequality to be proved is equivalent to

$$\frac{\pi}{4} + \sin x \cos y + \sin y \cos z > \sin x \cos x + \sin y \cos y + \sin z \cos z$$

$$\Leftrightarrow \frac{\pi}{4} > \sin x(\cos x - \cos y) + \sin y(\cos y - \cos z) + \sin z \cos z.$$

Construct a figure as shown by Figure 3.1. The circle O is a unit circle, S_1, S_2, and S_3 are the areas of the three small rectangles, and x, y, and z are the central angles of the corresponding circular sectors, respectively.

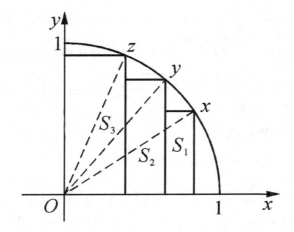

Figure 3.1 Figure for Example 8.

Then

$$S_1 = \sin x(\cos x - \cos y),$$
$$S_2 = \sin y(\cos y - \cos z),$$
$$S_3 = \sin z \cos z.$$

Since $S_1 + S_2 + S_3 < \frac{1}{4} \cdot \pi \cdot 1^2 = \frac{\pi}{4}$, so

$$\frac{\pi}{4} > \sin x(\cos x - \cos y) + \sin y(\cos y - \cos z) + \sin z \cos z.$$

Therefore, the original inequality is proved.   □

Remark. Quantitative relations of some algebraic inequalities may be concentrated on specific geometric figures. If such geometric figures can be constructed, then a rather pretty constructive proof may be obtained by means of the properties of the figures.

3. Exercises

Group A

I. Filling Problems

1. If two real numbers x and y satisfy $x^2 - 2xy + 5y^2 = 4$, then the range of the values of $x^2 + y^2$ is _____.
2. Let two real numbers x and y satisfy $x^2 + xy + y^2 = 1$. Then the sum of the maximum value and the minimum value of $F(x,y) = x^3y + xy^3$ is _____.
3. Suppose that $a > 1, b > 1, a^t + 1 \le b^t \le 2a^t$, and $t \in \mathbf{R}$. Then the maximum value of $\frac{b^t-2}{a^{2t}} + \frac{a^t-1}{b^{2t}}$ is _____.
4. Let $a, b, c \in \mathbf{R}_+$ satisfy $a + b + c = 1$. Then the minimum value of $\frac{a+3c}{a+2b+c} + \frac{4b}{a+b+2c} - \frac{8c}{a+b+3c}$ is _____.
5. Assume that $x, y, z \in \mathbf{R}_+$ and $\frac{x^2}{1+x^2} + \frac{y^2}{1+y^2} + \frac{z^2}{1+z^2} = 1$. Then the maximum value of xyz is _____.

II. Calculation Problems

6. Let a, b, and c be real numbers. Prove:
$$\frac{|a+b+c|}{1+|a+b+c|} \le \frac{|a|}{1+|a|} + \frac{|b|}{1+|b|} + \frac{|c|}{1+|c|}.$$

7. Assume that x_1, x_2, x_3, x_4, and x_5 are positive numbers. Prove:
$$(x_1 + x_2 + x_3 + x_4 + x_5)^2 > 4(x_1x_2 + x_2x_3 + x_3x_4 + x_4x_5 + x_5x_1).$$

8. Let $0 \le a \le 1, 0 \le b \le 1$, and $0 \le c \le 1$. Prove:

$$\frac{a}{b+c+1} + \frac{b}{c+a+1} + \frac{c}{a+b+1} + (1-a)(1-b)(1-c) \le 1.$$

Group B

9. Suppose that x, y, and z are positive numbers. Prove:

$$\frac{xy}{z} + \frac{yz}{x} + \frac{zx}{y} > 2\sqrt[3]{x^3 + y^3 + z^3}.$$

10. Let $x, y, z \le 0$ be such that $x + y + z = 1$. Prove:

$$\frac{1}{2} \le x^2 + y^2 + z^2 + 6xyz \le 1.$$

11. Assume that $\angle A, \angle B$, and $\angle C$ are the three angles of an acute triangle. Prove:

$$\sin A + \sin B + \sin C + \tan A + \tan B + \tan C \ge 2\pi.$$

12. Let a, b, and c be the lengths of the three sides of a triangle. Prove:

$$a^2 b(a - b) + b^2 c(b - c) + c^2 a(c - a) \ge 0,$$

and state a condition for the equality to be satisfied.

Chapter 4

Arithmetic and Geometric Sequences

1. Key Points of Knowledge and Basic Methods

Sequences are functions defined on the set of all positive integers (or integers). Arithmetic sequences and geometric sequences are the two most basic members in the family of sequences. Most problems related to sequences need to be dealt with by transforming them to arithmetic sequences or geometric sequences.

A. *Some basic properties of an arithmetic sequence* $\{a_n\}$

(1) For positive integers p, q, r, and s, if $p+s = q+r$, then $a_p+a_s = a_q+a_r$.
(2) For any real number b, the sequence $\{ba_n\}$ is an arithmetic sequence.
(3) If $\{b_n\}$ is also an arithmetic sequence, then $\{a_n + b_n\}$ is an arithmetic sequence.

B. *Let S_n be the sum of the first n terms of an arithmetic sequence $\{a_n\}$. One can prove*

(1) $S_{3m} = 3(S_{2m} - S_m)$.
(2) If $S_m = S_n$ $(m \neq n)$, then $S_{m+n} = 0$.
(3) If $S_p = q$ and $S_q = p$ with $p \neq q$, then $S_{p+q} = -(p + q)$.

C. *Some basic properties of a geometric sequence* $\{a_n\}$

(1) For positive integers p, q, r, and s, if $p + s = q + r$, then $a_p a_s = a_q a_r$.
(2) For any nonzero real number b, the sequence $\{ba_n\}$ is a geometric sequence.

(3) If $\{b_n\}$ is also a geometric sequence, then $\{a_nb_n\}$ is a geometric sequence.

D. *Ideas and methods to deal with sequence problems*

(1) Ideas from equations.

Sequence-related problems are often around expansions of a_n and S_n. In the problems of arithmetic sequences and geometric sequences, one mainly encounters the five quantities a_1, d (or q), n, a_n, and S_n, among which the whole sequence can be determined after the first two quantities are obtained. Thus, ideas from equations can be used to treat the problem.

(2) Ideas from functions.

For an arithmetic sequence, if it is not a constant one, a_n can be considered as a linear function of n and S_n can be considered as a quadratic function of n. Therefore, ideas from functions can be applied to solving the problem.

2. Illustrative Examples

Example 1. Suppose that each term of a sequence $\{a_n\}$ is a distinct positive number and their reciprocals form an arithmetic sequence. Find the value of $\frac{a_1a_2+a_2a_3+\cdots+a_{2014}a_{2015}}{a_1a_{2015}}$.

Solution. Let the common difference of the arithmetic sequence $\{\frac{1}{a_n}\}$ be $d \neq 0$. Then

$$\frac{1}{a_2} - \frac{1}{a_1} = \frac{1}{a_3} - \frac{1}{a_2} = \cdots = \frac{1}{a_{2015}} - \frac{1}{a_{2014}} = d, \qquad (4.1)$$

in other words,

$$\frac{a_1 - a_2}{a_1a_2} = \frac{a_2 - a_3}{a_2a_3} = \cdots = \frac{a_{2014} - a_{2015}}{a_{2014}a_{2015}} = d.$$

By the equal ratios theorem,

$$\frac{a_1 - a_{2015}}{a_1a_2 + a_2a_3 + \cdots + a_{2014}a_{2015}} = d. \qquad (4.2)$$

From (4.1), $\frac{1}{a_{2015}} - \frac{1}{a_1} = 2014d$, that is,

$$\frac{a_1 - a_{2015}}{a_1 a_{2015}} = 2014d. \tag{4.3}$$

It follows from (4.2) and (4.3) that $\frac{a_1 a_2 + a_2 a_3 + \cdots + a_{2014} a_{2015}}{a_1 a_{2015}} = 2014$. ☐

Example 2. Let $\{a_n\}$ be an arithmetic sequence with common difference $d \neq 0$, and assume that partial terms $a_{k_1}, a_{k_2}, \ldots, a_{k_n}, \ldots$ form a geometric sequence, where $k_1 = 1, k_2 = 5$, and $k_3 = 17$. Find k_{2017}.

Solution. By the condition, $a_5^2 = a_1 a_{17}$, namely

$$(a_1 + 4d)^2 = a_1(a_1 + 16d).$$

Then $a_1 d = 2d^2$.

Since $d \neq 0$, we have $a_1 = 2d$, or equivalently $d = \frac{a_1}{2}$.

Thus, $a_5 = a_1 + 4d = 3a_1$.

Consequently, in the geometric sequence $a_{k_1}, a_{k_2}, \ldots, a_{k_n}, \ldots$, the common ratio $q = \frac{a_5}{a_1} = 3$.

Since a_{k_n} is the nth term of the geometric sequence $\{a_{k_n}\}$, we see that $a_{k_n} = a_1 3^{n-1}$.

Since a_{k_n} is the k_nth term of the arithmetic sequence $\{a_n\}$,

$$a_{k_n} = a_1 + (n_k - 1)d = \frac{k_n + 1}{2} a_1.$$

Therefore, $a_1 3^{n-1} = \frac{k_n + 1}{2} a_1$, from which

$$k_n = 2 \cdot 3^{n-1} - 1.$$

Hence, $k_{2017} = 2 \cdot 3^{2016} - 1$. ☐

Example 3. Let the sum of the first n terms of a geometric sequence $\{a_n\}$ of real numbers be S_n. Suppose that $S_{10} = 10$ and $S_{30} = 70$. Find S_{40}.

Solution. Denote $b_1 = S_{10}, b_2 = S_{20} - S_{10}, b_3 = S_{30} - S_{20}$, and $b_4 = S_{40} - S_{30}$. Let q be the common ratio of $\{a_n\}$. Then b_1, b_2, b_3, and b_4 form a geometric sequence with common ratio $r = q^{10}$. Hence,

$$70 = S_{30} = b_1 + b_2 + b_3 = b_1(1 + r + r^2) = 10(1 + r + r^2),$$

that is, $r^2 - r + 6 = 0$, the solutions of which are $r = 2$ or $r = -3$.

Since $r = q^{10} > 0$, so $r = 2$, and thus

$$S_{40} = 10(1 + 2 + 2^2 + 2^3) = 150.$$ □

Example 4. Suppose that $\{a_n\}$ is an arithmetic sequence and a_1^2, a_2^2, and a_3^2 are also terms of the sequence $\{a_n\}$. Prove: For any $n \in \mathbf{Z}_+$, the term a_n is an integer.

Proof. Let the common difference of the arithmetic sequence $\{a_n\}$ be d. If $d = 0$, then $a_n = a_1$ for any $n \in \mathbf{Z}_+$. Since a_1^2 is one term of the sequence $\{a_n\}$, we have $a_1^2 = a_1$, from which $a_1 = 0$ or 1. Hence, $a_n = 0$ or 1 for any $n \in \mathbf{Z}_+$, so the proposition is satisfied.

If $d \neq 0$, then there exist $r, s, l \in \mathbf{N}$ such that

$$\begin{cases} a_1^2 = a_1 + rd, \\ a_2^2 = a_1 + sd = (a_1 + d)^2, \\ a_3^2 = a_1 + ld = (a_1 + 2d)^2. \end{cases}$$

It is easy to obtain

$$\begin{cases} 2a_1 + d = s - r, \\ 4a_1 + 4d = l - r, \end{cases}$$

and the solutions are

$$a_1 = \frac{4s - 3r - l}{4}, \quad d = \frac{r - 2s + l}{2}.$$

Since $r, s, l \in \mathbf{N}$, so $a_1, d \in \mathbf{Q}$. Also

$$a_1^2 = a_1 + rd = a_1 + \frac{r(r - 2s + l)}{2}$$

$$= a_1 + \frac{r(r - 2s + 4s - 3r - 4a_1)}{2}$$

$$= (1 - 2r)a_1 + r(s - r),$$

that is,

$$a_1^2 + (2r - 1)a_1 + r(s - r) = 0. \tag{4.4}$$

This is a quadratic equation for variable a_1 with integer coefficients.

Since the equation (4.4) has the quadratic term coefficient 1, its rational number solutions must be integers, from which $a_1 \in \mathbf{Z}$.

Thus, $d = s - r - 2a_1 \in \mathbf{Z}$, and so $a_n \in \mathbf{Z}$.

In summary, a_n is an integer for any positive integer n. □

Example 5. Let $\{a_n\}$ be a geometric sequence with every term a positive number, and let

$$S = a_1 + a_2 + \cdots + a_n, \quad T = \frac{1}{a_1} + \frac{1}{a_2} + \cdots + \frac{1}{a_n}.$$

Find the product of the first n terms of $\{a_n\}$.

Solution. Let the common ratio of the geometric sequence $\{a_n\}$ be q. Then by the hypothesis of the problem, $S = na_1$ and $T = \frac{n}{a_1}$ when $q = 1$. So $a_1^2 = \frac{S}{T}$, and consequently $a_1 a_2 \cdots a_n = a_1^n = \left(\frac{S}{T}\right)^{\frac{n}{2}}$.

When $q \neq 1$, we have $S = \frac{a_1 - a_n q}{1-q}$ and

$$T = \frac{\frac{1}{a_1} - \frac{1}{a_n}\frac{1}{q}}{1 - \frac{1}{q}} = \frac{a_1 - a_n q}{a_1 a_n(1-q)},$$

thus $a_1 a_n = \frac{S}{T}$.

For the geometric sequence, the product of two terms with the same distance to the first and last terms respectively equals the product of the first and last terms, in other words,

$$a_1 a_n = a_1 a_n, \ a_2 a_{n-1} = a_1 a_n, \ a_3 a_{n-2} = a_1 a_n, \ \ldots, \ a_n a_1 = a_1 a_n.$$

Multiplying the above n equalities leads to $(a_1 a_2 \cdots a_n)^2 = \left(\frac{S}{T}\right)^n$, therefore

$$a_1 a_2 \cdots a_n = \left(\frac{S}{T}\right)^{\frac{n}{2}}. \qquad \square$$

Example 6. Let $\{a_n\}$ be a geometric sequence consisting of positive numbers and let S_n be its sum of the first n terms.

(1) Prove: $\frac{\lg S_n + \lg S_{n+2}}{2} < \lg S_{n+1}$.

(2) Does there exist a constant $c > 0$ such that $\frac{\lg(S_n - c) + \lg(S_{n+2} - c)}{2} = \lg(S_{n+1} - c)$? Prove your conclusion.

Proof. (1) Let the common ratio of $\{a_n\}$ be q. From the condition, $a_1 > 0$ and $q > 0$.

(i) When $q = 1$, we have $S_n = na_1$, thus

$$S_n S_{n+2} - S_{n+1}^2 = na_1(n+2)a_1 - (n+1)^2 a_1^2 = -a_1^2 < 0.$$

(ii) When $q \neq 1$,

$$S_n = \frac{a_1(1-q^n)}{1-q},$$

so

$$S_n S_{n+2} - S_{n+1}^2 = \frac{a_1^2(1-q^n)(1-q^{n+2})}{(1-q)^2} - \frac{a_1^2(1-q^{n+1})^2}{(1-q)^2} = -a_1^2 q^n < 0.$$

In either (i) or (ii), $\frac{\log S_n + \log S_{n+2}}{2} < \log S_{n+1}$, and taking logarithm to the both sides proves the result.

(2) No, it does not exist such a constant c. The reason is as follows.

If $\frac{\lg(S_n - c) + \lg(S_{n+2} - c)}{2} = \lg(S_{n+1} - c)$, then

$$\begin{cases} (S_n - c)(S_{n+2} - c) = (S_{n+1} - c)^2, \\ S_n - c > 0. \end{cases}$$

We divide our discussion into the following two cases.

(i) When $q = 1$,

$$(S_n - c)(S_{n+2} - c) - (S_{n+1} - c)^2$$
$$= (na_1 - c)[(n+2)a_1 - c] - [(n+1)a_1 - c]^2$$
$$= -a_1^2 < 0.$$

In other words, there exists no constant $c > 0$ to satisfy the conclusion.

(ii) When $q \neq 1$, if $\frac{\log(S_n - c) + \log(S_{n+2} - c)}{2} = \log(S_{n+1} - c)$ is satisfied, then

$$(S_n - c)(S_{n+2} - c) - (S_{n+1} - c)^2$$
$$= \left[\frac{a_1(1-q^n)}{1-q} - c \right] \left[\frac{a_1(1-q^{n+2})}{1-q} - c \right] - \left[\frac{a_1(1-q^{n+1})}{1-q} - c \right]^2$$
$$= -a_1 q^n [a_1 - c(1-q)].$$

But since $a_1 q^n \neq 0$, so $a_1 - c(1-q) = 0$, namely $c = \frac{a_1}{1-q}$.

Now, since $c > 0$ and $a - 1 > 0$, we must have $0 < q < 1$. But when $0 < q < 1$,

$$S_n - \frac{a_1}{1-q} = \frac{-a_1 q^n}{1-q} < 0,$$

which does not satisfy $S_n - c > 0$. In other words, there does not exist a constant $c > 0$ that satisfies the requirement.

Combining (i) and (ii), we know that there is no constant $c > 0$ that satisfies the equality in (2). □

Example 7. Assume that n real numbers $a_1, a_2, \ldots, a_n$ ($n \geq 3$) are mutually distinct, and let the $\frac{n(n+1)}{2}$ sums $a_i + a_j$ ($1 \leq i \leq j \leq n$) form a new sequence according to the increasing order. If the new sequence is an arithmetic one with a nonzero common difference, find the value of n.

Solution. When $n = 3$, consider

$$(a_1, a_2, a_3) = (1, 2, 3).$$

The sums of any two of them are $3, 4$, and 5, forming an arithmetic sequence.

When $n = 4$, consider

$$(a_1, a_2, a_3, a_4) = (1, 3, 4, 5).$$

The sums of any two of them are $4, 5, 6, 7, 8$, and 9, forming an arithmetic sequence.

In the following, we assume that $n \geq 5$ and the sums of two of $a_1, a_2, \ldots, a_n$ form an arithmetic sequence according to the increasing order.

Without loss of generality, assume that $a_1 < a_2 < \cdots < a_n$. Let $d \neq 0$ be the common difference of the resulting sequence of the sums. Obviously, among the terms of the sums sequence, $a_1 + a_2$ is the minimum, $a_2 + a_3$ is the next to the minimum, $a_n + a_{n-2}$ is the next to the maximum, and $a_n + a_{n-1}$ is the maximum.

Hence, $a_3 - a_2 = d$ and $a_{n-1} - a_{n-2} = d$. As a result, $d > 0$ and

$$a_2 + a_{n-1} = (a_3 - d) + (a_{n-2} + d) = a_3 + a_{n-2}.$$

If $n \geq 6$, then the left hand side and the right hand side are respectively the sums of two different pairs of the numbers, which is contradictory to the fact that such sums of pairs form an arithmetic sequence of nonzero common difference. Therefore, there is no solution for $n \geq 6$.

When $n = 5$, we have $a_3 - a_2 = d$ and $a_4 - a_3 = d$.

In the sequence of the sums, the third minimal term is $a_1 + a_4$. From $a_1 + a_4 = (a_1 + a_3) + d$ and $a_2 + a_5 = a_3 + a_5 - d$, we see that the third maximal sum is $a_5 + a_2$.

On the other hand, since $a_2 + a_3 < a_2 + a_4 < a_3 + a_4$ and the differences of the two nearby terms of the sequence of the sums are all d, so $a_1 + a_5$ is the forth minimum term or the forth maximal term of the sums.

By symmetry, we may assume that $a_1 + a_5$ is the forth minimum term of the sums. Then

$$a_1 + a_2 < a_1 + a_3 < a_1 + a_4 < a_1 + a_5$$
$$< a_2 + a_3 < a_2 + a_4 < a_3 + a_4 < a_2 + a_5$$
$$< a_3 + a_5 < a_4 + a_5.$$

Thus, $(a_5 + a_2) - (a_3 + a_4) = d$, and so

$$a_5 - a_4 = d + a_3 - a_2 = 2d,$$

which contradicts $(a_1 + a_5) - (a_1 + a_4) = d$.

Hence, there is no solution for $n = 5$.

In summary, $n = 3$ and 4.   □

Example 8. Arrange a sequence $\{a_n\}$ of positive numbers into a triangular array as the following shows (there are n numbers in the nth row that is in the increasing order of the subscripts):

$$
\begin{array}{ccccccccc}
& & & & a_1 & & & & \\
& & & a_2 & & a_3 & & & \\
& & a_4 & & a_5 & & a_6 & & \\
& a_7 & & a_8 & & a_9 & & a_{10} & \\
\cdots & & \cdots & & \cdots & & \cdots & & \cdots
\end{array}
$$

Let b_n represent the number located at the nth row and the first column of the array. Suppose that $\{b_n\}$ is a geometric sequence, and from the third row on, each row is an arithmetic sequence of common difference d (the three numbers of the third row form an arithmetic sequence of common difference d, the four numbers of the fourth row form an arithmetic sequence of common difference d, and etc.). In addition, assume that $a_1 = 1$, $a_{12} = 17$, and $a_{18} = 34$.

(1) Find the number $A(m, n)$ (in terms of m and n) at the location of the mth row and the nth column of the array.
(2) Find the value of a_{2013}.
(3) Is 2013 in the array of the numbers? Give the reason of your answer.

Solution. (1) Let the common ratio of $\{b_n\}$ be q.

From the condition, a_{12} is the number at the location of the fifth row and the 2nd column of the array, and a_{18} is the number at the location of the sixth row and the third column of the array.

So, $b_1 = 1, b_n = q^{n-1}, a_{12} = q^4 + d = 17$, and $a_{18} = q^5 + 2d = 34$.
Thus, $q = 2, d = 1$, and $b_n = 2^{n-1}$.
Therefore, $A(m, n) = b_m + (n - 1)d = 2^{m-1} + n - 1$.

(2) From $1 + 2 + 3 + \cdots + 62 = 1953, 1 + 2 + 3 + \cdots + 62 + 63 = 2016$, and $2013 - 1953 = 60$, we know that a_{2013} is the number at the location of the 63rd row and the 60th column of the array.

So, $a_{2013} = 2^{62} + 59$.

(3) Suppose that 2013 is the number at the location of the mth row and the nth column of the array.

Since in the mth row, 2^{m-1} is the minimal number and $2^{m-1} + m - 1$ is the maximal number, so
$$2^{m-1} \le 2013 \le 2^{m-1} + m - 1. \qquad (4.5)$$
Since $2^{m-1} + m - 1 \le 2^{10} + 10 = 1031 < 2013$ when $m \le 11$, we see that $m \le 11$ does not satisfy (4.5).

Since $2^{m-1} \ge 2^{11} = 2048 > 2013$ when $m \ge 12$, it follows that $m \ge 12$ also does not satisfy (4.5).

Hence, the above inequality (4.5) has no positive integer solutions.

Therefore, there is no 2013 in the array of the numbers. $\qquad \square$

3. Exercises

Group A

I. Filling Problems

1. Let 7 positive numbers constitute a geometric sequence $a_1, a_2, \ldots, a_7$. If the sum of the first five terms is $\frac{62}{7\sqrt{2}-6}$ and the sum of the last five terms is $12 + 14\sqrt{2}$, then the value of the product of the seven terms is

   _____.

2. Let $\{a_n\}$ be an arithmetic sequence with $a_1 = 19$ and $a_{26} = -1$. Write $A = a_n + a_{n+1} + \cdots + a_{n+6}$, where n is a positive integer. Then the minimum value of $|A|$ is _____.

3. The sum of the first n terms of an arithmetic sequence $\{a_n\}$ is S_n. If $\frac{S_{25}}{a_{23}} = 5$ and $\frac{S_{45}}{a_{33}} = 25$, then the value of $\frac{S_{65}}{a_{43}}$ is _____.

4. In the sequences $\{a_n\}$ and $\{b_n\}$, if $a_1 = 2, 3a_{n+1} - a_n = 0$, and b_n is the arithmetic mean of a_n and a_{n+1} for any positive integer n, then the sum of all terms of $\{b_n\}$ is _____.

5. Let S_n be the sum of the first n terms of an arithmetic sequence $\{a_n\}$. Suppose that the geometric mean of $\frac{1}{3}S_3$ and $\frac{1}{4}S_4$ is $\frac{1}{5}S_5$, and the arithmetic mean of $\frac{1}{3}S_3$ and $\frac{1}{4}S_4$ is 1. Then the general term formula for $\{a_n\}$ is $a_n =$ _____ .

II. Calculation Problems

6. Let $\{a_n\}$ be a geometric sequence with the first term $a_1 > 1$ and the common ratio $q > 1$. Prove: The sequence $\{\log_{a_n} a_{n+1}\}$ is a decreasing sequence.

7. Let $\{a_n\}$ be a geometric sequence with common ratio q. Prove: A necessary and sufficient condition for the products of any two terms of $\{a_n\}$ still to be terms of the sequence is that there exists a nonnegative integer m such that $a_1 = q^m$.

8. Let $\{a_n\}$ and $\{b_n\}$ be an arithmetic sequence and a geometric sequence, respectively. If $a_1 = b_1 > 0$ and $a_2 = b_2 > 0$, please compare the magnitudes of a_n and b_n.

Group B

9. Given a positive integer n and a positive number M. For all arithmetic sequences $a_1, a_2, a_3, \ldots$ that satisfy the condition $a_1^2 + a_{n+1}^2 \leq M$, find the maximum value of $S = a_{n+1} + a_{n+2} + \cdots + a_{2n+1}$.

10. Let $\{a_n\}$ be an increasing arithmetic sequence of positive terms, and let $n, k \in \mathbf{N}_+$ with $k \geq 2$. Prove:

$$\sqrt[k]{\frac{a_{(n+1)k+1}}{a_{k+1}}} < \frac{a_{k+2}}{a_{k+1}} \frac{a_{2k+2}}{a_{2k+1}} \cdots \frac{a_{nk+2}}{a_{nk+1}} < \sqrt[k]{\frac{a_{nk+2}}{a_2}}.$$

11. Given $S = \{1, 2, 3, \ldots, n\}$. Let A be an arithmetic sequence of at least two terms with a positive common difference. Suppose that each term of A is in S, and putting any other elements of S into A cannot make an arithmetic sequence with the same common difference as A. Find the number of such sequences A (a sequence of only two terms is also considered as an arithmetic sequence).

12. Two positive sequences satisfy: One is an arithmetic sequence with r ($r > 0$) as its common difference, and the other is a geometric sequence with q ($q > 1$) as its common ratio. Here r and q are positive integers that are relatively prime. Prove: If there is one pair of equal corresponding terms for the two sequences, then there are infinitely many pairs of equal corresponding terms for them.

Chapter 5

Arithmetic Sequences of Higher Order

1. Key Points of Knowledge and Basic Methods

A. For a given sequence $\{a_n\}$, denote by b_n the difference $a_{n+1} - a_n$ of the consecutive terms a_{n+1} and a_n. Thus, we have a new sequence $\{b_n\}$, and the sequence $\{b_n\}$ is called the first order difference sequence of the original sequence $\{a_n\}$. If $c_n = b_{n+1} - b_n$ for $n = 1, 2, \ldots$, then the sequence $\{c_n\}$ is the first order difference sequence of $\{b_n\}$, and $\{c_n\}$ is called the second order difference sequence of $\{a_n\}$; and so on. In this way, we can obtain a pth order difference sequence of the sequence $\{a_n\}$, where $p \in \mathbf{N}_+$.

B. If the pth order difference sequence of a sequence $\{a_n\}$ is a nonzero constant sequence, then we call the sequence $\{a_n\}$ an arithmetic sequence of order p.

C. An arithmetic sequence of order 1 is just an arithmetic sequence (non-constant sequence of numbers) that we usually talk about.

D. Arithmetic sequences of higher order are the general name for arithmetic sequences of order 2 or greater than 2.

E. Properties of arithmetic sequences of higher order:

(1) If a sequence $\{a_n\}$ is an arithmetic sequence of order p, then its first order difference sequence is an arithmetic sequence of order $p - 1$.

43

(2) A necessary and sufficient condition for a sequence $\{a_n\}$ to be an arithmetic sequence of order p is: The general term of the sequence is a pth order polynomial of n.

(3) If a sequence $\{a_n\}$ is an arithmetic sequence of order p, then the sum S_n of its first n terms is a $(p+1)$st order polynomial of n.

Property (1) is obvious, and properties (2) and (3) can be proved by mathematical induction.

F. The most important and most often encountered problems for arithmetic sequences of higher order are to find the general term and the sum of the first n terms, and a deeper problem is to find solutions of difference equations. The basic methods to solve the problems include:

(1) Method of successive differences: The starting point is $a_n = a_1 + \sum_{k=1}^{n-1}(a_{k+1} - a_k)$.

(2) Method of undetermined coefficients: In the arithmetic sequence of known order, its general term a_n and the sum S_n of its first n terms are polynomials of n with given degrees. We first write down the polynomial, and then plug in the given conditions to solve the resulting equations, thus finding the coefficients of the polynomial.

(3) Method of elimination by splitting terms: The starting point is that a_n can be written in the form of

$$a_n = f(n+1) - f(n).$$

2. Illustrative Examples

Example 1. Suppose that an arithmetic sequence $\{a_n\}$ of order 2 satisfies $a_2 - a_1 = 110$, and the sum S_{11} of the first 11 terms of its first order difference sequence $\{b_n\}$ is the maximal number. Assume that when S_{11} is a maximum value, the maximum value of $\{a_n\}$ is 2017. Find the general term of $\{a_n\}$.

Solution. By the condition, $\{b_n\}$ is an arithmetic sequence and $b_1 = a_2 - a_1 = 110$. Let the common difference of $\{b_n\}$ be d. Then

$$\begin{cases} b_{11} = 110 + 10d \geq 0, \\ b_{12} = 110 + 11d \leq 0. \end{cases}$$

Solving the above inequalities, we get $-11 \leq d \leq -10$.

When S_{11} is a maximum value, $d = -10$. In this case,

$$S_{11} = \frac{2b_1 + 10d}{2} 11 = 660.$$

On the other hand,

$$a_{n+1} = a_1 + (a_2 - a_1) + (a_3 - a_2) + \cdots + (a_{n+1} - a_n)$$
$$= a_1 + b_1 + b_2 + \cdots + b_n$$
$$= a_1 + S_n.$$

Clearly, when S_n is a maximum value, a_{n+1} is a maximum value. So,

$$(a_{n+1})_{\max} = a_{12} = a_1 + S_{11} = a_1 + 660 = 2017,$$

from which $a_1 = 1357$.

Thus, $a_{n+1} = a_1 + \frac{2b_1 + (n-1)d}{2} n = -5n^2 + 115n + 1357$, and so $a_n = -5n^2 + 115n + 1237$. $\qquad\square$

Example 2. Let every term of the second order difference sequence of a sequence $\{a_n\}$ be 16, and $a_{63} = a_{89} = 10$. Find a_{51}.

Solution 1. Obviously, the first order difference sequence $\{b_n\}$ of the sequence $\{a_n\}$ is an arithmetic sequence of common difference 16. Suppose that its first term is a. Then $b_n = a + (n-1)16$, and so

$$a_n = a_1 + \sum_{k=1}^{n-1}(a_{k+1} - a_k)$$

$$= a_1 + \sum_{k=1}^{n-1} b_k$$

$$= a_1 + \frac{a + [a + (n-1)16]}{2}(n-1)$$

$$= a_1 + (n-1)a + 8(n-1)(n-2).$$

This is a quadratic polynomial of n, in which the coefficient of n^2 is 8. Since $a_{63} = a_{89} = 10$,

$$a_n = 8(n-63)(n-89) + 10.$$

Hence,

$$a_{51} = 8(51-63)(51-89) + 10 = 3658. \qquad\square$$

Solution 2. By the condition, the sequence $\{a_n\}$ is an arithmetic sequence of order 2, so its general term is a quadratic polynomial of n. Also from $a_{63} = a_{89} = 10$, we may assume that

$$a_n = A(n - 63)(n - 89) + 10.$$

Since the second order difference sequence of $\{a_n\}$ is the sequence of constant terms 16, so

$$(a_3 - a_2) - (a_2 - a_1) = 16,$$

namely

$$a_3 - 2a_2 + a_1 = 16.$$

Consequently,

$$A(3 - 63)(3 - 89) + 10 - 2[A(2 - 63)(2 - 89) + 10]$$
$$+ A(1 - 63)(1 - 89) + 10 = 16,$$

the solution of which is

$$A = 8.$$

Thus,

$$a_n = 8(n - 63)(n - 89) + 10,$$

and so $a_{51} = 3658.$ $\square$

Remark. Solution 1 used the method of successive differences and Solution 2 used the method of undetermined coefficients.

Example 3. The first four terms of an arithmetic sequence of order 3 are $30, 72, 140,$ and 240 in succession. Find a formula for its general term.

Solution. By property (2), a_n is a cubic polynomial of n, so we assume that $a_n = An^3 + Bn^2 + Cn + D$, where $A, B, C,$ and D are to be determined. Since $a_1 = 30, a_2 = 72, a_3 = 140,$ and $a_4 = 240$, we can set the system of equations

$$\begin{cases} A + B + C + D = 30, \\ 8A + 4B + 2C + D = 72, \\ 27A + 9B + 3C + D = 140, \\ 64A + 16B + 4C + D = 240, \end{cases}$$

whose solutions are $A = 1, B = 7, C = 14,$ and $D = 8$. Hence,

$$a_n = n^3 + 7n^2 + 14n + 8.$$ $\square$

Example 4. Observe that $\frac{1}{1} = \frac{1}{2} + \frac{1}{2}, \frac{1}{2} = \frac{1}{3} + \frac{1}{6}, \frac{1}{3} = \frac{1}{4} + \frac{1}{12}$, and $\frac{1}{4} = \frac{1}{5} + \frac{1}{20}$. Inspired by such examples, state a general rule and then prove it. Prove that for any integer n greater than 1, there exist two positive integers $i < j$ such that

$$\frac{1}{n} = \frac{1}{i(i+1)} + \frac{1}{(i+1)(i+2)} + \frac{1}{(i+2)(i+3)} + \cdots + \frac{1}{j(j+1)}.$$

Proof. The given examples imply the general rule: $\frac{1}{n} = \frac{1}{n+1} + \frac{1}{n(n+1)}$, namely $\frac{1}{n(n+1)} = \frac{1}{n} - \frac{1}{n+1}$ with $n = 1, 2, 3, \ldots$. Thus,

$$\frac{1}{i(i+1)} + \frac{1}{(i+1)(i+2)} + \cdots + \frac{1}{j(j+1)} = \frac{1}{i} - \frac{1}{j+1} = \frac{1}{n}.$$

Comparing the above equality with $\frac{1}{n} = \frac{1}{n-1} - \frac{1}{(n-1)n}$, we find that $i = n - 1$ and $j + 1 = (n-1)n$. This proves the result. $\qquad\square$

Example 5. Find the sum: $S_n = 1 \cdot 3 \cdot 2^2 + 2 \cdot 4 \cdot 3^2 + \cdots + n(n+2)(n+1)^2$.

Solution. Since $k(k+2)(k+1)^2 = k(k+1)(k+2)(k+3) - 2k(k+1)(k+2)$, so the problem of finding S_n can be reduced to finding $K_n = \sum_{k=1}^{n} k(k+1)(k+2)(k+3)$ and $T_n = \sum_{k=1}^{n} k(k+1)(k+2)$.

Since $k(k+1)(k+2)(k+3) = \frac{1}{5}[k(k+1)(k+2)(k+3)(k+4) - (k-1)k(k+1)(k+2)(k+3)]$, we see that

$$K_n = \frac{1}{5} \sum_{k=1}^{n} [k(k+1)(k+2)(k+3)(k+4) - (k-1)k(k+1)(k+2)(k+3)]$$

$$= \frac{1}{5}n(n+1)(n+2)(n+3)(n+4).$$

Similarly, we can get that

$$T_n = \frac{1}{4} \sum_{k=1}^{n} [k(k+1)(k+2)(k+3) - (k-1)k(k+1)(k+2)]$$

$$= \frac{1}{4}n(n+1)(n+2)(n+3).$$

Therefore,

$$S_n = K_n - 2T_n = \frac{1}{10}n(n+1)(n+2)(n+3)(2n+3). \qquad\square$$

Remark. To find K_n and T_n, we used the method of elimination by splitting terms.

Example 6. Assume that the sequence $\{a_n\}$ of integers satisfies the conditions:

(1) $a_{n+2} = 3a - n + 1 - 3a_n + a_{n-1}$ for $n = 2, 3, 4, \ldots$;
(2) $2a_2 = a_1 + a_3 - 2$;
(3) $a_5 - a_4 = 9$ and $a_1 = 1$.

Find the sum S_n of the first n terms of the sequence $\{a_n\}$.

Solution. Let $b_n = a_{n+1} - a_n$ and $c_n = b_{n+1} - b_n$. From condition (1),

$$c_n = b_{n+1} - b_n$$
$$= (a_{n+2} - a_{n+1}) - (a_{n+1} - a_n)$$
$$= a_{n+2} - 2a_{n+1} + a_n$$
$$= (3a_{n+1} - 3a_n + a_{n-1}) - 2a_{n+1} + a_n$$
$$= a_{n+1} - 2a_n + a_{n-1}$$
$$= c_{n-1} \ (n = 2, 3, 4, \ldots).$$

Thus, $\{c_n\}$ is a constant sequence.

From condition (2), $c_2 = a_3 - 2a_2 + a_1 = 2$, hence $c_n \equiv 2$. Note that

$$a_n = a_1 + \sum_{k=1}^{n-1} b_k$$
$$= a_1 + (n-1)b_1 + \frac{(n-1)(n-2)}{2} 2$$
$$= 1 + (n-1)b_1 + (n-1)(n-2).$$

Now, from condition (3), $b_4 = 9$, but $b_4 = b_1 + 3 \cdot 2$, so $b_1 = 3$. It follows that

$$a_n = 1 + 3(n-1) + (n-1)(n-2)$$
$$= 1 + 3n - 3 + n^2 - 3n + 2$$
$$= n^2.$$

Therefore, $S_n = 1^2 + 2^2 + 3^2 + \cdots + n^2 = \frac{1}{6}n(n+1)(2n+1)$. $\quad\square$

Example 7. Suppose that the second order difference sequence of a sequence $\{a_n\}$ is a geometric sequence, and $a_1 = 5, a_2 = 6, a_3 = 9$, and $a_4 = 16$. Find a formula for the general term of $\{a_n\}$.

Solution. It is easy to calculate out that the second order difference sequence $\{C_n\}$ of $\{a_n\}$ is a geometric sequence with 2 as its first term and 2 as its common ratio, so $C_n = 2^n$. Denote the first order difference sequence of $\{a_n\}$ as $\{b_n\}$. Then $b_1 = a_2 - a_1 = 1$, and

$$b_n = b_1 + \sum_{k=1}^{n-1}(b_{k+1} - b_k) = 1 + \sum_{k=1}^{n-1} C_k$$

$$= 1 + \sum_{k=1}^{n-1} 2^k = 1 + \frac{1 - 2^{n-1}}{1 - 2} 2$$

$$= 2^n - 1.$$

Consequently,

$$a_n = a_1 + \sum_{k=1}^{n-1}(a_{k+1} - a_k) = a_1 + \sum_{k=1}^{n-1}(2^k - 1)$$

$$= 5 + \frac{1 - 2^{n-1}}{1 - 2} 2 - (n - 1)$$

$$= 2^n - n + 4. \qquad \square$$

Example 8. Let $a_1 < a_2 < \cdots < a_m$ be real numbers. The sequence $a_1, a_2, \ldots, a_m$ is called a weakly arithmetic sequence of length m if there exist real numbers $x_0, x_1, \ldots, x_m$ and d such that $x_0 \le a_1 < x_1 \le a_2 < x_2 \le \cdots \le a_m \le x_m$ and $x_{i+1} - x_i = d$ for any $0 \le i \le m - 1$ (in other words, $x_0, x_1, \ldots, x_m$ constitute an arithmetic sequence of common difference d).

(1) Prove: If $a_1 < a_2 < a_3$, then the sequence a_1, a_2, a_3 is a weakly arithmetic sequence of length 3.
(2) Let A be a subset of the set $\{0, 1, 2, \ldots, 999\}$ that contains at least 730 elements. Prove: A contains a weakly arithmetic sequence of length 10.

Proof. (1) Consider the first order difference sequence d_1, d_2 of the sequence a_1, a_2, a_3. If $d_1 = d_2$, then by choosing $x_0 = a_1, x_1 = a_2, x_2 = a_3$, and $x_3 = a_3 + d_1$, we see that a_1, a_2, a_3 is a weakly arithmetic sequence. If $d_1 > d_2$, then by letting $d = d_1 + \frac{d_2}{2}$ and choosing $x_1 = a_1 + \frac{1}{3}d_2, x_0 = x_1 - d, x_2 = x_1 + d$, and $x_3 = x_1 + 2d$, we have $x_0 < a_1 < x_1 < a_2 < x_2 < a_3 < x_3$ and the conclusion is satisfied. If $d_1 < d_2$, then by letting $d = \frac{d_1}{2} + d_2$ and choosing $x_1 = a_1 + \frac{1}{3}d_2, x_0 = x_1 - d, x_2 = x_1 + d$, and $x_3 = x_1 + 2d$, we again obtain the conclusion. Hence, a_1, a_2, a_3 is a weakly arithmetic sequence.

(2) Suppose that there is no weakly arithmetic sequence of length 10. Consider the following sets

$$A_k = \{100k, 100k + 1, 100k + 2, \ldots, 100k + 99\}, \ k = 0, 1, 2, \ldots, 9.$$

Note that the sequence $0, 100, 200, \ldots, 1000$ is an arithmetic sequence of length 11. So, there exists an index k' such that $A_{k'}$ does not have an element that belongs to A.

For remaining indices k, consider the following sets

$$A_{k,j} = \{100k + 10j, 100k + 10j + 1, \ldots, 100k + 10j + 9\}, \ j = 0, 1, 2, \ldots, 9.$$

Since $100k + 10j, 100k + 10j + 10, \ldots, 100k + 10j + 100$ is an arithmetic sequence of length 10, there exists a j such that $A_{k,j}$ does not have an element that belongs to A. Also, all the 10 numbers of $A_{k,j}$ form an arithmetic sequence, so every $A_{k,j}$ has at least one element that does not belong to A.

The above discussion indicates that there are at least $100 + 9(10 + 9) = 271$ elements in the set $\{0, 1, 2, \ldots, 999\}$ that do not belong to A, but this contradicts the fact that A has at least 730 elements.

In summary, the proposition has been proved. □

3. Exercises

Group A

I. Filling Problems

1. Denote the first order difference sequence of a sequence $\{a_n\}$ as $\Delta\{a_n\}$. If all terms of the sequence $\Delta(\Delta\{a_n\})$ are 1 and $a_{19} = a_{92} = 0$, then the value of a_1 is _____.

2. The sum of the first n terms of the sequence $\{n^3\}$ is _____.

3. If the first six terms of a third order arithmetic sequence are $1, 2, 8, 22, 47,$ and 86, then its general term is _____.

4. The first order difference sequence of $\{a_n\}$ is $\{n(n+2)\}$. Then the general term of $\{a_n\}$ with $a_1 = 2$ is _____.

II. Calculation Problems

5. Suppose that a given sequence $\{a_n\}$ satisfies $a_n = n^4 - 2n^3 + 2n^2 - n + \frac{1}{5}$. Find the sum S_n of its first n terms.

6. The first term of a sequence $\{a_n\}$ is $a_1 = 1$, the first term of the first order difference sequence of $\{a_n\}$ is 7, and the second order difference sequence of $\{a_n\}$ is $\{6n + 6\}$. Find the general term formula of $\{a_n\}$.

7. Prove: $\frac{1}{1\cdot 2} + \frac{1}{3\cdot 4} + \cdots + \frac{1}{(2n-1)2n} = \frac{1}{n+1} + \frac{1}{n+2} + \cdots + \frac{1}{n+n}$.

Group B

8. If the first order difference sequence of $\{a_n\}$ is a geometric sequence with common ratio q ($\neq 1$), then prove: The general term of $\{a_n\}$ is
$$a_n = a_1 + (a_2 - a_1)\frac{1-q^{n-1}}{1-q}.$$

9. Find the sum $S_n = a + 4a^2 + 9a^3 + \cdots + n^2 a^n$ ($a \neq 1$).

10. Arrange positive odd numbers according to Table 5.1 and denote the number at the mth row from the top and the nth column from the left as a_{mn}. Find the general term formulas of the sequence $\{a_{m1}\}$ and the sequence $\{a_{1n}\}$.

Table 5.1 Table for Exercise 10.

1	3	7	13	21	31	
5	9	15	23			
11	17	25				
19	27					
29						

11. The first term of a sequence $\{a_n\}$ is $a_1 = 4$ and the first order difference sequence of $\{a_n\}$ is $\{n^2\}$. Find the general term of $\{a_n\}$.

12. Let $\{a_n\}$ be an arithmetic sequence of order 2 with initial values $a_1 = 1$ and $a_2 = 3$. If the sum of its first three terms is $S_3 = 13$, find a formula for the sum S_n of the first n terms of $\{a_n\}$.

Chapter 6

Sequence Summation

1. Key Points of Knowledge and Basic Methods

Sequence summation is an important aspect of sequence research. The sum of the first n terms is usually denoted as S_n, that is,

$$S_n = a_1 + a_2 + \cdots + a_n = \sum_{k=1}^{n} a_k.$$

A. *Important formulas*

(1) $1 + 2 + 3 + \cdots + n = \frac{1}{2}n(n+1)$.

(2) $1^2 + 2^2 + 3^2 + \cdots + n^2 = \frac{1}{6}n(n+1)(2n+1)$.

(3) $1^3 + 2^3 + 3^3 + \cdots + n^3 = (1 + 2 + 3 + \cdots + n)^2 = \frac{1}{4}n^2(n+1)^2$.

(4) For any arithmetic sequence, $S_{m+n} = S_m + S_n + mnd$, where $m, n \in \mathbf{N}_+$ and d is the common difference.

(5) For any geometric sequence, $S_{m+n} = S_n + q^n S_m = S_m + q^m S_n$, where $m, n \in \mathbf{N}_+$ and q is the common ratio.

(6) If the common ratio q of an infinite geometric sequence $\{a_n\}$ satisfies $|q| < 1$, then the sum of all the terms

$$S = \lim_{n \to \infty} S_n = \frac{a_1}{1 - q}.$$

B. *Basic methods*

(1) Idea of reduction.

"When solving a problem, we always use the problems that have been solved before, the results obtained or the methods adopted, or the experience that we have obtained from solving them" (quoted from Polya).

In the middle school algebra textbooks, only the summation formulas for arithmetic and geometric sequences are deduced, so reducing the summation problem of sequences to that of arithmetic or geometric sequences is an important idea in solving problems.

(2) Summation by splitting terms.

Split the general term formula of a sequence $\{a_n\}$ into the difference of two expressions in the form of $a_n = f(n+1) - f(n)$, where f is a function, and then eliminate the middle terms by mutual cancellation in the summation. This method of summation by means of "splitting first and elimination second" is called the summation by splitting terms. The method is the most frequently used one in sequence summation.

(3) Summation by combining terms.

Put some terms together to sum up, and then find S_n.

There are numerous methods of summation for sequences, and which method to choose depends on the concrete situation.

2. Illustrative Examples

Example 1. Find the sum $\sum_{n=1}^{2015}(-1)^n \frac{n^2+n+1}{n!}$.

Solution. For $n \in \mathbf{N}_+$, we have

$$\frac{n^2+n+1}{n!} = \frac{n+1}{n!} + \frac{n^2}{n!} = \frac{n+1}{n!} + \frac{n}{(n-1)!} \quad (0! = 1 \text{ by convention}),$$

so

$$(-1)^n \frac{n^2+n+1}{n!} = (-1)^n \frac{n+1}{n!} + (-1)^n \frac{n}{(n-1)!}$$

$$= (-1)^n \frac{n+1}{n!} - (-1)^{n-1} \frac{n}{(n-1)!}.$$

It follows that

$$\sum_{n=1}^{2015}(-1)^n \frac{n^2+n+1}{n!} = \sum_{n=1}^{2015}\left[(-1)^n \frac{n+1}{n!} - (-1)^{n-1} \frac{n}{(n-1)!}\right]$$

$$= -\frac{2016}{2015!} - 1. \qquad \square$$

Example 2. Assume that $a_k = \frac{2^k}{3^{2^k}+1}$ for all natural numbers k. Let $A = a_0 + a_1 + \cdots + a_9$ and $B = a_0 a_1 \cdots a_9$. Find $\frac{A}{B}$.

Solution. For any number m greater than 1, since

$$\frac{1}{m-1} - \frac{1}{m+1} = \frac{2}{m^2-1},$$

so

$$\frac{1}{m+1} = \frac{1}{m-1} - \frac{2}{m^2-1}.$$

Multiplying 2^k to the both sides and letting $m = 3^{2^k}$, we get

$$\frac{2^k}{3^{2^k}+1} = \frac{2^k}{3^{2^k}-1} - \frac{2^{k+1}}{3^{2^{k+1}}-1}. \tag{6.1}$$

Substituting $k = 0, 1, \ldots, 9$ into (6.1) and adding the equalities give

$$A = \frac{1}{2} - \frac{2^{10}}{3^{2^{10}}-1}. \tag{6.2}$$

On the other hand, denote

$$C = 2^0 2^1 2^2 \cdots 2^9 = 2^{45},$$

$$\begin{aligned}
D &= \left(3^{2^0}+1\right)\left(3^{2^1}+1\right)\left(3^{2^2}+1\right)\cdots\left(3^{2^9}+1\right)\\
&= \frac{1}{2}\left(3^{2^0}-1\right)\left(3^{2^0}+1\right)\left(3^{2^1}+1\right)\left(3^{2^2}+1\right)\cdots\left(3^{2^9}+1\right)\\
&= \frac{1}{2}\left(3^{2^1}-1\right)\left(3^{2^1}+1\right)\left(3^{2^2}+1\right)\cdots\left(3^{2^9}+1\right) = \cdots\\
&= \frac{1}{2}\left(3^{2^{10}}-1\right).
\end{aligned}$$

Then

$$\frac{1}{B} = \frac{D}{C} = \frac{3^{2^{10}}-1}{2^{46}}. \tag{6.3}$$

By (6.2) and (6.3), $\frac{A}{B} = \left(\frac{1}{2} - \frac{2^{10}}{3^{2^{10}}-1}\right)\frac{3^{2^{10}}-1}{2^{46}} = \frac{3^{2^{10}}-1}{2^{47}} - \frac{1}{2^{36}}$. $\qquad\square$

Example 3. Suppose that the general term of a sequence $\{a_n\}$ is $a_k = 2^k$ for $k = 1, 2, \ldots, n$. Find the sum of all possible products $a_i a_j$ ($1 \leq i \leq j \leq n$).

Solution. Since $\sum_{1 \le i \le j \le n} a_i a_j = \frac{1}{2}[a_1^2 + a_2^2 + \cdots + a_n^2 + (a_1 + a_2 + \cdots + a_n)^2]$, and

$$a_1 + a_2 + \cdots + a_n = \frac{2(2^n - 1)}{2 - 1} = 2(2^n - 1),$$

$$a_1^2 + a_2^2 + \cdots + a_n^2 = \frac{4(4^n - 1)}{4 - 1} = \frac{4(2^{2n} - 1)}{3},$$

so we have that

$$\sum_{1 \le i \le j \le n} a_i a_j = \frac{1}{2}\left[\frac{4(2^{2n} - 1)}{3} + (2(2^n - 1))^2\right]$$

$$= \frac{4}{3}(2^{2n+1} - 3 \cdot 2^n + 1). \qquad \square$$

Example 4. A sequence $\{x_n\}$ is defined as follows: $x_1 = \frac{1}{2}$ and $x_n = \frac{2n-3}{2n}x_{n-1}$ for $n = 2, 3, \ldots$. Prove: For any positive integer n, there holds that $x_1 + x_2 + \cdots + x_n < 1$.

Proof. From the given condition we know that $x_n > 0$ for any $n \in \mathbf{N}_+$, and

$$x_k = (2k - 3)x_{k-1} - (2k - 1)x_k, \ k = 2, 3, \ldots.$$

Thus, we have

$$\sum_{k=1}^{n} x_k = x_1 + \sum_{k=2}^{n}[(2k - 3)x_{k-1} - (2k - 1)x_k]$$

$$= x_1 + x_1 - (2n - 1)x_n$$

$$= 1 - (2n - 1)x_n$$

$$< 1.$$

Hence, the proposition is valid. $\qquad \square$

Example 5. Suppose that the sum of the first n terms of a sequence $\{a_n\}$ is $S_n = 2a_n - 1$ and the sequence $\{b_n\}$ satisfies $b_1 = 3$ and $b_{k+1} = a_k + b_k$ for $k = 1, 2, \ldots$. Find the sum of the first n terms of the sequence $\{b_n\}$.

Solution. Since $S_n = 2a_n - 1$, we have $a_1 = S_1 = 2a_1 - 1$, so $a_1 = 1$.
Also

$$a_k = S_k - S_{k-1} = (2a_k - 1) - (2a_{k-1} - 1)$$
$$= 2a_k - 2a_{k-1},$$

thus

$$a_k = 2a_{k-1}.$$

Hence, $\{a_n\}$ is a geometric sequence with first term 1 and common ratio 2.
Choosing $k = 1, 2, \ldots, n-1$ in $b_{k+1} = a_k + b_k$ gives

$$b_2 = a_1 + b_1,$$
$$b_3 = a_2 + b_2,$$
$$\cdots$$
$$b_n = a_{n-1} + b_{n-1}.$$

Adding up the both sides of the above equalities, we obtain

$$b_n = S_{n-1} + b_1 = \frac{2^{n-1} - 1}{2 - 1} + 3 = 2^{n-1} + 2.$$

Therefore, the sum of the first n terms of the sequence $\{b_n\}$ is

$$S'_n = 1 + 2 + 2^2 + \cdots + 2^{n-1} + 2n = 2^n + 2n - 1. \qquad \square$$

Example 6. The sum of all terms of the sequence

$$1, \ 1, \ 3, \ 3, \ 3^2, \ 3^2, \ \ldots, \ 3^{2002}, \ 3^{2002}$$

is denoted by S. For a given positive integer n, if one can choose different
terms of the sequence whose sum is exactly n, then it is called one option of
choice. The number of all options of choice with the same sum n is denoted
as a_n. Find $a_1 + a_2 + \cdots a_S$.

Solution. Note that $1 + 1 + 3 + 3 + 3^2 + 3^2 + \cdots + 3^{n-1} + 3^{n-1} = 3^n - 1$.
Let $P_n = a_1 + a_2 + \cdots + a_{3^n - 1}$, where $n \in \mathbf{N}_+$.
For the sequence

$$1, \ 1, \ 3, \ 3, \ \ldots, \ 3^{n-1}, \ 3^{n-1}, \ 3^n, \ 3^n \qquad (6.4)$$

of $2(n+1)$ terms, the number of options of choice to represent a natural
number from 1 to $3^n - 1$ from the first $2n$ terms is denoted as P_n. If the

number of options of choice to represent a natural number $m \in [1, 3^n - 1]$ from the first $2n$ terms of the sequence (6.4) is a_m, then there are $2a_m$ options of choice to represent the natural number $m + 3^n$ by the sequence (6.4). There are a_m options of choice to represent the natural number $m + 2 \cdot 3^n$ by the sequence (6.4). Besides, it is easy to obtain that $a_{s^n} = 2$ and $a_{2 \cdot 3^n} = 1$.

Therefore, we deduce that $P_{n+1} = P_n + 2P_n + P_n + 2 + 1 = 4P_n + 3$. Furthermore,

$$P_1 = a_1 + a_{3-1} = 2 + 1 = 3.$$

Thus, $P_n = 4P_{n-1} + 3 = 4(4P_{n-2} + 3) + 3 = \cdots = 4^{n-1}P_1 + 3(4^{n-2} + 4^{n-3} + \cdots + 4 + 1) = 4^{n-1}3 + (4^{n-1} - 1) = 4^n - 1$.

In this problem, $n = 2003$, so $a_1 + a_2 + \cdots + a_S = P_{2003} = 4^{2003} - 1$. $\square$

Remark. The problem can be extended to a general case: Suppose that an integer $k \geq 2$ and a sequence

$$1, 1, \ldots, 1, \quad k, k, \ldots, k, \quad \ldots, \quad k^{n-1}, k^{n-1}, \ldots, k^{n-1} \quad (n \in \mathbf{N}_+)$$

$$(6.5)$$

is given, in which there are $k-1$ repeating numbers for $1, k, \ldots, k^{n-1}$ respectively. The sum of all terms of the above sequence is denoted by S. For a given natural number $m \in [1, k^n - 1]$, the number of options of choice to represent m as the sum of different terms of the sequence (6.5) is denoted as a_m. Then $a_1 + a_2 + \cdots a_S = 2^{n(k-1)} - 1$. The solution method for the present problem is the recursive method.

Example 7. Assume that a sequence $\{a_n\}$ satisfies: $S_n = 1 - a_n$ $(n \in \mathbf{N}_+)$, where S_n is the sum of the first n terms of $\{a_n\}$.

(1) Find a formula for the general term of $\{a_n\}$.
(2) Let $c_n = \frac{1}{1+a_n} + \frac{1}{1-a_n}$ and denote by P_n the sum of the first n terms of $\{c_n\}$. Prove: $P_n > 2n - \frac{1}{5}$.

Solution. (1) Since

$$S_n = 1 - a_n,$$

$$(6.6)$$

so

$$S_{n+1} = 1 - a_{n+1}.$$

$$(6.7)$$

(6.7) $-$ (6.6) gives $a_{n+1} = -a_{n+1} + a_n$, thus $a_{n+1} = \frac{1}{2}a_n$ for $n \in \mathbf{N}_+$.

Also, $a_1 = \frac{1}{2}$ when $n = 1$, from which

$$a_n = \frac{1}{2}\left(\frac{1}{2}\right)^{n-1} = \left(\frac{1}{2}\right)^n, \ n \in \mathbf{N}_+.$$

(2) From the given condition,

$$
\begin{aligned}
P_n &= \frac{1}{1+a_1} + \sum_{i=2}^{n}\left(\frac{1}{1-a_i} - \frac{1}{1+a_i}\right) + \frac{1}{1-a_{n+1}} \\
&= \frac{2}{3} + \frac{1}{1-\left(\frac{1}{2}\right)^{n+1}} + 2\sum_{i=2}^{n}\frac{1}{1-a_i^2} \\
&= \frac{2}{3} + \frac{2^{n+1}}{2^{n+1}-1} + 2\sum_{i=2}^{n}\frac{4^i}{4^i-1} \\
&= \frac{2}{3} + \frac{2^{n+1}}{2^{n+1}-1} + 2\sum_{i=2}^{n}\left(1 + \frac{1}{4^i-1}\right) \\
&> \frac{2}{3} + 1 + 2\left(1 + \frac{1}{4^2-1}\right) + 2(n-2) = 2n - \frac{1}{5} \ (n \geq 2).
\end{aligned}
$$

When $n = 1$, we have $P_1 = 2 > 2 - \frac{1}{5}$. Therefore, the conclusion is true. □

Example 8. Let $f(n)$ be a function defined on the set of all positive integers such that $f(1) = 2$ and

$$f(n+1) = f(n)^2 - f(n) + 1, \ n = 1, 2, 3, \ldots.$$

Prove: For any integer $n > 1$, there holds that

$$1 - \frac{1}{1-2^{2n-1}} < \frac{1}{f(1)} + \frac{1}{f(2)} + \cdots + \frac{1}{f(n)} < 1 - \frac{1}{1-2^{2n}}.$$

Proof. Note that $f(n) > 1$ for any $n \in \mathbf{N}_+$. Then

$$\frac{1}{f(n+1)-1} = \frac{1}{f(n)^2 - f(n)} = \frac{1}{f(n)-1} - \frac{1}{f(n)}.$$

It follows that

$$\frac{1}{f(n)} = \frac{1}{f(n)-1} - \frac{1}{f(n+1)-1}, \ n = 1, 2, \ldots.$$

Therefore,

$$\sum_{k=1}^{n} \frac{1}{f(k)} = \sum_{k=1}^{n} \left[\frac{1}{f(k) - 1} - \frac{1}{f(k+1) - 1} \right]$$

$$= 1 - \frac{1}{f(n+1) - 1}. \tag{6.8}$$

On the other hand, from the recursive formula we know that $f(n + 1) - 1 = f(n)^2 - f(n) > [f(n) - 1]^2$, so $f(n+1) - 1 > [f(n) - 1]^2 > [f(n-1) - 1]^{2^2} > \cdots > [f(2) - 1]^{2^{n-1}} = 2^{2^{n-1}}$. Hence, by (6.8) we have $\sum_{k=1}^{n} \frac{1}{f(k)} > 1 - \frac{1}{2^{2^{n-1}}}$.

Again from the recursive formula, we are sure that $f(n) \in \mathbf{N}_+$, and consequently $f(n+1) = f(n)^2 - f(n) + 1 < f(n)^2$. This recursive relation implies that $f(n+1) < f(1)^{2^n} = 2^{2^n}$, from which

$$\frac{1}{f(n+1) - 1} > \frac{1}{2^{2^n} - 1} > \frac{1}{2^{2^n}}.$$

Therefore,

$$\sum_{k=1}^{n} \frac{1}{f(k)} < 1 - \frac{1}{2^{2^n}}.$$

To summarize, the proposition is proved. □

3. Exercises

Group A

I. Filling Problems

1. Let the general term of a given sequence $\{a_n\}$ be $a_n = \sqrt{1 + \frac{1}{n^2} + \frac{1}{(n+1)^2}}$, and let the sum of its first n terms be S_n. Then $\lfloor S_n \rfloor = $ _____. ($\lfloor S_n \rfloor$ represents the largest integral part of S_n.)

2. Suppose that for any positive integer n, the parabola $y = (n^2 + n)x^2 - (2n + 1)x + 1$ intersects the x-axis at two points A_n and B_n. If $|A_n B_n|$ represents the distance between the two points, then the value of $|A_1 B_1| + |A_2 B_2| + \cdots + |A_{2002} B_{2002}|$ is _____.

3. The sum $\frac{1}{2!} + \frac{2}{3!} + \cdots + \frac{3}{4!} + \cdots + \frac{n}{(n+1)!} = $ _____.

4. The value of $1! + 2 \cdot 2! + 3 \cdot 3! + 4 \cdot 4! + \cdots + nn! = $ _____.

5. If $a \in \mathbf{R}$ and $a \neq -1$, then $a - 2a^2 + 3a^3 - \cdots + (-1)^{n-1}na^n = $ _____.

6. The sum $\frac{4}{1\cdot2\cdot3} + \frac{5}{2\cdot3\cdot4} + \cdots + \frac{n+3}{n(n+1)(n+2)} =$ _____.

7. Suppose that the consecutive terms a_n and a_{n+1} of a sequence $\{a_n\}$ are the two roots of the equation $x^2 - c_n x + \left(\frac{1}{3}\right)^n = 0$, and $a_1 = 2$. Let the sum $c_1 + c_2 + \cdots + c_n + \cdots$ of the infinite sequence $\{c_n\}$ be S. Then $S =$ _____.

II. Calculation Problems

8. Calculate: $\arctan \frac{1}{1+1+1^2} + \arctan \frac{1}{1+2+2^2} + \cdots + \arctan \frac{1}{1+n+n^2}$.

9. Find the sum of the first n terms of $\frac{3}{1\cdot3}, \frac{7}{2\cdot4}3, \ldots, \frac{4n-1}{n(n+2)}3^{n-1}, \ldots$.

Group B

10. Let $x = 1 + \frac{1}{\sqrt{2}} + \frac{1}{\sqrt{3}} + \frac{1}{\sqrt{4}} + \cdots + \frac{1}{\sqrt{1000000}}$. Determine the value of $[x]$, where $[x]$ is the maximal integer not greater than x.

11. Let $f(n)$ be the integer that is the closest to $\sqrt[4]{n}$. Find $\sum_{k=1}^{2002} \frac{1}{f(k)}$.

12. Let S_n be the sum of the first n terms of a sequence $\{a_n\}$ that satisfies $4S_n = 3a_n + 2^{n+1}$ ($n \geq 0$ and $n \in \mathbf{Z}$).

    (1) Find a relation formula for a_n and a_{n-1}.
    (2) Find all the values of a_0 that make the sequence $a_0, a_1, a_2, \ldots$ an increasing one.

Chapter 7

Synthetic Problems for Sequences

Sequence problems in mathematics competitions are not restricted to the synthesis of sequence knowledge, but in most cases are related to the contents of functions, inequalities, number theory, and combinatorics. Consequently, we need to apply various knowledge and methods to solving problems.

1. Illustrative Examples

Example 1. Removing all perfect squares from the sequence $1, 2, \ldots$ of all positive integers, we obtain a new sequence $\{a_n\}$ according to the original order. Find the general term a_n.

Solution. Suppose that $k^2 < a_n < (k+1)^2$ with $k \in \mathbf{Z}_+$. Then $a_n = n+k$.

Since $k^2 < n+k < (k+1)^2$,

$$\begin{cases} k^2 - k - n < 0, \\ k^2 + k - n + 1 > 0. \end{cases}$$

By properties of integers, we have

$$\begin{cases} k^2 - k - n + \dfrac{1}{4} < 0, \\ k^2 + k - n + \dfrac{1}{4} > 0. \end{cases}$$

It follows that

$$\sqrt{n} - \frac{1}{2} < k < \sqrt{n} + \frac{1}{2},$$

hence $k = \left[\sqrt{n} + \frac{1}{2}\right]$.

Thus, the desired general term formula is $a_n = n + \left[\sqrt{n} + \frac{1}{2}\right]$.   □

Example 2. Assume that a monotonically increasing geometric sequence $\{a_n\}$ satisfies $a_2 + a_3 + a_4 = 28$, and $a_3 + 2$ is the arithmetic mean of a_2 and a_4.

(1) Find the general term formula of the sequence $\{a_n\}$.

(2) Let $b_n = a_n \log_{\frac{1}{2}} a_n$ and $S_n = b_1 + b_2 + \cdots + b_n$. Find the minimum value of the positive integers n that satisfy $S_n + n2^n > 50$.

Solution. (1) Let the common ratio of the geometric sequence be q ($q \neq 0$). By the hypothesis of the problem,

$$a_1 q + a_1 q^2 + a_1 q^3 = 28, \tag{7.1}$$

$$a_1 q + a_1 q^3 = 2(a_1 q^2 + 2). \tag{7.2}$$

After eliminating the constant terms in (7.1) and (7.2), we obtain

$$6a_1 q^3 - 15a_1 q^2 + 6a_1 q = 0,$$

that is

$$2q^2 - 5q + 2 = 0.$$

The solutions of the above are $q = 2$ or $q = \frac{1}{2}$. Substituting into (7.1) gives $a_1 = 2$ or $a_1 = 32$. Since $\{a_n\}$ is monotonically increasing, so $a_1 = 2$ and $q = 2$, Hence, the general term formula of the sequence $\{a_n\}$ is $a_n = 2^n$.

(2) From (1) we have $b_n = a_n \log_{\frac{1}{2}} a_n = -n2^n$. Then

$$S_n = b_1 + b_2 + \cdots + b_n = -(1 \cdot 2 + 2 \cdot 2^2 + 3 \cdot 2^3 + \cdots + n2^n).$$

Denote

$$T_n = 1 \cdot 2 + 2 \cdot 2^2 + 3 \cdot 2^3 + \cdots + n2^n. \tag{7.3}$$

Then

$$2T_n = 1 \cdot 2^2 + 2 \cdot 2^3 + 3 \cdot 2^4 + \cdots + (n-1)2^n + n2^{n+1}. \tag{7.4}$$

(7.3) $-$ (7.4) gives that

$$-T_n = 1 \cdot 2 + 1 \cdot 2^2 + \cdots + 1 \cdot 2^n - n2^{n+1}$$
$$= 2^{n+1} - 2 - n2^{n+1},$$

in other words,

$$S_n = -(n-1)2^{n+1} - 2.$$

Now, the inequality $S_n + n2^n > 50$ is the same as

$$-(n-1)2^{n+1} - 2 + n2^{n+1} > 50,$$

namely

$$2^n > 26.$$

Since $2^4 = 16 < 26 < 32 = 2^5$ and the function 2^x is a monotonically increasing one, it follows that the minimum value of the positive integers that satisfy the condition is 5.   □

Example 3. Suppose that a finite sequence $\{a_n\}$ has $2k$ terms (the integer $k \geq 2$) and the first term $a_1 = 2$. Let the sum of the first n terms be S_n, and assume that $a_{n+1} = (a-1)S_n + 2$ ($k = 1, 2, \ldots, 2k-1$), in which the constant $a > 1$.

(1) Prove: The sequence $\{a_n\}$ is a geometric sequence.
(2) If $a = 2^{\frac{2}{2k-1}}$ and a sequence $\{b_n\}$ satisfies $\frac{1}{n}\log_2(a_1 a_2 \cdots a_n)$ ($n = 1, 2, \ldots, 2k$), find the general term formula of the sequence $\{b_n\}$.
(3) If the sequence $\{b_n\}$ in (2) above satisfies the inequality

$$\left| b_1 - \frac{3}{2} \right| + \left| b_2 - \frac{3}{2} \right| + \cdots + \left| b_{2k-1} - \frac{3}{2} \right| + \left| b_{2k} - \frac{3}{2} \right| \leq 4,$$

find the value of k.

Solution. (1) First we see that $a_2 = 2$ when $n = 1$, thus $\frac{a_2}{a_1} = a$.
    When $2 \leq n \leq 2k - 1$,

$$a_{n+1} = (a-1)S_n + 2, \ a_n = (a-1)S_{n-1} + 2, \ a_{n+1} - a_n = (a-1)a_n,$$

so $\frac{a_{n+1}}{a_n} = a$. In other words, the sequence $\{a_n\}$ is a geometric sequence.

(2) From (1) we know that $a_n = 2a^{n-1}$, so

$$a_1 a_2 \cdots a_n = 2^n a^{1+2+\cdots+(n-1)} = 2^n a^{\frac{(n-1)n}{2}} = 2^{n+\frac{(n-1)n}{2k-1}},$$

$$b_n = \frac{1}{2}\left[n + \frac{(n-1)n}{2k-1}\right] = \frac{n-1}{2k-1} + 1, \quad n = 1, 2, \ldots, 2k.$$

(3) Suppose that $b_n \le \frac{3}{2}$, whose solution is $n \le k + \frac{1}{2}$. Since n is a positive integer, $b_n < \frac{3}{2}$ when $n \le k$, and $b_n > \frac{3}{2}$ when $n \ge k+1$. Now, the original expression

$$= \left(\frac{3}{2} - b_1\right) + \left(\frac{3}{2} - b_2\right) + \cdots + \left(\frac{3}{2} - b_k\right)$$

$$+ \left(b_{k+1} - \frac{3}{2}\right) + \cdots + \left(b_{2k} - \frac{3}{2}\right)$$

$$= (b_{k+1} + \cdots + b_{2k}) - (b_1 + \cdots + b_k)$$

$$= \left[\frac{\frac{1}{2}(k+2k-1)k}{2k-1} + k\right] - \left[\frac{\frac{1}{2}(0+k-1)k}{2k-1} + k\right]$$

$$= \frac{k^2}{2k-1}.$$

The inequality $\frac{k^2}{2k-1} \le 4$ is equivalent to $k^2 - 8k + 4 \le 0$, whose solution is $4 - 2\sqrt{3} \le k \le 4 + 2\sqrt{3}$.

Since $k \ge 2$, the original inequality is satisfied for $k = 2, 3, 4, 5, 6, 7$.   □

Example 4. Suppose that two sequences $\{x_n\}$ and $\{y_n\}$ satisfy $x_1 = x_2 = 1$ and $y_1 = y_2 = 2$. Assume that $\frac{x_{n+1}}{x_n} = \lambda \frac{x_n}{x_{n-1}}$ and $\frac{y_{n+1}}{y_n} \ge \lambda \frac{y_n}{y_{n-1}}$, where λ is a nonzero parameter and $n = 2, 3, 4, \ldots$.

(1) If x_1, x_2, and x_3 constitute a geometric sequence, find the value of the parameter λ.
(2) When $\lambda > 0$, prove: $\frac{x_{n+1}}{y_{n+1}} \le \frac{x_n}{y_n}$ ($n \in \mathbf{N}_+$).
(3) When $\lambda > 1$, prove:

$$\frac{x_1 - y_1}{x_2 - y_2} + \frac{x_2 - y_2}{x_3 - y_3} + \cdots + \frac{x_n - y_n}{x_{n+1} - y_{n+1}} < \frac{\lambda}{\lambda - 1} \quad (n \in \mathbf{N}_+).$$

Solution. (1) From the conditions $x_1 = x_2 = 1$ and $\frac{x_3}{x_2} = \lambda \frac{x_2}{x_1}$, we have $x_3 = \lambda$. Furthermore, $x_4 = \lambda^3$ from $\frac{x_4}{x_3} = \lambda \frac{x_3}{x_2}$, and $x_5 = \lambda^6$ from $\frac{x_5}{x_4} = \lambda \frac{x_4}{x_3}$.

If x_1, x_2, and x_3 constitute a geometric sequence, then $x_3^2 = x_1 x_5$, namely $\lambda^2 = \lambda^6$. Since $\lambda \neq 0$, the solutions are $\lambda = \pm 1$.

(2) By the hypothesis, $\lambda > 0, x_1 = x_2 = 1$, and $y_1 = y_2 = 2$. So, $x_n > 0$ and $y_n > 0$. From properties of inequalities, we have

$$\frac{y_{n+1}}{y_n} \geq \lambda \frac{y_n}{y_{n-1}} \geq \lambda^2 \frac{y_{n-1}}{y_{n-2}} \geq \cdots \geq \lambda^{n-1} \frac{y_2}{y_1} = \lambda^{n-1}.$$

On the other hand,

$$\frac{x_{n+1}}{x_n} = \lambda \frac{x_n}{x_{n-1}} = \lambda^2 \frac{x_{n-1}}{x_{n-2}} = \cdots = \lambda^{n-1} \frac{x_2}{x_1} = \lambda^{n-1}.$$

Hence,

$$\frac{y_{n+1}}{y_n} \geq \lambda^{n-1} = \frac{x_{n+1}}{x_n} \ (n \in \mathbf{N}_+),$$

and consequently,

$$\frac{x_{n+1}}{y_{n+1}} \leq \frac{x_n}{y_n} \ (n \in \mathbf{N}_+).$$

(3) When $\lambda > 1$, by (2) we know that $y_n > x_n \geq 1 \ (n \in \mathbf{N}_+)$. Also from (2), $\frac{x_{n+1}}{y_{n+1}} \leq \frac{x_n}{y_n} \ (n \in \mathbf{N}_+)$. Then

$$\frac{y_{n+1} - x_{n+1}}{x_{n+1}} \geq \frac{y_n - x_n}{x_n},$$

and so

$$\frac{y_{n+1} - x_{n+1}}{y_n - x_n} \geq \frac{x_{n+1}}{x_n} = \lambda^{n-1} \ (n \in \mathbf{N}_+).$$

It follows that

$$\frac{x_1 - y_1}{x_2 - y_2} + \frac{x_2 - y_2}{x_3 - y_3} + \cdots + \frac{x_n - y_n}{x_{n+1} - y_{n+1}}$$

$$\leq 1 + \frac{1}{\lambda} + \cdots + \left(\frac{1}{\lambda}\right)^{n-1} = \frac{1 - \left(\frac{1}{\lambda}\right)^n}{1 - \frac{1}{\lambda}} < \frac{\lambda}{\lambda - 1}. \qquad \square$$

Example 5. Let a sequence $\{a_n\} \ (a_n \geq 0)$ satisfy $a_1 = 0, a_2 = 1, a_3 = 9$, and $S_n^2 S_{n-2} = 10 S_{n-1}^3 \ (n > 3)$, where S_n is the sum of the first n terms of the sequence $\{a_n\}$. Find the expression of $a_n \ (n \geq 3)$.

Solution. Since $S_n^2 S_{n-2} = 10 S_{n-1}^3$, we have

$$\left(\frac{S_n}{S_{n-1}} \right)^2 = 10 \frac{S_{n-1}}{S_{n-2}}.$$

Let $b_n = \frac{S_n}{S_{n-1}}$ $(n \geq 3)$. Then $b_n = (10 b_{n-1})^{\frac{1}{2}}$, hence

$$b_n = 10^{\frac{1}{2}} \left(10^{\frac{1}{2}} b_{n-2}^{\frac{1}{2}} \right)^{\frac{1}{2}}$$

$$= 10^{\frac{1}{2} + \frac{1}{4}} b_{n-2}^{\frac{1}{4}}$$

$$\cdots$$

$$= 10^{\frac{1}{2} + \frac{1}{4} + \cdots + \frac{1}{2^{n-3}}} b_3^{\frac{1}{2^{n-3}}}$$

$$= 10.$$

Thus, $S_n = 10 S_{n-1}$, and so

$$a_n = 9 S_{n-1} = 9 a_{n-1} + 9 S_{n-2} = 10 a_{n-1}.$$

Since $a_3 = 9$, we see that $a_n = 9 \cdot 10^{n-3}$ $(n \geq 3)$. $\qquad \square$

Example 6. Suppose that the sum of the first n terms of a sequence $\{a_n\}$ is S_n and $a_1 = 1$. Assume that $S_n = n(a_n + a_{n-1})$ when $n \geq 2$. Find the value of $\left[\frac{a_{2017}}{10} \right]$.

Solution. From the given conditions we know that $S_1 = 1$ and $S_2 = 0$.

When $n \geq 3$, the condition $S_n = n(a_n + a_{n-1})$ together with the equalities $a_n = S_n - S_{n-1}$ and $a_{n-1} = S_{n-1} - S_{n-2}$ imply

$$S_n = n(S_n - S_{n-2}),$$

that is

$$n S_{n-2} = (n-1) S_n.$$

When n is an even integer, $S_n = 0$.
When n is an odd integer, $\frac{S_n}{S_{n-2}} = \frac{n}{n-1}$.
Therefore,

$$S_{2017} = \frac{S_{2017}}{S_{2015}} \frac{S_{2015}}{S_{2013}} \cdots \frac{S_3}{S_1} S_1 = \frac{2017}{2016} \frac{2015}{2014} \cdots \frac{3}{2},$$

and $a_{2017} = S_{2017}$.

Also, since $a_{2017}^2 < \frac{3}{2} \prod_{i=2}^{2016} \frac{i+1}{i} = 2017 \cdot \frac{3}{4} = 1512\frac{3}{4}$ and

$$a_{2017}^2 > \prod_{i=2}^{2017} \frac{i+1}{i} = 1009,$$

we see that $30 < a_{2017} < 40$.

Therefore, $\left[\frac{a_{2017}}{10}\right] = 3$. □

Example 7. Let a function $f(x)$ defined on $\mathbf{R}$ satisfy: $f(1) = \frac{10}{3}$, and for any real numbers x and y, it is always true that

$$f(x)f(y) = f(x+y) + f(x-y). \tag{7.5}$$

Suppose that a sequence $\{a_n\}$ satisfies $a_n = 3f(n) - f(n-1)$ for $n \in \mathbf{N}_+$.

(1) Find a formula for the general term of the sequence $\{a_n\}$.
(2) Let $b_n = \frac{24a_n}{(3a_n-8)^2}$ with $n \in \mathbf{N}_+$, and let S_n be the sum of the first n terms of the sequence $\{b_n\}$. Prove: $S_n < 1$.

Solution. (1) Letting $x = 1$ and $y = 0$ in (7.5) gives $f(1)f(0) = 2f(1)$. Since $f(1) = \frac{10}{3}$, we have $f(0) = 2$.

Substituting $x = n$ and $y = 1$ into (7.5) ensures that

$$f(n)f(1) = f(n+1) + f(n-1),$$

and consequently,

$$f(n+1) = \frac{10}{3}f(n) - f(n-1).$$

Therefore,

$$a_{n+1} = 3f(n+1) - f(n) = 9f(n) - 3f(n-1)$$

$$= 3[3f(n) - f(n-1)] = 3a_n.$$

On the other hand, $a_1 = 3f(1) - f(0)$, so $a_n = 8 \cdot 3^{n-1}$.

(2) By (1), $b_n = \frac{24a_n}{(3a_n-8)^2} = \frac{24 \cdot 8 \cdot 3^{n-1}}{(3 \cdot 8 \cdot 3^{n-1}-8)^2} = \frac{3^n}{(3^n-1)^2}$.

It is easy to show that $3^k - 1 \geq \frac{1}{4}(3^{k+1} - 1)$ is satisfied by all $k \in \mathbf{N}_+$. Thus,

$$b_k = \frac{3^k}{(3^k - 1)^2} \leq \frac{4 \cdot 3^k}{(3^k - 1)(3^{k+1} - 1)} = 2\left(\frac{1}{3^k - 1} - \frac{1}{3^{k+1} - 1}\right).$$

Hence,

$$S_n = b_1 + b_2 + \cdots + b_n$$

$$\leq 2\left[\left(\frac{1}{3^1 - 1} - \frac{1}{3^2 - 1}\right) + \left(\frac{1}{3^2 - 1} - \frac{1}{3^3 - 1}\right) + \cdots \right.$$

$$\left. + \left(\frac{1}{3^n - 1} - \frac{1}{3^{n+1} - 1}\right)\right]$$

$$= 2\left(\frac{1}{3^1 - 1} - \frac{1}{3^{n+1} - 1}\right) = 1 - \frac{2}{3^{n+1} - 1} < 1. \qquad \square$$

Example 8. Suppose that a given sequence $\{x_n\}$ satisfies:

$$x_1 > 0, \ x_{n+1} = \sqrt{5}x_n + 2\sqrt{x_n^2 + 1}, \ n \in \mathbf{N}_+.$$

Prove: Among $x_1, x_2, \ldots, x_{2016}$, there exist at least 672 irrational numbers.

Proof. By the method of mathematical induction, we can show that $\{x_n\}$ is a sequence of positive numbers, and

$$x_{n+1} = \sqrt{5}x_n + 2\sqrt{x_n^2 + 1} > x_n.$$

From the hypothesis of the problem,

$$(x_{n+1} - \sqrt{5}x_n)^2 = 4(x_n^2 + 1),$$

namely

$$x_{n+1}^2 + x_n^2 - 2\sqrt{5}x_n x_{n+1} = 4.$$

By the same token, we obtain

$$x_{n+2}^2 + x_{n+1}^2 - 2\sqrt{5}x_{n+1}x_{n+2} = 4.$$

Subtracting the above two equalities gives

$$x_{n+2}^2 - x_n^2 - 2\sqrt{5}x_{n+1}(x_{n+2} - x_n) = 0,$$

in other words,

$$(x_{n+2} - x_n)(x_{n+2} + x_n - 2\sqrt{5}x_{n+1}) = 0.$$

Since $x_{n+2} - x_n \neq 0$, it follows that $x_{n+2} + x_n - 2\sqrt{5}x_{n+1} = 0$, that is,

$$\frac{x_{n+2} + x_n}{x_{n+1}} = 2\sqrt{5}.$$

This shows that among the three terms x_n, x_{n+1}, and x_{n+2}, at least one term is an irrational number.

As a consequence, there are at least $\left[\frac{2016}{3}\right] = 672$ irrational numbers among $x_1, x_2, \ldots, x_{2016}$. $\qquad\Box$

2. Exercises

Group A

I. Filling Problems

1. Let three positive numbers a, b, and c constitute a geometric sequence with common ratio $q \neq 1$ such that $\log_a b, \log_b c$, and $\log_c a$ form an arithmetic sequence. Then the common difference $d = $ _____.
2. Group the terms of the sequence $1, 2, 3, \ldots$ of positive integers from the left to the right according to the following rule: The first group has $1 \cdot 2$ numbers, the second group has $2 \cdot 3$ numbers, $\ldots$, the nth group has $n(n+1)$ numbers, $\ldots$. Then the number 2014 belongs to _____ th group.
3. Given the function $f(x) = x^2 \cos \frac{\pi x}{2}$. Suppose that in a sequence $\{a_n\}$, the general term $a_n = f(n) + f(n+1)$ $(n \in \mathbf{N}_+)$. Then the sum S_{100} of the first 100 terms of the sequence $\{a_n\}$ is _____.

II. Calculation Problems

4. In an arithmetic sequence $\{a_n\}$, let $a_1 = 1$ and the sum S_n of the first n terms satisfy the condition

$$\frac{S_{2n}}{S_n} = \frac{4n+2}{n+1}, \quad n = 1, 2, \ldots.$$

   (1) Find a general term formula of the sequence $\{a_n\}$.
   (2) Denote $b_n = a_n p^{a_n}$ $(p > 0)$. Find the sum T_n of the first n terms of the sequence $\{b_n\}$.

5. Let the first term a_1 and the common difference d of an arithmetic sequence $\{a_n\}$ be both integers, and let the sum of the first n terms be S_n.

(1) If $a_{11} = 0$ and $a_{14} = 98$, find a general term formula of the sequence $\{a_n\}$.

(2) If $a_1 \geq 6, a_{11} > 0$, and $S_{14} \leq 77$, find general term formulas for all possible sequences $\{a_n\}$.

6. Suppose that the sum of the first n terms of a sequence $\{a_n\}$ is S_n and the points $\left(n, \frac{S_n}{n}\right)$ are all on the graph of the function $y = 3x - 2$.

(1) Find a general term formula of the sequence $\{a_n\}$.

(2) Let $b_n = \frac{3}{a_n a_{n+1}}$ and T_n the sum of the first n terms of the sequence $\{b_n\}$. Find the minimal positive integer m such that $T_n < \frac{m}{20}$ is satisfied for all $n \in \mathbf{N}_+$.

7. A sequence $\{a_n\}$ satisfies $a_1 = \frac{1}{2}$ and $a_{n+1} = a_n^2 + a_n$ for $n \in \mathbf{N}_+$. Let $b_n = \frac{1}{1+a_n}$ and

$$S_n = b_1 + b_2 + \cdots + b_n, \quad P_n = b_1 b_2 \cdots b_n.$$

Find the value of $2P_n + S_n$.

8. Assume that $a_1 = 2$ and the points (a_n, a_{n+1}) are on the graph of the function $f(x) = x^2 + 2x$ for $n = 1, 2, 3, \ldots$.

(1) Show that the sequence $\{\lg(1 + a_n)\}$ is a geometric sequence.

(2) Let $T_n = (1 + a_1)(1 + a_2) \cdots (1 + a_n)$. Find the general terms of the sequences $\{T_n\}$ and $\{a_n\}$.

(3) Denote $b_n = \frac{1}{a_n} + \frac{1}{a_n+2}$. Find the sum S_n of the first n terms of the sequence $\{b_n\}$, and show that

$$S_n + \frac{2}{3T_n - 1} = 1.$$

Group B

9. Let three sequences $\{a_n\}, \{b_n\}$, and $\{c_n\}$ satisfy:

$$b_n = a_n - a_{n+2}, \quad c_n = a_n + 2a_{n+1} + 3a_{n+2} \ (n = 1, 2, 3, \ldots).$$

Prove: A necessary and sufficient condition for the sequence $\{a_n\}$ to be an arithmetic sequence is that $\{c_n\}$ is an arithmetic sequence and $b_n \leq b_{n+1}$ for $n = 1, 2, 3, \ldots$.

10. Suppose that positive number sequences $\{x_n\}$ and $\{y_n\}$ satisfy the condition: For all positive integers n,

$$x_{n+2} = x_n + x_{n+1}^2, \quad y_{n+2} = y_n^2 + y_{n+1},$$

and x_1, x_2, y_1, and y_2 are all greater than 1. Prove: There exists a positive integer n such that $x_n > y_n$.

11. Let $a_1, a_2, \ldots$ be a sequence of integers, in which there are not only infinitely many positive integers but also infinitely many negative integers. If for each positive integer n, the n remainders of the integers $a_1, a_2, \ldots, a_n$ divided by n are mutually different, prove: Every integer appears just once in the sequence $a_1, a_2, \ldots$.

12. For the give function $f(x) = \frac{x^3+3x}{3x^2+1}$, a sequence $\{x_n\}$ satisfies:

$$x_1 = 2, \ x_{n+1} = f(x_n) \ (n \in \mathbf{N}_+).$$

Denote $b_n = \log_3 \frac{x_{n+1}-1}{x_{n+1}+1} \ (n \in \mathbf{N}_+)$.

(1) Prove: The sequence $\{b_n\}$ is a geometric one. And find a general term formula for the sequence $\{b_n\}$.

(2) Denote $c_n = -nb_n \ (n \in \mathbf{N}_+)$. Find a formula for the sum T_n of the first n terms of the sequence $\{c_n\}$.

Chapter 8

Coordinates Systems

1. Key Points of Knowledge and Basic Methods

A. In a rectangular coordinates plane, a point with the horizontal coordinate and the vertical coordinate both rational numbers (integers) is called a rational point (integral point). Problems concerning rational points can often be solved by reducing to problems with integral points.

B. The distance between two points $P_1(x_1, y_1)$ and $P_2(x_2, y_2)$ in a plane:

$$|P_1 P_2| = \sqrt{(x_1 - x_2)^2 + (y_1 - y_2)^2}.$$

C. *Coordinates of fixed proportion points*

Suppose that a point $P(x, y)$ is on the straight line that passes through two points $P_1(x_1, y_1)$ and $P_2(x_2, y_2)$ with $\overrightarrow{P_1 P} = \lambda \overrightarrow{P P_2}$. Then

$$\begin{cases} x = \dfrac{x_1 + \lambda x_2}{1 + \lambda}, \\ x = \dfrac{y_1 + \lambda y_2}{1 + \lambda}. \end{cases}$$

Corollary. *Three points $P_1(x_1, y_1), P_2(x_2, y_2)$, and $P_3(x_3, y_3)$ are co-linear (not parallel to the y-axis) if and only if*

$$\frac{y_1 - y_2}{x_1 - x_2} = \frac{y_2 - y_3}{x_2 - x_3}.$$

D. In a rectangular coordinates system, it is easy to get the following result from the coordinates formula for a fixed proportion point:

Theorem 1. *The coordinates of the center of gravity for a triangle with vertices $A(x_A, y_A), B(x_B, y_B)$, and $C(x_C, y_C)$ are $\left(\frac{x_A + x_B + x_C}{3}, \frac{y_A + y_B + y_C}{3}\right)$.*

This theorem can be extended to the arbitrary case of n points. That is, the coordinates of the center of gravity G_n of n points $P_i(x_i, y_i)$ ($i = 1, 2, \ldots, n$) are

$$\left(\frac{1}{n}\sum_{i=1}^{n} x_i, \frac{1}{n}\sum_{i=1}^{n} y_i\right).$$

Theorem 2. *The coordinates of the inner center of a triangle with vertices $A(x_A, y_A), B(x_B, y_B)$, and $C(x_C, y_C)$ are*

$$\left(\frac{ax_A + bx_B + cx_C}{a + b + c}, \frac{ay_A + by_B + cy_C}{a + b + b + c}\right),$$

where a, b, and c are the lengths of the three sides of the triangle $\triangle ABC$ opposite to the three vertices A, B, and C, respectively.

Theorem 3. *The coordinates of a parallelogram with successive vertices $A(x_A, y_A), B(x_B, y_B), C(x_C, y_C)$, and $D(x_D, y_D)$ satisfy*

$$\begin{cases} x_A + x_C = x_B + x_D, \\ y_A + y_C = y_B + y_D. \end{cases}$$

Theorem 4. *The directed area of $\triangle ABC$ can be expressed as*

$$\Delta = \frac{1}{2}\begin{vmatrix} 1 & x_A & y_A \\ 1 & x_B & y_B \\ 1 & x_C & y_C \end{vmatrix}.$$

Note that in the usual sense, the area of $\triangle ABC$ is the absolute value of the determinant Δ.

Corollary. *If and only if*

$$\begin{vmatrix} 1 & x_A & y_A \\ 1 & x_B & y_B \\ 1 & x_C & y_C \end{vmatrix} = 0,$$

the three points $A(x_A, y_A), B(x_B, y_B)$, and $C(x_C, y_C)$ are co-linear.

E. Basic methods

The combination of numbers and figures is an information transformation that is full of mathematical feature, and analytic geometry reflects this thought perfectly. By means of a perpendicular coordinates system, we can make a correspondence between pairs (x, y) of ordered numbers and points of the plane, so that a correspondence is established between the equality relation $f(x, y) = 0$ satisfied by pairs of ordered numbers and curves in the plane. Therefore, we can not only use algebraic methods to study shapes, sizes, and position relations of figures, but also use properties of figures to illustrate algebraic facts. Such mutual transformations and organic combinations between the information of algebraic expressions and the information of figures make us much more flexible when solving related problems.

2. Illustrative Examples

Example 1. Suppose that in a perpendicular coordinates system xOy, the coordinates of point $A(x_1, y_1)$ and point $B(x_2, y_2)$ are all positive integers of one digit, the angle of the line segment OA and the positive direction of the x-axis is greater than $45°$, the angle of the line segment OB and the positive direction of the x-axis is less than $45°$, the projection of B onto the x-axis is B', the projection of A onto the y-axis is A', and the area of $\triangle OB'B$ is 33.5 more than the area of $\triangle OA'A$. Let x_1, y_1, x_2, and y_2 form a number a four-digit positive integer $\overline{x_1 x_2 y_2 y_1} = x_1 10^3 + x_2 10^2 + y_2 10 + y_1$. Find all such four-digit numbers, and write out the solution process.

Analysis. Express the areas of $\triangle OB'B$ and $\triangle OA'A$ in terms of the coordinates of the points, and then make discussions according to the feature of the figure and properties of integers.

Solution. As shown in Figure 8.1, the area relation of $\triangle OB'B$ and $\triangle OA'A$ implies that

$$x_2 y_2 - x_1 y_1 = 67.$$

Since $x_1 > 0$ and $y_1 > 0$, so $x_2 y_2 > 67$. Also, since x_2 and y_2 are positive integers of one digit, we have

$$x_2 y_2 = 8 \cdot 9 = 72$$

or

$$x_2 y_2 = 9 \cdot 9 = 81.$$

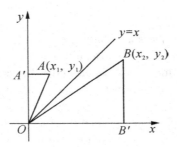

Figure 8.1   Figure for Example 1.

The condition $\angle BOB' < 45°$ implies that $x_2 > y_2$, from which $x_2 y_2 \neq 81$. Thus, $x_2 y_2 = 9 \cdot 8$, namely $x_2 = 9$ and $y_2 = 8$. Hence, $x_1 y_1 = x_2 y_2 - 67 = 5$. Since $\angle AOB' > 45°$, so $x_1 < y_1$. Again the fact that x_1 and y_1 are positive integers of one digit ensures the only possibility that $x_1 = 1$ and $y_1 = 5$. Therefore,

$$\overline{x_1 x_2 y_2 y_1} = 1985. \qquad \square$$

Remark. In general, some simple knowledge of elementary number theory is often used to deal with integral point problems.

Example 2. Pick three arbitrary points (lattice points) in a plane with all integral coordinates. Prove: They cannot constitute the three vertices of an equilateral triangle.

Analysis. If the three vertices of an equilateral triangle are all lattice points, then from the determinantal expression of its area, we see that the area of the triangle is a rational number. But the area in terms of the three sides of the triangle is an irrational number, which leads to a contradiction.

Proof. We use the method of contradiction. Let the coordinates of the three vertices of an equilateral $\triangle ABC$ be $A(x_1, y_1), B(x_2, y_2)$, and $C(x_3, y_3)$, where x_i and y_i are integers for $i = 1, 2$, and 3. Then from

$$S_{\triangle ABC} = \text{the absolute value of } \frac{1}{2} \begin{vmatrix} 1 & x_1 & y_1 \\ 1 & x_2 & y_2 \\ 1 & x_3 & y_3 \end{vmatrix},$$

we see that the area of the equilateral triangle is a rational number. On the other hand, by

$$S_{\triangle ABC} = \frac{\sqrt{3}}{4}|AB|^2 = \frac{\sqrt{3}}{4}\left[(x_2 - x_1)^2 + (y_2 - y_1)^2\right],$$

we know that the area of the equilateral triangle is an irrational number. This gives a contradiction, so three points whose coordinates are all integers cannot constitute an equilateral triangle. □

Example 3. Prove: There does not exist a closed broken line in a plane with an odd number of vertices and the length of each line segment 1, such that the coordinates of each vertex are both rational numbers.

Analysis. Suppose that such a closed broken line exists. Assume that the coordinates of all the vertices are $\left(\frac{a_i}{b_i}, \frac{c_i}{d_i}\right)$, where $\frac{a_i}{b_i}$ and $\frac{c_i}{d_i}$ are irreducible fractions for $i = 1, 2, \ldots, 2k+1$. Using the method of odd-even analysis and a relation of the coordinates for nearby vertices, we arrive at a contradiction.

Proof. Suppose that such a closed broken line exists. Without loss of generality, assume that the origin of a coordinates system is one vertex of the broken line, which is denoted as A_0. The coordinates of the other vertices are

$$A_1 = \left(\frac{a_1}{b_1}, \frac{c_1}{d_1}\right), \ldots, A_n = \left(\frac{a_n}{b_n}, \frac{c_n}{d_n}\right),$$

where $\frac{a_i}{b_i}, \frac{c_i}{d_i}$ are both irreducible fractions. We also denote $A_{n+1} = A_0$.

If the two integers p and q are both odd or even, we write $p \equiv q$; otherwise we write $p \not\equiv q$. In the following we use mathematical induction to prove

$$b_k \equiv 1, \; d_k \equiv 1 \; (k = 1, 2, \ldots, n),$$

$$a_k + c_k \not\equiv a_{k-1} + c_{k-1} \; (k = 1, 2, \ldots, n, n+1).$$

When $k = 1$, from $\left(\frac{a_1}{b_1}\right)^2 + \left(\frac{c_1}{d_1}\right)^2 = 1$, we obtain

$$\frac{a_1^2 d_1^2}{b_1^2} = d_1^2 - c_1^2.$$

Since a_1 and b_1 are relatively prime, so d_1 can be divided by b_1, and vice versa (that is, b_1 can be divided by d_1). Thus, $b_1 = \pm d_1$, and consequently

$$b_1^2 = d_1^2 = a_1^2 + c_1^2.$$

On the other hand, a_1 and c_1 cannot be both even (otherwise b_1 is also even, a contradiction); they cannot be both odd, since the sum of the squares of two odd numbers can be divided by 2, but cannot be divided by 4, thus cannot be a perfect square. It follows that

$$a_1 \not\equiv c_1, \ b_1 \equiv d_1 \equiv 1,$$

and furthermore

$$a_1 + c_1 \not\equiv 0 = a_0 + c_0.$$

Suppose that the conclusion is true for $k = 1, 2, \ldots, m - 1 \leq n$. Let

$$\frac{a_m}{b_m} - \frac{a_{m-1}}{b_{m-1}} = \frac{a}{b}, \ \frac{c_m}{d_m} - \frac{c_{m-1}}{d_{m-1}} = \frac{c}{d},$$

where $\frac{a}{b}$ and $\frac{c}{d}$ are irreducible fractions. Since the length of each line segment is 1, so

$$\left(\frac{a}{b}\right)^2 + \left(\frac{c}{d}\right)^2 = 1.$$

Similar to the case of $k = 1$, we have $a \not\equiv c$ and $d \equiv b \equiv 1$. Also, since

$$\frac{a_m}{b_m} = \frac{a}{b} + \frac{a_{m-1}}{b_{m-1}} = \frac{ab_{m-1} + ba_{m-1}}{bb_{m-1}},$$

the fraction $\frac{a_m}{b_m}$ is irreducible. Hence, b_m is a factor of bb_{m-1}, and thus $b_m \equiv 1$. By the same token, we see that $d_m \equiv 1$. Also $a_m \equiv ab_{m-1} + ba_{m-1}$ (similarly $c_m \equiv cd_{m-1} + dc_{m-1}$). Hence,

$$a_m + c_m - a_{m-1} - c_{m-1}$$

$$\equiv ab_{m-1} + ba_{m-1} + cd_{m-1} + dc_{m-1} - a_{m-1} - c_{m-1}$$

$$\equiv a_{m-1}(b - 1) + ab_{m-1} + c_{m-1}(d - 1) + cd_{m-1}$$

$$\equiv a + c$$

$$\equiv 1.$$

Therefore,

$$a_m + c_m \not\equiv a_{m-1} + c_{m-1},$$

and this is exactly what we have wanted to prove.

As a consequence, when the number $n + 1$ of the vertices is odd, from the above proof,

$$a_{n+1} + c_{n+1} \not\equiv a_0 + c_0,$$

so the broken line cannot be closed. In other words, the closed broken line desired in the problem does not exist. $\qquad\square$

Example 4. In a coordinates plane, a triangle with its three vertices all integral points is called an integral point triangle. Find the number of the integral point right triangles OAB with the point $I(2015, 7 \cdot 2015)$ as the inner center and the coordinates origin O as the right angle vertex.

Solution. See Figure 8.2. Without loss of generality, assume that the point A is in the first quadrant. Let $\angle xOL = \alpha$. Then $\tan \alpha = 7$, and the slope of the straight line OA is

$$k_{OA} = \tan \left(\alpha - \frac{\pi}{4} \right) = \frac{\tan \alpha - 1}{1 + \tan \alpha} = \frac{7 - 1}{1 + 7} = \frac{3}{4}.$$

Thus, $k_{OB} = -\frac{4}{3}$.

Since A and B are integral points, we write $A(4t_1, 3t_1)$ and $B(-3t_2, 4t_2)$, where t_1 and t_2 are positive integers. So, $|OA| = 5t_1$ and $|OB| = 5t_2$.

The radius of the inscribed circle of $\triangle OAB$ is

$$r = \frac{\sqrt{2}}{2} |OI| = \frac{\sqrt{2}}{2} \cdot 5\sqrt{2} \cdot 2015 = 5 \cdot 2015.$$

On the other hand, $r = \frac{|OA| + |OB| - |AB|}{2}$, namely

$$|AB| = |OA| + |OB| - 2r,$$

and then

$$|AB|^2 = (|OA| + |OB| - 2r)^2 = |OA|^2 + |OB|^2.$$

Hence,

$$(5t_1 + 5t_2 - 2 \cdot 5 \cdot 2015)^2 = 25t_1^2 + 25t_2^2.$$

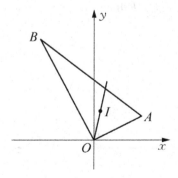

Figure 8.2 Figure for Example 4.

In other words,

$$(t_1 + t_2 - 2 \cdot 2015)^2 = t_1^2 + t_2^2.$$

Let $t_1 = x + 2015$ and $t_2 = y + 2015$. Then

$$(x + y)^2 = (x + 2015)^2 + (y + 2015)^2,$$

that is $xy = 2015x + 2015y + 2015^2$, namely

$$(x - 2015)(y - 2015) = 2 \cdot 2015^2 = 2 \cdot 5^2 \cdot 13^2 \cdot 31^2.$$

From $|OA| > 2r$ and $|OB| > 2r$, we see that $x - 2015$ and $y - 2015$ are positive integers. Also there are $2 \cdot 3 \cdot 3 \cdot 3 = 54$ positive factors of $2 \cdot 5^2 \cdot 13^2 \cdot 31^2$, therefore there are 54 pairs of (x, y) that satisfy the condition.

Hence, there are 54 triangles that satisfy the given requirement.   □

Example 5. Choose three points D, E, and F on the three sides BC, CA, and AB of $\triangle ABC$, respectively. Suppose that $\frac{|BD|}{|DC|} = \frac{|CE|}{|EA|} = \frac{|AF|}{|FB|}$. Prove: $\triangle ABC$ and $\triangle DEF$ have the same center of gravity.

Proof. Set up a rectangular coordinates system as in Figure 8.3. Let the coordinates of A, B, and C be $A(a, b), B(0, 0)$, and $C(c, 0)$. Then the formula for the coordinates of the center of gravity implies that the coordinates of the center of gravity of $\triangle ABC$ are $G\left(\frac{a+c}{3}, \frac{b}{3}\right)$.

Let $\frac{|BD|}{|DC|} = \frac{|CE|}{|EA|} = \frac{|AF|}{|FB|} = \lambda$. The formula for the coordinates of a fixed proportion point gives $D\left(\frac{\lambda c}{1+\lambda}, 0\right), E\left(\frac{c+\lambda a}{1+\lambda}, \frac{\lambda b}{1+\lambda}\right)$, and $F\left(\frac{a}{1+\lambda}, \frac{b}{1+\lambda}\right)$. Again from the formula for the coordinates of the center of gravity, we see that the coordinates of the center of gravity of $\triangle DEF$ are also $G\left(\frac{a+c}{3}, \frac{b}{3}\right)$. Therefore, $\triangle ABC$ and $\triangle DEF$ have the same center of gravity.   □

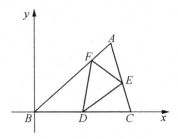

Figure 8.3   Figure for Example 5.

Remark. The method that by establishing a coordinates system in the plane of a plane figure to solve a geometric problem via the knowledge and methods of analytic geometry is called the **analytic method** (or the **method of coordinates**). Whether the process of the analytic method is complicated or simple depends on the choice of the coordinates system. Experiences tell us: According to the feature of the geometric figure, choosing one special point (such as the middle point of a line segment, center of a circle, etc.) as the origin of a coordinates system, and a special line (such as the symmetric axis of the figure or one side of a rectangle, etc.) as a coordinate axis, is a technique that avoids a complicated process of problem solving.

Example 6. Suppose that two points P and Q are on the side AB of $\triangle ABC$ and a point R is on the side AC. Assume that P, Q, and R divide the perimeter of $\triangle ABC$ into three equal parts, as Figure 8.4 shows. Prove:

$$\frac{S_{\triangle PQR}}{S_{\triangle ABC}} > \frac{2}{9}.$$

Proof. Set up a rectangular coordinates system with A as the origin and the straight line AB as the x-axis. Suppose that the sides of $\triangle ABC$ opposite to $\angle A, \angle B$, and $\angle C$ are a, b, and c, respectively, and let the horizontal coordinates of Q and P be q and p, respectively. Then

$$q - p = \frac{1}{3}(a + b + c),$$

$$|AR| = |PQ| - |AP| = q - 2p.$$

So

$$\frac{y_R}{y_C} = \frac{|AR|}{|AC|} = \frac{q - 2p}{b}.$$

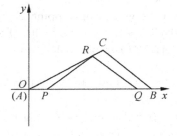

Figure 8.4 Figure for Example 6.

Since $2S_{\triangle PQR} = y_R(q - p)$ and $2S_{\triangle ABC} = x_B y_C$, we have

$$\frac{S_{\triangle PQR}}{S_{\triangle ABC}} = \frac{y_R(q - p)}{y_C x_B} = \frac{(q - p)(q - 2p)}{bc}.$$

By noting that

$$p = q - \frac{1}{3}(a + b + c) < c - \frac{1}{3}(a + b + c),$$

we get

$$q - 2p > \frac{2}{3}(a + b + c) - c$$

$$> \frac{2}{3}(a + b + c) - \frac{1}{2}(a + b + c)$$

$$= \frac{1}{6}(a + b + c).$$

It follows that

$$\frac{S_{\triangle PQR}}{S_{\triangle ABC}} > \frac{2}{9} \cdot \frac{(a + b + c)^2}{4bc} > \frac{2}{9} \cdot \frac{(b + c)^2}{4bc} > \frac{2}{9}. \qquad \square$$

Remark. The number $\frac{2}{9}$ in this problem is the optimal value: Choose $b = c$ and Q the same as B. Then $p \to \frac{1}{3}q$ as $a \to 0$, and the ratio of the areas approaches $\frac{2}{9}$.

Example 7. Please design a method to dye all the integral points: Every integral point has one of the white, red, or black colors, so that

(1) the points of one fixed color appear on infinitely many straight lines that are parallel to the horizontal axis;
(2) for any point A with the white color, point B with the red color, and point C with the black color, a point D with the red color can always be found such that $ABCD$ is a parallelogram.

Prove: You can design a method satisfying the above requirement.

Analysis. Classify the integral points according to the odd-even number structure of the coordinates. This will lead to a dying method satisfying the condition.

Proof. We divide all the integral points of the plane into three groups:

$$P = \{(x, y) \mid (x, y) = (\text{odd, odd})\},$$

$$Q = \{(x, y) \mid (x, y) = (\text{even, even})\},$$

$$R = \{(x, y) \mid (x, y) = (\text{odd, even}) \text{ or } (\text{even, odd})\}.$$

Dye all points of P with the white color, all points of Q with the black color, and all points of R with the red color.

Obviously, this dying method satisfies condition (1), and no three points with distinct colors are co-linear. Otherwise, suppose that white point $A(x_1, y_1)$, red point $B(x_2, y_2)$, and black point $C(x_3, y_3)$ are co-linear. Then

$$(y_2 - y_1)(x_3 - x_1) = (y_3 - y_1)(x_2 - x_1).$$

Since $x_3 - x_1$ and $y_3 - y_1$ are both odd numbers, $y_2 - y_1$ and $x_2 - x_1$ are one odd and one even. Then one side of the above equality is odd and the other side of the equality is even, which is impossible.

Suppose that the point D of the parallelogram $ABCD$ is (x_4, y_4). Then

$$\begin{cases} x_4 = x_1 + x_3 - x_2, \\ y_4 = y_1 + y_3 - y_2. \end{cases}$$

Clearly, x_4 and y_4 are integers. Since $x_1 + x_3$ and $y_1 + y_3$ are both odd numbers, while x_2 and y_2 are one odd and one even, thus x_4 and y_4 are one odd and one even, that is D is a red point. Hence, the condition (2) is also satisfied. $\qquad\square$

Remark. There are simpler dying methods for this problem. For example, Dye all integral points of the forms $(0, n - 1), (0, -n)$ $(n \in \mathbf{N}_+)$, and (p, q) $(p, q \in \mathbf{Z}, \text{and } p \neq 0)$ to the white color, black color, and red color, respectively.

Example 8. Suppose that integers n and k satisfy $n \geq 2$ and $k \geq \frac{5}{2}n - 1$. Prove: For any chosen k different integral points (x, y) $(1 \leq x, y \leq n)$, there always exists a circle that passes through at least four of such points.

Proof. Assume that there are a_i chosen points on the straight line $y = i$. Suppose that $a_i \geq 2$. Let the horizontal coordinates of the a_i points on the line $y = i$ be $x_1, x_2, \ldots, x_{a_i}$ and satisfy $x_1 < x_2 < \cdots < x_{a_i}$. From $x_1 + x_2 < x_1 + x_3 < x_2 + x_3 < x_2 + x_4 < x_3 + x_4 < \cdots < x_{a_i-1} < x_{a_i}$, we see that there are at least $2a_i - 3$ different values among all the sums of two distinct terms in $x_1, x_2, \ldots, x_{a_i}$. This is also true for the case $a_i < 2$.

Since there are exactly $2n - 3$ different values among all the sums of two distinct terms in $1, 2, \ldots, n$, and since

$$\sum_{i=1}^{n}(2a_i - 3) = 2k - 3n \geq 5n - 2 - 3n > 2n - 3,$$

by the drawer principle, there exist four different points

$$(x_1', y_1), \ (x_1'', y_1), \ (x_2', y_2), \ (x_2'', y_2)$$

such that $x_1' + x_1'' = x_2' + x_2''$ and $y_1 \neq y_2$.

Therefore, there exists a circle that passes through these four points. □

3. Exercises

Group A

I. Filling Problems

1. Suppose that D, E, and F are three points on the sides BC, CA, and AB of $\triangle ABC$, respectively, and $\frac{|BD|}{|DC|} = \frac{|CE|}{|EA|} = \frac{|AF|}{|FB|} = \frac{1}{2}$. Also, let line segments BE and CF intersect at L, line segments CF and AD intersect at M, and line segments AD and BE intersect at N. Then $\frac{S_{\triangle LMN}}{S_{\triangle ABC}}$ equals _____.

2. Assume that the coordinates of the three vertices A, B, and C of $\triangle ABC$ are $(-1, 5), (-2, -1)$, and $(5, 2)$, respectively. Choose three points P, Q, and R in the plane, so that the middle points of the line segments PQ, QR, and RP are C, A, and B, respectively. Then the coordinates of P are _____.

3. There are given three parallel lines l_1, l_2, and l_3 with l_2 between l_1 and l_3. Let the distance between l_1 and l_2 be a, and the distance between l_2 and l_3 be b. If the three vertices of an equilateral triangle ABC are on l_1, l_2, and l_3 respectively, then the side length of the equilateral triangle is _____.

II. Calculation Problems

4. For any natural number n, connect the origin O and the point $A_n(n, n+3)$. Let $f(n)$ represent the number of integral points on the line segment OA_n except for the end points. Find the value of $f(1) + f(2) + \cdots + f(1990)$.

5. Extend the sides AB and AC of $\triangle ABC$ to D and E so that $|BD| = |CE|$. Let the middle points of BC and DE be M and N, respectively, and let AT be the interior angular bisector of $\angle A$. Prove: $MN \parallel AT$.

6. Make eight marks on a chessboard such that each row and each column have just one mark. Prove: The number of marks in a black check is an even number.

7. There are two triangles ABC and $A'B'C'$ in the same plane. Suppose that the straight lines AA', BB', and CC' are mutually parallel. If $\triangle ABC$ also represents its signed area with appropriate $+$ or $-$ signs, and so do all other involved triangles, prove:

$$3(\triangle ABC + \triangle A'B'C')$$
$$= \triangle AB'C' + \triangle BC'A' + \triangle CA'B' + \triangle A'BC + \triangle B'CA + \triangle C'AB.$$

8. Let Q be the inner center of $\triangle ABC, |BC| = a, |CA| = b$, and $|AB| = c$. Prove: For any point P, it is always true that

$$a|PA|^2 + b|PB|^2 + c|PC|^2$$
$$= a|QA|^2 + b|QB|^2 + c|QC|^2 + (a + b + c)|QP|^2.$$

Group B

9. As in Figure 8.5, there are two fixed points P and Q on the same side of a fixed straight line l. Find the third point R such that $|PR| + |RQ| + |RS|$ achieves the minimum value. where $RS \perp l$ with S the foot of the perpendicular.

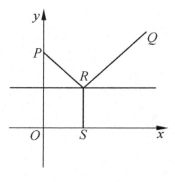

Figure 8.5 Figure for Exercise 9.

10. Prove: In a rectangular coordinates system, there is a set of n points, in which there are no three points that are co-linear, and the center of gravity of any subset of this set is an integral point.

11. In a coordinates plane, move points according to the following rule: A known point $P(x, y)$ can be moved to any one of the four points $P_H(x, y+2x)$, $P_D(x, y-2x)$, $P_L(x-2y, y)$, and $P_R(x+2y, y)$, but if we move P to Q, we cannot return to P from Q. Prove: If starting from the point $(1, \sqrt{2})$, then we are never able to come back to the original starting point.

12. Given a 100-side polygon P in a plane rectangular coordinates system that satisfies:

    (1) The coordinates of the vertices of P are all integers.
    (2) All sides of P are parallel to the coordinates axes.
    (3) All side lengths of P are odd numbers.

    Prove: The area of P is an odd number.

Chapter 9

Straight Lines

1. Key Points of Knowledge and Basic Methods

A. *Several forms of straight line equations*

(1) Point-slope form: $y - y_0 = k(x - x_0)$.

(2) Slope-intercept form: $y = kx + b$.

(3) Two-point form: $\frac{x-x_1}{x_2-x_1} = \frac{y-y_1}{y_2-y_1}$.

(4) Intercepts form: $\frac{x}{a} + \frac{y}{b} = 1$.

(5) General form: $Ax + By + C = 0$ (A and B are not both zero).

B. *Normal equation of a straight line*

Suppose that the distance of the origin O and a straight line l is p, the ray OD is perpendicular to l, and the angle of the rotation from the positive half x-axis to OD is α in the direction of counter clockwise. Then the equation of the straight line l is

$$x \cos \alpha + y \sin \alpha - p = 0.$$

We call the above equation the normal equation of the straight line l. If the equation of l is given by the general form

$$ax + by + c = 0 \ (c \leq 0),$$

then its normal equation is

$$\frac{ax + by + c}{\sqrt{a^2 + b^2}} = 0.$$

C. *Position relation of two straight lines*

(1) Suppose that the slopes of two straight lines both exist. Let the equations of the two straight lines be $l_1 : y = k_1 x + b_1$ and $l_2 : y = k_2 x + b_2$, respectively. Then:

If and only if $k_1 = k_2$ and $b_1 \neq b_2$, the two lines l_1 and l_2 are parallel;

if and only if $k_1 = k_2$ and $b_1 = b_2$, the two lines l_1 and l_2 coincide;

if and only if $k_1 \neq k_2$, the two lines l_1 and l_2 intersect, and in particular if and only if $k_1 k_2 = -1$, the two lines l_1 and l_2 are perpendicular.

When two lines intersect, by solving the system of their equations, we can find the intersection point.

(2) If the slopes of two intersecting straight lines l_1 and l_2 both exist, the angle of the rotation from the straight line l_1 to the straight line l_2 is α in the direction of counter clockwise, and $k_1 k_2 \neq -1$, then

$$\tan \alpha = \frac{k_2 - k_1}{1 + k_1 k_2}.$$

Let the angle between the two straight lines be θ. Then

$$\tan \theta = |\tan \alpha| = \left| \frac{k_2 - k_1}{1 + k_1 k_2} \right|.$$

If at least one slope of the two straight lines does not exist, then their position relation and their angle can be written out directly from the specialty of their graphs.

When two straight lines are parallel or coincident, their angle is $0°$. The range of the angles of two straight lines is $[0, 90°]$.

(3) Suppose that the general equations of the two straight lines are $l_1 : A_1 x + B_1 y + C_1 = 0$ and $l_2 : A_2 x + B_2 y + C_2 = 0$. Then their position relation can be reduced to the above cases for discussion, but it can also be determined directly via the ratios of their coefficients. If $A_2, B_2, C_2 \neq 0$, then

if and only if $\frac{A_1}{A_2} = \frac{B_1}{B_2} \neq \frac{C_1}{C_2}$, the two lines l_1 and l_2 are parallel;

if and only if $\frac{A_1}{A_2} = \frac{B_1}{B_2} = \frac{C_1}{C_2}$, the two lines l_1 and l_2 coincide;

if and only if $\frac{A_1}{A_2} \neq \frac{B_1}{B_2}$, the two lines l_1 and l_2 intersect, and in particular if and only if $A_1 A_2 + B_1 B_2 = 0$, the two lines l_1 and l_2 are perpendicular.

If one of A_2, B_2, and C_2 equals 0, then we can discuss it as a special case.

D. *Distance of a point to a straight line*

Suppose that the distance of a point $P(x_1, y_1)$ to a straight line $l : ax + by + c = 0$ is d. Then

$$d = \left| \frac{ax_1 + by_1 + c}{\sqrt{a^2 + b^2}} \right|.$$

If straight lines l_1 and l_2 are parallel, then their equations can be expressed as

$$l_1 : ax + by + c_1 = 0, \quad l_2 : ax + by + c_2 = 0,$$

and their distance equals

$$d = \frac{|c_1 - c_2|}{\sqrt{a^2 + b^2}}.$$

E. *Basic methods*

Equations of straight lines have various forms. When solving problems, we should consider the selection of the equation with an eye to the whole situation. For the equation of a straight line, it is important to master the meaning of the slope and intercepts, and to use them to determine the equation of the straight line and judge the position relation.

2. Illustrative Examples

Example 1. Suppose that the equation of a straight line m is $y = kx + 1$. Let A and B be two points on the straight line m, whose horizontal coordinates are just the two distinct negative real roots of the quadratic equation $(1 - k^2)x^2 - 2kx - 2 = 0$ of one variable x, let a straight line l pass through the point $P(-2, 0)$ and the middle point of the line segment AB, and let CD be a line segment on the y-axis. Consider all possible straight lines l that have no common point with CD. Does the maximum value of the length of CD exist? If it does exist, find the maximum value; if it does not exist, express your reason for the answer.

Analysis. This problem is actually to find the range of the intercept of the straight line l on the y-axis.

Solution. Write $A(x_1, y_1)$ and $B(x_2, y_2)$. Then

$$x_1 + x_2 = \frac{2k}{1 - k^2}.$$

Denote the middle point of the line segment AB by M. Then

$$x_M = \frac{x_1 + x_2}{2} = \frac{k}{1 - k^2}, \quad y_M = kx_M + 1 = \frac{1}{1 - k^2}.$$

Let the straight line l intersect the y-axis at $Q(0, b)$. From the fact that $P(-2, 0), Q(0, b)$, and $M = \left(\frac{k}{1-k^2}, \frac{1}{1-k^2}\right)$ are co-linear, we have

$$\frac{b - 0}{0 - (-2)} = \frac{\frac{1}{1-k^2} - 0}{\frac{k}{1-k^2} - (-2)},$$

thus

$$b = \frac{1}{-k^2 + \frac{k}{2} + 1}.$$

On the other hand, since x_1 and x_2 are the two distinct negative real roots of the quadratic equation $(1 - k^2)x^2 - 2kx - 2 = 0$ of x, the following is satisfied:

$$\begin{cases} x_1 + x_2 = \dfrac{2k}{1 - k^2} < 0, \\[2mm] x_1 x_2 = \dfrac{-2}{1 - k^2} > 0, \\[2mm] \Delta = 4k^2 + 8(1 - k^2) > 0. \end{cases}$$

The solution of the above system is

$$1 < k < \sqrt{2}.$$

Therefore,

$$b = \frac{1}{-k^2 + \frac{k}{2} + 1} \in (-\infty, -(2 + \sqrt{2})) \cup (2, +\infty).$$

Hence, the maximum value of the length of CD exists, and $|CD|_{\max} = 4 + \sqrt{2}$. $\qquad\square$

Example 2. In a right $\triangle ABC$, the altitude of the hypotenuse is AD, and the inner centers of $\triangle ABD$ and $\triangle ACD$ are M and N, respectively. Extend the line segment MN to intersect AB and AC at K and L, respectively. Prove: $S_{\triangle ABC} \geq 2S_{\triangle AKL}$.

Analysis. Set up a rectangular coordinates system with the straight line containing AC as the x-axis and the straight line containing AB as the

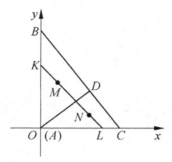

Figure 9.1   Figure for Example 2.

y-axis, and assume that $|AC| = a$ and $|AB| = b$. Then the key step of solving the problem is calculating $|AK|$ and $|AL|$, which are just the vertical and horizontal intercepts of the straight line MN. Computing the coordinates of M and N, we can find the equation of the straight line MN, thus proving the proposition.

Proof. Establish a rectangular coordinates system with the straight line containing AC as the x-axis and the straight line containing AB as the y-axis, and assume that $|AC| = a, |AB| = b$, and $|AD| = c$. Then $c = \frac{ab}{\sqrt{a^2+b^2}}$.

Let the radii of the inscribed circles of $\triangle ACD$ and $\triangle ABD$ be r_1 and r_2, respectively. Then the coordinates of N and M are $N(c - r_1, r_1)$ and $M(r_2, c - r_2)$. Thus, the slope of the straight line MN is

$$k_{MN} = \frac{c - r_2 - r_1}{r_2 - c + r_1} = -1.$$

This indicates that $\triangle AKL$ is an isosceles right triangle (see Figure 9.1), the equation of the straight line MN is $y - r_1 = -(x - c + r_1)$, and the vertical and horizontal intercepts of MN are both c. Consequently,

$$2S_{\triangle AKL} = c^2 = \frac{a^2b^2}{a^2 + b^2} \leq \frac{a^2b^2}{2ab} = \frac{1}{2}ab = S_{\triangle ABC}. \qquad \square$$

Example 3. Given two points $A(1,0)$ and $B(3,0)$. Find a point $P(x,y)$ on the straight line $l : x + y - 5 = 0$, so that the angle $\angle APB$ between the line segments PA and PB becomes maximal.

Solution. As Figure 9.2 shows, $|AB| = 2$ and P is a point on the line l. Draw AE and BF to be perpendicular to AB, which intersect l at E

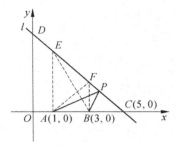

Figure 9.2   Figure for Example 3.

and F, respectively. It is obvious to see that

$$\angle AEB < \angle AFB$$

and

$$\angle AFB = \frac{\pi}{4}.$$

When the point P does not coincide with E or F, the slopes of the straight lines PA and PB are

$$k_{PA} = \frac{y}{x-1}, \quad k_{PB} = \frac{y}{x-3}.$$

Let $\angle APB = \alpha$, and denote $\tan \alpha = R$.

(1) When P is located in the upper half plane,

$$R = \frac{k_{PB} - k_{PA}}{1 + k_{PB}k_{PA}} = \frac{\frac{y}{x-3} - \frac{y}{x-1}}{1 + \frac{y}{x-3}\frac{y}{x-1}},$$

after a simplification of which we have

$$R(x^2 - 4x + 3 + y^2) = 2y. \tag{9.1}$$

Substituting $x = 5 - y$ into (9.1) gives

$$Ry^2 - (3R + 1)y + 4R = 0.$$

At this time, the discriminant of the above quadratic equation

$$\Delta = (3R + 1)^2 - 16R^2 \geq 0,$$

and its solution is $-\frac{1}{7} \leq R \leq 1$.

Clearly, $R > 0$, so $R_{\max} = 1$ and then $\alpha_{\max} = \frac{\pi}{4}$.

(2) When P is located in the lower half plane,

$$R = \frac{k_{PA} - k_{PB}}{1 + k_{PA}k_{PB}} = \frac{\frac{y}{x-1} - \frac{y}{x-3}}{1 + \frac{y}{x-1}\frac{y}{x-3}},$$

a manipulation of which and a substitution of $x = 5 - y$ give

$$Ry^2 - (3R - 1)y + 4R = 0.$$

Now,

$$\Delta = (3R - 1)^2 - 16R^2 \geq 0,$$

and its solution is $-1 \leq R \leq \frac{1}{7}$.

By the same reason, $R > 0$, so $R_{\max} = \frac{1}{7}$ with $\alpha_{\max} = \arctan \frac{1}{7}$.

In summary, the maximum value of $\angle APB$ is $\frac{\pi}{4}$, and the coordinates of the point P are $(3, 2)$. □

Remark. In fact, the solution of this problem is based on such a geometric fact: If the angle $\angle APB$ formed by a point P on a straight line l and two points A and B not on the straight line l is the maximum, then the circle ABP and the straight line l are tangent. Otherwise, the straight line l and the circle ABP have another common point, denoted as Q. Then any point M that is located in the line segment PQ must satisfy $\angle AMB > \angle APB$, leading to a contradiction.

For the present example, it is easy to see that l intersects the straight line AB at the point $C(5, 0)$ and intersects the y-axis at the point $D(0, 5)$. By the circular power theorem, we know that

$$|CP|^2 = |AC||BC| = 4 \cdot 2 = 8,$$

hence $|CP| = 2\sqrt{2}$.

It follows that $|CP| : |PD| = \frac{2}{3}$. From the coordinates formula of a fixed proportion point, it is easy to obtain the coordinates $(3, 2)$ of the point P.

Example 4. If $a\cos\theta + b\sin\theta = c$ and $a\cos\phi + b\sin\phi = c\left(\frac{\phi-\theta}{2} \neq k\pi$ with $k \in \mathbf{Z}\right)$, prove:

$$\frac{a}{\cos\frac{\theta+\phi}{2}} = \frac{b}{\sin\frac{\theta+\phi}{2}} = \frac{c}{\cos\frac{\theta-\phi}{2}}.$$

Proof. When $a = b = 0$, it is obviously true. For a and b not equal to 0 at the same time, the given conditions indicate that the two different points

$A(\cos\theta, \sin\theta)$ and $B(\cos\phi, \sin\phi)$ are on the straight line

$$ax + by = c. \tag{9.2}$$

Also the equation of the straight line determined by A and B is

$$(\sin\theta - \sin\phi)(x - \cos\phi) - (\cos\theta - \cos\phi)(y - \sin\phi) = 0,$$

in other words,

$$x\cos\frac{\theta+\phi}{2} + y\sin\frac{\theta+\phi}{2} = \cos\frac{\theta-\phi}{2}. \tag{9.3}$$

Since two distinct points determine a straight line uniquely, the equations (9.2) and (9.3) are the same, and consequently

$$\frac{a}{\cos\frac{\theta+\phi}{2}} = \frac{b}{\sin\frac{\theta+\phi}{2}} = \frac{c}{\cos\frac{\theta-\phi}{2}}. \qquad \square$$

Remark. Associating the given equations to the equation of a straight line, and then using the property of the coincidence relation for the straight line to solve the problem, we have transformed an algebraic problem to a problem of analytic geometry. We should grasp the essence of the problem as a whole, and figure out a solution method creatively. This is a necessary quality for the students to take part in the mathematics competition.

Example 5. Suppose that we are given a fixed point O and two fixed straight lines l_1 and l_2 that are mutually perpendicular. Make two moving straight lines passing through O and forming the right angle. One of them intersects l_2 at P and the other intersects l_1 at Q. Assume that the orthogonal projection of O onto PQ is R. Find the trajectory of R (see Figure 9.3).

Analysis. Since $OR \perp PQ$, we can write down the coordinates of R via the normal equation of the straight line PQ. And then by using $OP \perp OQ$, we are able to find the trajectory of R that satisfies the given condition.

Solution. Choosing O as the origin and the straight line passing through O and parallel to l_1 as the x-axis, we establish a rectangular coordinates system. Let the equations of l_1 and l_2 be

$$y = k, \tag{9.4}$$

$$x = h, \tag{9.5}$$

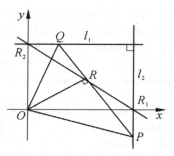

Figure 9.3 Figure for Example 5.

respectively, where k and h are known numbers. Let the normal equation of PQ be

$$x \cos \alpha + y \sin \alpha - p = 0. \tag{9.6}$$

Then the coordinates of R are

$$x = p \cos \alpha, \; y = p \sin \alpha. \tag{9.7}$$

From (9.4) and (9.6), the coordinates of Q are

$$\left(\frac{p - k \sin \alpha}{\cos \alpha}, k \right); \tag{9.8}$$

from (9.5) and (9.6), the coordinates of P are

$$\left(h, \frac{p - h \cos \alpha}{\sin \alpha} \right). \tag{9.9}$$

By $OP \perp OQ$, (9.8), and (9.9), we obtain

$$\frac{h(p - k \sin \alpha)}{\cos \alpha} + \frac{k(p - h \cos \alpha)}{\sin \alpha} = 0,$$

that is,

$$kp \cos \alpha + hp \sin \alpha = kh. \tag{9.10}$$

Substituting (9.7) into (9.10), we find that the equation of the trajectory of the point R is $kx + hy = kh$.

This is a straight line, whose intercepts on the x-axis and the y-axis are h and k, respectively. In other words, if we let R_1 be the intersection point of l_2 with the x-axis and R_2 be the intersection point of l_1 with the y-axis, then the straight line $R_1 R_2$ is the required trajectory. $\quad\square$

Remark. By choosing an appropriate coordinates system and selecting the normal equation for the straight line in a flexible way, we can reduce complicated computations. In addition, for problems concerning trajectories, we should not only point out the concrete figure of the trajectory, but also pay attention to the completeness and purity of the trajectory.

Example 6. Consider the point net consisting of all integral points of a rectangular coordinates plane. Try to prove the following conclusions for the straight lines of rational slopes:

(1) Such straight lines either do not pass through the lattice points of the net or pass through infinitely many lattice points of the net;
(2) for each of such straight lines, there exists a positive number d such that, except for possible lattice points on the straight line, there does not exist any lattice point of the net whose distance to the straight line is less than d.

Proof. (1) If l is a straight line with a rational slope, then its equation can be written as $ax + by + c = 0$, where a and b are integers that are not both 0. Let (x_1, y_1) be a point of the straight line, and let x_1 and y_1 be both integers. Since

$$a(x_1 + kb) + b(y_1 - ka) + c = ax_1 + by_1 + c = 0,$$

we know that all points of the form $(x_1 + kb, y_1 - ka)$ $(k = 0, \pm 1, \pm 2, \ldots)$ are on the same straight line. As a consequence, if a straight line with a rational slope contains integral points, it must have infinitely many integral points. Therefore, it is impossible for the line to contain only finitely many integral points.

(2) The distance from a point (p, q) to the straight line l is

$$d = \frac{|ap + bq + c|}{\sqrt{a^2 + b^2}}.$$

Since l has a rational slope, a and b can both be chosen to be integers. So, $ap + bq$ can only take integral values. Thus, if c is also an integer, then d is either equal to 0 or at least equal to $\frac{1}{\sqrt{a^2+b^2}}$. If c is not an integer, then d is at least equal to $\frac{\epsilon}{\sqrt{a^2+b^2}}$, where ϵ is the distance of c to the nearest integer. $\qquad\square$

Example 7. Prove: If $P_1(x_1, y_1)$ and $P_2(x_2, y_2)$ are two points that are on the different sides of a straight line $Ax + By + C = 0$, then $Ax_1 + By_1 + C$

and $Ax_2 + By_2 + C$ have different signs; if $P_1(x_1, y_1)$ and $P_2(x_2, y_2)$ are on the same side of the straight line $Ax + By + C = 0$, then $Ax_1 + By_1 + C$ and $Ax_2 + By_2 + C$ have the same sign.

Proof. Suppose that the straight line connecting P_1 and P_2 intersects the straight line $Ax + By + C = 0$ at a point $P(x, y)$. Let $|P_1P| : |PP_2| = \lambda$. Then, by the coordinates formula for a fixed proportion point, $x = \frac{x_1 + \lambda x_2}{1 + \lambda}$ and $y = \frac{y_1 + \lambda y_2}{1 + \lambda}$. Thus,

$$A\frac{x_1 + \lambda x_2}{1 + \lambda} + B\frac{y_1 + \lambda y_2}{1 + \lambda} + C = 0,$$

the solution of which is $\lambda = -\frac{Ax_1 + By_1 + C}{Ax_2 + By_2 + C}$.

If P_1 and P_2 are on the different sides of the straight line, then P is an interior division point of P_1P_2, namely $\lambda > 0$. So $Ax_1 + By_1 + C$ and $Ax_2 + By_2 + C$ have different signs.

Suppose that P_1 and P_2 are on the same side of the straight line. When P_1P_2 intersects the straight line, P is an exterior division point of P_1P_2, namely $\lambda < 0$. Then $Ax_1 + By_1 + C$ and $Ax_2 + By_2 + C$ have the same sign. When P_1P_2 is parallel to the straight line, we have

$$A(x_2 - x_1) + B(y_2 - y_1) = 0,$$

and consequently $Ax_1 + By_1 = Ax_2 + By_2$, so $Ax_1 + By_1 + C = Ax_2 + By_2 + C$, still the same sign for the both sides. $\square$

Example 8. In a coordinates plane, we ask whether there exists a family of infinitely many straight lines $l_1, l_2, \ldots, l_n, \ldots$ that satisfy the conditions:

(1) The point $(1, 1) \in l_n$ for $n = 1, 2, 3, \ldots$.
(2) $k_{n+1} = a_n - b_n$, where k_{n+1} is the slope of l_{n+1}, and a_n and b_n are the intercepts of l_n on the x-axis and the y-axis, respectively, $n = 1, 2, 3, \ldots$.
(3) $k_n k_{n+1} \geq 0$ with $n = 1, 2, 3, \ldots$.

Solution. No such a family of straight lines exists to satisfy (1), (2), and (3). Otherwise, let the equation of l_n be

$$y - 1 = k_n(x - 1).$$

Then the horizontal and vertical intercepts of l_n are

$$a_n = 1 - \frac{1}{k_n}, \quad b_n = 1 - k_n,$$

so

$$a_n - b_n = k_n - \frac{1}{k_n} = k_{n+1}$$

are all well defined for $n = 1, 2, 3, \ldots$. Consequently,

$$k_n \neq 0, \ n = 1, 2, 3, \ldots.$$

For $n \geq 1$, we have

$$k_{n+1} - k_n = -\frac{1}{k_n},$$

$$k_n - k_{n-1} = -\frac{1}{k_{n-1}},$$

$$\cdots$$

$$k_2 - k_1 = -\frac{1}{k_1}.$$

Adding all the above equalities, we get

$$k_{n+1} = k_1 - \left(\frac{1}{k_1} + \frac{1}{k_2} + \cdots + \frac{1}{k_n} \right).$$

From $k_n \neq 0$ and condition (3), $k_n k_{n+1} > 0$, so all k_n have the same sign.

If $k_n > 0$ for $n = 1, 2, 3, \ldots$, then from $k_{n+1} = k_n - \frac{1}{k_n} < k_n$, we see that $\frac{1}{k_{n+1}} > \frac{1}{k_n}$, hence

$$k_{n+1} = k_1 - \left(\frac{1}{k_1} + \frac{1}{k_2} + \cdots + \frac{1}{k_n} \right) < k_1 - \frac{n}{k_1}.$$

It follows that $k_{n+1} < 0$ when $n > k_1^2$, a contradiction to $k_{n+1} > 0$.

If $k_n < 0$ for $n = 1, 2, 3, \ldots$, then the same contradiction will come by the same reason.

Therefore, there does not exist an infinite family of straight lines satisfying (1), (2), and (3). $\qquad \square$

Remark. In this problem, we transformed the research question whether a family of straight lines exists or not to an algebraic problem that was then investigated by using the related knowledge of sequences and their limits. This method is widely used in analytic geometry.

3. Exercises

Group A

I. Filling Problems

1. If point $M\left(a, \frac{1}{b}\right)$ and point $N\left(b, \frac{1}{c}\right)$ are both on the straight line l_1 : $x + y = 1$, then the points among $P\left(\frac{1}{c}, b\right)$ and $Q\left(\frac{1}{a}, c\right)$ that also lie on l are _____.

2. Among all straight lines passing through the point $(\sqrt{2}, 0)$, the number of those that pass through two different rational points is _____.

3. Suppose that nonnegative real numbers x and y satisfy $3x + y = 1$. Then the minimum value of $x + \sqrt{x^2 + y^2}$ is _____.

4. A light ray that goes along the direction of the straight line $x - y + 3 = 0$ is reflected by the straight line $2x - y + 2 = 0$ when the ray hits it. Then the equation of the straight line that contains the reflected light ray is

_____.

5. Given the point $A(3, 5)$ and the straight line $l : x - 2y + 2 = 0$. Let B be a moving point on the y-axis and C a moving point on l. Then the minimum value of the perimeter of $\triangle ABC$ is _____.

6. Let $E = \{(x, y) \mid 0 \leq x \leq 2, 0 \leq y \leq 2\}$ and $F = \{(x, y) \mid x \leq 10, 2 \leq y \leq x - 4\}$ be two point sets in a plane of rectangular coordinates. Then the area of the point set

$$G = \left\{\left(\frac{x_1 + x_2}{2}, \frac{y_1 + y_2}{2}\right) \mid (x_1, y_1) \in E, (x_2, y_2) \in F\right\}$$

is _____.

7. Suppose that real numbers x and y satisfy $x \leq y, y \leq 6 - 2x$, and $x \geq 1$. Let $\vec{a} = (2x - y, m)$ and $\vec{b} = (-1, 1)$. If $\vec{a} \parallel \vec{b}$, then the minimum value of the real number m is _____.

II. Calculation Problems

8. Given the straight line $l : y = 4x$ and the point $R(6, 4)$, find a point Q on l such that the triangle bounded by the straight lines RQ, l, and the x-axis in the first quadrant has the minimal area.

9. In a plane rectangular coordinates system xOy, let a point set be given as

$$K = \{(x, y) \mid (|x| + |3y| - 6)(|3x| + |y| - 6) \leq 0\}.$$

Find the area of the plane region corresponding to the point set K.

10. If (x, y) is a moving point in the region given by $|x| + |y| \leq 1$. Find the maximum value of the function

$$f(x, y) = ax + y \ (a > 0).$$

Group B

11. In $\triangle ABC$, the length of the altitude AD over the side BC is $|AD| = 12$, and the length of the bisector AE of $\angle A$ is $|AE| = 13$. Let the length of the median line AF of the side BC be $|AF| = m$. What are the value ranges of m so that $\angle A$ is an acute angle, right angle, or an obtuse angle, respectively?

12. Given a circle O of radius R in a plane and a straight line a outside the circle. Pick two points M and N on a such that the circle O' with diameter MN is tangent to circle O from outside. Prove: There exists a point A in the plane such that all the angles $\angle MAN$ are the same.

Chapter 10

Circles

1. Key Points of Knowledge and Basic Methods

A. *Equation of a circle*

Standard equation: $(x-a)^2 + (y-b)^2 = r^2$, where (a, b) is the center of the circle and r $(r > 0)$ is the radius. The general equation: $x^2 + y^2 + Dx + Ey + F = 0$ $(D^2 + E^2 - 4F > 0)$. The center of the circle is $\left(-\frac{D}{2}, -\frac{E}{2}\right)$ and the radius is $\frac{1}{2}\sqrt{D^2 + E^2 - 4F}$.

B. *Tangent lines of circles*

The equation of a tangent line to a circle $x^2 + y^2 = r^2$ at a point $P(x_0, y_0)$ of the circle is $x_0 x + y_0 y = r^2$.

The equation of a tangent line to a circle $(x-a)^2 + (y-b)^2 = r^2$ at a point $P(x_0, y_0)$ of the circle is $(x_0 - a)(x - a) + (y_0 - b)(y - b) = r^2$.

The equation of a tangent line to a circle $x^2 + y^2 + Dx + Ey + F = 0$ at a point $P(x_0, y_0)$ of the circle is $x_0 x + y_0 y + D\frac{x+x_0}{2} + E\frac{y+y_0}{2} + F = 0$.

If we draw a tangent line to a circle that passes through a point $P(x_0, y_0)$ outside the circle, then:

When the equation of the circle is the standard one, the length of the tangent line segment between $P(x_0, y_0)$ and the tangent point is

$$\sqrt{(x_0 - a)^2 + (y_0 - b)^2 - r^2};$$

when the equation of the circle is the general one, the length of the tangent line segment between $P(x_0, y_0)$ and the tangent point is

$$\sqrt{x_0^2 + y_0^2 + Dx_0 + Ey_0 + F}.$$

C.   *Position relations of straight lines and circles*

(1) Suppose that the distance of the center of a circle C and a straight line l is d and the radius of the circle is r. When $d > r$, the circle C and the straight line l are disjoint; when $d = r$, the circle C and the straight line l are tangent; when $d < r$, the circle C and the straight line l intersect at two points.

(2) Put together the equations of a straight line and a circle, and eliminate one variable. Then, we obtain a quadratic equation of the other variable. If its discriminant $\Delta > 0$, then the circle and the straight line intersect at two points; if $\Delta = 0$, then the circle and the straight line are tangent; if $\Delta < 0$, then the circle and the straight line are disjoint.

D.   *Position relations of two circles*

Suppose that the radii of two circles are r_1 and r_2, respectively and the distance of the centers of the two circles is d.

When $d > r_1 + r_2$, the two circles are exterior to each other; when $d = r_1 + r_2$, the two circles are tangent from outside; when $|r_1 - r_2| < d < r_1 + r_2$, the two circles intersect at two points; when $d = |r_1 - r_2|$, the two circles are tangent from inside; when $d < |r_1 - r_2|$, one circle is inside the other.

E.   *Basic methods*

(1) The standard equation and the general equation of a circle can be transformed to each other, and we may choose one of them according to the actual situation, when solving problems.

(2) We should be good at transforming the problem on the position relation of a circle and a straight line or another circle to an algebraic relation, when solving a problem.

(3) Circles have symmetry. One can use the position relation of a chord and the center of a circle to establish an appropriate coordinates system for the convenience of solving problems.

2.   **Illustrative Examples**

Example 1. Suppose that the following are given: The straight line l : $y = \sqrt{3}x + 4$, a moving circle $O : x^2 + y^2 = r^2$ $(1 < r < 2)$, a rhombus $ABCD$ with the inner angle $60°$, the vertices A and B on the straight line

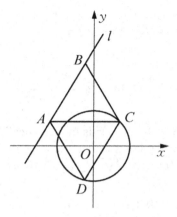

Figure 10.1   Figure for Example 1.

l, and the vertices C and D on the circle O. When r varies, find the range of the values for the area S of the rhombus $ABCD$.

Analysis. A geometric problem is always related to a figure. Draw Figure 10.1. Assume that the distance of the opposite sides of the rhombus is d. From the conditions and properties of a rhombus, $d = \frac{\sqrt{3}}{2}|CD|$. Thus, the following solution process looks natural.

Solution. Since one inner angle of the rhombus $ABCD$ is $60°$, so $\triangle ACD$ or $\triangle BCD$ is an equilateral triangle. Without loss of generality, assume that $\triangle ACD$ is an equilateral triangle, as Figure 10.1 shows.

Since the distance of the center of the circle O to the straight line l is $2 > r$, the straight line l and the circle O are disjoint.

Let the equation of the straight line CD be $y = \sqrt{3}x + b$. Then the distance between the straight lines l and CD is $d = \frac{|b-4|}{2}$.

Also the distance from the center of the circle O to the straight line CD is $\frac{|b|}{2}$, hence

$$|CD| = 2\sqrt{r^2 - \left(\frac{|b|}{2}\right)^2} = \sqrt{4r^2 - b^2}.$$

Since $d = \frac{\sqrt{3}}{2}|CD|$, we have

$$\frac{|b-4|}{2} = \frac{\sqrt{3}}{2}\sqrt{4r^2 - b^2}.$$

After a simplification we get $b^2 - 2b + 4 = 3r^2$.

The assumption that $1 < r < 2$ implies that $3 < b^2 - 2b + 4 < 12$.
The solutions are $-2 < b < 1$ or $1 < b < 4$.
On the other hand,

$$S = 2S_{\triangle ACD} = 2 \cdot \frac{\sqrt{3}}{4}|CD|^2 = 2 \cdot \frac{\sqrt{3}}{4}\left(\frac{2}{\sqrt{3}}\right)^2 d^2 = \frac{\sqrt{3}}{6}(b-4)^2.$$

Since the function $S = \frac{\sqrt{3}}{6}(b-4)^2$ is monotonically increasing on $(-2, 1)$ and $(1, 4)$, respectively, it follows that the range of the values for the area S of the rhombus $ABCD$ is $\left(0, \frac{3}{2}\sqrt{3}\right) \cup \left(\frac{3}{2}\sqrt{3}, 6\sqrt{3}\right)$. □

Example 2. As Figure 10.2 shows, in a rectangular coordinates system xOy of a plane, two circles O_1 and O_2 are both tangent to a straight line $l : y = kx$ and the positive half x-axis. Suppose that the product of the radii of the two circles is 2 and an intersection point of the two circles is $P(2, 2)$. Find the equation of the straight line l.

Solution. By the given conditions of the problem, the centers O_1 and O_2 of the circles are both on the angular bisector of the x-axis and the straight line l.

Suppose that the inclination angle of the straight line l is α. Then $k = \tan \alpha$. Denote $t = \tan \frac{\alpha}{2}$. Then $k = \frac{2t}{1-t^2}$.

Since the centers O_1 and O_2 of the circles are on the straight line $y = tx$, we may write them as $O_1(m, mt)$ and $O_2(n, nt)$.

Since the intersection point $P(2, 2)$ is in the first quadrant, there must be that $m > 0, n > 0$, and $t > 0$.

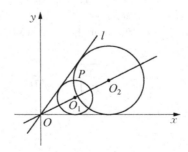

Figure 10.2　Figure for Example 2.

Thus,

the equation of the circle $O_1 : (x - m)^2 + (y - mt)^2 = (mt)^2$,

the equation of the circle $O_2 : (x - n)^2 + (y - nt)^2 = (nt)^2$.

Consequently,

$$\begin{cases} (2 - m)^2 + (2 - mt)^2 = (mt)^2, \\ (2 - n)^2 + (2 - nt)^2 = (nt)^2. \end{cases}$$

In other words,

$$\begin{cases} m^2 - (4 + 4t)m + 8 = 0, \\ n^2 - (4 + 4t)n + 8 = 0. \end{cases}$$

Hence, m and n are the two solutions of the equation $x^2 - (4+4t)x+8 = 0$, and so $mn = 8$.

Since the product of the radii of the circles is $mt \cdot nt = 2$, we find that $t^2 = \frac{1}{4}$, thus $t = \frac{1}{2}$.

Hence, $k = \frac{2t}{1-t^2} = \frac{1}{1-\frac{1}{4}} = \frac{4}{3}$. As a result, the equation of the straight line l is $y = \frac{4}{3}x$. $\qquad\square$

Example 3. As Figure 10.3 shows, in a rectangular coordinates system xOy of a plane, the diagonals AC and BD of the inscribed quadrilateral $ABCD$ to a circle M with an equation $x^2 + y^2 + Kx + Ey + F = 0$ are perpendicular to each other, and AC and BD are located in the x-axis and y-axis, respectively.

(1) Prove: $F < 0$.
(2) If the area of the quadrilateral $ABCD$ is 8, the length of the diagonal AC is 2, and $\overrightarrow{AB} \cdot \overrightarrow{AD} = 0$, find the value of $K^2 + E^2 - 4F$.

Solution. (1) **Method 1:** From the given conditions of the problem, the origin $O(0,0)$ is the intersection point of the chords AC and BD, so they must be inside the circle M. Consequently,

$$0^2 + 0^2 + K \cdot 0 + E \cdot 0 + F < 0,$$

that is, $F < 0$.

Method 2: By the given conditions, it is not difficult to find that the two points A and C are on the negative half axis and the positive half axis of

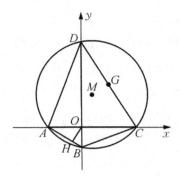

Figure 10.3   Figure for Example 3.

the x-axis, respectively. Suppose that the two points are written as $A(a, 0)$ and $C(c, 0)$. Then $ac < 0$.

For the equation $x^2 + y^2 + Kx + Ey + F = 0$ of the circle, when $y = 0$, we have $x^2 + Kx + F = 0$, the solutions x_A and x_C of which are just the horizontal coordinates of the points A and C, respectively. It follows that $x_A x_C = ac = F$. Since $ac < 0$, it is true that $F < 0$.

(2) The area of a quadrilateral $ABCD$ whose diagonals are perpendicular is $S = \frac{|AC||BD|}{2}$. Since $S = 8$ and $|AC| = 2$, we see that $|BD| = 8$.

Also, since $\overrightarrow{AB} \cdot \overrightarrow{AD} = 0$, so $\angle A$ is a right angle. It follows from $|BD| = 2r = 8$ that $r = 4$. For any circle given by the equation $x^2 + y^2 + Kx + Ey + F = 0$, the radius r of the circle satisfies $\frac{K^2}{4} + \frac{E^2}{4} - F = r^2$. Hence,

$$K^2 + E^2 - 4F = 4r^2 = 64. \qquad \square$$

Example 4. As Figure 10.4 indicates, in a rectangular coordinates system xOy of a plane, there are given two circles $C_1 : (x + 3)^2 + (y - 1)^2 = 4$ and $C_2 : (x - 4)^2 + (y - 5)^2 = 4$.

(1) If a straight line l passes through the point $P(4, 0)$ and the length of the chord formed by the intersection of the line and the circle C_1 is $2\sqrt{3}$. Find the equation of the line.

(2) Suppose that P is a point in the plane satisfying: There exist infinitely many pairs of mutually perpendicular straight lines l_1 and l_2 passing through P such that they intersect the circle C_1 and the circle C_2 respectively, and the length of the chord, which is formed from the intersection of the straight line l_1 and the circle C_1, and the length of

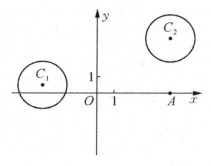

Figure 10.4 Figure for Example 4.

the chord, which is formed from the intersection of the straight line l_2 and the circle C_2, are equal. Find the coordinates of all points P that satisfy the condition.

Analysis. If the radius of a circle is known, then the length of a chord can be expressed in terms of the distance from the center of the circle to the chord.

Solution. (1) Let the equation of the straight line l be $y = k(x - 4)$, namely $kx - y - 4k = 0$. From the center-chord distance theorem, we see that the distance of the center of the circle C_1 to the straight line l is $d = \sqrt{2^2 - \left(\frac{2\sqrt{3}}{2}\right)^2} = 1$. Combined with the point-line distance formula, we get the equation $\frac{|-3k-1-4k|}{\sqrt{k^2+1}} = 1$, which can be simplified to

$$24k^2 + 7k = 0,$$

whose solutions are $k = 0$ or $k = -\frac{7}{24}$.

Hence, the desired equations of the straight lines l are $y = 0$ or $y = -\frac{7}{24}(x - 4)$, that is, $y = 0$ or $7x + 24y - 28 = 0$.

(2) Let the coordinates of the point P be (m, n), and let the equations of the straight lines l_1 and l_2 be

$$y - n = k(x - m), \quad y - n = -\frac{1}{k}(x - m),$$

respectively. In other words, $kx - y + n - km = 0$ and $-\frac{1}{k}x - y + n + \frac{1}{k}m = 0$.

Since the length of the chord formed by the intersection of the straight line l_1 and the circle C_1 and the length of the chord formed by the intersection of the straight line l_2 and the circle C_2 are equal, and since the two circles have the same radii, from the center-chord distance theorem, the distance of the center of the circle C_1 to the straight line l_1 equals the distance of the center of the circle C_2 to the straight line l_2. Hence,

$$\frac{|-3k-1+n-km|}{\sqrt{k^2+1}} = \frac{|-\frac{4}{k}-5+n+\frac{1}{k}m|}{\sqrt{\frac{1}{k^2}+1}},$$

a simplification of which gives $(2-m-n)k = m-n-3$ or $(m-n+8)k = m+n-5$.

There are infinitely many solutions for the variable k to the above linear equations when the coefficients and the right hand sides of the two linear equations are both zero. This observation implies that

$$\begin{cases} 2-m-n=0, \\ m-n-3=0, \end{cases} \text{ or } \begin{cases} m-n+8=0, \\ m+n-5=0. \end{cases}$$

Solving the above systems of equations gives that the coordinates of the point P are $\left(-\frac{3}{2}, \frac{13}{2}\right)$ or $\left(\frac{5}{2}, -\frac{1}{2}\right)$. $\square$

Example 5. Suppose that the equation of a circle O is $x^2+y^2=1$ and P is a moving point outside the circle O. Draw two tangent lines to the circle O that pass through P with tangent points A and B. Connect A and B, and the straight line AB intersects the straight lines $y=x$ and $y=-x$ at points M and N, respectively. If

$$\frac{5}{|OM|^2} + \frac{4}{|ON|^2} = 10,$$

find the equation of the trajectory of the point P.

Solution. At the point P, draw a perpendicular line PC to the straight line $y=x$ with the intersection point C and a perpendicular line PD to the straight line $y=-x$ with the intersection point D.

It is easy to deduce that the six points P, O, A, B, C, and D lie on the same circle.

Also, since $|PA|=|PB|$, so

$$\angle ACO = \angle OAM.$$

Thus, $\triangle OAM \sim \triangle OCA$ and $|OM| = \frac{|OA|^2}{|OC|} = \frac{1}{|OC|}$.

Similarly, $|ON| = \frac{1}{|OD|}$. Also, since the quadrilateral $PCOD$ is a rectangle, we have

$$\frac{5}{|OM|^2} + \frac{4}{|ON|^2} = 5|OC|^2 + 4|OD|^2$$

$$= 5|PD|^2 + 4|PC|^2$$

$$= 5\left(\frac{|x+y|}{\sqrt{2}}\right)^2 + 4\left(\frac{|x-y|}{\sqrt{2}}\right)^2 = 10.$$

Simplifying the above gives

$$\frac{9}{2}x^2 + \frac{9}{2}y^2 + xy = 10. \tag{10.1}$$

Since AB intersects both straight lines $y = x$ and $y = -1$,

$$(x, y) \neq (1, 1),\ (-1, -1),\ \left(\frac{\sqrt{5}}{2}, -\frac{\sqrt{5}}{2}\right),\ \left(-\frac{\sqrt{5}}{2}, \frac{\sqrt{5}}{2}\right).$$

To summarize, the demanded trajectory equation is (10.1) with the above four points excluded. □

Example 6. We are given a circle $O : x^2 + y^2 = r^2$ (O is the origin) and a moving straight line l that passes through a fixed point $(m, 0)$ $(m > r > 0)$, which does not coincide with the x-axis and intersects the circle l at two points P and Q (P and Q are allowed to coincide). Suppose that S is the symmetric point of P with respect to the x-axis.

(1) Prove: The straight line SQ passes through a fixed point; and find the coordinates of the fixed point.
(2) Find the maximum value of the area of $\triangle OSQ$.

Solution. (1) By symmetry we know that if the straight line l passes through a fixed point, then this fixed point must be on the x-axis. Suppose that $l : y = kx + m$. Substituting $x^2 + y^2 = r^2$ into the above equation gives

$$(1 + k^2)y^2 + 2kmy + m^2 - r^2 = 0. \tag{10.2}$$

Write $P = P(x_1, y_1)$ and $Q = Q(x_2, y_2)$. Then the coordinates of S are $(x_1, -y_1)$ and the equation of the straight line SQ is $y + y_1 = \frac{y_2 + y_1}{x_2 - x_1}(x - x_1)$.

So by letting $y = 0$, we obtain that

$$x = \frac{y_1(x_2 - x_1)}{y_1 + y_2} + x_1 = \frac{x_1 y_2 + x_2 y_1}{y_1 + y_2}$$

$$= \frac{(ky_1 + m)y_2 + (ky_2 + m)y_1}{y_1 + y_2}$$

$$= \frac{2ky_1 y_2}{y_1 + y_2} + m.$$

By applying Vieta's theorem to (10.2), we get $x = \frac{r^2}{m}$, in other words, SQ passes through the fixed point $B = \left(\frac{r^2}{m}, 0\right)$.

(2) From (1) we know that

$$S_{\triangle OSQ} = \frac{1}{2}|OB|(|y_1| + |y_2|) = \frac{1}{2}|OB||y_1 + y_2|$$

$$= \frac{1}{2}\frac{r^2}{m}\left|\frac{2km}{1 + k^2}\right| = \frac{r^2}{|k| + \frac{1}{|k|}}.$$

Also the straight line l intersects the circle O at two points, so the discriminant $\Delta = 4k^2m^2 - 4(1 + k^2)(m^2 - r^2) \geq 0$ for the equation (10.2). Thus, $|k| \geq \frac{\sqrt{m^2 - r^2}}{r}$, and consequently when $\frac{\sqrt{m^2 - r^2}}{r} > 1$, namely $m > \sqrt{2}r$,

$$S_{\triangle OSQ} \leq \frac{r^2}{\frac{\sqrt{m^2 - r^2}}{r} + \frac{r}{\sqrt{m^2 - r^2}}} = \frac{r^3\sqrt{m^2 - r^2}}{m^2};$$

when $\frac{\sqrt{m^2 - r^2}}{r} \leq 1$, namely $r < m \leq \sqrt{2}r$, we have $S_{\triangle OSQ} \leq \frac{r^2}{2}$. □

Example 7. In a coordinates plane, a circle that passes through the origin with r the radius is completely in the region $y \geq x^4$. Find the maximum value of r.

Analysis. See Figure 10.5. From the graph of the function $y = x^4$, we see that the region $y \geq x^4$ is the shaded part of the figure.

Solution. By symmetry, to find the maximum value of r, we may assume that the equation of the circle is

$$x^2 + (y - r)^2 = r^2.$$

By the given condition, for any $x \in [-r, r]$, there always holds that $r - \sqrt{r^2 - x^2} \geq x^4$.

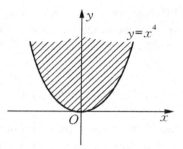

Figure 10.5 Figure for Example 7.

Let $x = r\cos\theta$. We only need to make the discussion for $0 \le \theta < \frac{\pi}{2}$. So, $r - \sqrt{r^2 - r^2\cos^2\theta} \ge r^4\cos^4\theta$ is always satisfied for any $\theta \in [0, \frac{\pi}{2})$.

Thus, $r - r\sin\theta \ge r^4\cos^4\theta$. Consequently,

$$r^3 \le \frac{1 - \sin\theta}{\cos^4\theta} = \frac{1}{(1 - \sin\theta)(1 + \sin\theta)^2}.$$

Note that

$$(1 - \sin\theta)(1 + \sin\theta)^2 = \frac{1}{2}(2 - 2\sin\theta)(1 + \sin\theta)(1 + \sin\theta) \le \frac{1}{2}\left(\frac{4}{3}\right)^3,$$

and the equality is achieved when $\sin\theta = \frac{1}{3}$. Since $r^3 \le \frac{1}{(1 - \sin\theta)(1 + \sin\theta)^2}$ is satisfied for any $\theta \in [0, \frac{\pi}{2}]$, it follows that $r^3 \le \frac{1}{\frac{1}{2}(\frac{4}{3})^3}$, and so $r \le \frac{3}{4}\sqrt[3]{2}$.

Conversely, when $r = \frac{3}{4}\sqrt[3]{2}$, it is always true that $r - \sqrt{r^2 - x^2} \ge x^4$ for any $x \in [-r, r]$. Thus, the circle $x^2 + (y - r)^2 = r^2$ is completely in the region defined by the inequality $y \ge x^4$.

In summary, the maximum value of r is $\frac{3}{4}\sqrt[3]{2}$. $\qquad\square$

Example 8. In a rectangular coordinates plane, find the number of the integral points on the circle centered as $(199, 0)$ with 199 as its radius.

Solution. Let $A(x, y)$ be an integral point on the circle O. The equation of the circle O is

$$y^2 + (x - 199)^2 = 199^2.$$

Obviously, $x = 0, y = 0; x = 199, y = 199; x = 199, y = -199;$ and $x = 398, y = 0$ are 4 groups of solutions of the equation. But y and 199 are relatively prime when $y \ne 0$ and $y \ne \pm 199$, so at this time $199, y,$ and

$|199 - x|$ are an array of Pythagorean numbers. Thus, 199 can be expressed as the square sum of two positive integers, namely $199 = m^2 + n^2$. On the other hand, $199 = 4 \cdot 49 + 3$, and by assuming $m = 2k$ and $n = 2l + 1$, we have

$$199 + 4k^2 + 4l^2 + 4l + 1 = 4(k^2 + l^2 + l) + 1,$$

which contradicts the fact that 199 is a prime number of the type $4d + 3$. Hence, there are exactly four integral points $(0, 0), (199, 199), (199, -199)$, and $(389, 0)$ on the circle. This means that the number of integral points on the circle is 4.   □

3. Exercises

Group A

I. Filling Problems

1. Let two point sets in a plane be $A = \left\{ (x, y) \mid (y - x) \left(y - \frac{1}{x} \right) \geq 0 \right\}$ and $B = \{ (x, y) \mid (x - 1)^2 + (y - 1)^2 \leq 1 \}$. Then the area of the figure represented by $A \cap B$ is _____.

2. In a rectangular coordinates system of a plane, give the point $A(0, 3)$ and the straight line $L : y = 2x - 4$. Let the radius of a circle C be 1 with the center of the circle on the straight line L. If there exists a point M on the circle C such that $|MA| = 2|MO|$, then the value range of the horizontal coordinate of the center of the circle C is _____.

3. Let a and b be two real roots of an equation $x^2 + \cot \theta \cdot x - \csc \theta = 0$. Then the location relation between the straight line passing through the two points $A(a, a^2)$ and $B(b, b^2)$ and the circle $x^2 + y^2 = 1$ is _____.

4. If there are two and only two points on a circle $(x - 3)^2 + (y + 5)^2 = r^2$ that have the distance 1 to the straight line $4x - 3y = 2$, then the range of the values for the radius of the circles is _____.

5. Draw two tangent lines from the point $M(3, 2)$ to the circle $x^2 + y^2 = 3$ with two tangent points A and B. Then the length of the minor arc with A and B as the end points is _____.

6. The equation of the trajectory for the middle point of a chord AB to a circle $x^2 + y^2 = r^2$ with $\angle ACB = 90°$ at a fixed point $C(c, 0)$ ($|c| < \sqrt{2}|r|$) is _____.

II. Calculation Problems

7. A light ray l starting from the point $A(-3, 3)$ reaches the x-axis and then is reflected by it. Suppose that the straight line that contains the reflected light ray is tangent to the circle $x^2 + y^2 - 4x - 4y + 7 = 0$. Find the equation of the straight line that contains the light ray l.

8. Let a circle satisfy:

   (1) The length of the chord formed by its intersection with the y-axis is 2;
   (2) it is divided into two arcs by the x-axis such that the ratio of the arc lengths is $3 : 1$.

   Among all the circles satisfying the above conditions, find the equation of the circle such that the distance of its center to the straight line $l : x - 2y = 0$ is minimal.

9. Suppose that the straight line $x + 2y - 3 = 0$ and a circle $x^2 + y^2 + x - 6y + F = 0$ intersect at points P and Q. Let O be the origin. For what values of F is $OP \perp OQ$ valid?

Group B

10. Let O be the origin of the x-y coordinates. Suppose that a straight line passing through the point O intersects the circles $C_1 : x^2 + y^2 + 2x + 2y = 0$ and $C_2 : x^2 + y^2 - 4x + 2y = 0$ at points Q and P respectively, excluding the point O. Find the equation for the trajectory of the middle point of PQ.

11. In a rectangular coordinates system of a plane, a circle centered at a point $C\left(t, \frac{2}{t}\right)$ passes through the coordinates origin O and intersects the x-axis and y-axis at points A and B that are different from the origin O.

    (1) Prove: The area S of $\triangle AOB$ is of fixed value.
    (2) Let the straight line $l : y = -2x + 4$ intersect the circle C at two different points M and N with $|OM| = |ON|$. Find the standard equation of the circle C.

12. There is a typhoon on the sea near some beach city. The current center of typhoon is located at P on the sea in the south-east direction

with angle θ $\left(\theta = \arctan \frac{\sqrt{2}}{10}\right)$ and has a 300 km distance to the city O (see Figure 10.6). It moves along the north-west direction with angle $45°$ in the speed of 20 km per hour. The current radius of the circular region influenced by the typhoon is 60 km, which is increasing at the rate of 10 km per hour. How many hours later will the city be hit by the typhoon?

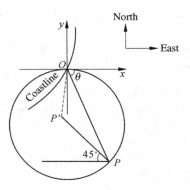

Figure 10.6   Figure for Exercise 12.

Chapter 11

Ellipses

1. Key Points of Knowledge and Basic Methods

A. *Standard equation and properties of ellipses*

Standard equation of an ellipse with the foci on the x-axis: Let the foci be $F_1(-c, 0)$ and $F_2(c, 0)$ $(c > 0)$, the length of the major axis be $2a$, and the length of the minor axis be $2b$ $(a > b)$. Then the standard equation of the ellipse is

$$\frac{x^2}{a^2} + \frac{y^2}{b^2} = 1, \tag{11.1}$$

where $a^2 = b^2 + c^2$. The ranges of x and y are $-a \le x \le a$ and $-b \le y \le b$, respectively; the symmetric axes are x-axis and y-axis; the vertices are $(-a, 0), (a, 0), (0, -b)$, and $(0, b)$.

The equations of the directrices are $x = \pm \frac{a^2}{c}$.

The eccentricity $e = \frac{c}{a} < 1$.

Similarly, one can write down the standard equation and the equations of the directrices for an ellipse whose foci are on the y-axis, and the corresponding properties.

B. *Second definition of ellipses (foci-directrix definition) and focal radii*

The trajectory of a point satisfying the following property is an ellipse: The ratio of the distance of the point to a fixed point (focus) and the distance of the point to a fixed straight line (directrix) is a positive constant (eccentricity) $e < 1$.

Suppose that $P(x_0, y_0)$ is a point on the ellipse (11.1). Then the left focal radius $|PF_1| = a + ex_0$ and the right focal radius $|PF_2| = a - ex_0$.

C. Optical properties of ellipses

Light rays emitted from a focus of an ellipse will go to the destination of the other focus after the refection from the ellipse.

D. Chord length formula of conic sections and midpoint of a chord

Substitute the equation $y = kx + m$ of a straight line into the equation $f(x, y) = 0$ of a conic section, and eliminate y (or x), resulting in a quadratic equation. If the straight line intersects the conic section at two points, then the discriminant of the quadratic equation is greater than 0. Suppose that the two intersection points are $A(x_1, y_1)$ and $B(x_2, y_2)$. Then by Vieta's theorem we can calculate out that

$$|AB| = \sqrt{1 + k^2}\sqrt{(x_1 + x_2)^2 - 4x_1x_2}$$

$$= \sqrt{1 + \frac{1}{k^2}}\sqrt{(y_1 + y_2)^2 - 4y_1y_2}.$$

Let the middle point of a chord AB be $M(x_0, y_0)$. Then $x_0 = \frac{x_1 + x_2}{2}$ and $y_0 = \frac{y_1 + y_2}{2}$.

In particular, the length of a focal chord of the ellipse (11.1), whose inclination angle is α, is $|AB| = \frac{2ab^2}{b^2 + c^2 \sin^2 \alpha}$.

E. Equation of the straight line containing a midpoint chord of a conic section

Let $M(x_0, y_0)$ be the middle point of a chord PQ of a conic section $Ax^2 + Cy^2 + Dx + Ey + F = 0$. Suppose that the slope of PQ is k. Then

$$k = -\frac{2Ax_0 + D}{2Cy_0 + E}.$$

The equation of the straight line PQ is $y - y_0 = -\frac{2Ax_0 + D}{2Cy_0 + E}(x - x_0)$.

F. Tangent lines and chords at contact of conic sections

(1) Suppose that $P(x', y')$ is a point on a conic section $L : Ax^2 + Cy^2 + Dx + Ey + F = 0$. Then the equation of the tangent line to the curve

L at the point P is

$$Ax'x + Cy'y + D\frac{x' + x}{2} + E\frac{y' + y}{2} + F = 0.$$

(2) Suppose that $P(x', y')$ is a point that is in the region of a conic section $L : Ax^2 + Cy^2 + Dx + Ey + F = 0$ that does not contain a focus (or the center of a circle). Then two tangent lines can be drawn to the curve L, which pass through the point P. Let their tangent points be Q and R. Then the equation of QR is

$$Ax'x + Cy'y + D\frac{x' + x}{2} + E\frac{y' + y}{2} + F = 0.$$

(3) Suppose that $P(x', y')$ is a point that is in the region of a conic section $L : Ax^2 + Cy^2 + Dx + Ey + F = 0$ that contains a focus (or the center of a circle), and a straight line that passes through the point P intersects the curve L at two points Q and R. The two tangent lines to the curve L at the two points Q and R intersect at a point G. Then the trajectory of the point G is the straight line

$$Ax'x + Cy'y + D\frac{x' + x}{2} + E\frac{y' + y}{2} + F = 0.$$

Here, we avoid introducing the concepts of poles and polar lines of conic sections.

G. *Basic methods*

When solving problems related to straight lines and conic sections, one often puts together the equation of the straight line and the equation of the conic section as a system, and then solve the system via algebraic methods.

Every point in the items D, E, F, G above is applicable to circles, ellipses, hyperbolas, and parabolas.

2. **Illustrative Examples**

Example 1. As Figure 11.1 shows, A and B are the two vertices of an ellipse $\frac{x^2}{a^2} + \frac{y^2}{b^2} = 1$ $(a > b > 0)$ along the x-axis, and P is a point on the ellipse that is different from A and B. Draw two lines l_1 and l_2 that are perpendicular to PA and PB, respectively. Suppose that l_1 and l_2 intersect at a point M. When the moving point P moves along the ellipse, find the equation of the trajectory of the point M.

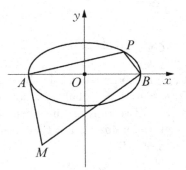

Figure 11.1   Figure 1 for Example 1.

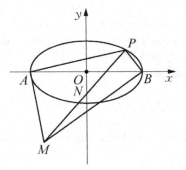

Figure 11.2   Figure 2 for Example 1.

Analysis. Since $l_1 \perp PA$ and $l_2 \perp PB$, from the symmetry of the ellipse, we see that the four points $A, P, B,$ and M are on a circle, and the center of the circle is the intersection point of the y-axis and PM.

Solution. By the given condition, $A, P, B,$ and M are on a circle, and the center N of the circle is on the y-axis, the perpendicular bisector of the chord AB, so we assume that the coordinates of N are $N(0, t)$, as Figure 11.2 shows. Let the coordinates of the points P and M be $P(x_0, y_0)$ and $M(x, y)$. Since N is the middle point of the diameter PM, so

$$x_0 = -x, \ y + y_0 = 2t,$$

and thus $y_0 = y - 2t$.

Since the point P is on the ellipse,

$$\frac{x_0^2}{a^2} + \frac{y_0^2}{b^2} = 1,$$

and consequently

$$\frac{x^2}{a^2} + \frac{(2t-y)^2}{b^2} = 1. \tag{11.2}$$

Also, the fact that $|NA| = |NM|$ implies

$$t^2 + a^2 = (t-y)^2 + x^2,$$

hence

$$t = \frac{x^2 + y^2 - a^2}{2y}. \tag{11.3}$$

From (11.2) and (11.3),

$$\frac{x^2}{a^2} + \left(\frac{x^2 - a^2}{by}\right)^2 = 1,$$

in other words,

$$\frac{x^2 - a^2}{a^2} + \frac{(x^2 - a^2)^2}{(by)^2} = 0. \tag{11.4}$$

Since $|x| = |x_0| < a$, we have $x^2 - a^2 \neq 0$, so by (11.4),

$$\frac{1}{a^2} + \frac{x^2 - a^2}{b^2 y^2} = 0,$$

namely

$$\frac{y^2}{a^2} + \frac{x^2}{b^2} = \frac{a^2}{b^2},$$

or in the standard form

$$\frac{x^2}{a^2} + \frac{y^2}{\left(\frac{a^2}{b}\right)^2} = 1 \ (|x| < a).$$

Therefore, the trajectory sought is an ellipse (excluding its two intersection points $(\pm a, 0)$ with the x-axis). $\qquad\square$

Example 2. As Figure 11.3 indicates, the left focus of the ellipse $E: \frac{x^2}{9} + \frac{y^2}{5} = 1$ is F_1, a straight line l that passes through F_1 intersects the ellipse E at two points A and B, and the coordinates of a point Q are $\left(-\frac{9}{2}, 0\right)$. If $\overrightarrow{QB} \perp \overrightarrow{AB}$, find the slopes of all possible straight lines l.

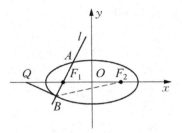

Figure 11.3   Figure for Example 2.

Solution. Let the right focus of the ellipse be F_2 and the inclination angle of the straight line l be $\angle F_2 F_1 A = \theta$. Then $\angle F_2 F_1 B = \pi - \theta$. Also let $|BF_1| = t$. Then by the given condition and the definition of ellipses,

$$|BF_2| = 6 - t.$$

In $\triangle F_1 B F_2$, by the law of cosines,

$$|BF_2|^2 = |BF_1|^2 + |F_1 F_2|^2 - 2|BF_1||F_1 F_2| \cos(\pi - \theta),$$

that is

$$(6 - t)^2 = t^2 + 4^2 + 8t \cos \theta,$$

the solution of which is

$$t = \frac{5}{3 + 2 \cos \theta}.$$

In the right $\triangle QBF_1$, we have $t = |BF_1| = |QF_1| \cos \theta = \frac{5}{2} \cos \theta$. It follows that

$$\frac{5}{3 + 2 \cos \theta} = \frac{5}{2} \cos \theta,$$

namely $2 \cos^2 \theta + 3 \cos \theta - 2 = 0$, that is,

$$(2 \cos \theta - 1)(\cos \theta + 2) = 0.$$

Since $\cos \theta + 2 > 0$, we must have $2 \cos \theta - 1 = 0$. Thus, $\cos \theta = \frac{1}{2}$, and so $\theta = \frac{\pi}{3}$.

By the symmetry of the ellipse, the slopes of the straight line l are $\pm\sqrt{3}$. □

Remark. Using the definition of conic sections and features of the corresponding figures can establish more directly the relation between the known and the unknown, avoid complicated manipulations and make things simpler, and sometimes prevent from possible mistakes in solving problems.

In the current example, if we use the polar coordinates equation of the ellipse, it may be more convenient to make the statements; if we use the parametric equations of the ellipse, it may be much simpler to solve the problem.

Example 3. Suppose that the left and right foci of an ellipse $C : \frac{x^2}{a^2} + \frac{y^2}{b^2} = 1$ $(a > b > 0)$ are F_1 and F_2, respectively, the eccentricity is e, the straight line $l : y = ex + a$ intersects the x-axis and the y-axis at A and B, respectively, and P is the symmetric point of the point F_1 with respect to the straight line l. Assume that $\overrightarrow{AM} = \lambda \overrightarrow{AB}$.

(1) Prove: $\lambda = 1 - e^2$.
(2) Determine the value of λ such that $\triangle PF_1F_2$ is an isosceles triangle.

Analysis. Combine the equations of l and C to find the common point M, and prove the conclusion of (1) from $\overrightarrow{AM} = \lambda \overrightarrow{AB}$; a condition to make $\triangle PF_1F_2$ an isosceles triangle is $\frac{1}{2}|PF_1| = c$.

Solution. (1) Since A and B are the intersection points of the straight line $l : y = ex + a$ with the x-axis and the y-axis respectively, so the coordinates of A and B are $\left(-\frac{a}{e}, 0\right)$ and $(0, a)$, respectively. From

$$\begin{cases} y = ex + a, \\ \dfrac{x^2}{a^2} + \dfrac{y^2}{b^2} = 1, \end{cases}$$

we have $x = -c$ and $y = \frac{b^2}{a}$ (in which $c = \sqrt{a^2 - b^2}$). Thus, the coordinates of the point M are $\left(-c, \frac{b^2}{a}\right)$. It follows from $\overrightarrow{AM} = \lambda \overrightarrow{AB}$ that $\left(-c + \frac{a}{e}, \frac{b^2}{a}\right) = \lambda \left(\frac{a}{e}, a\right)$, hence $\lambda = 1 - e^2$.

(2) Since $PF_1 \perp l$, so $\angle PF_1F_2 = 90° + \angle BAF_1$ is an obtuse angle. For $\triangle PF_1F_2$ to be an isosceles triangle, it is necessary that $|PF_1| = |F_1F_2|$, namely $\frac{1}{2}|PF_1| = c$.

Let the distance from F_1 to l be d. By $\frac{1}{2}|PF_1| = d = \frac{|e(-c)+0+a|}{1+e^2} = \frac{a-ce}{1+e^2} = c$, we obtain $e^2 = \frac{1}{3}$. Thus, $\lambda = 1 - e^2 = \frac{2}{3}$.

That is, $\triangle PF_1F_2$ is an isosceles triangle when $\lambda = \frac{2}{3}$. $\qquad\square$

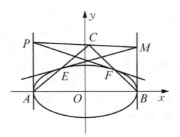

Figure 11.4   Figure for Example 4.

Example 4. As Figure 11.4 shows, let A and B be the left and right vertices respectively of an ellipse $\frac{x^2}{r^2} + y^2 = 1$ $(r > 0)$, and $PA \perp AB$ and $MB \perp AB$. Draw a tangent line to the ellipse that passes through the point P with the tangent point F different from A, and draw a tangent line to the ellipse that passes through the point M with the tangent point E different from B. Assume that AE and BF intersect at a point C. Prove: The three points P, C, and M are co-linear.

Proof. It is easy to see that the coordinates of A and B are $A(-r, 0)$ and $B(r, 0)$. Let the coordinates of P and M be $P(-r, t)$ and $M(r, s)$. Then the equation of the chord at contact at the point M with respect to the ellipse is $\frac{x}{r} + sy = 1$.

Substituting the above into the equation of the ellipse in the problem and after a manipulation, we get

$$(1 + s^2)y^2 - 2sy = 0.$$

It follows that

$$y_E = \frac{2s}{1 + s^2}.$$

Now, $x_E = r\left(1 - s\frac{2s}{1+s^2}\right) = \frac{1-s^2}{1+s^2}r$.

Thus, the coordinates of E are $E\left(\frac{1-s^2}{1+s^2}r, \frac{2s}{1+s^2}\right)$.

Similarly, the coordinates of F are $F\left(\frac{t^2-1}{1+t^2}r, \frac{2t}{1+t^2}\right)$.

From $k_{AE} = \frac{\frac{2s}{1+s^2}}{\frac{1-s^2}{1+s^2}r+r} = \frac{s}{r}$, we know that

$$l_{AE} : y = \frac{s}{r}(x + r). \tag{11.5}$$

Similarly,

$$l_{BF} : y = -\frac{t}{r}(x - r). \tag{11.6}$$

Eliminating y from (11.5) and (11.6), we obtain $x_C = \frac{t-s}{t+s}r$.

Consequently, $y_C = \frac{s}{r}\left(\frac{t-s}{t+s}r + r\right) = \frac{2ts}{t+s}$.

Hence, the coordinates of C are $C\left(\frac{t-s}{t+s}r, \frac{2ts}{t+s}\right)$.

As a result, $\frac{y_C - y_M}{x_C - x_M} = \frac{s-t}{2r}$ and $\frac{y_C - y_P}{x_C - x_P} = \frac{s-t}{2r}$, from which the three points P, C, and M are co-linear.   □

Example 5. Let a point O be the center of an ellipse, and let A be an arbitrary point on the ellipse that is different from any of its vertices. Draw a perpendicular line to the major axis of the ellipse at the point A with M the foot of the perpendicular. Connect A and O and extend AO to intersect the ellipse at another point B, and connect B and M and extend BM to intersect the ellipse at a point C. Does there exist an ellipse such that $BA \perp CA$?

Solution. Choose the center of the ellipse as the origin, the major axis of the ellipse as the x-axis, thus establishing a coordinates system. Let the equation of the ellipse be $\frac{x^2}{a^2} + \frac{y^2}{b^2} = 1$. If $a^2 = 2b^2$, then necessarily $BA \perp CA$.

Let the coordinates of A and C be $A(x_0, y_0)$ and $C(x_1, y_1)$. The conditions of the problem imply that the coordinates of M and B are $M(x_0, 0)$ and $B(-x_0, -y_0)$. Substituting the above into the equation of the ellipse and simplifying the resulting expression, we obtain

$$(a^2k^2 + b^2)x^2 - 2a^2k^2x_0x + a^2k^2x_0^2 - a^2k^2 = 0.$$

By Vieta's theorem,

$$x_1 - x_0 = \frac{2a^2k^2x_0}{a^2k^2 + b^2}.$$

Thus, from the equation $y = k(x - x_0)$ of the straight line BC, we get

$$y_1 = k(x - x_0) = k\frac{2a^2k^2x_0}{a^2k^2 + b^2} = \frac{2a^2k^3x_0}{a^2k^2 + b^2}.$$

Again from the conditions of the problem, the slope of the straight line AB is

$$k_{AB} = \frac{y_0}{x_0},$$

and the slope of the straight line AC is

$$k_{AC} = \frac{y_1 - y_0}{x_1 - x_0} = \frac{\frac{2a^2 k^3 x_0}{a^2 k^2 + b^2} - y_0}{\frac{2a^2 k^2 x_0}{a^2 k^2 + b^2}}.$$

The requirement $BA \perp CA$ implies $k_{AB} k_{AC} = -1$, and by noticing that $k = \frac{y_0}{2x_0}$, a manipulation gives the condition $a^2 = 2b^2$ for the existence of the demanded ellipse.   □

Example 6. As in Figure 11.5, an ellipse with its center as the origin, the foci on the x-axis, and the eccentricity $\frac{\sqrt{3}}{2}$ passes through the point $\left(\sqrt{2}, \frac{\sqrt{2}}{2}\right)$. Suppose that a straight line l that does not pass through the origin intersects the ellipse at two points P and Q, and the slopes of the straight lines OP, PQ, and OQ constitute a geometric sequence. Find the range of the values for the area of $\triangle OPQ$.

Solution. By the condition of the problem, let the equation of the ellipse be

$$\frac{x^2}{a^2} + \frac{y^2}{b^2} = 1 \ (a > b > 1).$$

From $\frac{c}{a} = \frac{\sqrt{3}}{2}$ and $\frac{2}{a^2} + \frac{1}{2b^2} = 1$, we have $a = 2$ and $b = 1$.

Thus, the equation of the ellipse is $\frac{x^2}{4} + y^2 = 1$.

The hypothesis of the problem guarantees that the slope of the straight line l exists and is not equal to 0, so we may assume that the equation of the straight line l is $y = kx + m \ (m \neq 0)$. The two points $P(x_1, y_1)$ and

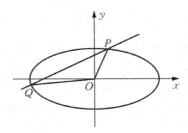

Figure 11.5   Figure for Example 6.

$Q(x_2, y_2)$ satisfy

$$\begin{cases} y = kx + m, \\ x^2 + 4y^2 - 4 = 0. \end{cases}$$

Eliminating y gives $(1 + 4k^2)x^2 + 8kmx + 4(m^2 - 1) = 0$. Its discriminant

$$\Delta = 64k^2m^2 - 16(1 + 4k^2)(m^2 - 1) = 16(4k^2 - m^2 + 1) > 0.$$

Also, $x_1 + x_2 = \frac{-8km}{1+4k^2}, x_1 x_2 = \frac{4(m^2-1)}{1+4k^2}$, and

$$y_1 y_2 = (kx_1 + m)(kx_2 + m) = k^2 x_1 x_2 + km(x_1 + x_2) + m^2.$$

Since the slopes of the straight lines OP, PQ, and OQ constitute a geometric sequence, so

$$\frac{y_1}{x_1} \frac{y_2}{x_2} = \frac{k^2 x_1 x_2 + km(x_1 + x_2) + m^2}{x_1 x_2} = k^2,$$

namely $\frac{-8k^2m^2}{1+4k^2} + m^2 = 0$. Since $m \neq 0$, we have $k^2 = \frac{1}{4}$, that is $k = \pm\frac{1}{2}$.

Since the slope of the straight line OQ exists and is nonzero, and since $\Delta > 0$, we find that $0 < m^2 < 2$ and $m^2 \neq 1$.

Let d be the distance of the point O to the straight line l. Then

$$S_{\triangle OPQ} = \frac{1}{2} d|PQ| = \frac{1}{2} - \frac{|m|}{\sqrt{1+k^2}} \sqrt{1+k^2}|x_1 - x_2|$$

$$= \frac{1}{2}|m|\sqrt{(x_1 + x_2)^2 - 4x_1 x_2} = \sqrt{m^2(2 - m^2)},$$

so the range of the values for the area of $\triangle OPQ$ is $(0, 1)$. □

Example 7. Draw two perpendicular chords AB and CD of the ellipse $\frac{x^2}{5} + \frac{y^2}{4} = 1$ that pass through the right focus F. Suppose that the middle points of AB and CD are M and N, respectively.

(1) Prove: The straight line MN must pass through a fixed point, and find the fixed point.
(2) If the slopes of AB and CD both exist, find the maximum value of the area of $\triangle FMN$.

Solution. (1) The condition of the problem implies that the coordinates of the focus F are $F(1, 0)$.

(i) When the slopes of the chords AB and CD both exist, suppose that the slope of AB is k and the slope of CD is $-\frac{1}{k}$.

Assume that the equation of l_{AB} is $y = k(x-1)$. Substituting it into the equation $\frac{x^2}{5} + \frac{y^2}{4} = 1$ of the ellipse gives

$$(5k^2 + 4)x^2 - 10k^2x + 5k^2 - 20 = 0.$$

Consequently,

$$x_M = \frac{x_A + x_B}{2} = \frac{5k^2}{5k^2 + 4},$$

$$y_M = k(x_M - 1) = \frac{-4k}{5k^2 + 4}.$$

Hence, the coordinates of M are $M\left(\frac{5k^2}{5k^2+4}, \frac{-4k}{5k^2+4}\right)$.

Since $CD \perp AB$, after replacing k with $-\frac{1}{k}$ in the coordinates of the point M, we obtain the point

$$N\left(\frac{5}{4k^2 + 5}, \frac{4k}{4k^2 + 5}\right).$$

When $k \neq \pm 1$,

$$k_{MN} = \frac{\frac{4k}{4k^2+5} + \frac{4k}{5k^2+4}}{\frac{5}{4k^2+5} - \frac{5k^2}{5k^2+4}} = \frac{36k(k^2+1)}{20 - 20k^4} = \frac{-9k}{5k^2 - 5},$$

and at this time the equation of l_{MN} is

$$y - \frac{4k}{4k^2 + 5} = \frac{-9k}{5k^2 - 5}\left(x - \frac{5}{4k^2 + 5}\right),$$

namely

$$y = \frac{-9k}{5k^2 - 5}\left(x - \frac{5}{9}\right).$$

Hence, the straight line l_{MN} passes through the fixed point $\left(\frac{5}{9}, 0\right)$.

When $k = \pm 1$, it is easy to obtain the equation $x = \frac{5}{9}$ of l_{MN}, which also passes through the same fixed point $\left(\frac{5}{9}, 0\right)$.

(ii) When the slope of the chord AB or the slope of the chord CD does not exist, it is easy to see that the straight line MN is the x-axis, which also passes through the fixed point $\left(\frac{5}{9}, 0\right)$.

To summarize, the straight line MN passes through the fixed point $E\left(\frac{5}{9}, 0\right)$.

(2) From (1), we know that the area of $\triangle FMN$ is

$$S = \frac{1}{2}|EF||y_M - y_N|$$

$$= \frac{2}{9}\left|\frac{-4k}{5k^2 + 4} - \frac{4k}{4k^2 + 5}\right|$$

$$= \left|\frac{8k(k^2 + 1)}{(5k^2 + 4)(4k^2 + 5)}\right|.$$

Without loss of generality, assume that $k > 0$. Then

$$S' = \frac{8(-20k^6 - 19k^4 + 19k^2 + 20}{(5k^2 + 4)^2(4k^2 + 5)^2}$$

$$= \frac{-8(20k^4 + 39k^2 + 20)(k^2 - 1)}{(5k^2 + 4)^2(4k^2 + 5)^2}.$$

Setting $S' = 0$ gives $k = 1$.

We see that $S' > 0$ when $k \in (0, 1)$ and $S' < 0$ when $k \in (1, \infty)$. It follows that S has the maximum value $\frac{16}{81}$ when $k = 1$.

Therefore, the maximum value of the area of $\triangle FMN$ is $\frac{16}{81}$. $\qquad\square$

Example 8. For a given ellipse $C_1 : \frac{x^2}{a^2} + \frac{y^2}{b^2} = 1$, a straight line l that does not pass through the origin intersects the ellipse at two points A and B.

(1) Find the maximum value of the area $S_{\triangle OAB}$ for $\triangle OAB$.

(2) Does there exist an ellipse C_2 such that for each tangent line to C_2 intersecting the ellipse C_1 at two points A and B, the area $S_{\triangle OAB}$ of $\triangle OAB$ always equals the maximum value obtained from (1)? If it exists, find the ellipse; otherwise, give your reason.

Solution. (1) Suppose that the slope of the straight line l exists. Let the equation of l be $y = kx + m$. By substituting $y = kx + m$ into the equation of the ellipse, we obtain

$$(a^2k^2 + b^2)x^2 + 2a^2kmx + a^2m^2 - a^2b^2 = 0.$$

Suppose that the points A and B are $A(x_1, y_1)$ and $B(x_2, y_2)$. Then

$$x_1 + x_2 = -\frac{2a^2km}{a^2k^2 + b^2}, \quad x_1x_2 = \frac{a^2m^2 - a^2b^2}{a^2k^2 + b^2}.$$

Consequently,

$$|AB| = \sqrt{(x_2 - x_1)^2 + (y_2 - y_1)^2}$$
$$= \sqrt{1 + k^2}\sqrt{(x_2 + x_1)^2 - 4x_1 x_2}$$
$$= \sqrt{1 + k^2}\frac{2ab}{a^2 k^2 + b^2}\sqrt{a^2 k^2 + b^2 - m^2}.$$

Let the height of $\triangle OAB$ with respect to the side AB be h. Then

$$h = \frac{|m|}{\sqrt{1 + k^2}}.$$

Thus, the area of $\triangle OAB$ is

$$S_{\triangle OAB} = \frac{1}{2}|AB|h = \frac{ab|m|}{a^2 k^2 + b^2}\sqrt{a^2 k^2 + b^2 - m^2}.$$

Fix k. Then

$$S_{\triangle OAB} = \frac{ab|m|}{a^2 k^2 + b^2}\sqrt{a^2 k^2 + b^2 - m^2} \le \frac{ab}{2}.$$

It follows that for any k, there holds that $S_{\triangle OAB} \le \frac{ab}{2}$, and the equality is satisfied if and only if $a^2 k^2 + b^2 = 2m^2$.

If the slope of the straight line l does not exist, then let $x = m$. It is easy to show that $S_{\triangle OAB} \le \frac{ab}{2}$, and the equality is satisfied if and only if $a^2 = 2m^2$.

Therefore, the maximum value of the area of $\triangle OAB$ is $\frac{ab}{2}$.

(2) There exists an ellipse $C_2 : \frac{x^2}{a^2} + \frac{y^2}{b^2} = \frac{1}{2}$, and any tangent line to this ellipse intersects the ellipse $C_1 : \frac{x^2}{a^2} + \frac{y^2}{b^2} = 1$ at two points A and B with the property that $S_{\triangle OAB} = \frac{ab}{2}$.

In fact, suppose that an ellipse that satisfies the requirement is $\frac{x^2}{a^2} + \frac{y^2}{b^2} = \lambda$ with some $\lambda \in (0, 1)$. Then the equation of a tangent line to the ellipse at any point (x_0, y_0) on it has the format of

$$\frac{x_0 x}{a^2} + \frac{y_0 y}{b^2} = \lambda.$$

Since $0 < \lambda < 1$, it is easy to see that the tangent line intersects the ellipse $C_1 : \frac{x^2}{a^2} + \frac{y^2}{b^2} = 1$ at two points A and B.

If $y_0 = 0$, then $x_0 = \pm a\sqrt{\lambda}$. From the equation of the tangent line, we see that $x = \frac{\lambda a^2}{x_0} = \pm a\sqrt{\lambda}$. By (1), the necessary and sufficient condition for $S_{\triangle OAB} = \frac{ab}{2}$ is $a^2 = 2(\sqrt{\lambda}a)^2 = 2\lambda a^2$, that is, $\lambda = \frac{1}{2}$.

In the following, we show that if $y_0 \neq 0$, it is still true that $S_{\triangle OAB} = \frac{ab}{2}$ when $\lambda = \frac{1}{2}$. This time, the equation of the tangent line at any point (x_0, y_0) of the ellipse $\frac{x^2}{a^2} + \frac{y^2}{b^2} = \frac{1}{2}$ is

$$y = -\frac{b^2 x_0}{a^2 y_0} x + \frac{b^2}{2 y_0}.$$

Let $k = -\frac{b^2 x_0}{a^2 y_0}$ and $m = \frac{b^2}{2 y_0}$. Then

$$a^2 k^2 + b^2 = a^2 \left(-\frac{b^2 x_0}{a^2 y_0} \right)^2 + b^2 = b^2 \left(1 + \frac{b^2 x_0^2}{a^2 y_0^2} \right).$$

Since $\frac{x_0^2}{a^2} + \frac{y_0^2}{b^2} = \frac{1}{2}$, we have

$$a^2 k^2 + b^2 = a^2 \left(-\frac{b^2 x_0}{a^2 y_0} \right)^2 + b^2$$

$$= b^2 \left(1 + \frac{b^2 x_0^2}{a^2 y_0^2} \right)$$

$$= \frac{b^4}{2 y_0^2} = 2m^2.$$

By (1), we obtain that $S_{\triangle OAB} = \frac{ab}{2}$.

In summary, there exists an ellipse $C_2 : \frac{x^2}{a^2} + \frac{y^2}{b^2} = \frac{1}{2}$, and any tangent line to this ellipse intersects the ellipse $C_1 : \frac{x^2}{a^2} + \frac{y^2}{b^2} = 1$ at two points A and B with $S_{\triangle OAB}$ being $\frac{ab}{2}$ exactly. $\qquad\square$

3. Exercises

Group A

I. Filling Problems

1. Let P be a point on the ellipse $\frac{x^2}{4} + \frac{y^2}{3} = 1$, and let $k = |PF_1||PF_2|$ (F_1 and F_2 are the two foci of the ellipse). Then the difference of the maximum value and the minimum value of k is _____.

2. Suppose that an ellipse $\frac{x^2}{a^2} + \frac{y^2}{b^2} = 1$ ($a > b > 0$) and the circle $x^2 + y^2 = \left(\frac{b}{2} + c \right)^2$ (c is the half focal distance) have four different intersection points. Then the value range for the eccentricity e of the ellipse is _____.

3. In an ellipse $\frac{x^2}{a^2} + \frac{y^2}{b^2} = 1$ ($a > b > 0$), denote the left focus by F, the right vertex by A, and the upper end point of the minor axis by B. If the eccentricity of the ellipse is $e = \frac{\sqrt{5}-1}{2}$, then $\angle ABF = $ _____.

4. Given an ellipse $\frac{x^2}{a^2} + \frac{y^2}{b^2} = 1$ $(a > b > 0)$. Draw a chord AB passing through the center O. Let F be the right focus of the ellipse. Then the maximum value of the area of $\triangle ABF$ is _____.

II. Calculation Problems

5. As Figure 11.6 shows, a circle centered at the origin O with radius R intersects the positive half y-axis at a point B. There is now a moving ellipse (the major axis and minor axis vary, but the x-axis and y-axis are always the symmetric ones) that intersects the circle O at A in the first quadrant, and the straight line BA intersects the x-axis at a point D. Prove: The product $|BA||BD|$ is of fixed value.

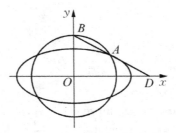

Figure 11.6   Figure for Exercise 5.

6. Let the two end points of the major axis of an ellipse $\frac{x^2}{a^2} + \frac{y^2}{b^2} = 1$ $(a > b > 0)$ be A and B. If there exists a point P such that $\angle APB = \frac{2\pi}{3}$, find the value range for the eccentricity of the ellipse.

7. Prove: Let P be a moving point on an ellipse $\frac{x^2}{a^2} + \frac{y^2}{b^2} = 1$ $(a > b > 0)$ that is not a vertex, O be the center of the ellipse, and F be the right focus of the ellipse. A tangent line l to the ellipse at a point M on the second quadrant part of the ellipse is parallel to OP, and MF intersects OP at a point N. Then $|MN| = a$.

8. Let the equation of an ellipse be $\frac{x^2}{a^2} + \frac{y^2}{b^2} = 1$ $(a > b > 0)$. A line segment PQ is a focal chord passing through the left focus F but not perpendicular to the x-axis. If there exists a point R on the left directrix such that $\triangle PQR$ is an equilateral triangle, find the value range for the eccentricity of the ellipse, and express the slope of the straight line PQ in terms of e.

Group B

9. A bird's view of the Stadium "Bird's Nest" for the 2008 Beijing Olympic Games is shown in Figure 11.7. The inner and outer steel skeletons are ellipses of the same eccentricity. From the outer vertices, draw tangent lines to the inner ellipse. If the product of the slopes of the tangent lines AC and BD is $-\frac{9}{16}$, find the eccentricity of the ellipses.

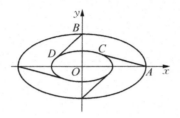

Figure 11.7 Figure for Example 9.

10. Suppose that the left vertex of an ellipse $C : \frac{x^2}{a^2} + \frac{y^2}{b^2} = 1 \ (a > b > 0)$ is $A(-a,0)$, two points M and N are on the ellipse, and $MA \perp NA$. Does the straight line MN pass through a fixed point? If so, give the coordinates of this point; if not so, express your reason.

11. We are given the circle $C_1 : x^2 + y^2 = 1$ and an ellipse $C_2 : \frac{x^2}{a^2} + \frac{y^2}{b^2} = 1 \ (a > b > 0)$. Question: If and only if what conditions do a and b satisfy for any point P on C_2, there always exists a parallelogram that has P as a vertex, is tangent to the circle C_1, and is inscribed into the ellipse C_2?

12. Let E be a given ellipse $\frac{x^2}{a^2} + \frac{y^2}{b^2} = 1 \ (a > b > 0)$, and let A_d be the set of all straight lines with distance d to the origin. Does there exist a constant $d \ (0 < d < b)$ such that for any $l \in A_d$, there always are $l_1, l_2 \in A_d$ with the property that l_1 and l_2 pass through two intersection points P and Q of l and the ellipse E, respectively, and $l_1 \parallel l_2$? Give your reason.

Chapter 12

Hyperbolas

1. Key Points of Knowledge and Basic Methods

A. *Standard equation and properties of hyperbolas*

Standard equation of a hyperbola with the foci on the x-axis: Let the foci be $F_1(-c, 0)$ and $F_2(c, 0)$ $(c > 0)$, the length of the real axis be $2a$, and the length of the imaginary axis be $2b$ $(a > b)$. Then the standard equation of the hyperbola is

$$\frac{x^2}{a^2} - \frac{y^2}{b^2} = 1, \qquad (12.1)$$

where $c^2 = a^2 + b^2$. The range of x is $x \leq -a$ or $x \geq a$; the symmetric axes are the x-axis and the y-axis; the vertices are $(-a, 0)$ and $(a, 0)$.

The equations of the directrices are $x = \pm \frac{a^2}{c}$.

The equations of the asymptotes are $y = \pm \frac{b}{a} x$.

The eccentricity $e = \frac{c}{a} > 1$.

Similarly, one can write down the standard equation and the equations of the directrices for a hyperbola whose foci are on the y-axis, and the corresponding properties.

B. *Second definition of hyperbolas (foci-directrix definition) and focal radii*

The trajectory of a point in a plane satisfying the following property is a hyperbola: The ratio of the distance of the point to a fixed point (focus) and the distance of the point to a fixed straight line (directrix), which does not pass through the fixed point, is a positive constant (eccentricity) $e > 1$.

Suppose that $P(x_0, y_0)$ is a point on the hypobola (12.1). Then the left focal radius $|PF_1| = |ex_0 + a|$ and the right focal radius $|PF_2| = |ex_0 - a|$.

Let the inclination angle of a chord AB of the hyperbola that passes through a focus be α. Then $|AB| = \frac{2ab^2}{|b^2 - c^2 \sin^2 \alpha|}$.

C. *Optical properties of hyperbolas*

The opposite direction extension of light rays emitted from a focus of a hyperbola will go to the destination of the other focus after the refection from the hyperbola.

D. *Special hyperbolas*

(1) Hyperbolas of equal axes: Hyperbolas with the same length of the real axis and imaginary axis.

   Standard equation of hyperbolas of equal axes with the foci on the x-axis: $\frac{x^2}{a^2} - \frac{y^2}{a^2} = 1 \ (a > 0)$.

   Equation of hyperbolas with the coordinates axes as the asymptotes: $xy = k \ (k \neq 0)$.

(2) Equation of the family of the hyperbolas with the same asymptotes $\frac{x}{a} \pm \frac{y}{b} = 0 : \frac{x^2}{a^2} - \frac{y^2}{b^2} = \lambda \ (\lambda \neq 0)$.

(3) Conjugate hyperbolas: The hyperbolas given by $\frac{x^2}{a^2} - \frac{y^2}{b^2} = 1$ and $\frac{x^2}{a^2} - \frac{y^2}{b^2} = -1$ respectively are called the mutually conjugate hyperbolas.

The other contents are referred to the previous chapter on ellipses.

2. Illustrative Examples

Example 1. Given the point $A(2, 1)$. Let the right focus of the hyperbola $x^2 - \frac{y^2}{4} = 1$ be F and let a point P be on the right branch of the hyperbola.

(1) Find the minimum value of $|AP| + |PF|$.
(2) Find the minimum value of $|AP| + \frac{\sqrt{5}}{5}|PF|$, and write down the corresponding coordinates of the point P.

Analysis. The question (1) can be transformed to another one if we can consider using the definition $|PF| = |PF'| - 2a$ (in which F' is the left focus), realizing the purpose that A and F' will be on the different sides of the curve. Compared to (1), the question (2) has an additional coefficient

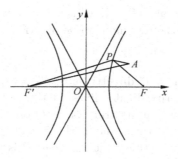

Figure 12.1 Figure 1 for Example 1.

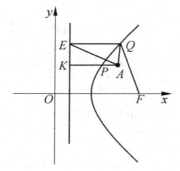

Figure 12.2 Figure 2 for Example 1.

$\frac{\sqrt{5}}{5}$. If we can observe that $e = \sqrt{5}$, then

$$|AP| + \frac{\sqrt{5}}{5}|PF| = |AP| + \frac{|PF|}{e}$$
$$= |AP| + d \ (d \text{ is the distance of } P \text{ to the directrix}).$$

As shown in Figure 12.2, the minimum value is $|AK|$.

Solution. (1) As Figure 12.1 shows, in the equation $x^2 - \frac{y^2}{4} = 1$ of the hyperbola, $a = 1, b = 2, c = \sqrt{5}$, and F' is the left focus. By the definition of a hyperbola, $|AP| + |PF| = |AP| + |PF'| - 2a \geq |AF'| - 2a$. Since the distance of $F'(-\sqrt{5}, 0)$ and $A(2, 1)$ is

$$|AF'| = \sqrt{(2 + \sqrt{5})^2 + 1^2} = \sqrt{10 + 4\sqrt{5}},$$

and the minimum value of $|AP| + |PF|$ is achieved when the point P is on the line segment AF', the minimum value of $|AP| + |PF|$ is $|AF'| - 2a = \sqrt{10 + 4\sqrt{5}} - 2$.

(2) See Figure 12.2. The right directrix is $x = \frac{a^2}{c} = \frac{\sqrt{5}}{5}$, and the eccentricity $e = \sqrt{5}$. Draw two perpendicular lines to the right directrix that pass through the point A and an arbitrary point Q on the right branch of the hyperbola, respectively, with K and E the feet of the perpendicular. Let KA intersect the curve at the point P. Then P is the desired point and $|AK|$ is the minimum value. We prove it in the following:

$$|QA| + \frac{\sqrt{5}}{5}|QF| = |QA| + \frac{|QA|}{e} = |QA| + |QE| \geq |EA| \geq |AK|$$

$$= |AP| + \frac{|PF|}{\sqrt{5}} = 2 - \frac{\sqrt{5}}{5},$$

and the equality is satisfied when $Q = P$. Thus, the minimum value is $2 - \frac{\sqrt{5}}{5}$ that is achieved at $P\left(\frac{\sqrt{5}}{2}, 1\right)$.   □

Remark. Sometimes, it is simple and direct to use the definition of conic sections for an extremal value.

Example 2. Suppose that the coordinates of the two foci of a hyperbola are $F_1(-2, 0)$ and $F_2(2, 0)$, and a tangent line to the hyperbola intersects the x-axis at $Q\left(\frac{1}{2}, 0\right)$ with slope 2.

(1) Find the equation of the hyperbola.
(2) Let the tangent point of the tangent line to the hyperbola be P. Prove: $\angle F_1 PQ = \angle F_2 PQ$.

Solution. (1) Assume that the equation of the hyperbola is $\frac{x^2}{a^2} - \frac{y^2}{b^2} = 1$. Since it is tangent to the straight line $y = 2\left(x - \frac{1}{2}\right)$, namely $y = 2x - 1$, so the system of the equations

$$\begin{cases} \dfrac{x^2}{a^2} - \dfrac{y^2}{b^2} = 1, \\ y = 2x - 1 \end{cases}$$

has only one solution, from which the equation of x,

$$\frac{x^2}{a^2} - \frac{(2x - 1)^2}{b^2} = 1, \tag{12.2}$$

has two equal real roots. Thus, its discriminant equals zero, and so

$$\left(\frac{4}{b^2}\right)^2 - 4\left(\frac{1}{a^2} - \frac{1}{b^2}\right)\left(-\frac{1}{b^2} - 1\right) = 0.$$

Simplifying the above gives $b^2 - 4a^2 + 1 = 0$.

Note that the coordinates of the two foci of the hyperbola are $(-2, 0)$ and $(2, 0)$. Hence, the half focal distance is $c = 2$, and thus $a^2 + b^2 = 4$. Combined with the previous equation, we see that $a^2 = 1$ and $b^2 = 3$. Therefore, the equation of the hyperbola is

$$x^2 - \frac{y^2}{3} = 1.$$

(2) Since $a^2 = 1$ and $b^2 = 3$, the equation (12.2) is $x^2 - \frac{(2x-1)^2}{3} = 1$, whose solution is $x = 2$. Substituting it into $y = 2x - 1$ gives $y = 3$. Thus, the coordinates of the tangent point P are $(2, 3)$.

Hence, the slope of $F_1 P$ is

$$k = \frac{3 - 0}{2 - (-2)} = \frac{3}{4}.$$

Also, the slope of the tangent line PQ is $t = 2$, so

$$\tan \angle F_1 P Q = \frac{t - k}{1 + kt} = \frac{1}{2}.$$

Note that the horizontal coordinates of F_2 and P are the same, and it follows that $F_2 P$ is parallel to the y-axis, from which $\tan \angle F_2 P Q = \frac{1}{t} = \frac{1}{2}$. Therefore, $\tan \angle F_1 P Q = \tan \angle F_2 P Q$, that is, $\angle F_1 P Q = \angle F_2 P Q$. $\square$

Example 3. As Figure 12.3 indicates, A and B are the common vertices of an ellipse $\frac{x^2}{a^2} + \frac{y^2}{b^2} = 1$ $(a > b > 0)$ and a hyperbola $\frac{x^2}{a^2} - \frac{y^2}{b^2} = 1$, P and Q are moving points on the hyperbola and ellipse respectively, which are different from A and B, and they satisfy $\overrightarrow{AP} + \overrightarrow{BP} = \lambda(\overrightarrow{AQ} + \overrightarrow{BQ})$ ($\lambda \in \mathbf{R}$ and $|\lambda| > 1$). Prove:

(1) The three points O, P, and Q are on the same straight line.
(2) If the slopes of the straight lines AP, BP, AQ, and BQ are k_1, k_2, k_3, and k_4 respectively, then $k_1 + k_2 + k_3 + k_4$ is a fixed value.

Proof. (1) Since O is the origin, so

$$\overrightarrow{AP} + \overrightarrow{BP} = 2\overrightarrow{OP},$$
$$\overrightarrow{AQ} + \overrightarrow{BQ} = 2\overrightarrow{OQ}.$$

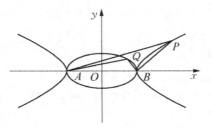

Figure 12.3 Figure for Example 3.

Also $\overrightarrow{OP} = \lambda \overrightarrow{OQ}$ from $\overrightarrow{AP} + \overrightarrow{BP} = \lambda(\overrightarrow{AQ} + \overrightarrow{BQ})$. Thus, the three points O, P, and Q are co-linear.

(2) Write $P(x_1, y_1)$ and $Q(x_2, y_2)$. Then $x_1^2 - a^2 = \frac{a^2}{b^2} y_1^2$ and $x_2^2 - a^2 = -\frac{a^2}{b^2} y_2^2$. Hence,

$$k_1 + k_2 = \frac{y_1}{x_1 + a} + \frac{y_1}{x_1 - a} = \frac{2x_1 y_1}{x_1^2 - a^2} = \frac{2b^2}{a^2} \frac{x_1}{y_1}. \tag{12.3}$$

By the same token,

$$k_3 + k_4 = -\frac{2b^2}{a^2} \frac{x_2}{y_2}. \tag{12.4}$$

From (1) we know that the three points O, P, and Q are co-linear, and consequently $\frac{x_1}{y_1} = \frac{x_2}{y_2}$.

Therefore, by (12.3) and (12.4) we have

$$k_1 + k_2 + k_3 + k_4 = \frac{2b^2}{a^2} \frac{x_1}{y_1} - \frac{2b^2}{a^2} \frac{x_2}{y_2}$$

$$= \frac{2b^2}{a^2} \left(\frac{x_1}{y_1} - \frac{x_2}{y_2} \right) = 0,$$

in other words, $k_1 + k_2 + k_3 + k_4$ is a fixed value. $\square$

Example 4. A straight line passing through an arbitrary point $P(x_0, y_0)$ on the right branch of the hyperbola $x^2 - \frac{y^2}{4} = 1$ intersects the two asymptotes of the hyperbola at two points A and B. If P is the middle point of AB, prove:

(1) The straight line l and the hyperbola have only one intersection point.
(2) The area of $\triangle OAB$ is a fixed value.

Proof. (1) The equations of the two asymptotes of the hyperbola are $y = \pm 2x$.

When $y_0 = 0$, it is easy to get the equation $x = x_0$ of the straight line l, and so there is only one intersection point of l and the hyperbola.

When $y_0 \neq 0$, clearly the straight line l has a slope, and thus assume that the equation of the straight line l is

$$y - y_0 = k(x - x_0).$$

Combining the above with $y = 2x$ to form a system of two equations, we find the solution that gives the coordinates $\left(\frac{kx_0 - y_0}{k-2}, \frac{2kx_0 - 2y_0}{k-2} \right)$ of the point A, and combining the same equation with $y = -2x$, we obtain the coordinates $\left(\frac{kx_0 - y_0}{k+2}, -\frac{2kx_0 - 2y_0}{k+2} \right)$ of the point B,

Since P is the middle point of AB, so $\frac{kx_0 - y_0}{k-2} + \frac{kx_0 - y_0}{k+2} = 2x_0$, from which $k = \frac{4x_0}{y_0}$. Thus, the equation of the straight line l is $y - y_0 = \frac{4x_0}{y_0}(x - x_0)$.

Combining the above equation of l with that of the hyperbola gives $4x^2 - \left[y_0 + \frac{4x_0}{y_0}(x - x_0) \right]^2 = 4$, that is,

$$4y_0^2 x^2 - \left[y_0 + \frac{4x_0}{y_0}(x - x_0) \right]^2 = 4y_0^2. \qquad (12.5)$$

Also, since $P(x_0, y_0)$ is on the hyperbola $x^2 - \frac{y^2}{4} = 1$, we have the equality $x_0^2 - \frac{y_0^2}{4} = 1$, namely $y_0^2 = 4x_0^2 - 4$, thus the equation (12.5) can be reduced to $4y_0^2 x^2 - (4x_0 x - 4)^2 = 4y_0^2$, whose simplified version is

$$-4x^2 + 8x_0 x - 4 - y_0^2 = 0.$$

Hence, the discriminant $\Delta = 64x_0^2 - 16(4 + y_0^2) = 64x_0^2 - 16y_0^2 - 64 = 0$. Equivalently, the straight line l and the hyperbola have only one intersection point.

(2) Since $S_{\triangle OAB} = \frac{1}{2}|OA||OB|\sin\angle AOB$, and since $\angle AOB$ has a fixed value, it is enough to prove that $|OA||OB|$ is of a fixed value.

When the slope k does not exist, it is easy to see that $|OA||OB| = 5$. When the slope k exists, from the two equalities

$$|OA|^2 = \left(\frac{kx_0 - y_0}{k - 2} \right)^2 + \left(\frac{2kx_0 - 2y_0}{k - 2} \right)^2 = \frac{20}{(2x_0 - y_0)^2},$$

$$|OB|^2 = \left(\frac{kx_0 - y_0}{k + 2} \right)^2 + \left(-\frac{2kx_0 - 2y_0}{k + 2} \right)^2 = \frac{20}{(2x_0 + y_0)^2},$$

we see that $|OA|^2|OB|^2 = \frac{400}{(4x_0^2 - y_0^2)^2} = 25$, namely $|OA||OB| = 5$. $\qquad \square$

Example 5. Let the left and right foci of the hyperbola $C : \frac{x^2}{4} - \frac{y^2}{5} = 1$ be F_1 and F_2, respectively, and let P be a moving point on the hyperbola in the first quadrant. Suppose that the center of gravity and the inner center of $\triangle PF_1F_2$ are G and I, respectively.

(1) Does there exist a point P such that $IG \parallel F_1F_2$? If so, find the point P. If not so, express your reason.

(2) Let A be the left vertex of the hyperbola C and let a straight line l passing through the right focus F_2 intersect the hyperbola C at points M and N. If the slopes of AM and AN are k_1 and k_2 respectively and $k_1 + k_2 = -\frac{1}{2}$, find the equation of the straight line l.

Solution. (1) Suppose that there exists a point $P(x_0, y_0)$ ($x_0 > 0$ and $y_0 > 0$) such that $IG \parallel F_1F_2$. Since G is the center of gravity of $\triangle PF_1F_2$, we can write $G\left(\frac{x_0}{3}, \frac{y_0}{3}\right)$.

Since I is the inner center of $\triangle PF_1F_2$, if the radius of the inscribed circle of $\triangle PF_1F_2$ is r, then

$$S_{\triangle PF_1F_2} = \frac{1}{2}|F_1F_2||y_0| = \frac{1}{2}(|PF_1| + |PF_2| + |F_1F_2|)r,$$

thus, $\frac{1}{2} \cdot 2c|y_0| = \frac{1}{2}(|PF_1| + |PF_2| + 2c)r$. That is,

$$r = \frac{2cy_0}{|PF_1| + |PF_2| + 2c}.$$

The condition $IG \parallel F_1F_2$ requires that

$$\frac{2cy_0}{|PF_1| + |PF_2| + 2c} = \frac{y_0}{3},$$

from which $|PF_1| + |PF_2| = 4c = 12$.

Also from $|PF_1| - |PF_2| = 2a = 4$ we get $|PF_2| = 4$.

Consequently,

$$\begin{cases} (x_0 - 3)^2 + y_0^2 = 16, \\ \dfrac{x_0^2}{4} - \dfrac{y_0^2}{5} = 1. \end{cases}$$

The solutions of the above system are $x_0 = 4$ and $y_0 = \sqrt{15}$ since the point P is in the first quadrant.

Hence, there exists a point $P(4, \sqrt{15})$ such that $IG \parallel F_1F_2$.

(2) By the condition of the problem, let the equation of the straight line l passing through $F_2(3,0)$ be $y = k(x-3)$, and the straight line and the hyperbola intersect at $M(x_1, y_1)$ and $N(x_2, y_2)$. Then from

$$\begin{cases} y = k(x-3), \\ 5x^2 - 4y^2 = 20, \end{cases}$$

we have

$$(5 - 4k^2)x^2 + 24k^2 x - 36k^2 - 20 = 0.$$

Vieta's theorem implies that

$$\begin{cases} x_1 + x_2 = \dfrac{24k^2}{4k^2 - 5}, \\[2mm] x_1 x_2 = \dfrac{36k^2 + 20}{4k^2 - 5}. \end{cases}$$

Also

$$k_1 + k_2 = \frac{y_1}{x_1 + 2} + \frac{y_2}{x_2 + 2} = k\left(\frac{x_1 - 3}{x_1 + 2} + \frac{x_2 - 3}{x_2 + 2}\right)$$

$$= k\left[2 - 5\left(\frac{1}{x_1 + 2} + \frac{1}{x_2 + 2}\right)\right],$$

while

$$\frac{1}{x_1 + 2} + \frac{1}{x_2 + 2} = \frac{x_1 + x_2 + 4}{x_1 x_2 + 2(x_1 + x_2) + 4}$$

$$= \frac{24k^2 + 4(4k^2 - 5)}{36k^2 + 20 + 48k^2 + 4(4k^2 - 5)}$$

$$= \frac{2k^2 - 1}{5k^2}.$$

It follows that

$$k_1 + k_2 = k\left(2 - 5\frac{2k^2 - 1}{5k^2}\right) = \frac{1}{k} = -\frac{1}{2},$$

in other words, $k = -2$.

Therefore, the desired equation of the straight line l is $y = -2x + 6$. $\square$

Example 6. As Figure 12.4 indicates, F_1 and F_2 are the left and right foci of the hyperbola $C : \frac{x^2}{4} - y^2 = 1$, and a moving point $P(x_0, y_0)$ $(y_0 \geq 1)$ is on the right branch of the hyperbola C. Suppose that the angular bisector

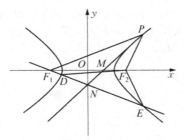

Figure 12.4　Figure for Example 6.

of $\angle F_1PF_2$ intersects the x-axis at a point $M(m,0)$ and the y-axis at a point N.

(1) Find the range of the values for m.
(2) If the straight line l passing through F_1 and N intersects the hyperbola at points D and E, find the maximum value for the area of $\triangle F_2DE$.

Solution. By the condition of the problem, the coordinates of the foci are $F_1(-\sqrt{5},0)$ and $(\sqrt{5},0)$. The equation of the straight line PF_1 is

$$y = \frac{y_0 - 0}{x_0 + \sqrt{5}}(x + \sqrt{5}),$$

namely

$$y_0x - (x_0 + \sqrt{5})y + \sqrt{5}y_0 = 0.$$

The equation of the straight line PF_2 is

$$y = \frac{y_0 - 0}{x_0 - \sqrt{5}}(x - \sqrt{5}),$$

namely

$$y_0x - (x_0 - \sqrt{5})y - \sqrt{5}y_0 = 0.$$

Since the point $M(m,0)$ in on the bisector of $\angle F_1PF_2$, we have

$$\frac{|y_0m + \sqrt{5}y_0|}{\sqrt{y_0^2 + (x_0 + \sqrt{5})^2}} = \frac{|y_0m - \sqrt{5}y_0|}{\sqrt{y_0^2 + (x_0 - \sqrt{5})^2}}.$$

From $y_0 \geq 1$ and $y_0^2 = \frac{1}{4}x_0^2 - 1$, we see that $x_0 \geq 2\sqrt{2}$ and

$$y_0^2 + (x_0 + \sqrt{5})^2 = \frac{5}{4}x_0^2 + 2\sqrt{5}x_0 + 4 = \left(\frac{\sqrt{5}}{2}x_0 + 2\right)^2,$$

$$y_0^2 + (x_0 - \sqrt{5})^2 = \left(\frac{\sqrt{5}}{2}x_0 - 2\right)^2.$$

It is easy to see that $-\sqrt{5} < m < \sqrt{5}$, so

$$\frac{m + \sqrt{5}}{\frac{\sqrt{5}}{2}x_0 + 2} = \frac{\sqrt{5} - m}{\frac{\sqrt{5}}{2}x_0 - 2}.$$

Solving the above equation gives $m = \frac{4}{x_0}$, and combined with $x_0 \geq 2\sqrt{2}$, we get $0 < \frac{4}{x_0} \leq \sqrt{2}$.

Therefore, the range of the values for m is $(0, \sqrt{2}]$.

(2) From (1), the equation of the straight line PM is

$$y = \frac{y_0 - 0}{x_0 - \frac{4}{x_0}}\left(x - \frac{4}{x_0}\right).$$

Letting $x = 0$ in the above gives $y = -\frac{4y_0}{x_0^2 - 4} = -\frac{1}{y_0}$.

Thus, the coordinates of the point N are $\left(0, -\frac{1}{y_0}\right)$.

Also, the slope of the straight line l is

$$k_1 = \frac{0 - \left(-\frac{1}{y_0}\right)}{-\sqrt{5} - 0} = -\frac{1}{\sqrt{5}y_0}.$$

Consequently, the equation of the straight line l is

$$y = -\frac{1}{\sqrt{5}y_0}(x + \sqrt{5}).$$

Eliminating x from the system

$$\begin{cases} y = -\dfrac{1}{\sqrt{5}y_0}(x + \sqrt{5}), \\ \dfrac{x^2}{4} - y^2 = 1, \end{cases}$$

we obtain

$$(5y_0^2 - 4)y^2 + 10y_0y + 1 = 0. \tag{12.6}$$

The discriminant of (12.6)

$$\Delta = 100y_0^2 - 4(5y_0^2 - 4) = 80y_0^2 + 16 > 0.$$

Write $D(x_1, y_1)$ and $E(x_2, y_2)$. Then

$$y_1 + y_2 = -\frac{10y_0}{5y_0^2 - 4}, \quad y_1 y_2 = \frac{1}{5y_0^2 - 4}.$$

Hence,

$$|y_1 - y_2| = \sqrt{(y_1 + y_2)^2 - 4y_1 y_2}$$

$$= \sqrt{\left(-\frac{10y_0}{5y_0^2 - 4}\right)^2 - 4 \cdot \frac{1}{5y_0^2 - 4}}$$

$$= \frac{4\sqrt{5y_0^2 + 1}}{|5y_0^2 - 4|}.$$

Since $y_0 \geq 1$,

$$y_1 + y_2 = -\frac{10y_0}{5y_0^2 - 4} < 0, \quad y_1 y_2 = \frac{1}{5y_0^2 - 4} > 0,$$

from which $y_1 < 0$ and $y_2 < 0$.

Consequently,

$$S_{\triangle F_2 DE} = \frac{1}{2}|F_1 F_2||y_1 - y_2|$$

$$= \frac{1}{2} \cdot 2\sqrt{5} \cdot \frac{4\sqrt{5y_0^2 + 1}}{5y_0^2 - 4}.$$

Denote $5y_0^2 - 4 = t$. Then $t \geq 1$, and

$$S_{\triangle F_2 DE} = 4\sqrt{5} \frac{\sqrt{t + 5}}{t} = 4\sqrt{5}\sqrt{\frac{5}{t^2} + \frac{1}{t}}$$

$$= 4\sqrt{5}\sqrt{5\left(\frac{1}{t} + \frac{1}{10}\right)^2 - \frac{1}{20}}.$$

Therefore, when $t = 1$, namely the point P is $P(2\sqrt{2}, 1)$, the area of $\triangle F_2 DE$ achieves the maximum value $4\sqrt{30}$. $\square$

Example 7. Let A and B be two points on the hyperbola $x^2 - \frac{y^2}{2} = \lambda$ ($\lambda \neq 0$), a point $N(1, 2)$ be the middle point of the line segment AB, and the perpendicular bisector of the line segment AB intersect the hyperbola at two points C and D.

(1) Determine the range of the values for λ.
(2) Try to judge whether the four points A, B, C, and D are on the same circle. Express your reason.

Solution. (1) By the condition of the problem, we may assume that the equation of the straight line AB is $y = k(x-1)+2$. Substituting it into the equation of the hyperbola and simplifying the resulting expression lead to

$$(2 - k^2)x^2 + 2k(k - 2)x - [(k - 2)^2 + 2\lambda] = 0. \tag{12.7}$$

Write the point A as $A(x_1, y_1)$ and the point B as $B(x_2, y_2)$. Then x_1 and x_2 are the two different real roots of the equation (12.7), and as a consequence,

$$\Delta = 4k^2(k - 2)^2 + 4(2 - k^2)[(k - 2)^2 + 2\lambda] > 0, \tag{12.8}$$

and $x_1 + x_2 = \frac{2k(2-k)}{2-k^2}$.

Since $N(1, 2)$ is the middle point of the line segment AB, so $\frac{k(2-k)}{2-k^2} = 1$, whose solution is $k = 1$. It follows that the equation of the straight line AB is $y = 1 \cdot (x - 1) + 2$, that is $y = x + 1$.

Substituting $k = 1$ into (12.8), we have $4 + 4(1 + 2\lambda) > 0$. Solving it gives $\lambda > -1$.

Also, since CD is the perpendicular bisector of the line segment AB, the equation of the straight line containing CD is

$$y - 2 = -(x - 1),$$

namely $y = -x + 3$. Substituting it into the equation of the hyperbola and after a manipulation, we obtain

$$x^2 + 6x - 2\lambda - 9 = 0. \tag{12.9}$$

From the hypothesis of the problem, the equation (12.9) has two different real roots, thus the discriminant of the above quadratic equation

$$\Delta_1 = 6^2 - 4(-2\lambda - 9) > 0,$$

the solution of which is $\lambda > -9$.
Since $\lambda \neq 0$, we see that the range of the values for λ is $(-1, 0) \cup (0, +\infty)$.

(2) Write C as $C(x_3, y_3)$, D as $D(x_4, y_4)$, and the middle point M of the line segment CD as $M(x_0, y_0)$. Then x_3 and x_4 are the two roots of the

equation (12.9). So, $x_3 + x_4 = -6$ and $x_3 x_4 = -2\lambda - 9$. Consequently,

$$x_0 = \frac{x_3 + x_4}{2} = -3, \ y_0 = -x_0 + 3 = 6.$$

Hence, by the chord length formula,

$$|CD| = \sqrt{1 + (-1)^2}|x_3 - x_4| = \sqrt{2}\sqrt{(x_3 + x_4)^2 - 4x_3 x_4}$$
$$= \sqrt{2}\sqrt{6^2 - 4(-2\lambda - 9)} = 4\sqrt{9 + \lambda}.$$

Simplifying the equation (12.7) results in $x^2 - 2x - 2\lambda - 1 = 0$, so by the same reason as above,

$$|AB| = \sqrt{1 + 1^2}\sqrt{(x_1 + x_2)^2 - 4x_1 x_2} = 4\sqrt{1 + \lambda}.$$

Obviously, $|AB| < |CD|$, and since CD is the perpendicular bisector of the line segment AB, if there exists

$$\lambda \in (-1, 0) \cup (0, +\infty)$$

such that the four points $A, B, C,$ and D are on the same circle, then CD must be the diameter of the circle with the point M as the center of the circle.

The distance of the point M to the straight line AB is

$$d = \frac{|x_0 - y_0 + 1|}{\sqrt{2}} = \frac{|-3 - 6 + 1|}{\sqrt{2}} = 4\sqrt{2},$$

and the Pythagorean theorem implies that

$$|MA|^2 = |MB|^2 = d^2 + \left(\frac{|AB|}{2}\right)^2$$
$$= (4\sqrt{2})^2 + (2\sqrt{1 + \lambda})^2$$
$$= 36 + 4\lambda.$$

On the other hand, $\left(\frac{|CD|}{2}\right)^2 = (2\sqrt{9 + \lambda})^2 = 36 + 4\lambda$, so

$$|MA|^2 = |MB|^2 = |MC|^2 = |MD|^2.$$

Therefore, when $\lambda \in (-1, 0) \cup (0, +\infty)$, the four points A, B, C, and D are on the circle centered at $M(-3, 6)$ with radius $2\sqrt{9 + \lambda}$.   □

Example 8. There are given an ellipse $\Gamma_1 : \frac{x^2}{a^2} + \frac{y^2}{b^2} = 1$ and a hyperbola $\Gamma_2 : \frac{x^2}{a^2} - \frac{y^2}{b^2} = 1$, where $a > b > 0$. Let the left and right foci of the ellipse be F_1 and F_2 respectively and the left and right foci of the hyperbola be F_3 and F_4 respectively. Draw a straight line passing through F_4, which intersects the ellipse at two different points A and B (the point B is between the points A and F_4). Suppose that the straight lines F_3A and F_2B intersect at C. If the three lines AF_2, BF_3, and CF_1 intersect at one point, find the value of $a^2 : b^2$.

Solution. Let $c_1 = \sqrt{a^2 - b^2}$ and $c_2 = \sqrt{a^2 + b^2}$. Then the four points F_1, F_2, F_3, and F_4 are $F_1(-c_1, 0), F_2(c_1, 0), F_3(-c_2, 0)$, and $F_4(c_2, 0)$.

Applying Menelaus' theorem to the straight line ABF_4 and the triangle CF_3F_2, we obtain

$$\frac{|CA|}{|AF_3|} \frac{|F_3F_4|}{|F_4F_2|} \frac{|F_2B|}{|BC|} = 1.$$

Also in $\triangle CF_3F_2$, the three lines AF_2, BF_3, and CF_1 intersect at one point, so by Cava's theorem,

$$\frac{|CA|}{|AF_3|} \frac{|F_3F_1|}{|F_1F_2|} \frac{|F_2B|}{|BC|} = 1.$$

It follows that $\frac{|F_3F_4|}{|F_4F_2|} = \frac{|F_3F_1|}{|F_1F_2|}$, from which $\frac{2c_2}{c_2 - c_1} = \frac{c_2 - c_1}{2c_1}$, that is

$$c_2 = (3 + 2\sqrt{2})c_1.$$

Therefore,

$$a^2 : b^2 = 3 : 2\sqrt{2}.$$ □

3. Exercises

Group A

I. Filling Problems

1. Suppose that F_1 and F_2 are the left and right foci of the hyperbola $C : x^2 - \frac{y^2}{24} = 1$, P is a point of the hyperbola C, and P is in the first quadrant. If $\frac{|PF_1|}{|PF_2|} = \frac{4}{3}$, then the radius of the inscribed circle of $\triangle PF_1F_2$ is _____.

2. A straight line is drawn to pass through the right focus of the hyperbola $x^2 - \frac{y^2}{2} = 1$ and to intersect the hyperbola at two points A and B. If a real number λ makes exactly three straight lines l satisfying $|AB| = \lambda$, then $\lambda = $ _____.

3. Let the focal distance of a hyperbola $\frac{x^2}{a^2} - \frac{y^2}{b^2} = 1$ $(a > 1, b > 0)$ be $2c$, a straight line l passes through the points $(a, 0)$ and $(0, b)$, and the sum of the distances from point $(1, 0)$ to the straight line l and from point $(-1, 0)$ to the straight line l be $s \geq \frac{4}{5}c$. Then the range of the values for the eccentricity e of the hyperbola is _____.

4. Suppose that the two end points of a line segment with a fixed length l $(l > \frac{2b^2}{a})$ are both on the right branch of a hyperbola $\frac{x^2}{a^2} - \frac{y^2}{b^2} = 1$ $(a > 0$ and $b > 0)$. Then the minimum value of the horizontal coordinate of the middle point of AB is _____.

5. Let P be a point on an equal axis hyperbola $x^2 - y^2 = a^2$ $(a > 0)$, and let F_1 and F_2 be its left and right foci. Then the range of the values of $\frac{|PF_1| + |PF_2|}{|PO|}$ is _____.

II. Calculation Problems

6. The center of a hyperbola $3x^2 - y^2 = k$ moves on the straight line $l : y = x$ and the symmetric axes of the hyperbola are kept parallel to the coordinates axes. Does there exist such a hyperbola during the translation such that the lengths of the chords obtained from its intersections with the straight line l and the y-axis are both equal to $2\sqrt{2}$? If there does, find the equation of the hyperbola; if there does not, explain the reason.

7. Let a straight line l passing through the point $P(0, 1)$ with slope k intersect the hyperbola $C : x^2 - \frac{y^2}{3} = 1$ at points A and B.

  (1) Find the value range of k.
  (2) If F_2 is the right focus of the hyperbola C and $|AF_2| + |BF_2| = 6$, find the value of k.

8. As in Figure 12.5, let $P(x_0, y_0)$ be a point on a straight line $x = m$ $(y \neq \pm m$ and $0 < m < 1)$. Draw two tangent lines PA and PB from P to the hyperbola $x^2 - y^2 = 1$ with tangent points A and B. Mark the fixed point $M\left(\frac{1}{m}, 0\right)$.

(1) Prove: The three points A, M, and B are co-linear.

(2) Draw the perpendicular line to the straight line $x - y = 0$ at the point A with N the foot of a perpendicular. Find the equation of a curve drawn from the center of gravity of $\triangle AMN$.

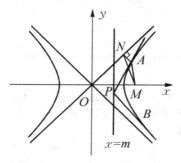

Figure 12.5 Figure for Exercise 8.

9. Let the eccentricity of a hyperbola $C : \frac{x^2}{a^2} - \frac{y^2}{b^2} = 1$ $(a > 0$ and $b > 0)$ be 2, and let a straight line l passing through a point $P(0, m)$ with slope 1 intersect the hyperbola C at two points A and B with $\overrightarrow{AP} = 3\overrightarrow{PB}$ and $\overrightarrow{OA} \cdot \overrightarrow{OB} = 3$.

(1) Find the equation of the hyperbola.

(2) Let Q be a moving point on the right branch of the hyperbola C and F the right focus of the hyperbola C. Does there exist a fixed point M on the negative half x-axis such that $\angle QFM = 2\angle QMF$? If there does, find the coordinates of the point M; if there does not, please give your reason.

Group B

10. As Figure 12.6 shows, a straight line $y = mx + b$ $(-1 < m < 1$ and $-1 < b < 1)$ intersects the circle $x^2 + y^2 = 1$ at two points P and Q and the hyperbola $x^2 - y^2 = 1$ at two points R and S. If P and Q divide the line segment RS into three equal parts, find the values of m and b.

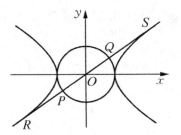

Figure 12.6 Figure for Exercise 10.

11. Let P be any point of a hyperbola, PQ be a tangent line to the hyperbola at the point P, and F_1 and F_2 be the foci of the hyperbola. Prove: PQ bisects $\angle F_1PF_2$.

12. Prove: From any point of a hyperbola, draw two tangent lines to another hyperbola with the same asymptotes. Then the area of the triangle bounded by the straight line passing through the two tangent points and the two asymptotes is a fixed value.

Chapter 13

Parabolas

1. Key Points of Knowledge and Basic Methods

A. *Standard equation and properties of parabolas*

Definition of parabolas: The trajectory of a point in a plane satisfying the following property is a parabola: The ratio of the distance of the point to a fixed point (focus) and the distance of the point to a fixed straight line (directrix), which does not pass through the fixed point, equals 1 (eccentricity).

Standard equation of a parabola with the focus on the positive semi-axis of the x-axis: The standard equation of the parabola with the focus $F\left(\frac{p}{2}, 0\right)$ $(p > 0)$ is

$$y^2 = 2px.$$

The range of x is $x \geq 0$; the symmetric axis is the x-axis; the vertex is $(0,0)$.

The equation of the directrix is $x = -\frac{p}{2}$.

The eccentricity $e = 1$.

Suppose that $P(x_0, y_0)$ is a point on the parabola. Then the focal radius $|FP| = x_0 + \frac{p}{2}$.

Let the inclination angle of a chord AB of the parabola that passes through the focus be α. Then $|AB| = \frac{2p}{\sin^2 \alpha}$.

Similarly, one can write down the standard equation and the equation of the directrix for a parabola whose focus is on the negative half x-axis or on the positive (negative) half y-axis, and the corresponding properties.

B.   *Optical properties of parabolas*

Light rays emitted from the focus of an parabola will go along the direction that is parallel to the symmetric axis of the parabola after the refection from the parabola.

C.   *Graph of a quadratic function of one variable*

The graph of a quadratic function of one variable is a parabola whose symmetric axis is parallel to (or coincides) the y-axis.

The other contents are referred to the chapter on ellipses.

2.   Illustrative Examples

Example 1. As Figure 13.1 shows, the vertex of a parabola $y = x^2$ is O, line segment AB is a chord of length 2 that passes through the focus F, and D is the intersection point of the perpendicular bisector of AB and the y-axis.

Find the area of the quadrilateral $AOBD$.

Solution. Let the coordinates of the points A and B be $A(x_1, y_1)$ and $B(x_2, y_2)$, the focus of the parabola be $F\left(0, \frac{1}{4}\right)$, and the equation of the directrix be $y = -\frac{1}{4}$. From the definition of a parabola, $|AF| = y_1 + \frac{1}{4}$ and $|BF| = y_2 + \frac{1}{4}$. So,

$$2 = |AF| + |BF| = y_1 + y_2 + \frac{1}{2}, \quad y_1 + y_2 = \frac{3}{2}.$$

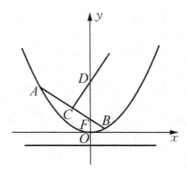

Figure 13.1   Figure for Example 1.

The middle point of AB is $C\left(\frac{x_1+x_2}{2}, \frac{y_1+y_2}{2}\right)$, and

$$k_{AB} = \frac{y_2 - y_1}{x_2 - x_1} = \frac{x_2^2 - x_1^2}{x_2 - x_1} = x_1 + x_2.$$

Thus, $k_{CD} = -\frac{1}{x_1+x_2}$ and the equation of CD is

$$y - \frac{y_1 + y_2}{2} = -\frac{1}{x_1 + x_2}\left(x - \frac{x_1 + x_2}{2}\right).$$

Letting $x = 0$ gives the coordinates $\left(0, \frac{y_1+y_2}{2} + \frac{1}{2}\right) = \left(0, \frac{5}{4}\right)$ of the point D. Hence, $|OD| = \frac{5}{4}$.

Since AB passes through the focus F, the equation of AB is $y - \frac{1}{4} = k_{AB}x$, namely $y = k_{AB}x + \frac{1}{4}$, from which $y_1 + y_2 = k_{AB}(x_1 + x_2) + \frac{1}{2} = (x_1 + x_2)^2 + \frac{1}{2}$. It follows that $(x_1 + x_2)^2 = 1$.

Consequently, $k_{AB} = x_1 + x_2 = 1$ or -1, so OD and AB intersect at the acute angle $45°$.

Therefore, the area of the quadrilateral $AOBD$ is

$$S = \frac{1}{2}|AB||OD|\sin 45° = \frac{5}{4}\frac{\sqrt{2}}{2} = \frac{5\sqrt{2}}{8}. \qquad \square$$

Example 2. A line segment AB of fixed length a is sliding on the parabola $y = x^2$. Find the minimum value of the vertical coordinate for the middle point M of AB.

Analysis. At the first glance the problem can be solved via the definition of a parabola. In Figure 13.2, l is the directrix. Passing through A, B, and M, draw three perpendicular lines to the directrix with A_1, B_1, and M_1

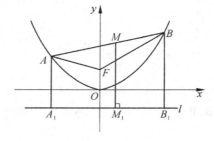

Figure 13.2 Figure for Example 2.

as the feet of the perpendicular, respectively. By the knowledge of plane geometry and the definition of a parabola, it is easy to get that

$$|MM_1| = \frac{1}{2}(|AA_1| + |BB_1|)$$

$$= \frac{1}{2}(|AF| + |BF|) \geq \frac{1}{2}|AB|$$

$$= \frac{1}{2}a.$$

Also, since $|MM_1| = y_M + \frac{1}{4}$, so the minimum value of y_M is expected to be $\frac{1}{2}a - \frac{1}{4}$. But it is not easy to solve the "equality" problem in the above inequality. Since $|AF| + |BF| \geq |AB|$, a condition for the inequality to become an equality is that the three points $A, F,$ and B are co-linear. Note that the minimal length of a chord that passes through the focus is $2p = 1$. So, the above approach works if $a \geq 1$. However, this approach does not work to solve the inequality issue in the case that $0 < a < 1$. Hence, we solve the problem by means of ideas from functions.

Solution. Let the coordinates of A and B be $A(x_1, x_1^2)$ and $B(x_2, x_2^2)$. By $|AB| = a$, we have

$$(x_1 - x_2)^2 + (x_1^2 - x_2^2)^2 = a^2,$$

that is $(x_1 - x_2)^2[1 + (x_1 + x_2)^2] = a^2$. So from

$$(x_1 + x_2)^2 = \frac{a^2}{(x_1 - x_2)^2} - 1, \qquad (13.1)$$

in other words,

$$2(x_1^2 + x_2^2) - (x_1 - x_2)^2 = \frac{a^2}{(x_1 - x_2)^2} - 1,$$

we see that

$$\frac{x_1^2 + x_2^2}{2} = \frac{1}{4}\left[\frac{a^2}{(x_1 - x_2)^2} + (x_1 - x_2)^2 - 1\right]$$

$$\geq \frac{1}{4}(2a - 1) = \frac{1}{2}a - \frac{1}{4},$$

and the equality holds if and only if $(x_1 - x_2)^2 = a$. But (13.1) implies that $\frac{a^2}{(x_1-x_2)^2} \geq 1$, namely

$$(x_1 - x_2)^2 \leq a^2.$$

(i) When $a \leq a^2$, that is $a \geq 1$, the minimum value $\frac{1}{2}a - \frac{1}{4}$ is obtained as $(x_1 - x_2)^2 = a$.

(ii) When $a > a^2$, that is $0 < a < 1$, we have $(x_1 - x_2)^2 \leq a^2 < a$. Note that the function

$$y = \frac{a^2}{x} + x \; (x \in (0, a^2])$$

is monotonically decreasing. It follows that the minimum value of the objective function is $\frac{1}{4}a^2$ as $(x_1 - x_2)^2 = a^2$. $\square$

Remark. When using the function idea to find extreme values, we should try to dig out the restriction condition; otherwise a mistake may occur.

Example 3. As Figure 13.3 shows, in a rectangular coordinates system xOy of a plane, P is a moving point not on the x-axis that satisfies the condition: One can draw two tangent lines passing through P to the parabola $y^2 = 4x$, and a line segment l_P connecting the two tangent points is perpendicular to PO. Assume that the straight line l_P intersects the straight line PO and the x-axis at points Q and R, respectively.

(1) Prove that R is a fixed point.
(2) Find the minimum value of $\frac{|PQ|}{|QR|}$.

Solution. (1) Let the coordinates of the point P be (a, b) $(b \neq 0)$. Clearly, $a \neq 0$. Denote the coordinates of the two tangent points A and B as

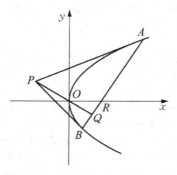

Figure 13.3 Figure for Example 3.

(x_1, y_1) and (x_2, y_2), respectively. Then the equations of PA and PB are respectively

$$y_1 y = 2(x + x_1), \tag{13.2}$$

$$y_2 y = 2(x + x_2), \tag{13.3}$$

while the coordinates (a, b) satisfy both (13.2) and (13.3). Thus, the coordinates (x_1, y_1) of A and (x_2, y_2) of B both satisfy the equation

$$by = 2(x + a). \tag{13.4}$$

So, (13.4) is just the equation of the straight line AB.

The slopes of the straight lines PO and AB are $\frac{b}{a}$ and $\frac{2}{b}$, respectively. From $PO \perp AB$, we see that $\frac{b}{a}\frac{2}{b} = -1$, hence $a = -2$.

It follows that (13.4) is $y = \frac{2}{b}(x - 2)$, therefore the intersection point R of AB and the x-axis is the fixed point $(2, 0)$.

(2) Since $a = -2$, the slope of the straight line PO is $k_1 = -\frac{b}{2}$ and the slope of the straight line PR is $k_2 = -\frac{b}{4}$. Denote $\angle OPR = \alpha$. Then α is an acute angle and

$$\frac{|PQ|}{|QR|} = \frac{1}{\tan \alpha} = \left| \frac{1 + k_1 k_2}{k_1 - k_2} \right|$$

$$= \left| \frac{1 + \left(-\frac{b}{2}\right)\left(-\frac{b}{4}\right)}{-\frac{b}{2} + \frac{b}{4}} \right|$$

$$= \frac{8 + b^2}{2|b|} \geq \frac{2\sqrt{8b^2}}{2|b|} = 2\sqrt{2}.$$

Consequently, the minimum value of $\frac{|PQ|}{|QR|}$ is $2\sqrt{2}$ obtained when $b = \pm 2\sqrt{2}$.
□

Remark. The line segment AB is the chord at contact of the point P with respect to the parabola $y = 4x$.

Example 4. Let a parabola $C : y^2 = 2px$ $(p > 0)$ be given. A straight line l and the parabola C intersect at two points A and B, the straight line connecting A and the vertex O of the parabola intersects the directrix at B', and the straight line connecting B and the vertex O of the parabola intersects the directrix at A'. Suppose that AA' and BB' are both parallel to the x-axis.

(1) Prove: The straight line l passes through a fixed point.

(2) Find the minimum value for the area of the quadrilateral $ABB'A'$.

Solution. (1) Let the equation of the straight line l be $y = kx + m$, and let the coordinates of A and B be $A(x_A, y_A)$ and (x_B, y_B), respectively. From the given condition of the problem,

$$\begin{cases} y = kx + m, \\ y^2 = 2px. \end{cases} \tag{13.5}$$

Substituting the first equation into the second one in (13.5), we obtain that

$$y^2 - \frac{2p}{k}y + \frac{2pm}{k} = 0.$$

By Vieta's theorem,

$$y_A + y_B = \frac{2p}{k}, \ y_A y_B = \frac{2pm}{k}.$$

So, $y_B = \frac{2pm}{k y_A}$.

The equation of the straight line AB' is $y = \frac{y_A}{x_A}x$. Since the point B' is on the directrix, the vertical coordinate of the point B' is

$$y_{B'} = \frac{y_A}{x_A}\left(-\frac{p}{2}\right) = -\frac{p y_A}{2 x_A}.$$

Since AA' and BB' are both parallel to the x-axis, so $y_B = y_{B'}$, which means that

$$\frac{2pm}{k y_A} = -\frac{p y_A}{2 x_A}. \tag{13.6}$$

Also, since the point A is on the parabola C, we have

$$y_A^2 = 2px_A. \tag{13.7}$$

If we substitute (13.7) into (13.6), then we obtain

$$m = -\frac{pk}{2}.$$

Thus, the equation of the straight line l is

$$y = k\left(x - \frac{p}{2}\right).$$

Hence, the straight line l passes through the fixed point $F\left(\frac{p}{2}, 0\right)$.

(2) From the definition and properties of parabolas, $|AA'| = |AF|$ and $|BB'| = |BF|$. Then

$$|AA'| = |BB'| = |AF| + |BF| = |AB|.$$

By (1) and the conditions of the problem, the quadrilateral $ABB'A'$ is a right-angled trapezoid, and so

$$
\begin{aligned}
S &= \frac{1}{2}|A'B'|(|AA'| + |BB'|) = \frac{1}{2}|A'B'||AB| \\
&= \frac{1}{2}(y_A - y_B)\sqrt{(x_A - x_B)^2 + (y_A - y_B)^2} \\
&= \frac{1}{2}(y_A - y_B)^2\sqrt{1 + \frac{1}{k^2}} \\
&= \frac{1}{2}\sqrt{1 + \frac{1}{k^2}}\left[(y_A + y_B)^2 - 4y_A y_B\right] \\
&= 2p^2\left(1 + \frac{1}{k^2}\right)^{\frac{3}{2}}.
\end{aligned}
$$

Hence, when the inclination angle of the straight line l is $\frac{\pi}{2}$, namely the quadrilateral $ABB'A'$ is a rectangle, its area has the minimum value, which is $2p^2$.   □

Example 5. Suppose that $E(m, n)$ is a fixed point inside a parabola $y^2 = 2px$ $(p > 0)$. Draw two straight lines passing through E with slopes k_1 and k_2 respectively, intersecting the parabola at A, B, C, and D. Assume that M and N are the middle points of the line segments AB and CD, respectively.

(1) When $n = 0$ and $k_1 k_2 = -1$, find the minimum value of the area of $\triangle EMN$.
(2) Let $k_1 + k_2 = \lambda$ ($\lambda \neq 0$ with λ a constant). Prove: The straight line MN passes through a fixed point.

Solution. Let the equation of the straight line containing AB be $x = t_1(y - n) + m$, where $t_1 = \frac{1}{k_1}$, and substitute it into $y^2 = 2px$. We then have

$$y^2 - 2pt_1 y + 2pt_1 n - 2pm = 0.$$

Denote $A(x_1, y_1)$ and $B(x_2, y_2)$. Then $y_1 + y_2 = 2pt_1$, and so

$$x_1 + x_2 = t_1(y_1 + y_2 - 2n) + 2m = t_1(2pt_1 - 2n) + 2m,$$

from which the coordinates of M are $M(pt_1^2 - nt_1 + m, pt_1)$.

Suppose that the equation of the straight line containing CD is $x = t_2(y - n) + m$, where $t_2 = \frac{1}{k_2}$. By the same reason, we have the coordinates of N as

$$N(pt_2^2 - nt_2 + m, pt_2).$$

(1) When $n = 0$, we have the points $E(m, 0), M(pt_1^2 + m, pt_1)$ and $N(pt_2^2 + m, pt_2)$, so $|EM| = |pt_1|\sqrt{1 + t_1^2}$ and $|EN| = |pt_2|\sqrt{1 + t_2^2}$.

The assumption $k_1 k_2 = -1$ implies $t_1 t_2 = -1$, from which the area of $\triangle EMN$ is

$$S = \frac{1}{2}|EM||EN|$$

$$= \frac{1}{2}|p^2 t_1 t_2|\sqrt{(1 + t_1^2)(1 + t_2^2)}$$

$$= \frac{p^2}{2}\sqrt{2 + t_1^2 + t_2^2}$$

$$\geq \frac{p^2}{2}\sqrt{4} = p^2,$$

in which the equality is valid if and only if $|t_1| = |t_2| = 1$.

Hence, the minimum value of the area of $\triangle EMN$ is p^2.

(2) The slope of MN is

$$k_{MN} = \frac{p(t_1 - t_2)}{p(t_1^2 - t_2^2) - n(t_1 - t_2)} = \frac{1}{(t_1 + t_2) - \frac{n}{p}}.$$

So, the equation of the straight line containing MN is

$$y - pt_1 = \frac{1}{(t_1 + t_2) - \frac{n}{p}}[x - (pt_1^2 - nt_1 + m)],$$

in other words,

$$y\left(t_1 + t_2 - \frac{n}{p}\right) - pt_1 t_2 = x - m.$$

Now, $k_1 + k_2 = \frac{1}{t_1} + \frac{1}{t_2} = \lambda$, namely $t_1 t_2 = \frac{t_1 + t_2}{\lambda}$. Substituting it into the above equality gives

$$y\left(t_1 + t_2 - \frac{n}{p}\right) - p\frac{t_1 + t_2}{\lambda} = x - m,$$

that is

$$(t_1 + t_2)\left(y - \frac{p}{\lambda}\right) = x + \frac{ny}{p} - m.$$

When $y - \frac{p}{\lambda} = 0$, we have $x + \frac{ny}{p} - m = 0$. Namely, $y = \frac{p}{\lambda}$ and $x = m - \frac{n}{\lambda}$ give a pair of solutions of the equation, hence the straight line passes through the fixed point $\left(m - \frac{n}{\lambda}, \frac{p}{\lambda}\right)$.   □

Example 6. As Figure 13.4 indicates, in a rectangular coordinates system xOy of a plane, F is a moving point on the positive semi-axis of the x-axis. Draw a parabola C with F as its focus and O as its vertex. Let P be a point on the parabola in the first quadrant and Q a point on the negative half x-axis such that PQ is a tangent line to C and $|PQ| = 2$. Two circles C_1 and C_2 are both tangent to the straight line OP with the tangent point P and also both tangent to the x-axis. Find the coordinates of the point F that makes the sum of the areas of the circles C_1 and C_2 the minimum value.

Solution. Suppose that the equation of the parabola C is $y^2 = 2px$ $(p > 0)$ and Q is $(-a, 0)$ $(a > 0)$. Assume that the centers of the circles C_1 and C_2 are $O_1(x_1, y_1)$ and $O_2(x_2, y_2)$, as Figure 13.5 shows.

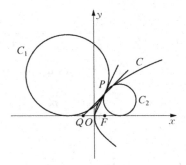

Figure 13.4   Figure 1 for Example 6.

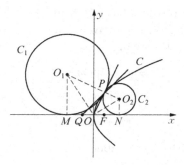

Figure 13.5   Figure 2 for Example 6.

Let the equation of the straight line PQ be $x = my - a$ $(m > 0)$. Combine it with the equation of C. After eliminating x from them, we get

$$y^2 - 2pmy + 2pa = 0.$$

Since PQ and C are tangent at the point P, so the discriminant of the quadratic equation is

$$\Delta = 4p^2m^2 - 4 \cdot 2pa = 0.$$

The solution of the above is $m = \sqrt{\frac{2a}{p}}$. It follows that the coordinates of P are

$$(x_P, y_P) = (a, \sqrt{2pa}).$$

Consequently,

$$|PQ| = \sqrt{1 + m^2}|y_P - 0|$$

$$= \sqrt{1 + \frac{2a}{p}} \sqrt{2pa}$$

$$= \sqrt{2a(p + 2a)}.$$

From $|PQ| = 2$, we have

$$4a^2 + 2pa = 4. \tag{13.8}$$

Note that OP is tangent to the circles C_1 and C_2 at the point P, so $OP \perp O_1O_2$. Let the circles C_1 and C_2 be tangent to the x-axis at two points M and N, respectively. Then OO_1 and OO_2 are the angular bisectors of $\angle POM$ and $\angle PON$, respectively, and as a result, $\angle O_1 O O_2 = 90°$. Thus, by the projection theorem,

$$y_1 y_2 = |O_1 M||O_2 N|$$
$$= |O_1 P||O_2 P| = |OP|^2$$
$$= x_P^2 + y_P^2 = a^2 + 2pa.$$

Combing the above with (13.8) implies that

$$y_1 y_2 = a^2 + 2pa = 4 - 3a^2. \tag{13.9}$$

Since O_1, P, and O_2 are co-linear,

$$\frac{y_1 - \sqrt{2pa}}{\sqrt{2pa} - y_2} = \frac{y_1 - y_P}{y_P - y_2} = \frac{|O_1 P|}{|PO_2|} = \frac{|O_1 M|}{|O_2 N|} = \frac{y_1}{y_2},$$

which can be simplified to

$$y_1 + y_2 = \frac{2}{\sqrt{2pa}} y_1 y_2. \tag{13.10}$$

Denote $T = y_1^2 + y_2^2$. Then the sum of the areas of the circles C_1 and C_2 is πT. By the condition of the problem, it is enough to consider the minimum value problem for T.

From (13.9) and (13.10), we know that

$$T = (y_1 + y_2)^2 - 2y_1 y_2 = \frac{4}{2pa} y_1^2 y_2^2 - 2y_1 y_2$$

$$= \frac{4}{4 - 4a^2}(4 - 3a^2)^2 - 2(4 - 3a^2)$$

$$= \frac{(4 - 3a^2)(2 - a^2)}{1 - a^2}.$$

Make the substitution $t = 1 - a^2$. Since

$$4t = 4 - 4a^2 = 2pa > 0,$$

we get $t > 0$, and so

$$T = \frac{(3t + 1)(t + 1)}{t} = 3t + \frac{1}{t} + 4$$

$$\geq 2\sqrt{3t\frac{1}{t}} + 4 = 2\sqrt{3} + 4.$$

The equality in the above inequality holds if and only if $t = \frac{\sqrt{3}}{3}$, and in this case

$$a = \sqrt{1 - t} = \sqrt{1 - \frac{1}{\sqrt{3}}}.$$

Combining the expression of a with (13.8), we finally get

$$\frac{p}{2} = \frac{1 - a^2}{a} = \frac{1}{\sqrt{1 - \frac{1}{\sqrt{3}}}} = \frac{\sqrt{3}t}{\sqrt{3 - \sqrt{3}}} = \frac{1}{\sqrt{3 - \sqrt{3}}},$$

and consequently, the coordinates of F are $\left(\frac{1}{\sqrt{3 - \sqrt{3}}}, 0 \right)$. $\qquad\square$

Example 7. From a point P outside a parabola $y^2 = 2px$ $(p > 0)$, draw two tangent lines to the parabola, with tangent points M and N. Let F be the focus of the parabola. Prove:

(1) $|PF|^2 = |MF||NF|$.
(2) $\angle PMF = \angle FPN$.

Proof. Let the given three points be $P(x_0, y_0), M(x_1, y_1)$, and $N(x_2, y_2)$. It is easy to find the equation $y_1 y = p(x + x_1)$ of the tangent line PM and the equation $y_2 y = p(x + x_2)$ of the tangent line PN.

Since the point P is on the two tangent lines, there hold that $y_1 y_0 = p(x_0 + x_1)$ and $y_2 y_0 = p(x_0 + x_2)$. Thus, the two points M and N are on the straight line $y_0 y == p(x + x_0)$, from which we know that the equation of the straight line MN is $y_0 y = p(x + x_0)$.

Putting the two equations $y_0 y = p(x + x_0)$ and $y^2 = 2px$ together, we have $[p(x + x_0)]^2 = 2py_0^2 x$, namely

$$x^2 + 2\left(x_0 - \frac{y_0^2}{p} \right) x + x_0^2 = 0.$$

By Vieta's theorem,

$$x_1 + x_2 = 2\left(\frac{y_0^2}{p} - x_0 \right), \quad x_1 x_2 = x_0^2.$$

(1) Since the focus of the parabola is $F\left(\frac{p}{2}, 0 \right)$, by the definition of parabolas,

$$|MF| = x_1 + \frac{p}{2}, \quad |NF| = x_2 + \frac{p}{2}.$$

Hence,

$$|MF||NF| = \left(x_1 + \frac{p}{2}\right)\left(x_2 + \frac{p}{2}\right)$$

$$= x_1 x_2 + \frac{p}{2}(x_1 + x_2) + \frac{p^2}{4}$$

$$= x_0^2 + \frac{p}{2} \cdot 2\left(\frac{y_0^2}{p} - x_0\right) + \frac{p^2}{4}$$

$$= x_0^2 + y_0^2 - px_0 + \frac{p^2}{4}$$

$$= \left(x_0 - \frac{p}{2}\right)^2 + y_0^2 = |PF|^2.$$

Therefore, $|PF|^2 = |MF||NF|$.

(2) Since $\left(x_0 - \frac{p}{2}, y_0\right), \left(x_1 - \frac{p}{2}, y_1\right)$, and $\left(x_2 - \frac{p}{2}, y_2\right)$ are the components of the vectors $\overrightarrow{FP}, \overrightarrow{FM}$, and $\overrightarrow{FN}$, respectively, using the dot product, we have

$$\overrightarrow{FP} \cdot \overrightarrow{FM} = \left(x_0 - \frac{p}{2}, y_0\right) \cdot \left(x_1 - \frac{p}{2}, y_1\right)$$

$$= x_0 x_1 - \frac{p}{2}(x_0 + x_1) + \frac{p^2}{4} + y_0 y_1$$

$$= x_0 x_1 - \frac{p}{2}(x_0 + x_1) + \frac{p^2}{4} + p(x_0 + x_1)$$

$$= x_0 x_1 + \frac{p}{2}(x_0 + x_1) + \frac{p^2}{4}$$

$$= \left(x_0 + \frac{p}{2}\right)\left(x_1 + \frac{p}{2}\right).$$

On the other hand,

$$|FM| = x_1 + \frac{p}{2}, \quad \overrightarrow{FP} \cdot \overrightarrow{FM} = |FP||FM|\cos\angle PFM,$$

from which

$$\cos\angle PFM = \frac{\overrightarrow{FP} \cdot \overrightarrow{FM}}{|FP||FM|} = \frac{\left(x_0 + \frac{p}{2}\right)\left(x_1 + \frac{p}{2}\right)}{|FP|\left(x_1 + \frac{p}{2}\right)} = \frac{x_0 + \frac{p}{2}}{|FP|}.$$

By the same argument, we have $\cos\angle PFN = \frac{x_0 + \frac{p}{2}}{|FP|}$.

Therefore, $\cos\angle PFM = \cos\angle PFN$, from which $\angle PFM = \angle PFN$.

Combining the above with the equality $|FP|^2 = |FM||FN|$, we see that $\triangle MFP \sim \triangle PFM$, hence $\angle PMF = \angle FPN$. $\square$

Example 8. Let A, B, and C be three different points on a parabola $y = x^2$ and let R be the radius of the circumcircle of $\triangle ABC$. Find the range of the values for R.

Solution. For any $a > 0$ choose $A(a, a^2), B(-a, a^2)$, and $O(0,0)$. Then the center of the circumcircle of $\triangle OAB$ is on the y-axis, which is assumed to be $Q(0, b)$. Since the equation of the perpendicular bisector of OA is

$$y - \frac{a^2}{2} = -\frac{1}{a}\left(x - \frac{a}{2}\right),$$

letting $x = 0$ in the above gives $b = \frac{1}{2} + \frac{a^2}{2}$. Thus, we know that the radius of the circumcircle of $\triangle OAB$ can take all values of the interval $\left(\frac{1}{2}, +\infty\right)$.

Now, assume that A, B, and C are three different points on the parabola $y = x^2$, and R is the radius of the circumcircle of $\triangle ABC$. Let (a, b) be the center of the circle. Then the equation

$$(x - a)^2 + (x^2 - b)^2 - R^2 = 0 \qquad (13.11)$$

has three distinct real roots, so it must have four real roots, denoted as x_1, x_2, x_3, and x_4. Since the equation (13.11) is

$$x^4 + (1 - 2b)x^2 - 2ax + a^2 + b^2 - R^2 = 0,$$

we have

$$x_1 + x_2 + x_3 + x_4 = 0.$$

It follows that

$$x_1^2 + x_2^2 + x_3^2 + x_4^2 + 2\sum_{i \neq j} x_i x_j = 0.$$

Also from $\sum_{i \neq j} x_i x_j = 1 - 2b < 0$, we find that $b > \frac{1}{2}$.
Since

$$2(2b - 1) = -2\sum_{i \neq j} x_i x_j = x_1^2 + x_2^2 + x_3^2 + x_4^2$$

$$\geq 4\sqrt[4]{x_1^2 x_2^2 x_3^2 x_4^2} = 4\sqrt{|x_1 x_2 x_3 x_4|}$$

$$= 4\sqrt{|a^2 + b^2 - R^2|},$$

we obtain the inequality

$$4b^2 - 4b + 1 \geq 4(a^2 + b^2 - R^2).$$

It follows that $R^2 - a^2 + \frac{1}{4} \geq b > \frac{1}{2}$.
Therefore, $R^2 > a^2 + \frac{1}{4}$, from which $R > \frac{1}{2}$. $\qquad\square$

3. Exercises

Group A

I. Filling Problems

1. Two parameters a and b of a parabola $y = ax^2 + bx + 1$ satisfy $8a^2 + 4ab = b^3$. When a and b vary, the equation of the trajectory for the vertex (s, t) of the parabola is _____.

2. Given the fixed point $B(-1, 0)$ and two moving points P and Q on the parabola $y = x^2 - 1$. When $BP \perp PQ$, the value range of the horizontal coordinate for the point Q is _____.

3. Let $P_1 P_2$ be a chord of the parabola $x^2 = y$. If the equation of the perpendicular bisector of $P_1 P_2$ is $x + y = 3$, then the equation of the straight line containing the chord $P_1 P_2$ is _____.

4. In a rectangular coordinates system xOy of a plane, two points A and B are on the parabola $y^2 = 2x$ with F the focus and satisfy $\overrightarrow{OA} \cdot \overrightarrow{OB} = -1$. Then the minimum value of $S_{\triangle OFA} + S_{\triangle OFB}$ is _____.

II. Calculation Problems

5. Given the circle $P : x^2 + y^2 = 2x$ and the parabola $S : y^2 = 4x$, and draw a straight line l passing through the center P of the circle that has four intersection points with the above two curves, denoted as A, B, C, and D according to the order from the top to the bottom. If the lengths of the line segments AB, BC, and CD constitute an arithmetic sequence with the same order, find the equation of the straight line l.

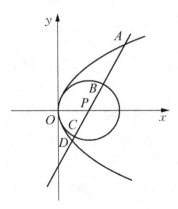

Figure 13.6 Figure for Exercise 5.

6. Let be given a parabola $y^2 = 2px$ $(p > 0)$, a point B the focus of the parabola, a point $C(c, 0)$ on the positive semi-axis of the x-axis, and a moving point A on the parabola. Question: In what range of the values of c is $\angle BAC$ always an acute angle for all the possible locations of the point A?

7. Suppose that $A(a, b)$ and $B(-a, 0)$ $(ab \neq 0)$ are fixed points on a parabola $y^2 = 2px$ $\left(p \neq \frac{b^2}{2a}\right)$. Let M be a point on the parabola. If straight lines AM and BM intersect the parabola at other points M_1 and M_2 respectively, prove: When the point M moves on the parabola (as long as M_1 and M_2 exist and $M_1 \neq M_2$), the straight line $M_1 M_2$ always passes through a fixed point. Also, find the coordinates of this fixed point.

8. As in Figure 13.7, P is a moving point on the parabola $y^2 = 2x$, two points B and C are on the y-axis, and the circle $(x - 1)^2 + y^2 = 1$ is inscribed into $\triangle PBC$. Find the minimum value of the area of $\triangle PBC$.

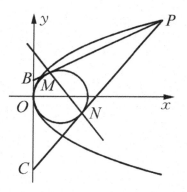

Figure 13.7 Figure for Exercise 8.

Group B

9. If there exists a symmetric point on the parabola $y = x^2$ with respect to a straight line $y = m(x - 3)$, find the value range of the real number m.

10. Let a straight line $y = x + b$ and a parabola $y^2 = 2px$ $(p > 0)$ intersect at two points A and B, and a circle passing through A and B intersect the same parabola $y^2 = 2px$ at two different points C and D. Prove: $AB \perp CD$.

11. Let F be a focus of the given parabola $C : y^2 = 4x$. And a straight line l from the point F intersects C at two points A and B.

   (1) If the slope of l is 1, find the angle between $\overrightarrow{OA}$ and $\overrightarrow{OB}$.

   (2) Suppose that $\overrightarrow{FB} = \lambda \overrightarrow{AF}$. If $\lambda \in [4, 9]$, find the range for the values of the intercept of l with the y-axis.

12. For a given integer $n \geq 2$, let $M_0(x_0, y_0)$ be an intersection point of the parabola $y^2 = nx - 1$ with the straight line $y = x$. Show that for any positive integer m, there must exist an integer $k \geq 2$ such that (x_0^m, y_0^m) is an intersection point of the parabola $y^2 = kx - 1$ with the straight line $y = x$.

Chapter 14

Parametric Equations

1. Key Points of Knowledge and Basic Methods

A. *Parametric equations commonly used for some curves*

(1) Parametric equations of straight lines:

$$\begin{cases} x = x_0 + t\cos\alpha, \\ y = y_0 + t\sin\alpha, \end{cases} \quad (t \text{ is a parameter}).$$

(2) Parametric equations of circles:

$$\begin{cases} x = a + r\cos\alpha, \\ y = b + r\sin\alpha, \end{cases} \quad (\alpha \text{ is a parameter}),$$

where the center of the circle is (a, b) and the radius of the circle is r $(r > 0)$.

(3) Parametric equations of ellipses:

$$\begin{cases} x = x_0 + a\cos\theta, \\ y = y_0 + b\sin\theta, \end{cases} \quad (\theta \text{ is a parameter}),$$

where the center of the ellipse is (x_0, y_0), the length of the semi-major axis is a, and the length of the semi-minor axis is b $(a > b > 0)$.

(4) Parametric equations of hyperbolas:

$$\begin{cases} x = x_0 + a\sec\theta, \\ y = y_0 + b\tan\theta, \end{cases} \quad (\theta \text{ is a parameter}),$$

where the center of the hyperbola is (x_0, y_0), the length of the real half axis is a $(a > 0)$, and the length of the imaginary half axis is b $(b > 0)$.

(5) Parametric equations of parabolas:

$$\begin{cases} x = 2pt^2, \\ y = 2pt, \end{cases} \quad (t \text{ is a parameter}),$$

where the focus of the parabola is $\left(\frac{p}{2}, 0\right)$ and the directrix is $x = -\frac{p}{2}$.

B. *Basic methods*

Parametric equations provide a bridge that connects the relationship of multi variables. In the process of solving a problem, by introducing a parameter or parametric equations, one can relate multi variables to a single variable for simpler computations, so that the purpose of solving the problem is easily realized.

Parameters or parametric equations play a prominent role in solving trajectory equations, extremal values, value ranges of variables, simplifying computations, and etc.

2. Illustrative Examples

Example 1. A point P is on the circle $A : x^2 + (y-1)^2 = 1$ and makes the counter clockwise rotation along the circle with a constant angular velocity. At the same time another point Q is on the circle $A : (x-3)^2 + (y-2)^2 = 1$ and makes the clockwise rotation along the circle with the same constant angular velocity. If the initial positions of the points P and Q are $(0,2)$ and $(2,2)$ respectively, find the maximum and minimum values of the distance between the two points P and Q.

Solution. As in Figure 14.1, let the point P come to a point M at a time moment and $\angle PAM = \theta$. Then the point Q moves to a point N and $\angle QBN = -\theta$, so the coordinates of the points M and N are

$$M\left(\cos\left(\frac{\pi}{2}+\theta\right), 1+\sin\left(\frac{\pi}{2}+\theta\right)\right),$$
$$N(3+\cos(\pi-\theta), 2+\sin(\pi-\theta)), \ \theta \in (-\pi, \pi),$$

respectively, and thus

$$|MN| = \sqrt{(-\sin\theta - 3 + \cos\theta)^2 + (1 + \cos\theta - 2 - \sin\theta)^2}$$
$$= \sqrt{2(\cos\theta - \sin\theta)^2 - 8(\cos\theta - \sin\theta) + 10}$$
$$= \sqrt{2(\cos\theta - \sin\theta - 2)^2 + 2}$$
$$= \sqrt{2\left[\sqrt{2}\sin\left(\frac{\pi}{4}-\theta\right) - 2\right]^2 + 2}.$$

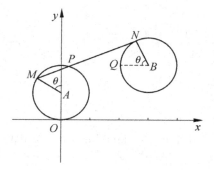

Figure 14.1   Figure for Example 1.

Clearly, when $\frac{\pi}{4} - \theta = -\frac{\pi}{2}$, that is $\theta = \frac{3\pi}{4}$, the distance between the two points P and Q has the maximum value $\sqrt{14 + 8\sqrt{2}}$; when $\frac{\pi}{4} - \theta = \frac{\pi}{2}$, namely $\theta = -\frac{\pi}{4}$, the distance between the two points P and Q has the minimum value $\sqrt{14 - 8\sqrt{2}}$. $\qquad\qquad\square$

Remark. Introducing the parametric equations of the circles, expressing the distance between two moving points in terms of the central angle θ, and using the feature of easy determination of the maximum value and the minimum value of trigonometric functions, made the solution to the problem easy to obtain. However, as in this problem, we should be careful about the value range of θ.

Example 2. A straight line passing through $M(2,1)$ intersects the ellipse $\frac{x^2}{16} + \frac{y^2}{4} = 1$ at two points A and B, so that a point M is the one-third point of the chord AB. Find the equation of the straight line.

Solution. Let the parametric equations of the straight line AB be

$$\begin{cases} x = 2 + t\cos\alpha, \\ y = 1 + t\sin\alpha, \end{cases} \quad (t \text{ is a parameter}).$$

Now, α can vary. Let the corresponding parameter values to the two points A and B be respectively t_1 and t_2. Then $t_1 = -2t_2$, and so $t_1 + t_2 = -t_2$ and $t_1 t_2 = -2t_2^2$. Thus, $t_1 t_2 = -2(t_1 + t_2)^2$.

Substituting the parametric equations of the straight line into $\frac{x^2}{16} + \frac{y^2}{4} = 1$, we get

$$(4\sin^2\alpha + \cos^2\alpha)t^2 + (8\sin\alpha + 4\cos\alpha)t - 8 = 0,$$

and then

$$t_1 + t_2 = \frac{-(8\sin\alpha + 4\cos\alpha)}{4\sin^2\alpha + \cos^2\alpha}, \quad t_1 t_2 = -\frac{8}{4\sin^2\alpha + \cos^2\alpha}.$$

Since $t_1 t_2 = -2(t_1 + t_2)^2$,

$$\tan\alpha = \frac{-4 \pm \sqrt{7}}{6}.$$

Hence, there are two desired straight lines whose equations are $y - 1 = \frac{-4+\sqrt{7}}{6}(x - 2)$ and $y - 1 = \frac{-4-\sqrt{7}}{6}(x - 2)$, respectively.   □

Remark. In the current example, the geometric meaning of the parameter in the parametric equations was combined with Vieta's theorem. If M is the middle point in the problem, then $t_1 + t_2 = 0$, and the length of the chord AB can be calculated via $|AB| = |t_1 - t_2| = \sqrt{(t_1 + t_2)^2 - 4t_1 t_2}$. Such strategies have many applications in parametric equations of straight lines.

Example 3. If there are two points on the hyperbola $C : x^2 - y^2 = 1$, which are symmetric with respect to a straight line $l : y = k(x + 4)$, find the range of the values for k.

Solution. Let $A(x_1, y_1)$ and $B(x_2, y_2)$ be the two points of the hyperbola that are symmetric with respect to the straight line l. Then $x_1^2 - y_1^2 = 1$ and $x_2^2 - y_2^2 = 1$. Subtracting one from the other gives

$$(x_1 + x_2)(x_1 - x_2) - (y_1 + y_2)(y_1 - y_2) = 0.$$

Let the middle point of AB be $M(x_0, y_0)$. By $k_{AB} = \frac{y_1 - y_2}{x_1 - x_2}$ ($x_1 \neq x_2$), $x_1 + x_2 = 2x_0$, and $y_1 + y_2 = 2y_0$, we obtain

$$2x_0 - 2y_0 k_{AB} = 0. \tag{14.1}$$

If $k \neq 0$, then the equality (14.1) is deformed to

$$2x_0 + 2y_0 \frac{1}{k} = 0. \tag{14.2}$$

Also the fact that $M(x_0, y_0)$ is on the straight line l implies

$$y_0 = k(x_0 + 4). \tag{14.3}$$

Solving (14.2) and (14.3) together gives the point $M(-2, 2k)$.

Let the parametric equations of the straight line AB be

$$\begin{cases} x = -2 + t \cos \alpha, \\ y = 2k + t \sin \alpha, \end{cases} \left(t \text{ is a parameter and } \tan \alpha = -\frac{1}{k} \right).$$

Substitute the above into the equation of the hyperbola. We then have

$$(\cos^2 \alpha - \sin^2 \alpha)t^2 - (4 \cos \alpha + 4k \sin \alpha)t + 3 - 4k^2 = 0. \qquad (14.4)$$

Since $\tan \alpha = -\frac{1}{k}$, we get $\cos \alpha = -k \sin \alpha$, and after substituting it into (14.4), we find that

$$t^2(k^2 - 1) \sin^2 \alpha + 3 - 4k^2 = 0.$$

It follows that $\frac{3-4k^2}{k^2-1} < 0$ from $t^2 > 0$ and $\sin^2 \alpha \neq 0$. In other words,

$$(k^2 - 1)(4k^2 - 3) > 0, \quad k \neq 0,$$

the solutions of which are $k \in (-\infty, 1) \cup \left(-\frac{\sqrt{3}}{2}, 0\right) \cup \left(0, \frac{\sqrt{3}}{2}\right) \cup (1, +\infty)$.

If $k = 0$, then the equation of the straight line l is $y = 0$. This time there exist points that are symmetric with respect to l on the hyperbola C. In summary, $k \in (-\infty, 1) \cup \left(-\frac{\sqrt{3}}{2}, \frac{\sqrt{3}}{2}\right) \cup (1, +\infty)$. $\qquad\square$

Example 4. Draw two mutually perpendicular chords OA and OB from the vertex of a parabola $y^2 = 2px$ $(p > 0)$. Find the equation of the trajectory for the middle point M of the line segment AB.

Solution 1. See Figure 14.2. Letting the equation of the straight line OA be $y = kx$, and solving the system of the equations

$$\begin{cases} y = kx, \\ y^2 = 2px, \end{cases}$$

we find the coordinates of the point A to be $\left(\frac{2p}{k^2}, \frac{2p}{k}\right)$.

Let the middle point of AB be $M(x, y)$. Then

$$\begin{cases} x = \dfrac{1}{2}\left(\dfrac{2p}{k^2} + 2pk^2\right) = p\left(\dfrac{1}{k^2} + k^2\right), \\ y = \dfrac{1}{2}\left(\dfrac{2p}{k} - 2pk\right) = p\left(\dfrac{1}{k} - k\right). \end{cases}$$

Eliminating k from the above gives that $y^2 = px - 2p^2$.

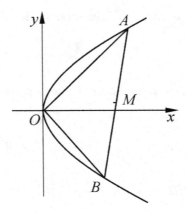

Figure 14.2　Figure for Example 4.

Solution 2. Write $A(2pt_1^2, 2pt_1)$ and $B(2pt_2^2, 2pt_2)$. Then

$$k_{OA}k_{OB} = \frac{2pt_1}{2pt_1^2}\frac{2pt_2}{2pt_2^2} = \frac{1}{t_1t_2} = -1,$$

namely $t_1t_2 = -1$.

For the middle point $M(x, y)$, we have

$$\begin{cases} x = p(t_1^2 + t_2^2), \\ y = p(t_1 + t_2), \end{cases}$$

and substituting $t_1t_2 = -1$ into the above and eliminating t_1 and t_2 ensure that

$$y^2 = px - 2p^2. \qquad \square$$

Remark. In Solution 1, using the slope k of OA as a parameter fastened the process of finding the result. In Solution 2, although two parameters t_1 and t_2 were chosen, actually there was only one parameter because of $t_1t_2 = -1$. So, the parameter elimination processes of the two solution methods have no essential difference, but Solution 2 is obviously more feasible and open.

Example 5. Suppose that there is a fixed pint A on an ellipse $C : \frac{x^2}{a^2} - \frac{y^2}{b^2} = 1$ $(a \neq b)$, and P and Q (both are different from A) are two arbitrary points on C satisfying $PA \perp QA$. Prove: The straight line passes through a fixed point.

Proof. Let the three points be $A(a \sec \theta, b \tan \theta)$, $P(a \sec \alpha_1, b \tan \alpha_1)$, and $Q(a \sec \alpha_2, b \tan \alpha_2)$. Then the equation of the straight line PQ is

$$(x - a \sec \alpha_1)(b \tan \alpha_1 - b \tan \alpha_2) = (y - b \tan \alpha_1)(a \sec \alpha_1 - a \sec \alpha_2),$$

which can be rewritten as

$$b \cos \frac{\alpha_1 - \alpha_2}{2} \cdot x - a \sin \frac{\alpha_1 + \alpha_2}{2} \cdot y - ab \cos \frac{\alpha_1 + \alpha_2}{2} = 0. \quad (14.5)$$

Since

$$k_{AP} = \frac{b \cos \frac{\alpha_1 - \theta}{2}}{a \sin \frac{\alpha_1 + \theta}{2}} \quad \text{and} \quad k_{AQ} = \frac{b \cos \frac{\alpha_2 - \theta}{2}}{a \sin \frac{\alpha_2 + \theta}{2}},$$

we see that

$$k_{AP} k_{AQ} = -1,$$

which is equivalent to

$$b^2 \cos \frac{\alpha_1 - \theta}{2} \cos \frac{\alpha_2 - \theta}{2} + a^2 \sin \frac{\alpha_1 + \theta}{2} \sin \frac{\alpha_2 + \theta}{2} = 0.$$

The above equality is valid if and only if

$$b^2 \left[\cos \left(\frac{\alpha_1 + \alpha_2}{2} - \theta \right) + \cos \frac{\alpha_1 - \alpha_2}{2} \right]$$

$$+ a^2 \left[\cos \frac{\alpha_1 - \alpha_2}{2} - \cos \left(\frac{\alpha_1 + \alpha_2}{2} + \theta \right) \right] = 0$$

$$\Leftrightarrow (a^2 + b^2) \cos \frac{\alpha_1 - \alpha_2}{2} + b^2 \left(\cos \frac{\alpha_1 + \alpha_2}{2} \cos \theta + \sin \frac{\alpha_1 + \alpha_2}{2} \sin \theta \right)$$

$$- a^2 \left(\cos \frac{\alpha_1 + \alpha_2}{2} \cos \theta - \sin \frac{\alpha_1 + \alpha_2}{2} \sin \theta \right) = 0$$

$$\Leftrightarrow \frac{a^2 + b^2}{b^2 - a^2} \frac{1}{\cos \theta} \cos \frac{\alpha_1 - \alpha_2}{2} - \frac{(a^2 + b^2) \sin \theta}{(a^2 - b^2) \cos \theta} \sin \frac{\alpha_1 + \alpha_2}{2}$$

$$+ \cos \frac{\alpha_1 + \alpha_2}{2} = 0.$$

The last equality can be written as

$$b \cos \frac{\alpha_1 - \alpha_2}{2} \cdot a \frac{a^2 + b^2}{a^2 - b^2} \sec \theta - a \sin \frac{\alpha_1 + \alpha_2}{2} \cdot b \frac{a^2 + b^2}{b^2 - a^2} \tan \theta$$

$$- ab \cos \frac{\alpha_1 + \alpha_2}{2} = 0,$$

the comparison of which to (14.5) implies that the straight line PQ always passes through the fixed point $\left(a\frac{a^2+b^2}{a^2-b^2} \sec \theta, b\frac{a^2+b^2}{b^2-a^2} \tan \theta \right)$. $\square$

Remark. Since $A, P,$ and Q are on the hyperbola, we only needed to introduce three angles (eccentric angles) to express the coordinates of the three points respectively by using the parametric equations.

Example 6. As Figure 14.3 shows, A and B are the left and right vertices of an ellipse $\frac{x^2}{a^2} + \frac{y^2}{b^2} = 1$ ($a > b > 0$), and P and Q are two points on the ellipse that are different from any vertex. Let the straight lines AP and QB intersect at a point M, and the straight lines PB and AQ intersect at a point N.

(1) Prove: $MN \perp AB$.
(2) If the chord PQ passes through the right focus F_2, find the equation of the straight line MN.

Solution. (1) Let $P(a\cos\alpha, b\sin\alpha)$ and $Q(a\cos\beta, b\sin\beta)$ be the coordinates of the points P and Q. Write $A(-a, 0)$ and $B(a, 0)$. Then

$$l_{AP} : a(1 + \cos\alpha)y = b\sin\alpha(x + a), \tag{14.6}$$

$$l_{QB} : a(\cos\beta - 1)y = b\sin\beta(x - a). \tag{14.7}$$

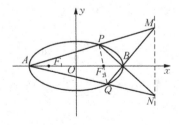

Figure 14.3   Figure for Example 6.

Eliminating y from (14.6) and (14.7) combined together, we see that

$\sin \alpha(\cos \beta - 1)(x + a) = \sin \beta(1 + \cos \alpha)(x - a)$

$\Leftrightarrow [\sin \alpha(\cos \beta - 1) - \sin \beta(1 + \cos \alpha)]x$

$\quad = a[\sin \alpha(1 - \cos \beta) - \sin \beta(1 + \cos \alpha)]$

$\Leftrightarrow [\sin(\alpha - \beta) - \sin \alpha - \sin \beta]x$

$\quad = a[\sin \alpha - \sin \beta - \sin(\beta + \alpha)]$

$\Leftrightarrow \cos \dfrac{\alpha - \beta}{2} \left(\sin \dfrac{\alpha - \beta}{2} - \dfrac{\alpha + \beta}{2} \right) x$

$\quad = a \cos \dfrac{\alpha + \beta}{2} \left(\sin \dfrac{\alpha - \beta}{2} - \sin \dfrac{\alpha + \beta}{2} \right)$

$\Leftrightarrow x_M = \dfrac{a \cos \frac{\alpha+\beta}{2}}{\cos \frac{\alpha-\beta}{2}}$

(since P and Q are different from the vertices).

By the same token, $x_N = \dfrac{a \cos \frac{\alpha+\beta}{2}}{\cos \frac{\alpha-\beta}{2}}$. Thus, $x_M = x_N$, and so $MN \perp AB$.

(2) Note that $\overrightarrow{F_2P} = (a \cos \alpha - c, b \sin \alpha)$ and $\overrightarrow{F_2Q} = (a \cos \beta - c, b \sin \beta)$. The fact that P, F_2, and Q are co-linear

$\Rightarrow \overrightarrow{F_2P}$ and $\overrightarrow{F_2Q}$ are co-linear

$\Rightarrow \sin \beta(a \cos \alpha - c) = \sin \alpha(a \cos \beta - c)$

$\Rightarrow a \sin(\alpha - \beta) = c(\sin \alpha - \sin \beta)$

$\Rightarrow a \sin \dfrac{\alpha - \beta}{2} \cos \dfrac{\alpha - \beta}{2} = c \cos \dfrac{\alpha + \beta}{2} \sin \dfrac{\alpha - \beta}{2}$

$\Rightarrow a \cos \dfrac{\alpha - \beta}{2} = c \cos \dfrac{\alpha + \beta}{2}$

$\Rightarrow x_M = x_N = \dfrac{a \cos \frac{\alpha+\beta}{2}}{\cos \frac{\alpha-\beta}{2}} = \dfrac{a^2}{c}$.

It follows that the equation of the straight line MN is $x = \frac{a^2}{c}$. $\quad\square$

Example 7. Assume that $\angle AOB = \theta$ (θ is a constant and $0 < \theta < \frac{\pi}{2}$), moving points P and Q are on the rays OA and OB respectively, and the area of $\triangle POQ$ is always 36. Let the center of gravity of $\triangle POQ$ be G, a point M be on the ray OG, and the condition $|OM| = \frac{3}{2}|OG|$ be satisfied. Find:

(1) The minimum value of $|OG|$.
(2) The equation of the trajectory for the moving point M.

Solution. (1) Establish a rectangular coordinates system with O as the origin and the angular bisector of $\angle AOB$ as the x-axis. Then we can write

$$P\left(a\cos\frac{\theta}{2}, a\sin\frac{\theta}{2}\right), \quad Q\left(b\cos\frac{\theta}{2}, -b\sin\frac{\theta}{2}\right).$$

So, the coordinates of the center of gravity $G(x_0, y_0)$ of $\triangle OPQ$ are

$$x_0 = \frac{1}{3}\left(a\cos\frac{\theta}{2} + b\cos\frac{\theta}{2} + 0\right) = \frac{1}{3}(a+b)\cos\frac{\theta}{2},$$

$$y_0 = \frac{1}{3}\left(a\sin\frac{\theta}{2} - b\sin\frac{\theta}{2} + 0\right) = \frac{1}{3}(a-b)\sin\frac{\theta}{2}.$$

Thus,

$$\begin{aligned}
|OG|^2 &= x_G^2 + y_G^2 \\
&= \frac{1}{9}(a^2+b^2) + \frac{2}{9}ab\left(\cos^2\frac{\theta}{2} - \sin^2\frac{\theta}{2}\right) \\
&= \frac{1}{9}(a^2+b^2) + \frac{2}{9}ab\cos\theta \\
&\geq \frac{1}{9}\cdot 2ab + \frac{2}{9}ab\cos\theta = \frac{4}{9}ab\cos^2\frac{\theta}{2}.
\end{aligned}$$

Also $S_{\triangle OPQ} = \frac{1}{2}ab\sin\theta = 36$, that is $ab = \frac{72}{\sin\theta}$, so

$$|OG| \geq \sqrt{\frac{4}{9}\frac{72}{\sin\theta}\cos^2\frac{\theta}{2}} = \sqrt{16\cot\frac{\theta}{2}} = 4\sqrt{\cot\frac{\theta}{2}}.$$

When $a = b = \sqrt{\frac{72}{\sin\theta}}$, the above inequality becomes equality, so $|OG|_{\min} = 4\sqrt{\cot\frac{\theta}{2}}$.

(2) Write $M(x, y)$. From $|OM| = \frac{3}{2}|OG|$, we have

$$x = \frac{3}{2}x_G = \frac{1}{2}(a+b)\cos\frac{\theta}{2} > 0,$$

$$y = \frac{3}{2}y_G = \frac{1}{2}(a-b)\sin\frac{\theta}{2} > 0,$$

the solutions of which are

$$a = \frac{x}{\cos \frac{\theta}{2}} + \frac{y}{\sin \frac{\theta}{2}}, \quad b = \frac{x}{\cos \frac{\theta}{2}} - \frac{y}{\sin \frac{\theta}{2}}.$$

Substituting the above into $ab = \frac{72}{\sin \theta}$ and simplifying the resulting expression, we obtain

$$\frac{x^2}{36 \cot \frac{\theta}{2}} - \frac{y^2}{36 \tan \frac{\theta}{2}} = 1 \ (x > 0),$$

which is the equation of the trajectory for the moving point M. □

Example 8. Given three points A, B, and C in a plane that are not co-linear, and with the line segment AB as its axis (major axis or minor axis) draw an ellipse that does not pass through the point C. Suppose that the ellipse intersects the two other line segments AC and BC at E and F, respectively. Draw the tangent lines to the ellipse from E and F, and assume that the two tangent lines intersect at a point C_0. Similarly, draw two other ellipses with the line segments BC and AC as an axis respectively, and obtain the intersection points A_0 and B_0 of the corresponding tangent lines respectively. Prove: No matter what length to choose the other axis of each ellipse, the three straight lines AA_0, BB_0, and CC_0 always pass through a fixed point.

Proof. We first consider the ellipse with the straight line containing the line segment AB as an axis. Establish a rectangular coordinates system, so that the equation of the ellipse is $\frac{x^2}{a^2} + \frac{y^2}{b^2} = 1 \ (a \neq b)$. Let the ellipse intersect the straight lines AC and BC at the points E and F respectively. Draw from E and F the tangent lines to the ellipse and they intersect at the point C_0.

Write $A(-a, 0), B(a, 0), E(a \cos \alpha, b \sin \alpha)$, and $F(a \cos \beta, b \sin \beta)$. Then

$$l_{AE} : \frac{y}{x+a} = \frac{b \sin \alpha}{a(1 + \cos \alpha)},$$

$$l_{BF} : \frac{y}{x-a} = \frac{-b \sin \beta}{a(1 - \cos \beta)}.$$

Solving the above two equations, we obtain

$$x_C = \frac{a \cos \frac{\alpha + \beta}{2}}{\cos \frac{\beta - \alpha}{2}}.$$

Thus, the equations of the tangent lines to the ellipse passing through E and F are

$$\frac{x\cos\alpha}{a} + \frac{y\sin\alpha}{b} = 1,$$

$$\frac{x\cos\beta}{a} + \frac{y\sin\beta}{b} = 1,$$

respectively, and solving them together gives

$$x_{C_0} = \frac{a\cos\frac{\alpha+\beta}{2}}{\cos\frac{\beta-\alpha}{2}}.$$

From $x_C = x_{C_0}$, we know that $CC_0 \perp AB$.

Similarly, $AA_0 \perp BC$ and $BB_0 \perp AC$.

Therefore, the straight lines AA_0, BB_0, and CC_0 coincide with the three altitudes of $\triangle ABC$, and consequently they all pass through the orthocenter of $\triangle ABC$.   □

3. Exercises

Group A

I. Filling Problems

1. Two vertices A and B of $\triangle ABC$ are the two foci of

$$\begin{cases} x = -1 + 2\sqrt{2}\cos\theta, \\ y = 2 + 2\sin\theta, \end{cases} \quad \theta \text{ is a parameter.}$$

   The vertex C moves on the circle $x^2 + y^2 = 64$. Then the trajectory equation for the center of gravity of $\triangle ABC$ is _____.

2. At a point M (not the origin) on the major axis of an ellipse $\frac{x^2}{a^2} + \frac{y^2}{b^2} = 1$ $(a > b > 0)$, draw a chord PP' that is perpendicular to the major axis AA'. Then the trajectory equation for the intersection point Q of $A'P'$ and PA is _____.

3. The intersection points of a straight line passing through the point $P(-2,3)$ with the inclination angle $\frac{3}{4}\pi$ and the circle $x^2 + y^2 = 25$ are A and B. Then the value of $|PA||PB|$ is _____.

4. There are two points P and Q on the ellipse $\frac{x^2}{16} + \frac{y^2}{4} = 1$ with center O. If the product of the slopes of OP and OQ is $-\frac{1}{4}$, then $|OP|^2 + |OQ|^2 = $ _____.

II. Calculation Problems

5. Let a moving straight line l be perpendicular to the x-axis and intersect the ellipse $\frac{x^2}{4} + \frac{y^2}{2} = 1$ at two points A and B. If P is a point on l satisfying $|PA||PB| = 1$, find the equation of the trajectory of the point P.

6. As Figure 14.4 shows, CD is a diameter of an ellipse $\frac{x^2}{a^2} + \frac{y^2}{b^2} = 1$. At the left end point A of the major axis of the ellipse, draw a parallel line to CD, intersecting the ellipse at another point N and intersecting the straight line containing the short axis at M. Prove: $|AM||AN| = |CO||CD|$.

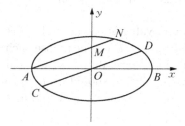

Figure 14.4   Figure for Exercise 6.

7. In a rectangular coordinates system xOy of a plane, the right focus of a hyperbola $C : \frac{x^2}{a^2} - \frac{y^2}{b^2} = 1$ is F. A straight line passing through F intersects the hyperbola C at two points A and B. If $|OF||AB| = |FA||FB|$, find the eccentricity e of the hyperbola C.

8. There is a point B outside a circle centered at A with radius $2\cos\theta$, where $0 < \theta < \frac{\pi}{2}$. Suppose that $|AB| = 2\sin\theta$, and M is the center of a circle that passes through B and is tangent to the circle A at a point T from outside.

   (1) When θ takes some value, what curve is the trajectory P of the point M?
   (2) The point M is a moving one on the trajectory P and N is a moving point on the circle A. Denote the minimum value of $|MN|$ by $f(\theta)$. Find the range for the values of $f(\theta)$ (you do not need to prove it).

Group B

9. Let P be any point on a hyperbola $\frac{x^2}{a^2} - \frac{y^2}{b^2} = 1$ ($a > 0$ and $b > 0$). At P draw two parallel lines to the asymptotes, each intersecting the other asymptote at Q and R, respectively. Prove:

   (1) $|PQ||PR| = \frac{1}{4}(a^2 + b^2)$.
   (2) The area of the parallelogram $PQOR$ is $\frac{1}{2}ab$.

10. As in Figure 14.5, at a moving point P on an ellipse $\frac{x^2}{a^2} + \frac{y^2}{b^2} = 1$ ($a > b > 0$), draw two tangent lines PA and PB to the circle $x^2 + y^2 = b^2$ with two tangent points A and B, respectively. Let the straight line AB intersect the x-axis and y-axis at M and N, respectively.

   (1) Find the minimum value of the area of $\triangle MON$.
   (2) Does there exist a point P on the ellipse such that the two tangent lines to the circle are mutually perpendicular? If there exists such a point P, what condition should a and b satisfy? Then also find the coordinates of the point P. If the point P does not exist, state the reason.

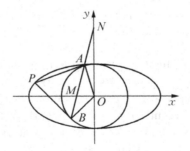

Figure 14.5   Figure for Exercise 10.

11. At any point P not on an ellipse $\frac{x^2}{a^2} + \frac{y^2}{b^2} = 1$, draw two straight lines l_1 and l_2 that intersect the ellipse at two points A, B and C, D, respectively. If the inclination angles of l_1 and l_2 are α and β, respectively and $\alpha + \beta = \pi$, prove: The four points A, B, C, and D are on the same circle.

12. Let the left vertex of a hyperbola $\frac{x^2}{a^2} - \frac{y^2}{b^2} = 1$ $(a, b > 0)$ be A and the right focus be F. From F draw any straight line that intersects the right branch of the hyperbola at two points M and N. Suppose that AM and AN intersect the right directrix at two points P and Q. Prove: The circle with PQ as diameter passes through the point F.

Chapter 15

Families of Curves

1. Key Points of Knowledge and Basic Methods

A collection of curves with some common properties is called a family of curves.

A. *Families of straight lines*

(1) The family of the straight lines passing through a fixed point (x_0, y_0) is

$$\lambda_1(y - y_0) + \lambda_2(x - x_0) = 0,$$

where λ_1 and λ_2 are parameters.

(2) The family of the straight lines that are parallel to a straight line $Ax + By + C = 0$ is

$$Ax + By + \lambda = 0,$$

where $\lambda \neq C$, and λ is a parameter.

(3) The family of the straight lines that are perpendicular to a straight line $Ax + By + C = 0$ is

$$Bx - Ay + \lambda = 0,$$

where λ is a parameter.

(4) When the general forms of two straight lines l_1 and l_2 are $f_1(x, y) = 0$ and $f_2(x, y) = 0$ respectively, the family of the straight lines

$$\lambda_1 f_1(x, y) + \lambda_2 f_2(x, y) = 0 \ (\lambda_1, \lambda_2 \text{ are parameters})$$

(i) represents all the straight lines that pass through the intersection point of l_1 and l_2 if l_1 and l_2 intersect;

(ii) represents all the straight lines that are parallel to l_1 and l_2 (or imaginary straight lines) if $l_1 \parallel l_2$.

(5) The family of the straight lines with sum a of the x and y intercepts is

$$\frac{x}{\lambda} + \frac{y}{a - \lambda} = 1 \ (\lambda \text{ is a parameter}).$$

(6) The family of the straight lines whose distance to the origin equals $r \ (r > 0)$ is

$$x \cos \theta + y \sin \theta = r \ (\theta \text{ is a parameter}).$$

B. Families of conic sections generated by straight lines

Let $f_i(x, y) = A_i x + B_i y + C_i \ (i = 1, 2, \ldots)$.

(1) If the equations of the three sides of a triangle are $f_i(x, y) = 0 \ (i = 1, 2, 3)$, then the equation of the family of the conic sections passing through the three vertices of the triangle is

$$f_1(x, y)f_2(x, y) + \lambda f_2(x, y)f_3(x, y) + \mu f_3(x, y)f_1(x, y) = 0,$$

where λ and μ are parameters.

(2) If the equations of the four sides of a quadrilateral are $f_i(x, y) = 0 \ (i = 1, 2, 3, 4)$, then the equation of the family of the conic sections passing through the four vertices of the quadrilateral is

$$f_1(x, y)f_3(x, y) + \lambda f_2(x, y)f_4(x, y) = 0 \ (\lambda \text{ is a parameter}),$$

where $f_1(x, y) = 0$ and $f_3(x, y) = 0$, and $f_2(x, y) = 0$ and $f_4(x, y) = 0$ are the equations of the opposite sides of the quadrilateral, respectively.

(3) The equation of the family of the conic sections that are tangent to two given straight lines $f_1(x, y) = 0$ and $f_2(x, y) = 0$ at given points M_1 and M_2 is

$$f_1(x, y)f_2(x, y) + \lambda f_3(x, y)f_3(x, y) = 0 \ (\lambda \text{ is a parameter}),$$

where $f_3(x, y) = 0$ is the equation of the straight line that passes through M_1 and M_2.

(4) The equation of the family of the conic sections that pass through the four intersection points of two straight lines $f_1(x, y) = 0$ and $f_2(x, y) = 0$ with a conic section $F(x, y) = 0$ is

$$F(x, y) + \lambda f_1(x, y)f_2(x, y) = 0 \ (\lambda \text{ is a parameter}).$$

C. *Families of circles*

Using families of circles is an important method to find equations of circles, and also a simple and direct way of proving co-circle of four points.

For two circles

$$C_i(x, y) \equiv x^2 + y^2 + D_i x + E_i y + F_i = 0 \ (i = 1, 2)$$

with distinct centers, the equation

$$C_1(x, y) + \lambda C_2(x, y) = 0$$

represents the family of the circles with the common axis. When $\lambda \neq -1$, it gives a circle; when $\lambda = -1$, it is reduced to the straight line

$$(D_1 - D_2)x + (E_1 - E_2)y + (F_1 - F_2) = 0,$$

and this straight line is called the root axis of the two circles.

For a given circle $C_1(x, y) = 0$ and a point (m, n) on the circle, the equation

$$C_1(x, y) + \lambda[(x - m)^2 + (y - n)^2] = 0 \ (\lambda \text{ is a parameter})$$

represents the family of the circles that are tangent to the circle $C_1(x, y) = 0$ at the point (m, n).

D. *Families of centered conics*

When $\lambda \mu \neq 0$ and λ, μ are not both negative, the standard equation of the family of the centered conics can be written uniformly as

$$\frac{x^2}{\lambda} + \frac{y^2}{\mu} = 1.$$

(1) The equation of the family of the centered conics with common foci is

$$\frac{x^2}{m^2 - \lambda} + \frac{y^2}{n^2 - \lambda} = 1,$$

where m and n are constants, the half focal distance is $c = \sqrt{|m^2 - n^2|}$, and λ is a parameter.

(2) The equation of the family of the centered conics with two common vertices is

$$\frac{x^2}{a^2} + \frac{y^2}{\lambda} = 1,$$

where $(\pm a, 0)$ are the fixed vertices, and λ is a parameter.

(3) The equation of the family of the ellipses with a common eccentricity is

$$\frac{x^2}{a^2} + \frac{y^2}{b^2} = \lambda,$$

where a and b are constants, and $\lambda > 0$ is a parameter.

(4) The equation of the family of the hyperbolas with common asymptotes is

$$\frac{x^2}{a^2} - \frac{y^2}{b^2} = \lambda,$$

where a and b are constants, and λ ($\neq 0$) is a parameter.

(5) The equation of the family of the conic sections that pass through the intersection points of two conic sections

$$\frac{x^2}{A_1} + \frac{y^2}{B_1} = 1 \text{ and } \frac{x^2}{A_2} + \frac{y^2}{B_2} = 1$$

is

$$\left(\frac{x^2}{A_1} + \frac{y^2}{B_1} - 1\right) + \lambda \left(\frac{x^2}{A_2} + \frac{y^2}{B_2} - 1\right) = 0,$$

where λ ($\neq 0$) is a parameter. This equation does not contain $\frac{x^2}{A_2} + \frac{y^2}{B_2} = 1$.

In general, if the equations of two conic sections are $f_1(x, y) = 0$ and $f_2(x, y) = 0$ respectively, then the equation of the family of the conic sections that pass through the intersection points of the two conic sections is

$$\lambda_1 f_1(x, y) + \lambda_2 f_2(x, y) = 0,$$

where λ_1 and λ_2 are parameters.

E. *Basic methods*

Using a family of curves to solve problems is basically to choose characteristic quantities of the equation of the curves (such as the slope of the equation of a straight line, intercept b, the radius of a circle, a and b of a centered conic, etc.) as variables to obtain a family of the curves. According to the known quantity, using the method of undetermined coefficients, one can realize the goal of solving the problem. This method often reflects the mathematical viewpoint of parameter transformations and solution strategies of global treatments. The types of problems include finding the coordinates of a point, the equation of a curve, properties of a geometric figure, and so on.

2. Illustrative Examples

Example 1. Find the equation of a circle that is tangent to the circle $x^2 + y^2 - 4x - 8y + 15 = 0$ at the point $P(3,6)$ and passes through the other point $Q(5,6)$.

Solution. The tangent point $(3,6)$ is on the known circle, so it can be thought as the "point circle": $(x-3)^2 + (y-6)^2 = 0$, so we establish the equation of the family of the circles

$$x^2 + y^2 - 4x - 8y + 15 + \lambda[(x-3)^2 + (y-6)^2] = 0.$$

Substituting the point $Q(5,6)$ into the above equation, we find that $\lambda = -2$. Hence, the equation of the desired circle is $x^2 + y^2 - 8x - 16y + 75 = 0$.

□.

Remark. When trying to find the equation of a circle that is tangent to a known straight line or a known circle at a given point, it is often to treat the tangent point as a "point circle" and then to use the equation of the family of the circles with a common point to solve the problem. This is an important method and technique.

Example 2. Suppose that $0 < a < b$. Passing through fixed points $A(a, 0)$ and $B(b, 0)$ respectively, draw straight lines l and m so that they intersect the parabola $y^2 = x$ at four different points. When the four points are on the same circle, find the trajectory of the intersection point P of the straight lines l and m.

Solution. Let the equations of the straight lines l and m be $y = k_1(x-a)$ and $y = k_2(x-b)$, respectively.

Then the coordinates of all points on the straight lines l and m satisfy the equation

$$(k_1 x - y - k_1 a)(k_2 x - y - k_2 b) = 0.$$

Thus, their intersection points with the parabola satisfy the system of the equations

$$\begin{cases} y^2 = x, \\ (k_1 x - y - k_1 a)(k_2 x - y - k_2 b) = 0. \end{cases}$$

If the four intersection points are on the same circle, then the equation of the circle can be written as

$$(k_1 x - y - k_1 a)(k_2 x - y - k_2 b) + \lambda(y^2 - x) = 0.$$

In this equation, the term that contains xy must be zero, hence $k_1 = -k_2$. Assume that $k_1 = -k_2 = k \neq 0$. Then

straight line $l : y = k(x - a)$, straight line $m : y = -k(x - b)$.

Eliminating k from the above two equations gives $2x - (a + b) = 0$ ($y \neq 0$), and this is the sought equation of the trajectory. $\square$

Example 3. The ellipse $x^2 + 2y^2 - 2 = 0$ and the straight line $x + 2y - 1 = 0$ intersect at two points B and C. Given $A(2, 2)$, find the equation of the circle that passes through the three points A, B, and C.

Analysis. For this problem, we could find the coordinates of B and C, and then find the equation of the circle passing through the known three points. But here we can consider the problem from the angle of families of conic sections. Every curve of the family of the conic sections

$$\lambda(x^2 + 2y^2 - 2) + \mu(x + 2y - 1) = 0$$

passes through B and C. However, very obviously the family does not contain all the conic sections that pass through B and C, since all the circles passing through B and C are not in the family. But we can construct another family of the conic sections

$$\lambda(x^2 + 2y^2 - 2) + \mu(x + 2y - 1)(x - 2y + m) = 0,$$

which contains the circles passing through B and C.

Solution. Let the equation of the desired circle be

$$\lambda(x^2 + 2y^2 - 2) + \mu(x + 2y - 1)(x - 2y + m) = 0,$$

which becomes

$$(\lambda + \mu)x^2 + (2\lambda - 4\mu)y^2 + \mu(m - 1)x + 2\mu(m + 1)y - m\mu - 2\lambda = 0 \tag{15.1}$$

after expansion. Let $\lambda + \mu = 2\lambda - 4\mu$ and choose $\mu = 1$. Then $\lambda = 5$. Substituting it into (15.1) results in

$$6x^2 + 6y^2 + (m - 1)x + 2(m + 1)y - m - 10 = 0. \tag{15.2}$$

Putting $A(2,2)$ into (15.2) gives $m = -8$, and so (15.2) is

$$6x^2 + 6y^2 - 9x - 14y - 2 = 0. \qquad \square$$

Remark. If two curves $C_1 : f_1(x,y) = 0$ and $C_2 : f_2(x,y) = 0$ have intersection points, then the curve $\lambda f_1(x,y) + \mu f_2(x,y) = 0$ must pass through the same points, but curves that pass through the intersection points may not be of the above form. When the two curves are conic sections and intersect at four points that are not co-linear, the curve $\lambda f_1(x,y) + \mu f_2(x,y) = 0$ can represent all the conic sections that pass through the four points. Hence, in the process of using the family of the conic sections, one should consider the feature of the equation.

Example 4. Let a be a real number, and let the two parabolas $y = x^2 + x + a$ and $x = 4y^2 + 3y + a$ intersect at four points.

(1) Find the range of the values for a.
(2) Prove that the four intersection points are on the same circle, and find the coordinates of the center of the circle.

Solution. Let $\lambda \in \mathbf{R}$. The equation of the conic sections that pass through the intersection points of the two parabolas is

$$x^2 + x - y + a + \lambda(4y^2 - x + 3y + a) = 0. \qquad (15.3)$$

(1) Let $\lambda = -1$ in (15.3). Factorizing the resulting expression gives

$$(x - 2y)(x + 2y + 2) = 0.$$

The two parabolas have four intersection points if and only if the system of the equations

$$\text{(I)} \quad \begin{cases} x - 2y = 0, \\ y = x^2 + x + a \end{cases}$$

and the system of the equations

$$\text{(II)} \quad \begin{cases} x + 2y + 2 = 0, \\ y = x^2 + x + a \end{cases}$$

have four groups of different solutions.

Eliminating x from (I) gives

$$4y^2 + y + a = 0.$$

By requiring the discriminant to be greater than 0, we get $1 - 16a > 0$, so $a < \frac{1}{16}$.

Eliminating x from (II) gives

$$4y^2 + 5y + a + 2 = 0.$$

Requiring the discriminant to be greater than 0 gives $25 - 16(a + 2) > 0$, so $a < -\frac{7}{16}$.

If the systems (I) and (II) of the equations have common solutions, then there exist real numbers x and y that satisfy

$$x - 2y = 0, \ x + 2y + 2 = 0, \ y = x^2 + x + a,$$

the solutions of which are $x = -1$ and $y = -\frac{1}{2}$, with $a = -\frac{1}{2}$.

To summarize, the range of the values for a is $\left(-\infty, -\frac{1}{2}\right) \cup \left(-\frac{1}{2}, -\frac{7}{16}\right)$.

(2) Letting $\lambda = \frac{1}{4}$ in (15.3) gives

$$x^2 + y^2 + \frac{3}{4}x - \frac{1}{4}y + \frac{5}{4}a = 0,$$

namely

$$\left(x + \frac{3}{8}\right)^2 + \left(y - \frac{1}{8}\right)^2 = \frac{5}{32} - \frac{5a}{4} \ (> 0).$$

This indicates that the four intersection points of the two parabolas lie on the circle with $\left(-\frac{3}{8}, \frac{1}{8}\right)$ as the center. $\qquad\square$

Example 5. Suppose that A and B are two intersection points of a straight line $y = x + b$ and the ellipse $3x^2 + y^2 = 1$, and O is the origin. If $\triangle OAB$ is an acute triangle, find the range of the values for b.

Solution. Let a straight line $y = x + b$ intersect the ellipse $3x^2 + y^2 = 1$ at two points A_1 and B_1 such that $\angle A_1 O B_1$ is the right angle. Clearly, the equation of the straight line $A_1' B_1'$ that is symmetric to the straight line $A_1 B_1$ with respect to the origin is

$$y = x - b.$$

Then the equation of the family of the curves that pass through the two straight lines $A_1 B_1$ and $A_1' B_1'$ and the ellipse $3x^2 + y^2 = 1$ is

$$\lambda(y - x - b)(y - x + b) + \mu(3x^2 + y^2 - 1) = 0,$$

that is

$$(\lambda + 3\mu)x^2 - 2\lambda xy + (\lambda + \mu)y^2 - (\lambda b^2 + \mu) = 0. \qquad (15.4)$$

If $A_1 A_1' \perp B_1 B_1'$, then the equations of $A_1 A_1'$ and $B_1 B_1'$ can be assumed to be

$$x + ky = 0, \quad kx - y = 0,$$

respectively. Then

$$(x + ky)(kx - y) = 0,$$

namely

$$kx^2 + (k^2 - 1)xy - ky^2 = 0. \qquad (15.5)$$

If the family of the curves (15.4) contains the family of the curves (15.5), then we may let

$$k = \lambda + 3\mu, \ k^2 - 1 = -2\lambda, \ -k = \lambda + \mu, \ \lambda b^2 + \mu = 0,$$

the solutions of which are $k = 2 \pm \sqrt{5}$ (choose $k = 2 + \sqrt{5}$), $\lambda = -4 - 2\sqrt{5}$, $\mu = 2 + \sqrt{5}$, and $b = \pm \frac{\sqrt{2}}{2}$.

As Figure 15.1 shows, this means that the desired straight line $y = x + b$ satisfying the condition of the problem is above the straight line $A_1 B_1 \left(y = x + \frac{\sqrt{2}}{2} \right)$ or below the straight line $A_1' B_1' \left(y = x - \frac{\sqrt{2}}{2} \right)$. That is, $b > \frac{\sqrt{2}}{2}$ or $b < -\frac{\sqrt{2}}{2}$.

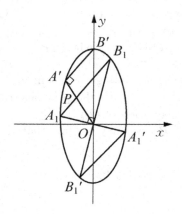

Figure 15.1 Figure for Example 5.

Without loss of generality, assume that the straight line AB intersects the ellipse at the two points A' and B' above A_1B_1. Since $|OB'| > |OA'|$, it is enough to consider that $\angle OA'B'$ is also an acute angle.

Eliminating y from $y = x + b$ and $3x^2 + y^2 = 1$, we obtain

$$4x^2 + 2bx + b^2 - 1 = 0. \tag{15.6}$$

Write $A'(x_1, y_1)$ and $B'(x_2, y_2)$ with $x_1 < x_2$. From (15.6),

$$x_1 = \frac{-b - \sqrt{4 - 3b^2}}{4} \quad \left(\frac{\sqrt{2}}{2} < b < \frac{2\sqrt{3}}{3} \right).$$

Consequently,

$$y_1 = \frac{3b - \sqrt{4 - 3b^2}}{4}.$$

Since the slope of the straight line $y = x + b$ is 1, it is sufficient to have $k_{OA'} > -1$, namely

$$\frac{3b - \sqrt{4 - 3b^2}}{-b - \sqrt{4 - 3b^2}} > -1,$$

the solution of which is $b < 1$.

In summary, and using the symmetry of the ellipse, we find that the range of b is $\left(-1, -\frac{\sqrt{2}}{2} \right) \cup \left(\frac{\sqrt{2}}{2}, 1 \right)$. $\square$

Remark. (1) There is a more concise method concerning that $\angle OA'B'$ is also acute: Since $A'B'$ is oriented, that is, its slope is 1. Assume that $OA' \perp A'B'$. Then the equation of OA' is $y = -x$. Combining it with the equation of the ellipse to solve, we get the solutions $x = -\frac{1}{2}$ and $y = \frac{1}{2}$. Now, $b = y - x = 1$, so $\angle OA'B' < \frac{\pi}{2}$ when $b < 1$.

(2) We can also solve the acute angle problem for $\angle OA'B'$ by the equation of the curve family, similar to the previous examples: Assume that $OA' \perp A'B'$. Then the equation of OA' is $y = -x$. Considering the existence of the slope of OB', we may assume that the equation of OB' is $x = ty$. Then

$$(x + y)(x - ty) = 0,$$

in other words,

$$x^2 + (1 - t)xy - ty^2 = 0. \tag{15.7}$$

Comparing (15.7) and (15.4) gives

$$1 = \lambda + 3\mu, \; 1 - t = -2\lambda, \; -t = \lambda + \mu, \; \lambda b^2 + \mu = 0,$$

the solutions of which are $\lambda = -\mu = -\frac{1}{2}$ and $b = 1$.

Since the slope of $A'B'$ is 1, we know that $\angle OA'B' < \frac{\pi}{2}$ when $b < 1$.

Example 6. As Figure 15.2 shows, let two circles be intersected by two straight lines. The two straight lines intersect the first circle at A', B', C', and D' that form the two chords $A'C'$ and $B'D'$, and the second circle at A, B, C, and D that form the two chords AB and DC. The four chords intersect at four points. Prove: The four points are on the same circle that has the same root axis as the original circles.

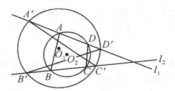

Figure 15.2 Figure for Example 6.

Proof. Let the equations of the circles O_1 and O_2 be $C_1(x, y) = 0$ and $C_2(x, y) = 0$, respectively, and let the equations of the straight lines l_1 and l_2 be $L_1(x, y) = 0$ and $L_2(x, y) = 0$ respectively. Then the equations of AB and DC are of the form $C_2(x, y) + \lambda_1 L_1(x, y) L_2(x, y) = 0$ (λ_1 is a parameter); the equations of $A'C'$ and $B'D'$ are of the form $C_1(x, y) + \lambda_2 L_1(x, y) L_2(x, y) = 0$ (λ_2 is a parameter). Thus, the four intersection points of AB, DC and $A'C', B'D'$ are on the curve $\lambda_2 C_2(x, y) - \lambda_1 C_1(x, y) = 0$. Clearly, this is the equation of a circle (denoted as the circle O), namely the four points are on the same circle.

On the one hand, we know that the equation of the root axis for the circles O_1 and O_2 is $C_1(x, y) - C_2(x, y) = 0$. On the other hand, the equation of the root axis for the circles O_1 and O is $(\lambda_2 - \lambda_1) C_1(x, y) - \lambda_2 C_2(x, y) + \lambda_1 C_1(x, y) = \lambda_2 [C_1(x, y) - C_2(x, y)] = 0$. Therefore, the circles O, O_1, and O_2 have the common root axis. □

Example 7. Let MN be a chord of a circle O and R be the middle point of the chord MN. Draw two chords AB and CD at the point R. A conic section that passes through A, B, C, and D intersects MN at P and Q. Prove: R is the middle point of PQ.

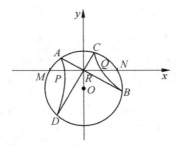

Figure 15.3   Figure for Example 7.

Analysis. The conic sections that pass through A, B, C, and D can be viewed as the family of the conic sections passing through the intersection points of the circle with l_1 and l_2.

Proof. Set up a rectangular coordinates system with R as the origin and MN as the x-axis; see Figure 15.3.

Suppose that the coordinates of the center O of the circle are $(0, a)$ and the radius is r. Then the equation of the circle is

$$x^2 + (y - a)^2 = r^2. \tag{15.8}$$

Let the equations of AB and CD be $y = k_1 x$ and $y = k_2 x$, respectively. Combine them together to form the equation

$$(y - k_1 x)(y - k_2 x) = 0. \tag{15.9}$$

Then the equation of the family of the curves that pass through the four intersection points A, B, C, and D of (15.8) and (15.9) is

$$\lambda(y - k_1 x)(y - k_2 x) + \mu[x^2 + (y - a)^2 - r^2] = 0. \tag{15.10}$$

Letting $y = 0$ gives

$$(\lambda k_1 k_2 + \mu)x^2 + \mu(a^2 - r^2) = 0. \tag{15.11}$$

The two roots of (15.11) are the horizontal coordinates of the intersection points P and Q of the conic section and MN. By Vieta's theorem, $x_P + x_Q = 0$, so R is the middle point of PQ.   □

Remark. This example is essentially a generalization of the butterfly theorem in plane geometry.

Example 8. Suppose that the three pairs of the opposite sides of the inscribed hexagon of a conic section are all non-parallel. Prove: The intersection points of the three pairs of the two straight lines containing the opposite sides of the hexagon are co-linear.

Proof. Let the equation of the conic section be $F(x, y) = 0$, let the equations of the six sides AB, BC, CD, DE, EF, and FA be $l_i(x, y) = 0$ ($i = 1, 2, \ldots, 6$) respectively, and let the equation of the diagonal AD be $m(x, y) = 0$. Then the equation of the conic sections that pass through the four points A, B, C, and D is $l_2(x, y)m(x, y) + \lambda_1 l_1(x, y)l_3(x, y) = 0$. Properly choose λ_1 so that the above equation coincides with $F(x, y) = 0$. Let this value of λ_1 be λ_1'. Then

$$l_2(x, y)m(x, y) + \lambda_1' l_1(x, y)l_3(x, y) = \mu_1 F(x, y) = 0. \qquad (15.12)$$

By the same reason, the equation of the conic sections that pass through the four points A, D, E, and F is $l_5(x, y)m(x, y) + \lambda_2 l_4(x, y)l_6(x, y) = 0$. Properly choose λ_2 so that the above equation coincides with $F(x, y) = 0$. Let this value of λ_2 be λ_2'. Then

$$l_5(x, y)m(x, y) + \lambda_2' l_4(x, y)l_6(x, y) = \mu_2 F(x, y) = 0. \qquad (15.13)$$

Eliminating $m(x, y)$ from (15.12) and (15.13), we obtain

$$\lambda_1' l_1(x, y)l_3(x, y)l_5(x, y) - \lambda_2' l_2(x, y)l_4(x, y)l_6(x, y)$$

$$= [\mu_1 l_5(x, y) - \mu_2 l_2(x, y)]F(x, y) = 0. \qquad (15.14)$$

Clearly, the coordinates of A satisfy $l_1(x, y) = 0$ and $l_6(x, y) = 0$, so the point A is on the curve represented by the formula (15.14); by the same reason, B, D, C, E, and F are also on the curve represented by (15.14).

Denote the three intersection points of the three opposite sides by L, M, and N. The coordinates of the point L satisfy $l_1(x, y) = 0$ and $l_4(x, y) = 0$, so the point L is on the curve represented by the formula (15.14); by the same reason, M and N are also on the curve represented by (15.14).

Since L, M, and N are not on the curve represented by $F(x, y) = 0$, so the three points must be on the straight line

$$\mu_1 l_5(x, y) - \mu_2 l_2(x, y) = 0.$$

That is, L, M, and N are co-linear. □

Remark. The above conclusion is nothing but Pascal's theorem.

3. Exercises

Group A

I. Filling Problems

1. The equation of a straight line that passes through the intersection point of the straight lines $2x - 3y = 1$ and $3x + 2y = 2$ and is parallel to the straight line $y + 3x = 0$ is _____.

2. A circle is tangent to the straight line $x + 3y - 26 = 0$ at $A(8, 6)$ and passes through the point $B(-2, -4)$. Then the equation of this circle is

_____.

3. The equation of a centered conic with the foci $(-2, 0)$ and $(2, 0)$ that passes through the point $P\left(\frac{5}{2}, -\frac{3}{2}\right)$ is _____.

4. The long and short axes of an ellipse are parallel to the coordinates axes, respectively. Suppose that the ellipse is tangent to the straight line $2x + y = 11$ at the point $P(4, 3)$ and passes through the points $Q(0, -1)$ and $R(1, \sqrt{10} + 1)$. Then the equation of the ellipse is _____.

II. Calculation Problems

5. Given an ellipse $\frac{x^2}{a^2} + \frac{y^2}{b^2} = 1$ $(a > b > 0)$. Find the hyperbola with the same foci as the ellipse such that the area of the quadrilateral with their intersection points as its vertices is maximal. And find the coordinates of the vertices of the quadrilateral.

6. From the vertex A of $\triangle ABC$, draw the perpendicular line to BC, with D as the foot of the perpendicular. Choose any point H on AD. Suppose that the straight lines BH and AC intersect at E, and the straight lines CH and AB intersect at F, Prove: AD bisects the angle formed by ED and DF.

7. Pick any point M in a line segment AB, and construct two squares with AM and BM as one side on the same side of AB, respectively. Let the circumscribed circles of the two squares intersect at points M and N. Prove: No matter how M is chosen, MN always passes through a fixed point.

Group B

8. Let A and B be two points on an ellipse $3x^2 + y^2 = \lambda$, let $N(1, 3)$ be the middle point of the line segment AB, and let the perpendicular bisector of the line segment AB intersect the ellipse at C and D. Does

there exist a real number λ such that the four points A, B, C, and D are on the same circle? Give your reason.

9. Given a family of curves $2(2\sin\theta - \cos\theta + 3)x^2 - (8\sin\theta + \cos\theta + 1)y = 0$ with θ being a parameter, find the maximum value for the length of the chord obtained from the intersection of the curve family with the straight line $y = 2x$.

10. Assumption: Every two of three given circles intersect. Prove: The straight lines that contain a resulting chord each from the intersections intersect at one point, or are mutually parallel.

11. Draw a tangent line to a circle M from any outside point E with tangent point F, and draw a secant line EAB with intersection points A and B to the circle. Connect the middle point O of EF and B, intersecting the circle M at D. Also ED intersects the circle M at C. Prove: $AC \parallel EF$.

12. In a quadrilateral $ABCD$, the diagonal AC bisects $\angle BAD$. Pick a point E on CD, and BE and AC intersect at F. Extend DF to intersect BC at G. Prove: $\angle GAC = \angle EAC$.

Chapter 16

Derivatives

1. Key Points of Knowledge and Basic Methods

A. *Meaning of derivatives*

The derivative of a function $y = f(x)$ at $x = x_0$ is the instantaneous rate of change of the function $f(x)$ at the point $x = x_0$.

B. *Geometric meaning of derivatives*

The derivative of a function $y = f(x)$ at $x = x_0$ is the slope of the tangent line to the graph of the function $f(x)$ at the point $x = x_0$.

C. *Derivative formulas of basic elementary functions*

(1) $(c)' = 0$ (c is a constant);   (2) $(x^\alpha)' = \alpha x^{\alpha-1}$ (α is a constant);

(3) $(\sin x)' = \cos x$;   (4) $(\cos x)' = -\sin x$;

(5) $(a^x)' = a^x \ln a$;   (6) $(e^x)' = e^x$;

(7) $(\log_a x)' = \frac{1}{x \ln a}$;   (8) $(\ln x)' = \frac{1}{x}$.

D. *Operational rules of derivatives*

If $u(x)$ and $v(x)$ have derivatives at x, then

$$[u(x) \pm v(x)]' = u'(x) \pm v'(x);$$

$$[u(x)v(x)]' = u'(x)v(x) + u(x)v'(x);$$

$$\left[\frac{u(x)}{v(x)}\right]' = \frac{u'(x)v(x) - u(x)v'(x)}{v(x)^2} \quad (v(x) \neq 0).$$

In particular, if c is a constant, then $[cf(x)]' = cf'(x)$.

E. *Derivatives of composite functions*

We call the function $y = f(g(x))$ the composite function of functions $y = f(u)$ and $u = g(x)$. If the function $f(u)$ has the derivative $f'(u)$ at $u = g(x)$ and the function $u = g(x)$ has the derivative at x, then the composite function $y = f(g(x))$ has the derivative at x, and

$$y'_x = y'_u u'_x = f'(g(x))g'(x).$$

F. *Applications of derivatives*

(1) Monotonicity of functions

In some interval $D = (a, b)$ in the domain of a function $f(x)$, if $f'(x) > 0$, then the function $f(x)$ is monotonically increasing in D; if $f'(x) < 0$, then the function $f(x)$ is monotonically decreasing in D; if $f'(x) = 0$, then $f(x)$ is a constant function in D.

(2) Local extreme values of functions

Let $D = (a, b)$ be an interval in the domain of a function $f(x)$. Assume that $f'(x_0) = 0$ for some $a < x_0 < b$.

If $f'(x) > 0$ in the interval (a, x_0) and $f'(x) < 0$ in the interval (x_0, b), then $f(x_0)$ is a local maximum value of $f(x)$.

If $f'(x) < 0$ in the interval (a, x_0) and $f'(x) > 0$ in the interval (x_0, b), then $f(x_0)$ is a local minimum value of $f(x)$.

(3) Absolute maximum and minimum values of functions

Suppose that a function $f(x)$ is defined on $[a, b]$. Then among the local extreme values and the function values $f(a)$ and $f(b)$ at the end points of $[a, b]$, the maximal (minimal) number is the absolute maximum (minimum) value of $f(x)$ on $[a, b]$. The absolute maximum and minimum values of $f(x)$ on $[a, b]$ are also called the maximum and minimum values of $f(x)$ on $[a, b]$, respectively.

G. *Basic methods*

Monotonicity of functions, local extreme values, and the absolute maximum or minimum values are all related to the signs and zeros of the derivative of the function on some interval. Hence, the main work in using the derivative tools to determine the monotonicity of the function, and find the local extreme values and the absolute maximum (minimum) value is to find the zeros of the derivative and determine the signs of the derivative function on the subintervals partitioned via the zeros of the derivative.

Using the geometric meaning of the derivative, one can find the slope of the tangent line to a curve. Let $P(x_0, y_0)$ be a point on the curve $y = f(x)$. Then the equation of the tangent line to the curve $y = f(x)$ at the point P is

$$y - y_0 = f'(x_0)(x - x_0).$$

2. Illustrative Examples

Example 1. Suppose that a function $f(x) = \ln x - ax$ with $a > 0$ and let $g(x) = f(x) + f'(x)$.

(1) If the maximum value of the function $f(x)$ is -4 for $1 \leq x \leq e$, find the expression of the function $f(x)$.
(2) Find the range of the values for a so that the function $g(x)$ is a monotone function on the interval $(0, +\infty)$.

Solution. (1) Since $f'(x) = \frac{1}{x} - a$, so $f(x)$ is monotonically increasing on $\left(0, \frac{1}{a}\right)$ and $f(x)$ is monotonically decreasing on $\left(\frac{1}{a}, +\infty\right)$. Thus, $f(x)$ has the maximum value when $x = \frac{1}{a}$.

(i) When $1 < \frac{1}{a} < 1$, namely $a > 1$, we have $f(x)_{\max} = f(1) = -4$. The solution is $a = 4$, which satisfies the condition.
(ii) When $1 \leq \frac{1}{a} \leq e$, namely $\frac{1}{e} \leq a \leq 1$, we have $f(x)_{\max} = f\left(\frac{1}{a}\right) = -4$. The solution is $a = e^3 > 1$, which does not satisfy the condition and so is omitted.
(iii) When $\frac{1}{a} > e$, namely $0 < a < \frac{1}{e}$, we have $f(x)_{\max} = f(e) = -4$. The solution is $a = \frac{5}{e}$, which does not satisfy the condition and so is omitted.

In summary, $f(x) = \ln x - 4x$.

(2) Since $g(x) = \ln x - ax + \frac{1}{x} - a$,

$$g'(x) = \frac{1}{x} - a - \frac{1}{x^2} = -\left(\frac{1}{x} - \frac{2}{2}\right)^2 + \frac{1}{4} - a.$$

(i) When $a \geq \frac{1}{4}$, we have $g'(x) \leq 0$, and $g'(x) = 0$ only when $x = 2$. Hence, $g(x)$ is monotonically decreasing on $(0, +\infty)$.

(ii) When $0 < a < \frac{1}{4}$, we have $g'(x) < 0$ if $\frac{1}{x} \in \left(0, \frac{1}{2} - \sqrt{\frac{1}{4} - a}\right)$.

When $\frac{1}{x} \in \left(\frac{1}{2} - \sqrt{\frac{1}{4} - a}, \frac{1}{2} + \sqrt{\frac{1}{4} - a}\right)$, it holds that $g'(x) > 0$.

When $\frac{1}{x} \in \left(\frac{1}{2} + \sqrt{\frac{1}{4} - a}, +\infty\right)$, we have $g'(x) < 0$.

Hence, $g(x)$ is not monotonic on $(0, +\infty)$.

To summarize, the range of the values for a is $\left[\frac{1}{4}, +\infty\right)$.   □

Example 2. Suppose that the maximum value of the function $f_n(x) = x^n(1-x)^2$ on $\left[\frac{1}{2}, 1\right]$ is a_n $(n = 1, 2, \ldots)$. Find the general term formula for the sequence $\{a_n\}$.

Analysis. Obviously, we should find the maximum value of $f_n(x)$ on $\left[\frac{1}{2}, 1\right]$ respectively for different positive integers n. This requires that for different n, we need to find out $f_n\left(\frac{1}{2}\right)$, $f_n(1)$, and the local extreme values of $f_n(x)$ inside $\left(\frac{1}{2}, 1\right)$ respectively, and then make comparisons of them.

Solution. First we calculate out that $f_n'(x) = nx^{n-1}(1-x)^2 - 2x^n(1-x) = x^{n-1}(1-x)[n(1-x) - 2x]$.

When $x \in \left[\frac{1}{2}, 1\right]$, from $f_n'(x) = 0$ we know that $x = 1$ or $x = \frac{n}{n+2}$. Also $f_n(1) = 0$.

When $n = 1$, we find $\frac{n}{n+2} = \frac{1}{3} \notin \left[\frac{1}{2}, 1\right]$. Since $f_1\left(\frac{1}{2}\right) = \frac{1}{8}$, so $a_1 = \frac{1}{8}$.

When $n = 2$, we see that $\frac{n}{n+2} = \frac{1}{2} \in \left[\frac{1}{2}, 1\right]$. Since $f_2\left(\frac{1}{2}\right) = \frac{1}{16}$, so $a_2 = \frac{1}{16}$.

When $n \geq 3$, clearly $\frac{n}{n+2} \in \left[\frac{1}{2}, 1\right]$.

Since $f_n'(x) > 0$ for $x \in \left[\frac{1}{2}, \frac{n}{n+2}\right)$ and $f_n'(x) < 0$ for $x \in \left(\frac{n}{n+2}, 1\right)$, hence $f_n(x)$ achieves the maximum value at $x = \frac{n}{n+2}$. In other words,

$$a_n = \left(\frac{n}{n+2}\right)^n \left(\frac{2}{n+2}\right)^2 = \frac{4n^2}{(n+2)^{n+2}}.$$

In summary,

$$a_n = \begin{cases} \dfrac{1}{8}, & n = 1, \\[2mm] \dfrac{4n^2}{(n+2)^{n+2}}, & n \geq 2. \end{cases}$$ □

In the following, we look at several problems that are solved by means of derivatives.

Example 3. Suppose that two real numbers x and y satisfy $2^x + 3^y = 4^x + 9^y$. Find the range of the values of $U = 8^x + 27^y$.

Solution. Let $a = 2^x$ and $b = 3^y$. Then the known condition is reduced to

$$a + b = a^2 + b^2, \text{ and } a, b > 0.$$

In other words,

$$\left(a - \frac{1}{2}\right)^2 + \left(b - \frac{1}{2}\right)^2 = \frac{1}{2}.$$

From the graph in an aOb plane, $t = a + b \in (1, 2]$. Also

$$ab = \frac{(a + b)^2 - (a^2 + b^2)}{2} = \frac{t^2 - t}{2}.$$

So, it follows that

$$\begin{aligned}
U = 8^x + 27^y &= a^3 + b^3 \\
&= (a + b)^3 - 3ab(a + b) \\
&= t^3 - 3\frac{t^2 - t}{2}t \\
&= -\frac{1}{2}t^3 + \frac{3}{2}t^2.
\end{aligned}$$

Denote $f(t) = -\frac{1}{2}t^3 + \frac{3}{2}t^2$. Then for $t \in (1, 2]$,

$$f'(t) = -\frac{3}{2}t^2 + 3t = -\frac{3}{2}t(t - 2) \geq 0.$$

Thus, $f(t)$ is monotonically increasing on $(1, 2]$. It is easy to see that $f(t) \in (1, 2]$ when $t \in (1, 2]$.

To summarize, the range of the values of $U = 8^x + 27^y$ is $(1, 2]$. □

Example 4. We know that π is an irrational number. The question is whether

$$\sin \pi x + \sin x \ (x \in \mathbf{R})$$

is a periodic function.

Solution. Suppose that $f(x) = \sin \pi x + \sin x$ is a periodic function. Assume that $T > 0$ is a period of $f(x)$.

Taking derivative to the both sides of $f(x + T) = f(t)$ gives

$$f'(x + T) = \pi \cos(\pi x + \pi T) + \cos(\pi + T) = \pi \cos \pi x + \cos x = f'(x).$$

$$(16.1)$$

Let $x = 0$ in (16.1). We have

$$\pi \cos \pi T + \cos T = f'(T) = f'(0) = \pi + 1,$$

and thus

$$\pi T = 2m\pi, \quad T = 2n\pi \ (m, n \in \mathbf{Z}_+),$$

namely

$$\pi = \frac{m}{n}.$$

This contradicts the fact that π is an irrational number.

Hence, $f(x)$ is not a periodic function. □

Example 5. Prove: (1) The given equation $x^3 - x - 1 = 0$ has exactly one real root r, and r is an irrational number. (2) The number r is not a root of any quadratic equation $ax^2 + bx + c = 0 \ (a \neq 0)$ with integral coefficients.

Proof. (1) Let $f(x) = x^3 - x - 1$. Then $f'(x) = 3x^2 - 1$. The function $f(x)$ is monotonically increasing on $\left(-\infty, -\frac{1}{\sqrt{3}}\right)$, monotonically decreasing on $\left(-\frac{1}{\sqrt{3}}, \frac{1}{\sqrt{3}}\right)$, and monotonically increasing on $\left(\frac{1}{\sqrt{3}}, +\infty\right)$, so $f(x)$ obtains a local maximum value $\frac{2}{3\sqrt{3}} - 1 \ (< 0)$ at $x = -\frac{1}{\sqrt{3}}$ and a local minimum value $-\frac{2}{3\sqrt{3}} - 1 \ (< 0)$ at $x = \frac{1}{\sqrt{3}}$. Also from $f(1) = -1 < 0$ and $f(2) = 5 > 0$, we see that the equation $f(x) = 0$ has a unique real root $r \in (1, 2)$. It is easy to see that r is also a unique root of $f(x) = 0$.

Suppose that $r = \frac{m}{n}$, where m and n are positive integers that are relatively prime. Then $m^3 = n^2(m + n)$, so from $n^2 \mid m^3$ we know that $n = 1$. In other words r is an integer, which is a contradiction to $r \in (1, 2)$. Hence, r is an irrational number.

(2) Suppose that r satisfies $ax^2 + bx + c = 0 \ (a, b, c \in \mathbf{Z}$ and $a \neq 0)$. Then

$$\begin{cases} ar^2 + br + c = 0, \\ r^3 - r - 1 = 0. \end{cases} \tag{16.2}$$

Multiplying the first equality by r and then subtracting the second one multiplied by a in (16.2), we obtain

$$br^2 + (a + c)r + a = 0. \tag{16.3}$$

Multiplying (16.3) by a and then subtracting the first equality of (16.2) multiplied by b, we arrive at

$$(a^2 + ac - b^2)r + (a^2 - bc) = 0.$$

Since r is an irrational number, it follows that

$$\begin{cases} a^2 + ac - b^2 = 0, \\ a^2 - bc = 0. \end{cases}$$

The assumption $a \neq 0$ implies $bc \neq 0$. Substituting $b = \frac{a^2}{c}$ into $a^2 + ac - b^2 = 0$ gives $\frac{a^3}{c^3} = \frac{a}{c} + 1$. Thus, $r = \frac{a}{c}$, which contradicts the fact that r is an irrational number. $\qquad\square$

Example 6. For two different points $A(x_1, y_1)$ and $B(x_2, y_2)$ on the graph of a function, if there exists a point $M(x_0, y_0)$, where $x_0 \in (x_1, x_2)$, such that the tangent line l at the point M is parallel to AB, then we say that AB has a "dependent tangent." Particularly, when $x_0 = \frac{x_1 + x_2}{2}$, we say that AB has a "mean value dependent tangent."

Let a function $f(x) = \ln x - \frac{1}{2}ax^2 + bx$ $(a > 0)$ be given with the condition that $f'(1) = 0$.

(1) Express b as a function of a, and find the intervals of monotonicity of $f(x)$.
(2) We ask whether there exist two points A and B on the graph of $f(x)$ such that AB has a "mean value dependent tangent." If they exist, find the coordinates of A and B; if they do not exist, give the reason.

Analysis. By the definition of a "dependent tangent," we should start from the equality relation between the slopes of AB and the tangent line to the curve, set up the equation and determine whether it has a solution, and transform the problem to that of finding a zero of a function.

Solution. (1) Since $f'(x) = \frac{1}{x} - ax + b \, (x > 0)$, we have $f'(1) = 1 - a + b = 0$, so $b = a - 1$. Thus,

$$f'(x) = \frac{-(ax + 1)(x - 1)}{x} \quad (x > 0).$$

Also, since $a > 0$ and $x > 0$, we see that $ax + 1 > 0$.

Consequently, $f'(x) > 0$ when $x \in (0,1)$ and $f'(x) < 0$ when $x \in (1,+\infty)$.

Hence, the interval of increase for $f(x)$ is $(0,1)$ and the interval of decrease for $f(x)$ is $(1,+\infty)$.

(2) Let $A(x_1, y_1)$ and $B(x_2, y_2)$ be two points with $0 < x_1 < x_2$ on the graph of $f(x)$. Then

$$y_1 = \ln x_1 - \frac{1}{2}ax_1^2 + (a-1)x_1, \quad y_2 = \ln x_2 - \frac{1}{2}ax_2^2 + (a-1)x_2,$$

$$k_{AB} = \frac{y_1 - y_2}{x_1 - x_2} = \frac{\ln x_1 - \ln x_2}{x_1 - x_2} - \frac{1}{2}a(x_1 + x_2) + (a-1).$$

The slope of the tangent line to the function graph at $x_0 = \frac{x_1 + x_2}{2}$ is

$$k = f'(x_0) = f'\left(\frac{x_1 + x_2}{2}\right) = \frac{2}{x_1 + x_2} - \frac{1}{2}a(x_1 + x_2) + (a-1).$$

The condition $k_{AB} = k$ implies that $\frac{\ln x_1 - \ln x_2}{x_1 - x_2} = \frac{2}{x_1 + x_2}$, namely

$$\ln \frac{x_1}{x_2} = \frac{2(x_1 - x_2)}{x_1 + x_2} = \frac{2\left(\frac{x_1}{x_2} - 1\right)}{\frac{x_1}{x_2} + 1}.$$

Denote $\frac{x_1}{x_2} = t$. Then $0 < t < 1$, and the above equality can be reduced to $\ln t = \frac{2(t-1)}{t+1} = 2 - \frac{4}{t+1}$, that is,

$$\ln t + \frac{4}{t+1} = 2.$$

Define $g(t) = \ln t + \frac{4}{t+1}$ $(0 < t \le 1)$. Then $g'(t) = \frac{1}{t} - \frac{4}{(t+1)^2} = \frac{(t-1)^2}{t(t+1)^2}$.

Hence, $g'(t) > 0$ when $t \in (0,1)$, so $g(t)$ is monotonically increasing, and it follows that

$$g(t) < g(1) = 2 \ (0 < t < 1).$$

This means that the equation $\ln t + \frac{4}{t+1} = 2$ has no solution in $(0,1)$.

In summary, there do not exist two points A and B on the graph of the function $f(x)$ such that AB has a "mean value dependent tangent." $\quad \square$

Remark. This problem gave a new concept, but after a transformation it was reduced to a familiar problem of solving equation zeros. The technique of introducing the parameter $t = \frac{x_1}{x_2}$ was used cleverly and flexibly in the solution process. That changed an unfamiliar situation to a familiar one, and made a complicated case to a simple one.

Example 7. Suppose that a function $f(x) = ax^2 + bx + c$ ($a, b, c \in \mathbf{R}$ and $a \neq 0$) satisfies the inequality $|f(x)| \leq 1$ for all $|x| \leq 1$. Let the maximum value of $|f'(x)|$ with $|x| \leq 1$ be K. Try to find all functions $f(x)$ satisfying $|f'(x_0)| = K$ for some $x_0 \in [-1, 1]$.

Solution. First, $|f(0)| = |c| \leq 1$. Since one of b and $-b$ has the same sign as $a + c$, so from

$$|f(1)| = |a + b + c| \leq 1,$$
$$|f(-1)| = |a - b + c| \leq 1,$$

we know that $|a + c| + |b| \leq 1$, namely $|a + c| \leq 1 - |b|$. Thus,

$$|a| - |c| \leq |a + c| \leq 1 - |b|,$$

and consequently $|a| \leq 1 + |c| - |b| \leq 2 - |b|$.

When $|x| \leq 1$, we have

$$|f'(x)| = |2ax + b| \leq 2|a| + |b| \leq 4 - |b| \leq 4.$$

In fact, with the choice $f(x) = 2x^2 - 1$, it holds that $|f'(1)| = 4$, so $K = 4$.

For $f(x)$ with an $x_0 \in [-1, 1]$ such that $|f'(x_0)| = 4$, from $4 = |f'(x_0)| \leq 4 - |b|$ we see that $|b| = 0$ and $|a| = 2$.

When $a = 2$, the function is $f(x) = 2x^2 + c$. From $|f(1)| = |2 + c| \leq 1$ we have $-3 \leq c \leq 3$. Also $c = -1$ since $|c| \leq 1$, hence $f(x) = 2x^2 - 1$.

When $a = -2$, the same argument gives that $f(x) = -2x^2 + 1$ satisfies the requirement.

To summarize, the functions that satisfy the given condition are $f(x) = \pm(2x^2 - 1)$. $\qquad\square$

Example 8. We are given a function $y = f(x)$ and three points A, B, and C on a straight line l. Suppose that

$$\overrightarrow{OA} + (y - ax^2e^x - 1)\overrightarrow{OB} - [(x - 1)e^x + f'(0)]\overrightarrow{OC} = \overrightarrow{0},$$

where e is the base of the natural logarithm, $O \notin l$, and $a \in \mathbf{R}$.

(1) If $a < 0$, find the function $f(x)$ and its intervals of monotonicity.
(2) If $a = -1$ and the graphs of the function $f(x)$ and a function $g(x) = \frac{1}{3}x^3 + \frac{1}{2}x^2 + m$ have three different intersection points, find the range of the values for m.

Solution. Since the three points A, B, and C are co-linear,

$$\overrightarrow{OA} = -(y - ax^2 e^x - 1)\overrightarrow{OB} + [(x-1)e^x + f'(0)]\overrightarrow{OC},$$

in which

$$-(y - ax^2 e^x - 1) + [(x-1)e^x + f'(0)] = 1.$$

It follows that

$$y = ax^2 e^x + (x-1)e^x + f'(0),$$

$$y' = [ax^2 + (2a+1)x]e^x,$$

and so $f'(0) = 0$, in other words $f(x) = (ax^2 + x - 1)e^x$.

(1) Now, $f'(x) = (2ax+1)e^x + (ax^2 + x - 1)e^x = x(ax + 2a + 1)e^x$.

(i) If $-\frac{1}{2} < a < 0$, then $f'(x) < 0$ when $x < 0$ or $x > -\frac{2a+1}{a}$, and $f'(x) > 0$ when $0 < x < -\frac{2a+1}{a}$.

   Thus, for $f(x)$, the intervals of decrease are $(-\infty, 0]$ and $[-\frac{2a+1}{a}, = \infty)$ and the interval of increase is $[0, -\frac{2a+1}{a}]$.

(ii) If $a = -\frac{1}{2}$, then the interval of decrease for $f(x)$ is $(-\infty, +\infty)$ since $f'(x) = -\frac{1}{2}x^2 e^x \le 0$.

(iii) If $a < -\frac{1}{2}$, then $f'(x) < 0$ when $x < -\frac{2a+1}{a}$ or $x > 0$, and $f'(x) > 0$ when $-\frac{2a+1}{a} < x < 0$.

Hence, for $f(x)$, the intervals of decrease are $\left(-\infty, -\frac{2a+1}{a}\right]$ and $[0, +\infty)$ and the interval of increase is $\left[-\frac{2a+1}{a}, 0\right]$.

(2) The condition that the graphs of the function $f(x)$ and the function $g(x) = \frac{1}{3}x^3 + \frac{1}{2}x^2 + m$ have three different intersection points is equivalent to the property that the function $h(x) = f(x) - g(x)$ has three different zeros.

   When $a = -1$, From (1)(iii), $f(x) = (-x^3 + x - 1)e^x$ is monotonically decreasing on $(-\infty, -1]$, monotonically increasing on $[-1, 0]$, and monotonically decreasing on $[0, +\infty)$. From $g(x) = \frac{1}{3}x^3 + \frac{1}{2}x^2 + m$, we have $g'(x) = x^2 + x$.

   Clearly, $g'(x) > 0$ whenever $x < -1$ or $x > 0$, and $g'(x) < 0$ whenever $-1 < x < 0$.

   Therefore, $g(x)$ is monotonically increasing on $(-\infty, -1]$, monotonically decreasing on $[-1, 0]$, and monotonically increasing on $[0, +\infty)$.

Hence, $h(x) = f(x) - g(x)$ is monotonically decreasing on $(-\infty, -1]$, monotonically increasing on $[-1, 0]$, and monotonically decreasing on $[0, +\infty)$. So, $h(x)$ achieves a local minimum value

$$h(-1) = -\frac{3}{e} - \frac{1}{6} - m$$

at $x = -1$, and achieves a local maximum value $h(0) = -1 - m$ at $x = 0$.

Since $h(x) \to +\infty$ as $x \to -\infty$ and $h(x) \to -\infty$ as $x \to +\infty$, we see that $h(x)$ has three different zeros if and only if

$$\begin{cases} h(-1) = -\dfrac{3}{e} - \dfrac{1}{6} - m < 0, \\ h(0) = -1 - m > 0. \end{cases}$$

Hence, $-\frac{3}{e} - \frac{1}{6} < m < -1$, which is the range of the values for m. $\qquad\square$

3. Exercises

Group A

I. Filling Problems

1. The maximum value of function $y = x(1 + \sqrt{1 - x^2})$ is _____.
2. Let a function be $f(x) = 2x - \cos x$, $\{a_n\}$ an arithmetic sequence with common difference $\frac{\pi}{8}$, and

$$f(a_1) + f(a_2) + \cdots + f(a_5) = 5\pi.$$

   Then the value of $f(a_3)^2 - a_1 a_5$ equals _____.
3. Suppose that a point P is on the curve $y = e^x$ and a point Q is on the curve $y = \ln x$. Then the minimum value of $|PQ|$ is _____.
4. A function $f(x)$ defined on $\mathbf{R}$ satisfies $f(1) = 1$ and $f'(x) < \frac{1}{2}$ for any $x \in \mathbf{R}$. Then the solution set of the inequality $f(\log_2 x) > \frac{\log_2 x + 1}{2}$ is _____.

II. Calculation Problems

5. Let a function $f(x) = 1 - e^{-x}$.

   (1) Prove: When $x > 0$, it holds that $f(x) > \frac{x}{x+1}$.
   (2) Let a sequence $\{a_n\}$ satisfy $a_1 = \frac{1}{3}$ and $a_n e^{-a_{n+1}} = f(a_n)$ for $n \geq 1$. Prove: The sequence $\{a_n\}$ is monotonically decreasing, and $a_n < \frac{1}{2^n}$ for any n.

6. Given an integer $n > 1$. Prove: There exists a positive integer m greater than n^n such that $\frac{n^m - m^n}{n + m}$ is a positive integer.

7. We are given a function $f(x) = \ln(1 + x) + 1 - 2x + \frac{1}{2}ax^2$ $(a \geq 1)$.

    (1) Let the tangent line at the point $P(0, 1)$ to the curve $C : y = f(x)$ be l. If the straight line l and the curve C have only one common point, find all the possible values of a.

    (2) Prove: There is an interval $[x_1, x_2]$ $(x_1 < x_2)$ on which the function is monotonically decreasing. Also find the value range for $x_2 - x_1$.

8. Let $b > 0$ and be given the equation $\frac{x^2}{2b^2} + \frac{y^2}{b^2} = 1$ of an ellipse and the equation $x^2 = 8(y - b)$ of a parabola, as shown by Figure 16.1. Draw a parallel line to the x-axis at $F(0, b + 2)$, whose intersection point with the parabola in the first quadrant is G. Suppose that the tangent line to the parabola at G passes through the right focus F_1 of the ellipse.

    (1) Find the equations of the ellipse and parabola satisfying the conditions.

    (2) Let A and B be the end points of the long axis of the ellipse, respectively. Investigate whether there exists a point P on the parabola such that $\triangle ABC$ is a right triangle. If the answer is yes, indicate how many of such points exist (you do not need to find the coordinates of these points).

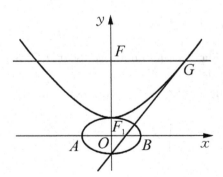

Figure 16.1   Figure for Exercise 8.

Group B

9. Give a function

$$f(x) = \ln x + a\left(\frac{1}{x} - 1\right), \ a \in \mathbf{R},$$

and assume that the minimum value of $f(x)$ is 0.

(1) Find the value of a.
(2) Let a sequence $\{a_n\}$ satisfy $a_1 = 1$ and $a_{n+1} = f(a_n) + 2$ $(n \in \mathbf{N}_+)$. Denote

$$S_n = [a_1] + [a_2] + \cdots + [a_n],$$

where $[m]$ represents the maximal integer that does not exceed the real number m. Find S_n.

10. Suppose that a function $f(x) = [\ln(1+x)]^2 - \frac{x^2}{1+x}$.

(1) Find the intervals of monotonicity for the function $f(x)$.
(2) If an inequality $\left(1 + \frac{1}{n}\right)^{n+a} \le e$ is satisfied for all $n \in \mathbf{N}_+$, where e is the base of the natural logarithm, find the maximum value of a.

11. We have a function $f(x) = \ln(x+1) + \frac{2}{x+1} + ax - 2$, where $a > 0$.

(1) When $a = 1$, find the minimum value of $f(x)$.
(2) If $f(x) \ge 0$ for $x \in [0, 2]$, find the range of the values for the number a.

12. Let $f(x) = \frac{1+a^x}{1-a^x}$ ($a > 0$ and $a \ne 1$), and let $g(x)$ be the inverse function of $f(x)$.

(1) If an equation $\log_a \frac{t}{(x^2-1)(7-x)} = g(x)$ about x has a real number solution on the interval $[2, 6]$, find the value range for t.
(2) When $a = e$ (e is the base of the natural logarithm), prove:

$$\sum_{k=2}^{n} g(k) > \frac{2 - n - n^2}{\sqrt{2n(n+1)}}.$$

(3) When $0 < a \le \frac{1}{2}$, compare the magnitudes of $\left|\sum_{k=1}^{n} f(k) - n\right|$ and 4, and express your argument.

Chapter 17

Mathematical Induction (I)

1. Key Points of Knowledge and Basic Methods

Usually those propositions that are related to positive integers directly or indirectly can be considered to be proved by mathematical induction.

A. *Fundamental forms of mathematical induction*

(1) First form of mathematical induction: Let $P(n)$ be a proposition related to all positive integers n. Suppose that

    $1°$ $P(1)$ is true (foundation);
    $2°$ if $P(k)$ is true, then it is deduced that $P(k + 1)$ is also true (induction).

    Then $P(n)$ is true for all positive integers n.
      If $P(n)$ is defined on the set $\mathbf{N}\backslash\{0, 1, \ldots, r - 1\}$, then "$P(1)$ is true" in $1°$ should be replaced by "$P(r)$ is true."

(2) Jump form of mathematical induction: Let $P(n)$ be a proposition related to all positive integers n. Suppose that

    $1°$ $P(1), P(2), \ldots, P(l)$ are true;
    $2°$ if $P(k)$ is true, then it is deduced that $P(k + l)$ is also true.

    Then $P(n)$ is true for all positive integers n.

(3) Second form of mathematical induction: Let $P(n)$ be a proposition related to all positive integers n. Suppose that

1° $P(1)$ is true;
2° if $P(n)$ $(1 \leq n \leq k)$ is true for any positive integer k, then it is deduced that $P(k+1)$ is also true.

Then $P(n)$ is true for all positive integers n.

Each of the above forms of mathematical induction consists of two steps: "foundation" and "induction," and neither can be omitted. In the process of "induction," the indispensable premise of "inductive hypothesis" must be used.

B. *Proof techniques of mathematical induction*

(1) "Moving starting point forward" or "moving starting point backward": In some propositions $P(n)$ of all positive integers, it is relatively difficult to verify $P(1)$, or $P(1), P(2), \ldots, P(r-1)$ cannot be unified to the process of induction. At this time one may consider moving the starting point forward to $P(0)$ (if it is meaningful) or moving the starting point backward to $P(r)$ (and then proving $P(1), P(2), \ldots, P(r-1)$ separately).

(2) Increase the "span": For propositions $P(n)$ defined on $\mathbf{M} = \{n_0, n_0 + r, n_0 + 2r, \ldots, n_0 + mr, \ldots\}$ $(n_0, r, m \in \mathbf{N})$, one should consider the method of increasing the "span" when using mathematical induction, that is, verify $P(n_0)$ in the first step, and prove that $P(k+r)$ is true under the assumption that $P(k)$ $(k \in \mathbf{M})$ is true in the second step.

(3) Strengthen the proposition: Some propositions that are not easy to prove by mathematical induction can be more conveniently proved by using mathematical induction via strengthening the proposition. There are usually two methods to strengthen the proposition: One is to make the proposition general and the other is to strengthen the conclusion. As for to what appropriate extent the conclusion of the proposition should be "strengthened," it can be determined only by catching the features of the proposition, making a careful exploration, and giving a bold conjecture. In this way one can find a proper scheme to solve the problem.

2. Illustrative Examples

Example 1. Try to prove that any postage of n ($n > 7$ with $n \in \mathbf{N}$) cents can be paid with stamps of 3 cents and 5 cents.

Proof. 1° When $n = 8$, the conclusion is obviously true.

2° Suppose that the proposition is true for $n = k$ ($k > 7$ with $n \in \mathbf{N}$).

If the k-cent postage is paid with all stamps of 3 cents, then there are at least 3 of them. Then by replacing three 3 cents stamps by 2 stamps of 5 cents, the postage of $k + 1$ cents can be paid. If there is at least one stamp of 5 cents, then it is enough to replace the 5 cents stamp with 2 stamps of 3 cents, so that the postage of $k + 1$ cents can also be paid. Thus, the proposition is also true when $n = k + 1$.

In summary, the proposition is true for all natural numbers $n > 7$. □

Remark. The key of the above proof is how to move to $P(k + 1)$ from the inductive hypothesis. Here, we adopted the method of classification for the discussion. We can also use the jump form of mathematical induction to prove the proposition.

Proof. 1° When $n = 8, 9$, and 10, the proposition is true from $8 = 3 + 5$, $9 = 3 + 3 + 3$, and $10 = 5 + 5$.

2° Suppose that the proposition is true for $n = k$ ($k > 7$ with $k \in \mathbf{N}$). Then for $n = k + 3$, the proposition is also true from 1° and the inductive hypothesis. This completes the proof. □

Remark. The above proof actually used the technique of increasing the "span"; the "span" is 3.

Example 2. Label the prime numbers with ordered numbers from the smallest to the largest: The first prime number is represented by 2, the second prime number by 3, and so on. Prove: The nth prime number $P_n < 2^{2^n}$.

Proof. 1° When $n = 1$, we have $P_1 = 2 < 2^{2^1}$, so the conclusion is valid.

$2°$ Suppose that the proposition is true for $n \leq k$, that is

$$P_i < 2^{2^i} \ (i = 1, 2, \ldots, k).$$

Multiplying the both sides of the k inequalities gives

$$P_1 P_2 \cdots P_k < 2^{2^1 + 2^2 + \cdots + 2^k} = 2^{2^{k+1} - 2}.$$

So,

$$P_1 P_2 \cdots P_k + 1 \leq 2^{2^{k+1} - 2} < 2^{2^{k+1}}.$$

Since $P_1, P_2, \ldots, P_k$ cannot divide $P_1 P_2 \cdots P_k + 1$, it follows that any prime factor q of $P_1 P_2 \cdots P_k + 1$ cannot be any of $P_1, P_2, \ldots, P_k$, and thus must be greater than or equal to P_{k+1}. Hence,

$$P_{k+1} \leq q \leq P_1 P_2 \cdots P_k + 1 < 2^{2^{k+1}},$$

in other words

$$P_{k+1} < 2^{2^{k+1}}.$$

To summarize, $P_n < 2^{2^n}$ for any natural number n. $\qquad\square$

Remark. Here, we used the second form of mathematical induction: The premise for $P_{k+1} < 2^{2^{k+1}}$ to be true is that $P_1 < 2^{2^1}, P_2 < 2^{2^2}, \ldots, P_k < 2^{2^k}$ are all true.

Example 3. Let a_i and b_i $(i = 1, 2, \ldots, n)$ be all positive numbers, and let $A = \sum_{i=1}^{n} a_i$ and $B = \sum_{i=1}^{n} b_i$. Prove:

$$\sum_{i=1}^{n} \frac{a_i b_i}{a_i + b_i} \leq \frac{AB}{A + B}.$$

Proof. When $n = 1$ and 2, it is easy to verify the conclusion.

Assume that the inequality is true for $n = k$. Then

$$\sum_{i=1}^{k+1} \frac{a_i b_i}{a_i + b_i} = \sum_{i=1}^{k} \frac{a_i b_i}{a_i + b_i} + \frac{a_{k+1} b_{k+1}}{a_{k+1} + b_{k+1}}$$

$$\leq \frac{A' B'}{A' + B'} + \frac{a_{k+1} b_{k+1}}{a_{k+1} + b_{k+1}}$$

$$\leq \frac{(A' + a_{k+1})(B' + b_{k+1})}{(A' + a_{k+1}) + (B' + b_{k+1})} = \frac{AB}{A + B},$$

where $A' = \sum_{i=1}^{k} a_i, A = A' + a_{k+1}, B' = \sum_{i=1}^{k} b_i$, and $B = B' + b_{k+1}$. This means that the inequality is also valid for $n = k + 1$.

By mathematical induction, the inequality is true for $n \in \mathbf{N}$. $\qquad\square$

Remark. In the above, we verified the inequality for the first two natural numbers. This is because that in the process of proving $P(k) \Rightarrow P(k+1)$, besides using the truth of $P(1)$, we also used the truth of $P(2)$.

Example 4. Suppose that a sequence $a_1, a_2, \ldots$ of positive real numbers satisfies that for each positive integer k,

$$a_{k+1} \geq \frac{ka_k}{a_k^2 + k - 1}. \tag{17.1}$$

Prove: For each positive integer $n \geq 2$,

$$a_1 + a_2 + \cdots + a_n \geq n.$$

Proof. From (17.1) we have

$$\frac{k}{a_{k+1}} \leq \frac{a_k^2 + k - 1}{a_k} = a_k + \frac{k-1}{a_k},$$

from which $a_k \geq \frac{k}{a_{k+1}} - \frac{k-1}{a_k}$. Thus,

$$\sum_{k=1}^{m} a_k \geq \sum_{k=1}^{m} \left(\frac{k}{a_{k+1}} - \frac{k-1}{a_k} \right) = \frac{m}{a_{m+1}}. \tag{17.2}$$

In the following, for $n \geq 2$ we use mathematical induction to prove

$$a_1 + a_2 + \cdots + a_n \geq n.$$

When $n = 2$, letting $k = 1$ in the inequality (17.1) gives $a_2 \geq \frac{1}{a_1}$, so $a_1 + a_2 \geq a_1 + \frac{1}{a_1} \geq 2$.

Assume that the conclusion is true for a positive integer $n \geq 2$.

For the positive integer $n+1$, if $a_{n+1} \geq 1$, then the inductive hypothesis implies that

$$a_1 + a_2 + \cdots + a_n + a_{n+1} \geq n + 1.$$

If $a_{n+1} < 1$, by the inequality (17.2) we obtain that

$$a_1 + a_2 + \cdots + a_n + a_{n+1}$$

$$\geq \frac{n}{a_{n+1}} + a_{n+1} = \frac{n-1}{a_{n+1}} + \frac{1}{a_{n+1}} + a_{n+1}$$

$$> n - 1 + 2 = n + 1.$$

That is, the conclusion is valid for $n + 1$.

To summarize, the conclusion is true for $n \geq 2$. $\qquad\square$

Example 5. Suppose that a sequence $\{a_n\}$ satisfies

$$a_1 = 2, \quad a_2 = 7, \quad a_n = 3a_{n-1} + 2a_{n-2} \quad (n \geq 3).$$

Prove: For any positive integer n, one can express a_{2n-1} as the square sum of two positive integers.

Proof. Write down the first odd numbered terms of the sequence:

$$a_1 = 2 = 1^2 + 1^2, \quad a_3 = 25 = 3^2 + 4^2,$$

$$a_5 = 317 = 11^2 + 14^2, \quad a_7 = 4021 = 39^2 + 50^2.$$

Conjecture: For the subsequence $a_1, a_3, \ldots, a_{2n-1}, a_{2n+1}, \ldots$ of the odd numbered terms of the sequence $\{a_n\}$, any two consecutive terms can be expressed as

$$a_{2n-1} = \alpha^2 + \beta^2,$$

$$a_{2n+1} = (\alpha + 2\beta)^2 + (2\alpha + 2\beta)^2.$$

We prove the above conjecture by using mathematical induction as follows:

From $a_1 = 2 = 1^2 + 1^2, a_3 = 25 = (1 + 2 \cdot 1)^2 + (2 \cdot 1 + 2 \cdot 1)^2$, and $a_5 = 317 = (3 + 2 \cdot 4)^2 + (2 \cdot 3 + 2 \cdot 4)^2$, we know that the proposition is true for $n = 1, 2$, and 3.

Assume that the conjecture is valid for $n \leq k$. Denote $\alpha + 2\beta = \alpha'$ and $2\alpha + 2\beta = \beta'$. By the recursive relation and the inductive hypothesis, we have that

$$a_{2k+3} = 3a_{2k+2} + 2a_{2k+1}$$

$$= 13a_{2k+1} - 4a_{2k-1}$$

$$= 13[(\alpha + 2\beta)^2 + (2\alpha + 2\beta)^2] - 4(\alpha^2 + \beta^2)$$

$$= [\alpha + 2\beta + 2(2\alpha + 2\beta)]^2 + [2(\alpha + 2\beta) + 2(2\alpha + 2\beta)]^2$$

$$= (\alpha' + 2\beta')^2 + (2\alpha' + 2\beta')^2,$$

in other words, the conjecture is true when $n = k + 1$.

In summary, a_{2n-1} can be expressed as the sum of the squares of two positive integers for $n \in \mathbf{N}_+$. □

Remark. Here, we used the method of "conjecture first-proof second": Find a general rule through an observation to several special cases, and then give a rigorous proof.

Example 6. Prove: There exists an infinite sequence $\{a_n\}$ of positive integers $a_1 < a_2 < a_3 < \cdots$ such that $a_1^2 + a_2^2 + \cdots + a_n^2$ are perfect squares for all natural numbers n.

Analysis. We suppose to use mathematical induction to prove it. The key point is to show that $P(k) \Rightarrow P(k+1)$. Assume that $a_1^2 + a_2^2 + \cdots + a_k^2 = x^2$ and $a_{k+1}^2 = y^2$. Then $a_1^2 + a_2^2 + \cdots + a_{k+1}^2 = x^2 + y^2$. If the conclusion is true, then there exists z such that $x^2 + y^2 = z^2$. By the Pythagorean theorem, it is not difficult to think it out that $a_1^2 + a_2^2 + \cdots + a_n^2$ being "perfect squares" should be strengthened to "squares of odd numbers."

Proof. We strengthen the conclusion by proving that there exists an infinite sequence $\{a_n\}$ of positive integers $a_1 < a_2 < a_3 < \cdots$ such that $a_1^2 + a_2^2 + \cdots + a_n^2$ are squares of odd numbers for all natural numbers n.

When $n = 1$, it is enough to choose $a_1 = 5$.

Suppose that the conclusion is true for $n = k$, that is, there exist k positive numbers $a_1 < a_2 < \cdots < a_k$ such that $a_1^2 + a_2^2 + \cdots + a_k^2 = (2m+1)^2$ for some $m \in \mathbf{N}$.

Choose $a_{k+1} = 2m^2 + 2m$. Then

$$a_1^2 + a_2^2 + \cdots + a_k^2 + a_{k+1}^2 = (2m+1)^2 + (2m^2 + 2m)^2$$
$$= (2m^2 + 2m + 1)^2,$$

which is the square of an odd number. Also, since $a_1 \geq 5$ and $a_k > a_1$, so

$$2a_{k+1} = a_1^2 + a_2^2 + \cdots + a_k^2 - 1 \geq a_k^2 - 1 > 2a_k,$$

namely $a_{k+1} > a_k$. Thus, the conclusion is also true for $n = k + 1$. Hence, the conclusion is valid for all natural numbers n. □

Remark. The technique of strengthened proposition was adopted for the current example. Strengthening a proposition can get a stronger inductive hypothesis, and sometimes can make it easier to go though the path from $P(k)$ to $P(k+1)$.

Example 7. Let $\{a_k\}$ be an infinite sequence of real numbers. Define

$$a'_k = \frac{a_k + a_{k+1}}{2} \ (k = 1, 2, \ldots).$$

We call $\{a'_k\}$ the mean value sequence of $\{a_k\}$. By the same way, one can define $\{a''_k\}$ as the mean value sequence of $\{a'_k\}$, which is called the second grade mean value sequence of $\{a_k\}$, and so on. If $\{a_k\}$ and the corresponding ascending grades mean value sequences $\{a'_k\}, \{a''_k\}, \ldots$ are all sequences of integers, then we say that $\{a_k\}$ is a "good" sequence. Prove: If $\{a_k\}$ is a good sequence, then $\{a_k^2\}$ is also a good sequence.

Proof. If $\{a_k\}$ and its first, second, $\ldots$, and up to mth grade mean value sequences are all sequences of integers, we call $\{a_k\}$ an "mth grade good sequence." We prove that if $\{a_k\}$ is a good sequence, then $\{a_k^2\}$ is always an mth grade good sequence for any nonnegative integer m. This is actually the conclusion of the current problem.

When $m = 0$, since $\{a_k\}$ is a good sequence, so $\{a_k^2\}$ is a sequence of integers, namely $\{a_k^2\}$ is a 0th grade good sequence, the conclusion being true.

Assume that $\{a_k^2\}$ is an mth grade good sequence. In the following, we prove that $\{a_k^2\}$ must be an $(m+1)$st grade good sequence. Obviously,

$$\frac{a_k^2 + a_{k+1}^2}{2} = \left(\frac{a_k + a_{k+1}}{2}\right)^2 + \left(\frac{a_k + a_{k+1}}{2} - a_{k+1}\right)^2.$$

Since the sequences with the general terms $\frac{a_k + a_{k+1}}{2}$ and $\frac{a_k + a_{k+1}}{2} - a_{k+1}$ respectively are both good sequences, by the inductive hypothesis, the sequences with the general terms $\frac{a_k + a_{k+1}}{2}$ and $\frac{a_k + a_{k+1}}{2} - a_{k+1}$, respectively are both mth grade good sequences. Consequently, $\{a_k^2\}$ is an $(m+1)$st good sequence. This completes the proof. $\qquad\square$

Remark. In the above proof, we changed the original conclusion with an equivalent form to express the question, for the convenience of proof by using mathematical induction.

Example 8. For arbitrary $2n-1$ subsets with two elements of $\{1, 2, \ldots, n\}$, prove: It is always possible to select n among them such that the number of elements in their union is not more than $\frac{2}{3}n + 1$.

Proof. We only need to show that it is always possible to select $3k$ ($k \le \frac{2n-1}{3}$) subsets of two elements among the $2n - 1$ subsets of two elements, such that the number of elements in the union of the remaining subsets is not more than $n - k$.

We use induction on k.

When $k = 0$, it is obviously satisfied.

Assume that for $k \ge 1$ there can be selected $3(k - 1)$ subsets of two elements such that the number of elements in the union of the remaining subsets is not more than $n - k + 1$.

By the drawer principle, from

$$2n - 1 - 3(k - 1) < 2(n - k + 1),$$

we know that there exists an element x_k in the remaining subsets, which belongs to three remaining subsets at most.

Thus, it is possible to select additional three subsets of two elements such that the remaining $2n - 1 - 3k$ subsets do not contain x_k, and so the conclusion is true for $k+1$. Therefore, the conclusion is valid for all positive integers $k \le \frac{2n-1}{3}$.

Let $k = \left[\frac{n-1}{3}\right]$ in the example. Then $n - \left[\frac{n-1}{3}\right] \le \frac{2}{3}n + 1$. $\qquad\qquad$ □

3. Exercises

Group A

I. Filling Problems

1. If a sequence $\{a_n\}$ is defined as $a_1 = 3$ and $a_{n+1} = 3^{a_n}$ for $n \ge 1$, then the last digit of a_{1993} is _____.
2. Let $f(n) = 2n + 1$, and the sequence $\{g(n)\}$ satisfy $g(1) = 3$ and $g(n + 1) = f(g(n))$ for $n \ge 1$. Then $g(n) = $ _____.
3. Assume that $a_1 = \frac{1}{3}$ and $a_{n+1} = a_n^2 + 6a_n + 4$ for $n \ge 1$. Then the general term formula of a_n is _____.

II. Calculation Problems

4. Suppose that a sequence $\{a_n\}$ satisfies

$$a_1 = 1, \quad a_2 = x, \quad a_{n+1} = \frac{a_n a_{n-1} + 1}{a_n + a_{n-1}} + (n + 2), \quad n \ge 2.$$

For what real numbers of x is the sequence $\{a_n\}$ that of positive integers?

5. All the centers of n semicircles are on a straight line l, every two of the n semicircles intersect to each other, and are all on the same side of l. Question: How many pieces of the arcs from these semicircles can be obtained at most from all the intersection points?

6. If $\{a_n\}$ is a sequence of all positive terms such that $a_{n+1} \leq a_n - a_n^2$ for all $n \geq 1$, prove: For all $n \geq 2$, it is always true that

$$a_n \leq \frac{1}{n+2}.$$

7. Suppose that 2^n balls are divided into several piles. Choose arbitrary two piles A and B and make changes according to the following rule: If the number p of balls in A is not less than the number q of balls in B, then move q balls from A to B. This is considered one shift. Prove: After finitely many shifts, all the balls can be moved to one pile.

8. Write numbers $1, 2, \ldots, 2016$ as the first line, and then each of the next lines will be obtained by adding the two consecutive numbers in the current line, for example, the numbers of the second line are $3, 5, \ldots, 4031$. If the last line has one number, find this number.

Group B

9. Prove: Every positive proper fraction $\frac{m}{n}$ can be represented as the sum of the reciprocals of different positive integers.

10. Find all the real numbers $x \geq -1$ such that for all $a_i \geq 1$ ($i = 1, 2, \ldots, n$ with $n \geq 2$), it is always true that

$$\prod_{i=1}^{n} \frac{a_i + x}{2} \leq \frac{a_1 a_2 \cdots a_n + x}{2}.$$

11. Prove: Any equilateral triangle can be partitioned into n ($n \geq 3$ with $n \in \mathbf{N}$) isosceles triangles.

12. On a table with 1993×1993 checks, remove an arbitrary check. Prove: The remaining figure can be exactly covered by several L-shaped figures (the union of 3 checks).

Chapter 18

Complex Numbers

1. Key Points of Knowledge and Basic Methods

A. *Representation forms of complex numbers*

Algebraic form: $z = a + bi$ $(a, b \in \mathbf{R})$.

Geometric form: Points $Z(a, b)$ in a complex plane or vectors $\overrightarrow{OZ}$ starting from the origin.

Trigonometric form: $z = r(\cos\theta + i\sin\theta)$ $(r \geq 0$ and $\theta \in \mathbf{R})$.

Exponential form: $z = re^{i\theta}$.

B. *Conditions for the equality of complex numbers*

A necessary and sufficient condition for two complex numbers $z_1 = a_1 + b_1 i$ and $z_2 = a_2 + b_2 i$ to be equal to each other is

$$a_1 = a_2 \quad \text{and} \quad b_1 = b_2.$$

A necessary and sufficient condition for two nonzero complex numbers $z_1 = r_1(\cos\theta_1 + i\sin\theta_1)$ and $z_2 = r_2(\cos\theta_2 + i\sin\theta_2)$ to be equal to each other is

$$r_1 = r_2 \quad \text{and} \quad \theta_1 - \theta_2 = 2k\pi \quad (k \in \mathbf{Z}).$$

C. *Operational rules of complex numbers*

Addition and subtraction: $(a + bi) \pm (c + di) = (a \pm c) + (b \pm d)i$.

Multiplication: $(a + bi)(c + di) = (ac - bd) + (bc + ad)i$ and

$$r_1(\cos\theta_1 + i\sin\theta_1) \cdot r_2(\cos\theta_2 + i\sin\theta_2) = r_1 r_2[\cos(\theta_1 + \theta_2) + i\sin(\theta_1 + \theta_2)].$$

Division: $\frac{a+bi}{c+di} = \frac{ac+bd}{c^2+d^2} + \frac{bc-ad}{c^2+d^2}i \ (c + di \neq 0)$ and

$$\frac{r_1(\cos\theta_1 + i\sin\theta_1)}{r_2(\cos\theta_2 + i\sin\theta_2)} = \frac{r_1}{r_2}[\cos(\theta_1 - \theta_2) + i\sin(\theta_1 - \theta_2)] \quad (r_2 \neq 0).$$

Powers: $[r(\cos\theta + i\sin\theta)]^n = r^n(\cos n\theta + i\sin n\theta) \ (n \in \mathbf{N})$.

Roots: The nth roots of a complex number $r(\cos\theta + i\sin\theta)$ are

$$\sqrt[n]{r}\left(\cos\frac{\theta + 2k\pi}{n} + i\sin\frac{\theta + 2k\pi}{n}\right) \quad (k = 0, 1, \ldots, n-1).$$

D. Moduli of complex numbers and conjugate complex numbers

Properties of conjugate complex numbers:

(1) $z\bar{z} = |z|^2 = |\bar{z}|^2$;

(2) $z + \bar{z} = 2\Re(z), z - \bar{z} = 2i\Im(z)$ ($\Re(z)$ and $\Im(z)$ represent the real part and the imaginary part of the complex number z respectively);

(3) $\bar{\bar{z}} = z$;

(4) $\overline{z_1 \pm z_2} = \overline{z_1} \pm \overline{z_2}$;

(5) $\overline{z_1 z_2} = \overline{z_1}\ \overline{z_2}$;

(6) $\overline{\left(\frac{z_1}{z_2}\right)} = \frac{\overline{z_1}}{\overline{z_2}} \ (z_2 \neq 0)$;

(7) a necessary and sufficient condition for z to be a real number is $\bar{z} = z$, and a necessary and sufficient condition for z to be a purely imaginary number is $\bar{z} = -z$.

Properties of moduli of complex numbers:

(1) $|z| \geq |\Re(z)|$ and $|z| \geq |\Im(z)|$;

(2) $|z_1 z_2 \cdots z_n| = |z_1||z_2| \cdots |z_n|$;

(3) $\left|\frac{z_1}{z_2}\right| = \frac{|z_1|}{|z_2|} \ (z_2 \neq 0)$;

(4) $||z_1| - |z_2|| \leq |z_1 + z_2|$, and the equality holds when the vectors $\overrightarrow{OZ_1}$ and $\overrightarrow{OZ_2}$ corresponding to the complex numbers z_1 and z_2, respectively have opposite directions;

(5) $|z_1 + z_2 + \cdots + z_n| \leq |z_1| + |z_2| + \cdots + |z_n|$, and the equality holds when the vectors $\overrightarrow{OZ_1}, \overrightarrow{OZ_2}, \ldots, \overrightarrow{OZ_n}$ corresponding to the complex numbers $z_1, z_2, \ldots, z_n$ respectively have the same direction.

E. *Basic methods*

Using a condition for the equality of complex numbers, we can reduce the complex number problem to the real number problem, so applications of the knowledge and methods of real numbers can help solve complex number problems. Conversely, one can also transform a real number problem to a complex number problem for an investigation.

The modulus of complex numbers is also one of the efficient methods to reduce complex number problems to real number ones. Being good at using the properties of moduli is a prominent aspect in the modulus operations.

The various forms of complex representations make many connections of complex numbers to other branches of mathematics, so complex numbers not only have synthetic applications in every branch of algebra, but also provide powerful solution tools for the subjects of trigonometry, geometry, etc. Constructing complex numbers according to the feature of the problem and solving it by using the related knowledge of complex numbers will provide new ideas towards solving problems in algebra, trigonometry, and geometry, etc.

2. Illustrative Examples

A. *Equality of complex numbers*

Example 1. Suppose that a complex number z satisfies $z^2 + 2z = \bar{z} \neq z$. Find the product of all possible values of z.

Solution. Let $z = a + bi$ $(a, b \in \mathbf{R})$. From $z^2 + 2z = \bar{z}$, we have

$$a^2 - b^2 + 2abi + 2a + 2bi = a - bi.$$

Comparing the real and imaginary parts gives

$$\begin{cases} a^2 - b^2 + a = 0, \\ 2ab + 3b = 0. \end{cases}$$

Also from $\bar{z} \neq z$ we know that $b \neq 0$, and so $2a + 3 = 0$, namely $a = -\frac{3}{2}$. Consequently,

$$b = \pm\sqrt{a^2 + a} = \pm\frac{\sqrt{3}}{2}.$$

Therefore, the product of the values of the complex numbers that satisfy the condition is $\left(-\frac{3}{2} + \frac{\sqrt{3}}{2}i\right)\left(-\frac{3}{2} - \frac{\sqrt{3}}{2}i\right) = 3.$ □

Remark. Solving complex number problems also needs to master and apply the following conclusions skillfully:

1. Necessary and sufficient conditions for a complex number to be a real number: (1) $z = a + bi \in \mathbf{R} \Leftrightarrow b = 0$ with $a, b \in \mathbf{R}$; (2) $z \in \mathbf{R} \Leftrightarrow z = \overline{z}$; (3) $z \in \mathbf{R} \Leftrightarrow z^2 \geq 0$.
2. Necessary and sufficient conditions for a complex number to be a purely imaginary number: (1) $z = a + bi$ is a purely imaginary number $\Leftrightarrow a = 0$ and $b \neq 0$ $(a, b \in \mathbf{R})$; (2) z is a purely imaginary number $\Leftrightarrow z + \overline{z} = 0$ $(z \neq 0)$; (3) z is a purely imaginary number $\Leftrightarrow z^2 < 0$. Any restriction to z should be investigated from the two aspects of the real and imaginary parts.

Example 2. Suppose that $n \leq 1999$ with $n \in \mathbf{N}$ and there exists a θ satisfying

$$(\sin \theta + i \cos \theta)^n = \sin n\theta + i \cos n\theta.$$

What is the total number of such n?

Analysis. If $z = r(\cos \theta + i \sin \theta)$, then

$$z^n = r^n(\cos n\theta + i \sin n\theta) \ (n \in \mathbf{N}),$$

which is de Moivre's formula. The premise of using de Moivre's formula is expressing a complex number z in its trigonometric form.

Solution. Note that

$$
\begin{aligned}
(\sin \theta + i \cos \theta)^n &= \left[\cos \left(\frac{\pi}{2} - \theta \right) + i \sin \left(\frac{\pi}{2} - \theta \right) \right]^n \\
&= \cos \left(\frac{n\pi}{2} - n\theta \right) + i \sin \left(\frac{n\pi}{2} - n\theta \right) \quad (18.1)
\end{aligned}
$$

and

$$\sin n\theta + i \cos n\theta = \cos \left(\frac{\pi}{2} - n\theta \right) + i \sin \left(\frac{\pi}{2} - n\theta \right). \quad (18.2)$$

By the necessary and sufficient conditions for the equality of complex numbers, (18.1) and (18.2) imply that

$$\frac{n\pi}{2} - n\theta = 2k\pi + \frac{\pi}{2} - n\theta \ (k \in \mathbf{Z}),$$

the solutions of which are $n = 4k + 1$ $(k \in \mathbf{Z})$.

Since $1 \leq n \leq 1999$, so $0 \leq k \leq 499\frac{1}{2}$.

Hence, the number of n that satisfy the condition of the problem is 500. $\qquad \square$

B. *Basic operations of complex numbers*

Reasonable computation is a problem that needs a great attention in complex number operations. Using different forms of complex numbers flexibly may finish the operations fast and precisely. At the same time, one should pay attentions to using the imaginary unit i and the operational properties of two cubic roots of 1: $\omega = -\frac{1}{2} + \frac{\sqrt{3}}{2}i$ and $\omega^2 = -\frac{1}{2} - \frac{\sqrt{3}}{2}i$.

Example 3. Prove:

$$[(2a - b - c) + (b - c)\sqrt{3}i]^3 = [(2b - c - a) + (c - a)\sqrt{3}i]^3.$$

Analysis. If we expand the two sides of the equality, then the operations are too complicated. We may consider the cubic root $\omega = -\frac{1}{2} + \frac{\sqrt{3}}{2}i$ of 1 to simplify the operations.

Proof. The left hand side of the equality

$$= [2a + b(-1 + \sqrt{3}i) + c(-1 - \sqrt{3}i)]^3$$
$$= (2a + 2b\omega + 2c\omega^2)^3,$$

where $\omega = -\frac{1}{2} + \frac{\sqrt{3}}{2}i$ and $\omega^2 = -\frac{1}{2} - \frac{\sqrt{3}}{2}i$.

The right hand side of the equality

$$= [a(-1 - \sqrt{3}i) + 2b + c(-1 + \sqrt{3}i)]^3$$
$$= (2a\omega^2 + 2b + 2c\omega)^3$$
$$= [(2a + 2b\omega + 2c\omega^2)\omega^2]^3$$
$$= (2a + 2b\omega + 2c\omega^2)^3 = \text{the left hand side of the equality,}$$

so the equality is valid. □

Remark. The number ω is a quantity that is very frequently seen in complex number problems and its related properties include: $\omega^3 = 1, 1 + \omega + \omega^2 = 0$, and $\omega^2 = \bar{\omega}$. These properties are very useful in the process of solving problems.

Example 4. Suppose that two complex numbers z_1 and z_2 satisfy the following conditions:

$$|z_1| = 2, \quad |z_2| = 3, \quad 3z_1 - 2z_2 = 2 - i.$$

Find the value of $z_1 z_2$.

Analysis. The way of trying to solve z_1 and z_2 out from the conditions (the system of the equations for z_1 and z_2) will meet a difficulty. Note the relation of conjugate complex numbers and moduli of complex numbers. From $|z_1| = 2$ and $|z_2| = 3$, we see that $2^2 = z_1\overline{z_1}$ and $3^2 = z_2\overline{z_2}$, and substituting into the third equality will find z_1z_2.

Solution. By the conditions, we know that $z_1\overline{z_1} = 4$ and $z_2\overline{z_2} = 9$. Then

$$3z_1 - 2z_2 = \frac{1}{3}z_1z_2\overline{z_2} - \frac{1}{2}z_2z_1\overline{z_1} = \frac{1}{6}z_1z_2(2\overline{z_2} - 3\overline{z_1}).$$

Thus, we have

$$z_1z_2 = \frac{6(3z_1 - 2z_2)}{2\overline{z_2} - 3\overline{z_1}} = -\frac{6(3z_1 - 2z_2)}{\overline{3z_1 - 2z_2}} = -\frac{6(2 - i)}{2 + i}$$

$$= -\frac{6}{5}(2 - i)^2 = -\frac{18}{5} + \frac{24}{5}i. \qquad \square$$

C.   *Moduli of complex numbers and conjugate complex numbers*

Example 5. Let complex numbers z and w satisfy $|z| = 3$ and $(z+\overline{w})(\overline{z} - w) = 7 + 4i$. Find the modulus of $(z + 2\overline{w})(\overline{z} - 2w)$.

Solution. From the properties of complex number operations,

$$7 + 4i = (z + \overline{w})(\overline{z} - w) = |z|^2 - |w|^2 - (zw - \overline{zw}).$$

Since $|z|^2$ and $|w|^2$ are real numbers and $\Re(zw - \overline{zw}) = 0$, so $|z|^2 - |w|^2 = 7$ and $zw - \overline{zw} = -4i$, Also $|w|^2 = 2$ from $|z| = 3$. It follows that

$$(z + 2\overline{w})(\overline{z} - 2w) = |z|^2 - 4|w|^2 - 2(zw - \overline{zw}) = 9 - 8 + 8i = 1 + 8i.$$

Therefore, the modulus of $(z + 2\overline{w})(\overline{z} - 2w)$ is $\sqrt{1^2 + 8^2} = \sqrt{65}$. $\qquad \square$

Example 6. Suppose that $z_1, z_2, \ldots z_n$ are complex numbers and satisfy $|z_1| + |z_2| + \cdots + |z_n| = 1$. Prove: There must be several complex numbers among the above n complex numbers such that the sum of their moduli is not less than $\frac{1}{6}$.

Proof. For the complex number $z_k = a_k + b_k i$ $(a_k, b_k \in \mathbf{R})$ for $k = 1, 2, \ldots, n$, there holds that $|z_k| \leq |a_k| + |b_k|$. So

$$1 = \sum_{k=1}^{n} |z_k| \leq \sum_{k=1}^{n} |a_k| + \sum_{k=1}^{n} |b_k|. \tag{18.3}$$

Also,

$$\left| \sum_{k=1}^{n} z_k \right| = \sqrt{\left(\sum_{k=1}^{n} a_k \right)^2 + \left(\sum_{k=1}^{n} b_k \right)^2} \geq \max \left\{ \left| \sum_{k=1}^{n} a_k \right|, \left| \sum_{k=1}^{n} b_k \right| \right\}. \tag{18.4}$$

For the comparison purpose of (18.3) and (18.4), rewrite the right hand side of (18.3) as

$$\sum_{a_k \geq 0} |a_k| + \sum_{a_k < 0} |a_k| + \sum_{b_k \geq 0} |b_k| + \sum_{b_k < 0} |b_k|$$

$$= \left| \sum_{a_k \geq 0} a_k \right| + \left| \sum_{a_k < 0} a_k \right| + \left| \sum_{b_k \geq 0} b_k \right| + \left| \sum_{b_k < 0} b_k \right|.$$

So, in the above four terms, there must be one term that is not less than $\frac{1}{4}$, say

$$\left| \sum_{a_k \geq 0} a_k \right| \geq \frac{1}{4}$$

without loss of generality. Consequently,

$$\left| \sum_{a_k \geq 0} z_k \right| \geq \left| \sum_{a_k \geq 0} a_k \right| \geq \frac{1}{4} > \frac{1}{6}. \qquad \square$$

Remark. In the following we rethink the problem, starting from another angle. First,

$$\left| \sum_{k=1}^{n} z_k \right| = \sqrt{\left(\sum_{k=1}^{n} a_k \right)^2 + \left(\sum_{k=1}^{n} b_k \right)^2} \geq \max \left\{ \left| \sum_{k=1}^{n} a_k \right|, \left| \sum_{k=1}^{n} b_k \right| \right\}.$$

Without loss of generality, assume that $\left| \sum_{k=1}^{n} a_k \right| \geq \left| \sum_{k=1}^{n} b_k \right|$. Then

$$\left| \sum_{k=1}^{n} z_k \right| \geq \left| \sum_{k=1}^{n} a_k \right|.$$

Using the above inequality directly may lead to too much decrease from the left hand side, so we adopt a novel idea as follows.

Divide the complex plane into four regions via the straight lines $y = x$ and $y = -x$. Since

$$|z_1| + |z_2| + \cdots + |z_n| = 1,$$

so at least one of the four regions has the property that those complex numbers among $z_1, z_2, \ldots, z_n$ that are inside this region must have the sum of their moduli not less than $\frac{1}{4}$. For the simplicity of presentation, we assume that this region is the one that contains the positive direction of the x-axis (since only moduli of the complex numbers are considered, otherwise we may make a rotation).

Suppose that the involved complex numbers are $z_{k_j} = a_{k_j} + b_{k_j} i$ ($a_{k_j} > 0$) with $j = 1, 2, \ldots, s$. Then

$$\sum_{j=1}^{s} |z_{k_j}| \geq \frac{1}{4} \quad (1 \leq s \leq n \text{ with } s \in \mathbf{N}),$$

from which we have

$$\left| \sum_{j=1}^{s} z_{k_j} \right| = \sqrt{\left(\sum_{j=1}^{s} a_{k_j} \right)^2 + \left(\sum_{j=1}^{s} b_{k_j} \right)^2}$$

$$\geq \left| \sum_{j=1}^{s} a_{k_j} \right| = \sum_{j=1}^{s} a_{k_j} \geq \frac{1}{\sqrt{2}} \sum_{j=1}^{s} \sqrt{a_{k_j}^2 + b_{k_j}^2}$$

$$= \frac{1}{\sqrt{2}} \sum_{j=1}^{s} |z_{k_j}| \geq \frac{1}{4\sqrt{2}} > \frac{1}{6}.$$

Thus, the problem is completely solved. □

D. *Method of complex numbers*

Example 7. Given eight nonzero real numbers $a_1, a_2, \ldots, a_8$. Prove: Among the following six numbers $a_1 a_3 + a_2 a_4, a_1 a_5 + a_2 a_6, a_1 a_7 + a_2 a_8, a_3 a_5 + a_4 a_6, a_3 a_7 + a_4 a_8$, and $a_5 a_7 + a_6 a_8$, at least one is nonnegative.

Analysis. In the given six numbers, $a_1, a_2; a_3, a_4; a_5, a_6; a_7, a_8$ appear in pairs. so we may construct four complex numbers $z_1 = a_1 + a_2 i, z_2 = a_3 + a_4 i, z_3 = a_5 + a_6 i$, and $z_4 = a_7 + a_8 i$. The moduli of the differences of all pairs of two complex numbers appear in the corresponding forms of the six numbers.

Proof. Let $z_1 = a_1 + a_2 i$, $z_2 = a_3 + a_4 i$, $z_3 = a_5 + a_6 i$, and $z_4 = a_7 + a_8 i$. Then

$$|z_1 - z_2|^2 = (a_1 - a_3)^2 + (a_2 - a_4)^2$$
$$= a_1^2 + a_2^2 + a_3^2 + a_4^2 - 2(a_1 a_3 + a_2 a_4).$$

Thus,

$$|z_1|^2 + |z_2|^2 - |z_1 - z_2|^2 = 2(a_1 a_3 + a_2 a_4).$$

Similarly, we obtain

$$|z_1|^2 + |z_3|^2 - |z_1 - z_3|^2 = 2(a_1 a_5 + a_2 a_6),$$
$$|z_1|^2 + |z_4|^2 - |z_1 - z_4|^2 = 2(a_1 a_7 + a_2 a_8),$$
$$|z_2|^2 + |z_3|^2 - |z_2 - z_3|^2 = 2(a_3 a_5 + a_4 a_6),$$
$$|z_2|^2 + |z_4|^2 - |z_2 - z_4|^2 = 2(a_3 a_7 + a_4 a_8),$$
$$|z_3|^2 + |z_4|^2 - |z_3 - z_4|^2 = 2(a_5 a_7 + a_6 a_8).$$

The remaining thing to do is only to show that among the four vectors $\overrightarrow{OZ_1}, \overrightarrow{OZ_2}, \overrightarrow{OZ_3}$, and $\overrightarrow{OZ_4}$ corresponding to the complex numbers z_1, z_2, z_3, and z_4 respectively, the angle of at least one pair of two vectors is less than or equal to $\frac{\pi}{2}$, which is obviously correct. Hence, the proposition is proved. □

Example 8. Suppose that $\cos \alpha + \cos \beta + \cos \gamma = \sin \alpha + \sin \beta + \sin \gamma = 0$. Prove:

$$\cos 2\alpha + \cos 2\beta + \cos 2\gamma = \sin 2\alpha + \sin 2\beta + \sin 2\gamma = 0.$$

Analysis. If we consider the formula $(\cos \alpha + i \sin \alpha)^2 = \cos 2\alpha + i \sin 2\alpha$ from the trigonometric form of complex numbers, then we can construct complex numbers so that we can solve the problem by means of the square operation of complex numbers.

Proof. Let $z_\theta = \cos \theta + i \sin \theta$. Then

$$z_\theta \overline{z_\theta} = |z_\theta|^2 = 1,$$
$$z_\theta^2 = \cos 2\theta + i \sin 2\theta \ (\theta = \alpha, \beta, \text{ and } \gamma).$$

So, $z_\alpha + z_\beta + z_\gamma = \cos\alpha + \cos\beta + \cos\gamma + i(\sin\alpha + \sin\beta + \sin\gamma) = 0$ and

$$z_\alpha^2 + z_\beta^2 + z_\gamma^2 = \cos 2\alpha + \cos 2\beta + \cos 2\gamma + i(\sin 2\alpha + \sin 2\beta + \sin 2\gamma).$$

$$(18.5)$$

Since $z_\alpha + z_\beta + z_\gamma = 0$, so $\overline{z_\alpha + z_\beta + z_\gamma} = 0$, and consequently

$$\frac{1}{z_\alpha} + \frac{1}{z_\beta} + \frac{1}{z_\gamma} = 0,$$

namely

$$\frac{z_\alpha z_\beta + z_\beta z_\gamma + z_\gamma z_\alpha}{z_\alpha z_\beta z_\gamma} = 0.$$

Hence,

$$z_\alpha z_\beta + z_\beta z_\gamma + z_\gamma z_\alpha = 0.$$

It follows that

$$z_\alpha^2 + z_\beta^2 + z_\gamma^2 = (z_\alpha + z_\beta + z_\gamma)^2 - 2(z_\alpha z_\beta + z_\beta z_\gamma + z_\gamma z_\alpha) = 0.$$

By (18.5),

$$\cos 2\alpha + \cos 2\beta + \cos 2\gamma = \sin 2\alpha + \sin 2\beta + \sin 2\gamma = 0. \qquad \square$$

3. Exercises

Group A

I. Filling Problems

1. Let a complex number z satisfy $z + 9 = 10\bar{z} + 22i$. Then the value of $|z|$ is _____.

2. A given sequence $\{z_n\}$ of complex numbers satisfies $z_1 = 1$ and $z_{n+1} = \overline{z_n} + 1 + ni$ $(n = 1, 2, \ldots)$. Then the value of z_{2015} is _____.

3. If two complex numbers z_1 and z_2 satisfy the condition $|z_1 - \overline{z_2}|^2 = |1 - \overline{z_1 z_2}|^2$, then $(|z_1| - 1)(|z_2| - 1) =$ _____.

4. Assume that $\arg(z + 1) = \frac{\pi}{6}$ and $\arg(z - 1) = \frac{2\pi}{3}$. Then the complex number $z =$ _____.

5. Given a geometric sequence $\{z_n\}$, where $z_1 = 1, z_2 = a + bi$, and $z_3 = b + ai$ $(a, b \in \mathbf{R}$ and $a > 0)$, when the condition $z_1 + z_2 + \cdots + z_n = 0$ is satisfied, the minimum value of the natural number n is _____.

6. Let complex numbers z_1 and z_2 satisfy $|z_1| = |z_1 + z_2| = 3$ and $|z_1 - z_2| = 3\sqrt{3}$. Then $|(z_1\overline{z_2})^{2000} + (\overline{z_1}z_2)^{2000}| =$ _____.

II. Calculation Problems

7. Let z_1 and z_2 be a pair of conjugate complex numbers. If $|z_1 - z_2| = \sqrt{6}$ and $\frac{z_1}{z_2^2}$ is a real number, find $|z_1|$.

8. Given complex numbers $z_1 = 1 + \cos\alpha + i\sin\alpha$ and $z_2 = 1 - \cos\beta + i\sin\beta$ $(0 < \alpha < \pi < \beta < 2\pi)$ such that $\arg z_1 + \arg z_2 = \frac{13\pi}{6}$ and $|z_1||z_2| = \sqrt{3}$, find α and β.

9. Let complex numbers α, β, and γ satisfy $|\alpha| = |\beta| = |\gamma| = 1$. Prove: $\frac{(\alpha+\beta)(\beta+\gamma)(\gamma+\alpha)}{\alpha\beta\gamma}$ is a real number.

Group B

10. Let $a, b, c \in \mathbf{N}$ satisfy the equality $c = (a + bi)^3 - 107i$. Find the value of c.

11. Given three real numbers a, b, and c, suppose that complex numbers z_1, z_2, and z_3 satisfy $|z_1| = |z_2| = |z_3| = 1$ and $\frac{z_1}{z_2} + \frac{z_2}{z_3} + \frac{z_3}{z_1} = 1$. Find the value of $|az_1 + bz_2 + cz_3|$.

12. Assume that $x^2 + y^2 \leq 1$ and $a^2 + b^2 \leq 2$ $(x, y, a, b \in \mathbf{R})$. Use the complex number method to prove:

$$\left| b(x^2 - y^2) + 2axy \right| \leq \sqrt{2}.$$

Chapter 19

Geometric Meaning
of Complex Operations

1. Key Points of Knowledge and Basic Methods

A. *Geometric meaning of complex numbers*

From the geometric representation of complex numbers, we know that the complex numbers $z = a + bi$ $(a, b \in \mathbf{R})$ and the points $Z(a, b)$ or the vectors $\overrightarrow{OZ}$ (O is the origin of the coordinates system) of the complex plane both constitute a one-one correspondence. We thus have $|z| = |\overrightarrow{OZ}| = \sqrt{a^2 + b^2}$.

Note that equal vectors represent the same complex number.

B. *Geometric meaning of addition and subtraction of complex numbers*

The addition and subtraction of complex numbers can be done according to the addition and subtraction of vectors, that is, they follow the parallelogram rule (or the triangle rule).

As in Figure 19.1, with vectors $\overrightarrow{OZ_1}$ and $\overrightarrow{OZ_2}$ corresponding to complex numbers $z_1 = a + bi$ and $z_2 = c + di$, respectively as adjacent sides, construct a parallelogram. Let complex number z correspond to the vector $\overrightarrow{OZ}$ represented by the diagonal of the parallelogram with O as the starting point. Then

$$z = (a + c) + (b + d)i = z_1 + z_2.$$

As in Figure 19.2, with vector $\overrightarrow{OZ_2}$ corresponding to complex number $z_2 = c + di$ as one side, and vector $\overrightarrow{OZ_1}$ corresponding to complex number

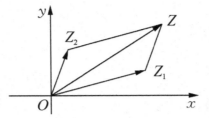

Figure 19.1   Addition of Complex Numbers.

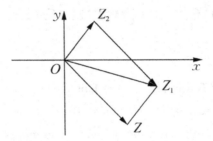

Figure 19.2   Subtraction of Complex Numbers.

$z_1 = a + bi$ as the diagonal, construct a parallelogram OZ_2Z_1Z. Let complex number z correspond to the vector $\overrightarrow{OZ}$. Then

$$z = (a - c) + (b - d)i = z_1 - z_2.$$

C.   Geometric meaning of multiplication and division of complex numbers

The geometric meaning of multiplication of complex numbers: Let $z_1 = r_1(\cos\theta_1 + i\sin\theta_1)$ and $z_2 = r_2(\cos\theta_2 + i\sin\theta_2)$. Then the vector corresponding to the product z_1z_2 of the two complex numbers is obtained via rotating the vector $\overrightarrow{OZ_1}$ corresponding to the complex number z_1 the angle θ_2 counter clockwise (if $\theta_2 < 0$, then rotating $\overrightarrow{OZ_1}$ the angle $|\theta_2|$ clockwise), followed by changing its modulus to the r_2 multiple of the original one.

The geometric meaning of division of complex numbers: Let $z_1 = r_1(\cos\theta_1 + i\sin\theta_1)$ and $z_2 = r_2(\cos\theta_2 + i\sin\theta_2)$ $(z_2 \neq 0)$. Then the vector corresponding to the quotient $\frac{z_1}{z_2}$ of the two complex numbers is obtained via rotating the vector $\overrightarrow{OZ_1}$ corresponding to the complex number z_1 the angle

θ_2 clockwise (if $\theta_2 < 0$, then rotating $\overrightarrow{OZ_1}$ the angle $|\theta_2|$ counter clockwise), followed by changing its modulus to the $\frac{1}{r_2}$ multiple of the original one.

D. *Geometric meaning of taking roots of complex numbers*

Let z_1 be a root of the binomial equation $x^n = z$, and let $\overrightarrow{OZ_1}$ be the corresponding vector (O is the origin). Then the vectors corresponding to the n roots of the equation $x^n = z$ all have O as their starting point and have their terminal points distributed uniformly on the circle with the origin as the center and $|\overrightarrow{OZ_1}| = \sqrt[n]{z}$ as the radius. In other words, their terminal points are just the vertices of a regular n-polygon. The principal values of the arguments of the two complex numbers corresponding to two nearby vectors differ by $\frac{2\pi}{n}$.

E. *Distance formula for two points of the complex plane*

Let the complex numbers correspond to two points Z_1 and Z_2 in the complex plane be $z_1 = a + bi$ and $z_2 = c + di$, respectively. Then the distance between the two points Z_1 and Z_2 is $d = |Z_1 Z_2| = |z_1 - z_2| = \sqrt{(a - c)^2 + (b - d)^2}$.

F. *Equations of commonly seen curves in the complex plane*

(1) The trajectory of the point corresponding to all the complex numbers z satisfying $|z - z_1| = |z - z_2|$ is the perpendicular bisector of the line segment $Z_1 Z_2$.

(2) The trajectory of the point corresponding to all the complex numbers z satisfying $|z - z_1| = r$ is the circle with the point Z_1 as the center and r as the radius.

(3) The trajectory of the point corresponding to all the complex numbers z satisfying $|z - z_1| + |z - z_2| = 2a$ ($|Z_1 Z_2| < 2a$) is the ellipse with the foci Z_1 and Z_2 and $2a$ as the length of the long axis.

(4) The trajectory of the point corresponding to all the complex numbers z satisfying $||z - z_1| - |z - z_2|| = 2a$ ($|Z_1 Z_2| > 2a$) is the hyperbola with the foci Z_1 and Z_2 and $2a$ as the length of the real axis.

(5) The trajectory of the point corresponding to all the complex numbers z satisfying $\arg z = \theta$ is the ray with the origin as its starting point (the angle with the positive half x-axis as the starting side and the ray as the terminal side is θ).

(6) The solutions of a system of complex number equations are the complex numbers corresponding to the intersection points of the trajectories of the points that satisfy one equation of the system respectively, and etc.

G. Basic methods

Using the geometric meaning of complex numbers and their operations to give geometric explanations of quantitative relations and then transform the problem of the numerical relation to the problem of the graphic properties, one can use figures to solve problems. The idea and method of combining numbers and figures not only strengthen the intuitive feature of the problem, but also play a role of turning difficulty into ease and turning complexity into simplicity.

2. Illustrative Examples

A. *Problems related to moduli*

Example 1. Let the complex numbers corresponding to three points A, B, and C of the complex plane be z_1, z_2, and z_3, respectively. Suppose that $|z_1| = 1$, $z_2 = z_1 z$, and $z_3 = z_2 z$, where $z = \frac{3}{2}(1 + \sqrt{3}i)$. Find the area of the quadrilateral $OABC$.

Analysis. Rewrite $z = \frac{3}{2}(1 + \sqrt{3}i) = 3\left(\cos\frac{\pi}{3} + i\sin\frac{\pi}{3}\right)$ in the trigonometric form. Then we know that $z_2 = z_1 z$ represents rotating the vector $\overrightarrow{OA}$ corresponding to z_1 the angle of $\frac{\pi}{3}$ in the direction of counter clockwise around the origin, and then followed by extending its modulus to three times, arriving at the location of the point B. By the same reason we can get the location of the point C, and in this way we can decompose the quadrilateral into two triangles, the areas of which can be calculated respectively.

Solution. Since $z = \frac{3}{2}(1 + \sqrt{3}i) = 3\left(\cos\frac{\pi}{3} + i\sin\frac{\pi}{3}\right)$ and the point A corresponds to $z_1 = \cos\theta + i\sin\theta$ with the corresponding vector $\overrightarrow{OA}$, we have $z_2 = z_1 z = z_1 \cdot 3\left(\cos\frac{\pi}{3} + i\sin\frac{\pi}{3}\right)$. This indicates that the point B corresponds to z_2, and the corresponding vector $\overrightarrow{OB}$ is obtained via rotating $\overrightarrow{OA}$ the angle of $\frac{\pi}{3}$ counter clockwise around the point O and then changing the original modulus to three times:

$$|OB| = 3 \quad \text{and} \quad \angle AOB = \frac{\pi}{3}.$$

Also $z_3 = z_2 z = z_2 \cdot 3 \left(\cos \frac{\pi}{3} + i \sin \frac{\pi}{3} \right)$. This means that the point C corresponds to z_3, and the corresponding vector $\overrightarrow{OC}$ is obtained via rotating $\overrightarrow{OB}$ the angle of $\frac{\pi}{3}$ counter clockwise around the point O and then changing the original modulus to three times:

$$|OC| = 9 \quad \text{and} \quad \angle BOC = \frac{\pi}{3}.$$

Hence,

$$
\begin{aligned}
S_{OABC} &= S_{\triangle OAB} + S_{\triangle OBC} \\
&= \frac{1}{2}|OA||OB| \sin \frac{\pi}{3} + \frac{1}{2}|OB||OC| \sin \frac{\pi}{3} \\
&= \frac{15}{2}\sqrt{3}.
\end{aligned}
$$

$\square$

Remark. This is a direct application of the geometric meaning of complex number multiplications. Here, it should be noted: (1) The object of rotation; (2) the rotation direction (determined by the sign of θ); (3) the multiple of change in the modulus.

Example 2. Assume that a complex number z satisfies $\arg(z + 3) = \frac{3}{4}\pi$. For what value of z, the maximum value of $u = \frac{1}{|z+6|+|z-3i|}$ can be obtained? Also find the maximum value.

Analysis. The equality $\arg(z + 3) = \frac{3}{4}\pi$ indicates that the curve represented by $z + 3$ with z satisfying the equation is the ray starting at the origin with the inclination angle $\frac{3}{4}\pi$. The curve represented by $z + 3$ is obtained by translating the curve represented by z to the left by 3 units. From this we can find the trajectory of the point Z, and by using the geometric meaning of the modulus of a complex number, we can transform the problem to that of finding a point in a straight line that has the minimal sum of the distances to two fixed points.

Solution. In the complex plane, the curve represented by the equation

$$\arg(z + 3) = \frac{3}{4}\pi$$

is the ray AE with $(-3, 0)$ as its starting point and $\frac{3}{4}\pi$ as its inclination angle (see Figure 19.3). Thus, the problem is reduced to that of finding a point Z (corresponding to the complex number z) on the ray AE such that the value of $|z + 6| + |z - 3i|$ is minimal. The number $|z + 6| + |z - 3i|$

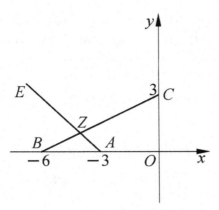

Figure 19.3   Figure for Example 2.

expresses the sum of the distances of the point Z on the ray to the points $B(-6, 0)$ and $C(0, 3)$, respectively.

From Figure 19.3 we see that when the point Z is on the line segment BC, the value of $|z + 6| + |z - 3i|$ is minimal, and the minimum value is $\sqrt{(-6)^2 + 3^2} = 3\sqrt{5}$. Thus,

$$u_{\max} = \frac{1}{3\sqrt{5}} = \frac{\sqrt{5}}{15}.$$

When u achieves the maximum value, let $z = x + yi$ $(x, y \in \mathbf{R})$. The above reduction idea transforms the problem to that of finding the intersection point of the straight lines AE and BC. By the knowledge of plane geometry,

the equation of the straight line AE : $\quad x + y + 3 = 0;$ (19.1)

the equation of the straight line BC : $\quad x - 2y + 6 = 0.$ (19.2)

The solution of the system of (19.1) and (19.2) is $x = -4$ and $y = 1$, from which $z = -4 + i$.

In summary, $u_{\max} = \frac{\sqrt{5}}{15}$ when $z = -4 + i$.   □

Remark. Understanding the geometric meaning of the operations, such as addition, subtraction, multiplication, division, finding the modulus, and arguments of complex numbers, and utilizing the method of combing figures and numbers, can often achieve an unexpected effect.

Example 3. Given a positive real number a, if $z = \frac{1}{a+ti}$ (here t is a real parameter) achieves the maximum value 2 of $|z - i|$, find the value of a.

Solution 1. Since

$$z - \frac{1}{2a} = \frac{1}{a+ti} - \frac{1}{2a} = \frac{a - ti}{2a(a + ti)},$$

$$\left| z - \frac{1}{2a} \right| = \frac{|a - ti|}{2a|a + ti|}.$$

Thus,

$$\left| z - \frac{1}{2a} \right| = \frac{1}{2a}.$$

Consequently, in the complex plane, the point Z corresponding to the complex number z lies on the circle (not containing the origin) centered at $\left(\frac{1}{2a}, 0\right)$ with radius $\frac{1}{2a}$. Denote $A\left(\frac{1}{2a}, 0\right)$ and $B(0, 1)$. Then

$$|z - i| = |ZB| \le |ZA| + |AB| = \frac{1}{2a} + \sqrt{\left(\frac{1}{2a}\right)^2 + 1},$$

and the equality is valid when the three points Z, A, and B are co-linear. It follows that

$$\frac{1}{2a} + \sqrt{\left(\frac{1}{2a}\right)^2 + 1} = 2$$

from the condition of the problem. The solution of the above is $a = \frac{2}{3}$. $\square$

Solution 2. Since $z - i = \frac{1+t-ai}{a+ti}$,

$$|z - i|^2 = \frac{|1 + t - ai|^2}{|a + ti|^2} = \frac{(1+t)^2 + a^2}{a^2 + t^2} = 1 + \frac{2t + 1}{a^2 + t^2}.$$

Since the maximum value of $|z - i|$ is 2, the maximum value of $|z - i|^2$ is 4. Thus, the maximum value of $\frac{2t+1}{a^2+t^2}$ is 3.

Let $u = 2t + 1$. Then $u > 0$, and so

$$\frac{2t + 1}{a^2 + t^2} = \frac{u}{a^2 + \frac{u^2 - 2u + 1}{4}} = \frac{4u}{4a^2 + 1 + u^2 - 2u}$$

$$= \frac{4}{\frac{4a^2+1}{u} + u - 2} \le \frac{4}{2\sqrt{4a^2 + 1} - 2} = \frac{2}{\sqrt{4a^2 + 1} - 1},$$

with the equality being satisfied if and only if $\frac{4a^2+1}{u} = u$, that is $u = \sqrt{4a^2+1}$. Hence,

$$\frac{2}{\sqrt{4a^2+1}-1} = 3,$$

and its solution is $a = \frac{2}{3}$ (since $a > 0$).   □

B.   *Problems with principal values of arguments*

The combination of numbers and figures is an efficient means to deal with problems concerning the principal value of arguments.

Example 4. Let a complex number z satisfy $\left|\arg\frac{z+1}{z+2}\right| = \frac{\pi}{6}$. Find the range of the values for $\arg z$.

Analysis. Let the point that corresponds to the vector z be Z. According to the geometric meaning of divisions of complex numbers, we may write $z+1 = r(z+2)(\cos\theta+i\sin\theta)$, where $z+1$ represents the vector with $(-1,0)$ as its starting point and Z as its terminal point, and $z+2$ represents the vector with $(-2,0)$ as its starting point and Z as its terminal point. Rotate the vector represented by $z+2$ the angle of $|\theta|$ counter clockwise ($\theta > 0$) or clockwise ($\theta < 0$). Then it will be parallel to the vector represented by $z+1$, and $\angle AZB$ is the fixed value $|\theta|$. This means that the point Z moves on the circle with AB as a chord and the corresponding central angle $2|\theta|$, from which the trajectory of the point Z can be obtained. This will give the solution.

Solution. Let $\left|\arg\frac{z+1}{z+2}\right| = \theta$. Then $z+1 = \lambda(z+2)(\cos\theta + i\sin\theta)$ with $\lambda \geq 0$.

As in Figure 19.4, denote $A(-2,0)$ and $B(-1,0)$. Then $\overrightarrow{BZ} = \lambda\overrightarrow{AZ}(\cos\theta + i\sin\theta)$. When $\theta = \frac{\pi}{6}$, the vector $\overrightarrow{AZ}$ will be parallel to $\overrightarrow{BZ}$ after it is rotated the angle of θ counter clockwise, so

$$\angle AZB = \frac{\pi}{6}.$$

Consequently, part of the trajectory of Z is the major arc AB (containing B but not A) of the circle centered at $O_1\left(-\frac{3}{2}, \frac{\sqrt{3}}{2}\right)$ with radius 1. When $\theta = -\frac{\pi}{6}$, by the symmetry we know that another part of the trajectory of

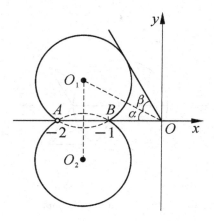

Figure 19.4   Figure for Example 4.

Z is the major arc AB (containing B but not A) of the circle centered at $O_2\left(-\frac{3}{2}, -\frac{\sqrt{3}}{2}\right)$ with radius 1. It can be calculated out easily that

$$\tan\alpha = \frac{\sqrt{3}}{3}, \quad \alpha = \frac{\pi}{6},$$

$$\sin\beta = \frac{\sqrt{3}}{3}, \quad \beta = \arcsin\frac{\sqrt{3}}{3}.$$

Therefore, $\arg z \in \left[\frac{5}{6}\pi - \arcsin\frac{\sqrt{3}}{3}, \pi\right) \cup \left(\pi, \frac{7}{6}\pi + \arcsin\frac{\sqrt{3}}{3}\right]$.   □

Remark. In general, the argument equation

$$\arg\frac{z - z_1}{z - z_2} = \theta \Leftrightarrow \overrightarrow{Z_1Z} = \lambda\overrightarrow{Z_2Z}(\cos\theta + i\sin\theta)\ (\lambda > 0),$$

which means that the trajectory of the point Z is the arc opposite to the chord Z_1Z_2.

Example 5. Suppose that two complex numbers $z_1 = 1 + \cos\alpha + i\sin\alpha$ $(0 < \alpha < \pi)$ and $z_2 = 1 - \cos\beta + i\sin\beta$ $(0 < \beta < 2\pi)$ satisfy $\arg z_1 + \arg z_2 = \frac{13}{6}\pi$. Find $\sin\frac{\alpha - \beta}{4}$.

Analysis. Construct the new complex numbers $z = \cos\alpha + i\sin\alpha$ and $z' = -\cos\beta + i\sin\beta$. Then the corresponding points Z and Z' move on the unit circle centered at the origin. Since $z_1 = 1 + z$ and $z_2 = 1 + z'$, one can determine the ranges of the arguments for z_1 and z_2 from the figure.

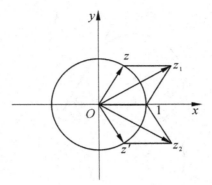

Figure 19.5   Figure for Example 5.

Solution. Let $z = \cos\alpha + i\sin\alpha$ $(0 < \alpha < \pi)$. Then $z_1 = 1 + z$. From Figure 19.5, we know that $\arg z_1 = \frac{\alpha}{2}$.

Let $z' = -\cos\beta + i\sin\beta = \cos(\pi - \beta) + \sin(\pi - \beta)$. Then $z_2 = 1 + z'$. Since $-\pi < \pi - \beta < 0$, Figure 19.5 tells us that $\arg z_2 = 2\pi + \frac{\pi - \beta}{2}$.

Also, since $\arg z_1 + \arg z_2 = \frac{\alpha}{2} + \frac{5}{2}\pi - \frac{\beta}{2} = \frac{13}{6}\pi$, we obtain $\frac{\alpha - \beta}{4} = -\frac{\pi}{6}$. It follows that

$$\sin\frac{\alpha - \beta}{4} = -\frac{1}{2}. \qquad \square$$

C.   Trajectory problems in the complex plane

Example 6. Suppose that the complex numbers corresponding to two points A and B in the complex plane are 1 and i, respectively. Let the complex number that corresponds to a moving point on the line segment AB be z. Find the trajectory of the point corresponding to the complex number z^2 in the complex plane.

Analysis. Let $z = a + bi$. Then $z^2 = a^2 - b^2 + 2abi$. Eliminating the parameters a and b gives the trajectory of the point corresponding to the complex number z^2.

Solution. Let the complex number corresponding to the moving point on the line segment AB be $z = a + bi$ $(a, b \in \mathbf{R})$. By the condition of the problem, the coordinates of the two points A and B are $(1, 0)$ and $(0, 1)$, respectively. So, the equation of the straight line that contains the line segment AB is $x + y = 1$. Since the point corresponding to $z = a + bi$ is on the line segment AB, we have $a + b = 1$ with $0 \le a \le 1$ and $0 \le b \le 1$. Also,

$$z^2 = (a + bi)^2 = a^2 - b^2 + 2abi.$$

Let the coordinates of the point corresponding to z^2 be (x, y). Then

$$\begin{cases} x = a^2 - b^2, \\ y = 2ab. \end{cases}$$

Since $a + b = 1$, we have $b = 1 - a$, thus

$$\begin{cases} x = a^2 - (1 - a)^2, \\ y = 2a(1 - a). \end{cases}$$

In other words,

$$\begin{cases} x = 2a - 1, \\ y = 2a - 2a^2. \end{cases}$$

Eliminating a from the above and simplifying the resulting expression, we find that $x^2 = -2\left(y - \frac{1}{2}\right)$.

Also, since $x = 2a - 1$ with $0 \le a \le 1$, we have $-1 \le x \le 1$.

Therefore, the trajectory of the point corresponding to the complex number z^2 is the part of the parabola $x^2 = -2\left(y - \frac{1}{2}\right)$ that corresponds to $-1 \le x \le 1$, namely the solid curve part of the parabola as shown in Figure 19.6. □

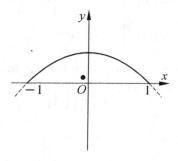

Figure 19.6   Figure for Example 6.

Remark. Using the algebraic form of complex numbers $z = x + yi$ ($x, y \in \mathbf{R}$), finding an expression of the relation between x and y, eliminating the parameter in the process, and finally obtaining an equation about x and y, give basic ideas towards solving the trajectory problem.

Example 7. For a complex number $z = t + 3 + \sqrt{3}i$, where t is a complex number that makes $\frac{t+3}{t-3}$ a purely imaginary number, (1) find the trajectory equation for the point corresponding to z in the complex plane; (2) find the range of the values for $\arg z$; (3) find the minimum value of $|z-1|^2 + |z+1|^2$ and the complex numbers z that make the minimum value.

Analysis. From the given information, we have $|z - (3 + \sqrt{3}i)| = |t|$, and so if we can find $|t|$, we can obtain the answer to the trajectory question.

Solution. (1) Since $\frac{t+3}{t-3}$ is purely imaginary, from $\frac{t+3}{t-3} = -\frac{\bar{t}+3}{\bar{t}-3}$, we get $t\bar{t} = 9$, namely $|t| = 3$ ($t \neq \pm 3$).

The equality $z = t + (3 + \sqrt{3}i)$ implies that

$$t = z - (3 + \sqrt{3}i) \ (z \neq 6 + \sqrt{3}i \text{ and } z \neq \sqrt{3}i).$$

Consequently,

$$|z - (3 + \sqrt{3}i)| = |t| = 3.$$

Hence, the trajectory of the point P corresponding to z is the circle centered at $(3, \sqrt{3})$ with radius 3, excluding the points $(6, \sqrt{3})$ and $(0, \sqrt{3})$.

(2) Let one tangent line to the above circle that passes through the origin be $y = kx$. The other tangent line is the y-axis. Then

$$\frac{|3k - \sqrt{3}|}{\sqrt{1 + k^2}} = 3, \quad k = -\frac{\sqrt{3}}{3}.$$

Also, since $z \neq \sqrt{3}i$, it does not reach the y-axis, so

$$\arg z \in \left[\frac{11}{6}\pi, 2\pi \right) \bigcup \left[0, \frac{\pi}{2} \right).$$

(3) We first calculate

$$|z - 1|^2 + |z + 1|^2 = (z - 1)(\bar{z} - 1) + (z + 1)(\bar{z} + 1)$$
$$= 2(|z|^2 + 1).$$

Since $|z|_{\min} = \sqrt{3}$, it follows that

$$(|z-1|^2 + |z+1|^2)_{\min} = 8.$$

Now, $z = \sqrt{3}i$ or $z = \sqrt{3}\left(\cos\frac{11}{6}\pi + i\sin\frac{11}{6}\pi\right) = \frac{3}{2} - \frac{\sqrt{3}}{2}i.$ $\qquad\square$

Remark. The form of the trajectory equation in the complex plane is rather special, and it often appears in the form of the complex number modulus. Hence, one often uses the method of the complex number modulus to solve a trajectory problem.

Example 8. Let complex numbers z_1 and z_2 satisfy $\Re(z_1) > 0, \Re(z_2) > 0$, and $\Re(z_1^2) = \Re(z_2^2) = 2$.

(1) Find the minimum value of $\Re(z_1 z_2)$.
(2) Find the minimum value of $|z_1 + 2| + |\overline{z_2} + 2| - |\overline{z_1} - z_2|$.

Solution. (1) For $k = 1$ and 2, let $z_k = x_k + y_k i$ $(x_k, y_k \in \mathbf{R})$. By the given conditions,

$$x_k = \Re(z_k) > 0, \; x_k^2 - y_k^2 = \Re(z_k^2) = 2.$$

Thus,

$$\Re(z_1 z_2) = \Re[(x_1 + y_1 i)(x_2 + y_2 i)] = x_1 x_2 - y_1 y_2$$

$$= \sqrt{(y_1^2 + 2)(y_2^2 + 2)} - y_1 y_2$$

$$\geq (|y_1 y_2| + 2) - y_1 y_2 \geq 2.$$

Also, when $z_1 = z_2 = \sqrt{2}$, we have $\Re(z_1 z_2) = 2$. This means that the minimum value of $\Re(z_1 z_2)$ is 2.

(2) For $k = 1$ and 2, let the point corresponding to z_k in the rectangular coordinates system xOy of the complex plane be $P_k(x_k, y_k)$, and denote the symmetric point of P_2 with respect to the x-axis to be P_2'. Then P_1 and P_2' are both on the right branch of the hyperbola $C : x^2 - y^2 = 2$.

Let $F_1(-2, 0)$ and $F_2(2, 0)$ be the left and right foci of C, respectively.

According to the definition of a hyperbola, we have $|P_1F_1| = |P_1F_2| + 2\sqrt{2}$ and $|P_2'F_1| = |P_2'F_2| + 2\sqrt{2}$. Consequently,

$$|z_1 + 2| + |\overline{z_2} + 2| - |\overline{z_1} - z_2| = |z_1 + 2| + |\overline{z_2} + 2| - |z_1 - \overline{z_2}|$$
$$= |P_1F_1| + |P_2'F_1| - |P_1P_2'|$$
$$= 4\sqrt{2} + |P_1F_2| + |P_2'F_2| - |P_1P_2'| \geq 4\sqrt{2},$$

with the equality being valid if and only if F_2 is on the line segment P_1P_2' (for example, F_2 is just the middle point of P_1P_2' when $z_1 = z_2 = 2 + \sqrt{2}i$).

In summary, the minimum value of $|z_1 + 2| + |\overline{z_2} + 2| - |\overline{z_1} - z_2|$ is $4\sqrt{2}$. □

3. Exercises

Group A

I. Filling Problems

1. If $|z_1| = |z_2| = 1$ and $|z_1 + z_2| = \sqrt{2}$, then $|z_1 - z_2|$ equals _____.

2. Let $\omega = z + ai$, where $a \in \mathbf{R}$ and $z = \frac{(1-4i)(1+i)+2+4i}{3+4i}$. If $|\omega| \leq \sqrt{2}$, then the range of the principal values for the argument of ω is _____.

3. Let complex numbers $z_1 = -3 - \sqrt{3}i, z_2 = \sqrt{3} + i$, and $z = \sqrt{3}\sin\theta + i(\sqrt{3}\cos\theta + 2)$. Then the minimum value of $|z - z_1| + |z - z_2|$ is _____.

4. Suppose that a is a real number and there exists a complex number z such that $|z + \sqrt{2}| = \sqrt{a^2 - 3a + 2}$ and $|z + \sqrt{2}i| < a$. Then the range of the values for a is _____.

5. Let the complex numbers corresponding to the 20 vertices of the inscribed regular 20-polygon to the unit circle in the complex plane be $z_1, z_2, \ldots, z_{20}$ in succession. Then the number of the points that correspond to the complex numbers $z_1^{1995}, z_2^{1995}, \ldots, z_{20}^{1995}$ is _____.

II. Calculation Problems

6. Let a complex number z_1 satisfy $|z_1 - 8i| = 2$ and another complex number z_2 satisfy $|z_2| = 4$. Denote $w = z_1 - z_2$. Find the area of the figure corresponding to the set M of the complex numbers w.

7. Suppose that $\triangle ABC$ is an isosceles right triangle with the vertices arranged in the clockwise direction, where A is a fixed point, B is a moving point on a fixed circle, and C is the vertex of the right angle. Find the trajectory of the point C.

8. Given two complex number sets

$$A = \{z \mid |z - 2| \leq 2\},$$

$$B = \left\{z \mid z = \frac{iz_1}{2 + b}, z_1 \in A, \text{ and } b \in \mathbf{R}\right\}.$$

   (1) If $A \cap B = \emptyset$, find the range of b.
   (2) If $A \cap B = B$, find the values of b.

Group B

9. The arguments of complex numbers z_1, z_2, and z_3 are α, β, and γ, respectively, $|z_1| = 1, |z_2| = k, |z_3| = 2 - k$, and $z_1 + z_2 + z_3 = 0$. For what values of k does $\cos(\beta - \gamma)$ achieve its maximum value and minimum value, respectively? Also, find the maximum value and minimum value.

10. Suppose that a and b are real numbers, the two imaginary roots of $x^2 + 2ax + b = 0$ are z_1 and z_2 with corresponding points Z_1 and Z_2 in the complex plane, and the circle with Z_1 and Z_2 as the two end points of its diameter includes the origin in its interior. Find the range for the existence of the complex numbers $z = a + bi$, and represent it with a figure.

11. Let O be the origin of the complex plane, and let Z_1 and Z_2 be two moving points in the complex plane that satisfy: (1) The arguments of the complex numbers corresponding to Z_1 and Z_2 are fixed values of θ and $-\theta$ $\left(0 < \theta < \frac{\pi}{2}\right)$ respectively; (2) the area of $\triangle OZ_1Z_2$ is a fixed value of S. Find the trajectory equation for the center of gravity Z of $\triangle OZ_1Z_1$ and the minimum value of the modulus of the complex number corresponding to the point Z.

12. Let nonzero complex numbers a_1, a_2, a_3, a_4, and a_5 satisfy $\frac{a_2}{a_1} = \frac{a_3}{a_2} = \frac{a_4}{a_3} = \frac{a_5}{a_4}$ and $a_1 + a_2 + a_3 + a_4 + a_5 = 4\left(\frac{1}{a_1} + \frac{1}{a_2} + \frac{1}{a_3} + \frac{1}{a_4} + \frac{1}{a_5}\right) = S$, where S is a real number and $|S| \leq 2$. Prove: The points in the complex plane that correspond to the complex numbers a_1, a_2, a_3, a_4, and a_5, respectively lie on the same circle.

Chapter 20

Mean Value Inequalities

1. Key Points of Knowledge and Basic Methods

Let $a_1, a_2, \ldots, a_n$ be n positive real numbers. Denote

$$H_n = \frac{n}{\frac{1}{a_1} + \frac{1}{a_2} + \cdots + \frac{1}{a_n}},$$

$$G_n = \sqrt[n]{a_1 a_2 \cdots a_n},$$

$$A_n = \frac{a_1 + a_2 + \cdots + a_n}{n},$$

$$Q_n = \sqrt{\frac{a_1^2 + a_2^2 + \cdots + a_n^2}{n}},$$

which are called respectively the harmonic mean, geometric mean, arithmetic mean, and square mean of the n positive numbers.

The four mean values have the following relation:

$$H_n \leq G_n \leq A_n \leq Q_n,$$

and any equality holds if and only if $a_1 = a_2 = \cdots = a_n$.

Proof. (1) We first prove $A_n \geq G_n$, using mathematical induction.

When $n = 1$, since $a_1 = a_1$, the inequality is true.

When $n = 2$, since

$$\frac{(a_1 + a_2)^2}{4} = \frac{(a_1 - a_2)^2}{4} + a_1 a_2 \geq a_1 a_2,$$

namely $\frac{a_1 + a_2}{2} \geq \sqrt{a_1 a_2}$, the inequality is true.

Assume that the inequality is true for $n = k$ ($k \geq 2$ with $k \in \mathbf{N}$). Then for $n = k+1$,

$$A_{k+1} = \frac{a_1 + a_2 + \cdots + a_k + a_{k+1}}{k+1}$$

$$= \frac{a_1 + a_2 + \cdots + a_{k+1} + (k-1)A_{k+1}}{2k}$$

$$= \frac{1}{2}\left(\frac{a_1 + a_2 + \cdots + a_k}{k} + \frac{a_{k+1} + (k-1)A_{k+1}}{k}\right)$$

$$\geq \frac{1}{2}\left(\sqrt[k]{a_1 a_2 \cdots a_k} + \sqrt[k]{a_{k+1} A_{k+1}^{k-1}}\right)$$

$$\geq \sqrt{\sqrt[k]{a_1 a_2 \cdots a_k} \cdot \sqrt[k]{a_{k+1} A_{k+1}^{k-1}}}.$$

After simplifying the above inequality, we obtain

$$A_{k+1} \geq \sqrt[k+1]{a_1 a_2 \cdots a_k a_{k+1}},$$

and the equality is valid if and only if $a_1 = a_2 = \cdots = a_k = a_{k+1} = A_{k+1}$.

(2) The inequality $A_n \geq G_n$ ensures that

$$\frac{1}{a_1} + \frac{1}{a_2} + \cdots + \frac{1}{a_n} \geq n \sqrt[n]{\frac{1}{a_1 a_2 \cdots a_n}} = \frac{n}{G_n},$$

that is

$$G_n \geq \frac{n}{\frac{1}{a_1} + \frac{1}{a_2} + \cdots + \frac{1}{a_n}} = H_n.$$

(3) The inequality $Q_n \geq A_n$ is equivalent to

$$n(a_1^2 + a_2^2 + \cdots + a_n^2) - (a_1 + a_2 + \cdots + a_n)^2 \geq 0.$$

But the left hand side expression of the above

$$= (a_1 - a_2)^2 + (a_1 - a_3)^2 + \cdots + (a_1 - a_n)^2 + (a_2 - a_3)^2 + \cdots$$

$$+ (a_2 - a_n)^2 + \cdots + (a_{n-1} - a_n)^2 \geq 0,$$

the inequality in which becomes an equality if and only if $a_1 = a_2 = \cdots = a_n$.

To summarize, $H_n \leq G_n \leq A_n \leq Q_n$, and equalities hold if and only if $a_1 = a_2 = \cdots = a_n$. $\qquad \square$

Here, we used a frequently used method for proving propositions related to positive integers: mathematical induction. The mean value inequalities can be proved by many methods.

The mean value inequalities involve the sum and product of n positive numbers, so whether one can view some parts of the expressions in the problem to be proved or solved as a sum or product of some n positive numbers, will be the key point whether using this inequality can finish the task of solving the problem. When using mean value inequalities to solve problems, one should be good at deforming the structure of the expression properly. In particular, one should pay an attention to performing multi-direction and invertible imaginations, thus obtaining successful information.

2. Illustrative Examples

Example 1. Let $a, b, c \in \mathbf{R}_+$ and $abc \leq 1$. Prove:

$$\frac{1}{a} + \frac{1}{b} + \frac{1}{c} \geq 1 + \frac{6}{a + b + c}.$$

Solution. The original inequality is equivalent to

$$\left(\frac{1}{a} + \frac{1}{b} + \frac{1}{c}\right)(a + b + c) \geq a + b + c + 6.$$

Since $abc \leq 1$ and $a, b, c \in \mathbf{R}_+$, we have

$$\frac{1}{a} + \frac{1}{b} + \frac{1}{c} \geq 3\sqrt[3]{\frac{1}{abc}} \geq 3.$$

Consequently,

$$\frac{1}{3}\left(\frac{1}{a} + \frac{1}{b} + \frac{1}{c}\right)(a + b + c) \geq \frac{1}{3} \cdot 3(a + b + c) = a + b + c.$$

On the other hand,

$$\frac{2}{3}\left(\frac{1}{a} + \frac{1}{b} + \frac{1}{c}\right)(a + b + c) \geq \frac{2}{3} \cdot 3\sqrt[3]{\frac{1}{abc}} \cdot 3\sqrt[3]{abc} = 6. \tag{20.1}$$

Hence,

$$\left(\frac{1}{a} + \frac{1}{b} + \frac{1}{c}\right)(a + b + c) \geq a + b + c + 6. \qquad \square$$

Remark. The inequality (20.1) can be proved by the Cauchy inequality.

Example 2. Assume that four real numbers a, b, c, and d satisfy $abcd > a^2 + b^2 + c^2 + d^2$. Prove:

$$abcd > a + b + c + d + 8.$$

Proof. From

$$abcd > a^2 + b^2 + c^2 + d^2 \geq 4\sqrt[4]{a^2 b^2 c^2 d^2},$$

we get $\sqrt[4]{abcd} > 2$.

Suppose that $abcd \leq a + b + c + d + 8$. Then

$$a^2 + b^2 + c^2 + d^2 < a + b + c + d + 8,$$

that is,

$$\left(a - \frac{1}{2}\right)^2 + \left(b - \frac{1}{2}\right)^2 + \left(c - \frac{1}{2}\right)^2 + \left(d - \frac{1}{2}\right)^2 < 9.$$

It follows that

$$\frac{\left(a - \frac{1}{2}\right) + \left(b - \frac{1}{2}\right) + \left(c - \frac{1}{2}\right) + \left(d - \frac{1}{2}\right)}{4}$$

$$\leq \sqrt{\frac{\left(a - \frac{1}{2}\right)^2 + \left(b - \frac{1}{2}\right)^2 + \left(c - \frac{1}{2}\right)^2 + \left(d - \frac{1}{2}\right)^2}{4}}$$

$$< \sqrt{\frac{9}{4}} = \frac{3}{2},$$

in other words,

$$a + b + c + d < 8.$$

Thus,

$$2 < \sqrt[4]{abcd} \leq \frac{a + b + c + d}{4} < \frac{8}{4} = 2,$$

a contradiction.

Hence, $abcd > a + b + c + d + 8$. $\qquad\square$

Remark. The arithmetic mean-geometric mean inequality implied $a^2 + b^2 + c^2 + d^2 \geq 4\sqrt[4]{a^2 b^2 c^2 d^2}$, and from the known inequality one got $\sqrt[4]{abcd} > 2$ further. This gave a preparation for making a contradiction by using the method of contradiction later.

Example 3. Suppose that $x_1, x_2, \ldots, x_{n+1}$ are positive real numbers. Prove:

$$\frac{1}{x_1} + \frac{x_1}{x_2} + \frac{x_1 x_2}{x_3} + \frac{x_1 x_2 x_3}{x_4} + \cdots + \frac{x_1 x_2 \cdots x_n}{x_{n+1}} \geq 4(1 - x_1 x_2 \cdots x_{n+1}).$$

$$(20.2)$$

Proof. Let $a, b > 0$. By the arithmetic mean-geometric mean inequality,

$$4ab + \frac{a}{b} \geq 2\sqrt{4ab \cdot \frac{a}{b}} = 4a,$$

namely

$$\frac{a}{b} \geq 4(a - ab).$$

Consequently,

$$\frac{x_1 x_2 \cdots x_k}{x_{k+1}} \geq 4(x_1 x_2 \cdots x_k - x_1 x_2 \cdots x_{k+1}) \quad (k = 1, 2, \ldots, n+1),$$

from which the left hand side of (20.2)

$$\geq 4(1 - x_1) + 4(x_1 - x_1 x_2) + \cdots + 4(x_1 x_2 \cdots x_n - x_1 x_2 \cdots x_{n+1})$$

$$= 4(1 - x_1 x_2 \cdots x_{n+1}) = \text{the right hand side of (20.2)}. \qquad \square$$

Example 4. Let real numbers $a_1, a_2, \ldots, a_{2016}$ satisfy $9a_i > 11a_{i+1}^2$ ($i = 1, 2, \ldots, 2015$). Find the maximum value of

$$(a_1 - a_2^2)(a_2 - a_3^2) \cdots (a_{2015} - a_{2016}^2)(a_{2016} - a_1^2).$$

Solution. Denote $P = (a_1 - a_2^2)(a_2 - a_3^2) \cdots (a_{2015} - a_{2016}^2)(a_{2016} - a_1^2)$. From the given condition, for $i = 1, 2, \ldots, 2015$,

$$a_i - a_{i+1}^2 > \frac{11}{9}a_{i+1}^2 - a_{i+1}^2 \geq 0.$$

If $a_{2016} - a_1^2 \leq 0$, then $P \leq 0$.

In the following, we consider the case that $a_{2016} - a_1^2 > 0$. Denote $a_{2017} = a_1$. By the arithmetic mean-geometric mean inequality,

$$P^{\frac{1}{2016}} \leq \frac{1}{2016} \sum_{i=1}^{2016} (a_i - a_{i+1}^2) = \frac{1}{2016} \left(\sum_{i=1}^{2016} a_i - \sum_{i=1}^{2016} a_{i+1}^2 \right)$$

$$= \frac{1}{2016} \left(\sum_{i=1}^{2016} a_i - \sum_{i=1}^{2016} a_i^2 \right) = \frac{1}{2016} \sum_{i=1}^{2016} a_i(1 - a_i)$$

$$\leq \frac{1}{2016} \sum_{i=1}^{2016} \left[\frac{a_i + (1 - a_i)}{2} \right]^2$$

$$= \frac{1}{2016} \cdot 2016 \cdot \frac{1}{4} = \frac{1}{4},$$

so

$$P \leq \frac{1}{4^{2016}}.$$

When $a_1 = a_2 = \cdots = a_{2016} = \frac{1}{2}$, the equality in the above inequality holds and the condition

$$9a_i > 11a_{i+1}^2 \ (i = 1, 2, \ldots, 2015)$$

is satisfied. Now, $P = \frac{1}{4^{2016}}$. □

Example 5. Assume that $a, b,$ and c are positive real numbers and $a + b + c = 1$. Prove:

$$\frac{a}{a + b^2} + \frac{b}{b + c^2} + \frac{c}{c + a^2} \leq \frac{1}{4} \left(\frac{1}{a} + \frac{1}{b} + \frac{1}{c} \right).$$

Proof. From the given condition and the arithmetic mean-geometric mean inequality,

$$\frac{a}{a + b^2} = \frac{a}{a(a + b + c) + b^2}$$

$$= \frac{a}{a^2 + b^2 + ab + ac}$$

$$\leq \frac{a}{2ab + ab + ac} = \frac{1}{3b + c}.$$

By the arithmetic mean-harmonic mean inequality,

$$\frac{4}{\frac{3}{b} + \frac{1}{c}} = \frac{4}{\frac{1}{b} + \frac{1}{b} + \frac{1}{b} + \frac{1}{c}} \leq \frac{b + b + b + c}{4} = \frac{3b + c}{4}.$$

So,

$$\frac{a}{a+b^2} \le \frac{1}{3b+c} \le \frac{1}{16}\left(\frac{3}{b}+\frac{1}{c}\right).$$

Similarly,

$$\frac{b}{b+c^2} \le \frac{1}{16}\left(\frac{3}{c}+\frac{1}{a}\right),$$

$$\frac{c}{c+a^2} \le \frac{1}{16}\left(\frac{3}{a}+\frac{1}{b}\right).$$

Summing up the above three inequalities gives

$$\frac{a}{a+b^2} + \frac{b}{b+c^2} + \frac{c}{c+a^2} \le \frac{1}{16}\left(\frac{4}{a}+\frac{4}{b}+\frac{4}{c}\right) = \frac{1}{4}\left(\frac{1}{a}+\frac{1}{b}+\frac{1}{c}\right). \qquad \square$$

Example 6. Suppose that $a, b,$ and c are positive real numbers satisfying $abc = 8$. Prove:

$$\frac{a^2}{\sqrt{(1+b^3)(1+c^3)}} + \frac{b^2}{\sqrt{(1+c^3)(1+a^3)}} + \frac{c^2}{\sqrt{(1+a^3)(1+b^3)}} \ge \frac{4}{3}.$$

Proof. Let $a = 2x, b = 2y,$ and $c = 2z$. Then $xyz = 1$, and the original inequality is equivalent to

$$\frac{x^2}{\sqrt{(1+8y^3)(1+8z^3)}} + \frac{y^2}{\sqrt{(1+8z^3)(1+8x^3)}} + \frac{z^2}{\sqrt{(1+8x^3)(1+8y^3)}} \ge \frac{1}{3}.$$

Since $1 + y^2 \ge 2y$, so $y^2 + y^4 \ge 2y^3$, and thus

$$4y^4 + 4y^2 + 1 \ge 8y^3 + 1,$$

namely

$$1 + 8y^3 \le (1 + 2y^2)^2.$$

Hence, it is enough to show that

$$\frac{x^2}{(1+2y^2)(1+2z^2)} + \frac{y^2}{(1+2z^2)(1+2x^2)} + \frac{z^2}{(1+2x^2)(1+2y^2)} \ge \frac{1}{3}.$$

The above inequality is equivalent to

$$3x^2(1+2x^2) + 3y^2(1+2y^2) + 3z^2(1+2z^2)$$
$$\ge (1+2x^2)(1+2y^2)(1+2z^2),$$

which is equivalent to

$$x^2 + y^2 + z^2 + 6(x^4 + y^4 + z^4) \geq 9 + 4(x^2y^2 + y^2z^2 + z^2x^2).$$

Since $x^4 + y^4 + z^4 \geq x^2y^2 + y^2z^2 + z^2x^2$, so it is sufficient to show:

$$x^2 + y^2 + z^2 + 2(x^4 + y^4 + z^4) \geq 9.$$

By the arithmetic mean-geometric mean inequality,

$$x^2 + y^2 + z^2 \geq 3\sqrt[3]{x^2y^2z^2} = 3,$$

$$x^4 + y^4 + z^4 \geq 3\sqrt[3]{x^4y^4z^4} = 3,$$

and consequently $x^2 + y^2 + z^2 + 2(x^4 + y^4 + z^4) \geq 3 + 2 \cdot 3 = 9$. That is, the original inequality is satisfied. $\square$

Example 7. Assume that positive integers $a_1, a_2, \ldots, a_n$ satisfy

$$\sum_{i=1}^{n} ia_i = 6n, \qquad \sum_{i=1}^{n} \frac{i}{a_i} = 2 + \frac{1}{n}.$$

Find the value of the positive integer n.

Solution. The arithmetic mean-harmonic mean inequality implies

$$\frac{6n}{\frac{1}{2}n(n+1)} = \frac{a_1 + 2a_2 + \cdots + na_n}{\frac{1}{2}n(n+1)}$$

$$\geq \frac{\frac{1}{2}n(n+1)}{\frac{1}{a_1} + \frac{2}{a_2} + \cdots + \frac{n}{a_n}} = \frac{\frac{1}{2}n(n+1)}{2 + \frac{1}{n}}.$$

Note that

$$\frac{6n}{\frac{1}{2}n(n+1)} = \frac{12}{n+1} < \frac{12}{n},$$

$$\frac{\frac{1}{2}n(n+1)}{2 + \frac{1}{n}} = \frac{\frac{1}{2}n^2(n+1)}{2n+1} > \frac{\frac{1}{2}n^2(n+1)}{2n+2} = \frac{1}{4}n^2.$$

It follows that $\frac{12}{n} > \frac{1}{4}n^2$. In other words, $n^3 < 48$, namely $n \leq 3$.

When $n = 1$, we have $a_1 = 6$ and $\frac{1}{a_1} = 3$, a contradiction. So, $n \neq 1$.

When $n = 2$, the equations $a_1 + 2a_2 = 12$ and $\frac{1}{a_1} + \frac{2}{a_2} = 2 + \frac{1}{2}$ have no positive integer solutions, and thus $n \neq 2$.

When $n = 3$, we obtain $a_1 = 6, a_2 = 3$, and $a_3 = 2$, therefore $n = 3$ satisfies the condition.

In summary, $n = 3$ is the unique solution. $\square$

The mean value inequalities can be generalized further. Let $x_1, x_2, \ldots, x_n$ be n positive real numbers and a real number $r \neq 0$. Then the number

$$M_r = \left(\frac{x_1^r + x_2^r + \cdots + x_n^r}{n} \right)^{\frac{1}{r}}$$

is called the rth power mean of $x_1, x_2, \ldots, x_n$.

Theorem. *If α and β are nonzero real numbers and $\alpha > \beta$, then $M_\alpha \geq M_\beta$, that is,*

$$\left(\frac{x_1^\alpha + x_2^\alpha + \cdots + x_n^\alpha}{n} \right)^{\frac{1}{\alpha}} \geq \left(\frac{x_1^\beta + x_2^\beta + \cdots + x_n^\beta}{n} \right)^{\frac{1}{\beta}},$$

and the equality holds if and only if $x_1 = x_2 = \cdots = x_n$. This inequality is called the power mean inequality.

Example 8. Suppose that the three sides of a triangle are $a, b,$ and c respectively and its area is S. Prove:

$$a^n + b^n + c^n \geq 2^n 3^{\frac{4-n}{4}} S^{\frac{n}{2}}, \quad n \in \mathbf{N}_+.$$

Proof. When $n = 1$, the fact $a + b + c \geq 2\sqrt{3\sqrt{3}S}$ is a frequently seen geometric inequality, so the conclusion is true for $n = 1$.

Assume that the conclusion is true for $n = k$, namely

$$a^k + b^k + c^k \geq 2^k \cdot 3^{\frac{4-k}{4}} \cdot S^{\frac{k}{2}}.$$

Then for $n = k + 1$, by the power mean inequality,

$$\left(\frac{a^{k+1} + b^{k+1} + c^{k+1}}{3} \right)^{\frac{1}{k+1}} \geq \left(\frac{a^k + b^k + c^k}{3} \right)^{\frac{1}{k}} \geq \left(\frac{2^k}{3} \cdot 3^{\frac{4-k}{4}} \cdot S^{\frac{k}{2}} \right)^{\frac{1}{k}},$$

and thus

$$a^{k+1} + b^{k+1} + c^{k+1} \geq 3 \left(\frac{2^k}{3} \cdot 3^{\frac{4-k}{4}} \cdot S^{\frac{k}{2}} \right)^{\frac{k+1}{k}} = 2^{k+1} \cdot 3^{\frac{4-(k+1)}{4}} \cdot S^{\frac{k+1}{2}}.$$

In other words, the original inequality is satisfied for $n = k + 1$. $\square$

Remark. Since the general power mean inequality is rarely used in competitions, we do not give more discussions here.

3. Exercises

Group A

I. Filling Problems

1. Two positive real numbers x and y satisfy $\frac{(x+2)^2}{y} + \frac{(y+2)^2}{x} = 16$. Then the value of $x + y$ is _____.

2. Let $a = \sqrt{3x+1} + \sqrt{3y+1} + \sqrt{3z+1}$, where
$$x + y + z = 1 \quad \text{and} \quad x, y, z \geq 0.$$
Then $[a] =$ _____.

3. If $a, d \geq 0, b, c > 0$, and $b + c \geq a + d$, then the minimum value of $\frac{b}{c+d} + \frac{c}{a+b}$ is _____.

II. Calculation Problems

4. Let $a, b, c \in \mathbf{R}_+$. Prove:
$$a + \sqrt{ab} + \sqrt[3]{abc} \leq \frac{4}{3}(a + b + c).$$

5. Suppose that $x, y, z, w \in \mathbf{R}_+$. Solve the system of equations
$$\begin{cases} x + y + z + w = 4, \\ \dfrac{1}{x} + \dfrac{1}{y} + \dfrac{1}{z} + \dfrac{1}{w} = 5 - \dfrac{1}{xyzw}. \end{cases}$$

6. Assume that x, y, and z are all positive real numbers and $xyz = 1$. Prove:
$$\frac{x^6 + 2}{x^3} + \frac{y^6 + 2}{y^3} + \frac{z^6 + 2}{z^3} \geq 3\left(\frac{x}{y} + \frac{y}{z} + \frac{z}{x}\right).$$

7. Let $n \in \mathbf{N}_+$. Prove: $n\left[(1+n)^{\frac{1}{n}} - 1\right] < 1 + \frac{1}{2} + \frac{1}{3} + \cdots + \frac{1}{n}$.

8. Assume that a, b, and c are positive real numbers and satisfy $a^2 + b^2 + c^2 = \frac{1}{2}$. Prove:
$$\frac{1 - a^2 + c^2}{c(a + 2b)} + \frac{1 - b^2 + a^2}{a(b + 2c)} + \frac{1 - c^2 + b^2}{b(c + 2a)} \geq 6.$$

Group B

9. Let x, y, and z be positive numbers and $x^2 + y^2 + z^2 = 1$. Find the minimum value of
$$S = \frac{xy}{z} + \frac{yz}{x} + \frac{zx}{y}.$$

10. If positive numbers a, b, and c satisfy $a + b + c = 1$, prove:

$$\left(a + \frac{1}{a}\right)\left(b + \frac{1}{b}\right)\left(c + \frac{1}{c}\right) \geq \frac{1000}{27}.$$

11. Given a positive integer k, if $x^k + y^k + z^k = 1$, find the minimum value of $x^{k+1} + y^{k+1} + z^{k+1}$.

12. Let n be any positive integer greater than 1 and let $a_1, a_2, \ldots, a_n$ be positive numbers. Find the maximal positive number λ such that

$$\frac{\sqrt{a_1^{n-1}}}{\sqrt{a_1^{n-1} + (n^2 - 1)a_2 a_3 \cdots a_n}} + \frac{\sqrt{a_2^{n-1}}}{\sqrt{a_2^{n-1} + (n^2 - 1)a_1 a_3 \cdots a_n}}$$

$$+ \cdots + \frac{\sqrt{a_n^{n-1}}}{\sqrt{a_n^{n-1} + (n^2 - 1)a_1 a_2 \cdots a_{n-1}}} \geq \lambda$$

is always satisfied.

Chapter 21

Cauchy Inequality

1. Key Points of Knowledge and Basic Methods

Cauchy inequality. Let $a_i, b_i \in \mathbf{R}$ $(i = 1, 2, \ldots, n)$. Then

$$\left(\sum_{i=1}^{n} a_i b_i \right)^2 \leq \left(\sum_{i=1}^{n} a_i^2 \right) \left(\sum_{i=1}^{n} b_i^2 \right),$$

and when $a_1, a_2, \ldots, a_n$ are not all zero, the equality in the above inequality holds if and only if $b_i = \lambda a_i$ $(1 \leq i \leq n)$ for some real number λ.

Proof. If $\sum_{i=1}^{n} a_i^2 = 0$, then $a_1 = a_2 = \cdots = a_n = 0$, and the inequality is obviously true.

When $\sum_{i=1}^{n} a_i^2 \neq 0$, construct the quadratic function

$$f(x) = \left(\sum_{i=1}^{n} a_i^2 \right) \cdot x^2 - 2 \left(\sum_{i=1}^{n} a_i b_i \right) \cdot x + \sum_{i=1}^{n} b_i^2.$$

It is easy to see that $f(x) = \sum_{i=1}^{n} (a_i x - b_i)^2 \geq 0$, so the discriminant of the quadratic function $\Delta \leq 0$, that is

$$4 \left(\sum_{i=1}^{n} a_i b_i \right)^2 - 4 \left(\sum_{i=1}^{n} a_i^2 \right) \left(\sum_{i=1}^{n} b_i^2 \right) \leq 0.$$

In other words,

$$\left(\sum_{i=1}^{n} a_i b_i \right)^2 \leq \left(\sum_{i=1}^{n} a_i^2 \right) \left(\sum_{i=1}^{n} b_i^2 \right),$$

and the equality is valid if and only if $a_i x - b_i = 0$ for all i, namely $b_i = \lambda a_i$ $(1 \leq i \leq n)$ for some real number λ. $\qquad \square$

There are many proofs of the Cauchy inequality; here we chose constructing a quadratic function to prove it.

Making appropriate substitutions for the number arrays a_i ($i = 1, 2, \ldots, n$) and b_i ($i = 1, 2, \ldots, n$) in the Cauchy inequality produces the following commonly used corollaries:

Corollary 1. *Let $a_i \in \mathbf{R}$ ($i = 1, 2, \ldots, n$). Then $n \sum_{i=1}^{n} a_i^2 \geq \left(\sum_{i=1}^{n} a_i \right)^2$, where the equality holds if and only if $a_1 = a_2 = \cdots = a_n$.*

Corollary 2. *Let $a_i \in \mathbf{R}$ ($i = 1, 2, \ldots, n$). Then $\left(\sum_{i=1}^{n} a_i \right) \left(\sum_{i=1}^{n} \frac{1}{a_i} \right) \geq n^2$, where the equality holds if and only if $a_1 = a_2 = \cdots = a_n$.*

Corollary 3. *Let $a_i, b_i \in \mathbf{R}$ ($i = 1, 2, \ldots, n$) be such that $b_i \neq 0$ and $\frac{a_i}{b_i} \geq 0$ for $i = 1, 2, \ldots, n$. Then $\left(\sum_{i=1}^{n} a_i b_i \right) \left(\sum_{i=1}^{n} \frac{a_i}{b_i} \right) \geq \left(\sum_{i=1}^{n} a_i \right)^2$, where the equality holds if and only if $b_1^2 = b_2^2 = \cdots = b_n^2$.*

Corollary 4. *Let $a_i, b_i \in \mathbf{R}$ ($i = 1, 2, \ldots, n$) be such that $b_i \neq 0$. Then $\left(\sum_{i=1}^{n} b_i \right) \left(\sum_{i=1}^{n} \frac{a_i^2}{b_i} \right) \geq \left(\sum_{i=1}^{n} a_i \right)^2$, where the equality holds if and only if $a_i^2 = \lambda b_i^2$ and $b_i > 0$ for $i = 1, 2, \ldots, n$ with λ some real number.*

The key point of applying the Cauchy inequality to solve problems is to find or construct two groups of numbers a_i ($i = 1, 2, \ldots, n$) and b_i ($i = 1, 2, \ldots, n$) in the expressions already known or to be solved.

2. Illustrative Examples

Example 1. Let $a, b, c \in \mathbf{R}_+$ and $a + b + c = 3$. Prove:
$$\sqrt{\frac{b}{a^2 + 3}} + \sqrt{\frac{c}{b^2 + 3}} + \sqrt{\frac{a}{c^2 + 3}} \leq \frac{3}{2} \sqrt[4]{\frac{1}{abc}}.$$

Proof. Denote $u = \left(\frac{1}{\sqrt{a^2+3}}, \frac{1}{\sqrt{b^2+3}}, \frac{1}{\sqrt{c^2+3}} \right)$ and $v = (\sqrt{b}, \sqrt{c}, \sqrt{a})$. Then the Cauchy inequality and the given condition imply that

$$\left(\sqrt{\frac{b}{a^2 + 3}} + \sqrt{\frac{c}{b^2 + 3}} + \sqrt{\frac{a}{c^2 + 3}} \right)^2$$

$$\leq \left(\frac{1}{a^2 + 3} + \frac{1}{b^2 + 3} + \frac{1}{c^2 + 3} \right) (b + c + a)$$

$$= 3 \left(\frac{1}{a^2 + 3} + \frac{1}{b^2 + 3} + \frac{1}{c^2 + 3} \right).$$

By the arithmetic mean-geometric mean inequality,

$$a^2 + 3 = a^2 + 1 + 1 + 1 \geq 4\sqrt[4]{a^2} = 4\sqrt{a},$$

and similarly $b^2 + 3 \geq 4\sqrt{b}$ and $c^2 + 3 \geq 4\sqrt{c}$. Thus,

$$\frac{1}{a^2+3} + \frac{1}{b^2+3} + \frac{1}{c^2+3} \leq \frac{1}{4}\left(\frac{1}{\sqrt{a}} + \frac{1}{\sqrt{b}} + \frac{1}{\sqrt{b}}\right)$$

$$= \frac{\sqrt{ab} + \sqrt{bc} + \sqrt{ca}}{4\sqrt{abc}} \leq \frac{\frac{a+b}{2} + \frac{b+c}{2} + \frac{c+a}{2}}{4\sqrt{abc}} = \frac{a+b+c}{4\sqrt{abc}},$$

and so

$$\left(\sqrt{\frac{b}{a^2+3}} + \sqrt{\frac{c}{b^2+3}} + \sqrt{\frac{a}{c^2+3}}\right)^2 \leq \frac{3(a+b+c)}{4\sqrt{abc}} = \frac{9}{4\sqrt{abc}}.$$

When $a = b = c = 1$, the above inequality becomes equality. $\qquad\square$

Remark. The proof here cannot be made one step to finish. The application of the Cauchy inequality is a key step.

Example 2. Suppose that x, y, and z are positive numbers and $x + y + z \geq 3$. Prove:

$$\frac{1}{x+y+z^2} + \frac{1}{y+z+x^2} + \frac{1}{z+x+y^2} \leq 1,$$

and express a condition for the equality to be satisfied.

Proof. By the Cauchy inequality, we have

$$(x + y + z^2)(x + y + 1) \geq (x + y + z)^2, \tag{21.1}$$

namely

$$\frac{1}{x+y+z^2} \leq \frac{x+y+1}{(x+y+z)^2}. \tag{21.2}$$

Hence,

$$\frac{x+y+1}{(x+y+z)^2} + \frac{y+z+1}{(y+z+x)^2} + \frac{z+x+1}{(z+x+y)^2}$$

$$= \frac{2(x+y+z)+3}{(x+y+z)^2} \leq 1$$

$$\Leftrightarrow (x+y+z)^2 - 2(x+y+z) - 3 \geq 0.$$

When $x + y + z \geq 3$, the above inequality is obviously correct.

The equality holds in (21.1) $\Leftrightarrow (x, y, z^2) = (x, y, 1)$, from which $z^2 = 1$, namely $z = 1$.

By symmetry, the original inequality becomes equality if and only if $x = y = z = 1$. $\qquad\square$

Remark. Here, we first used the Cauchy inequality to prove the "local" inequality (21.2).

Example 3. Let a, b, and c be the lengths of the three sides of a triangle. Prove:

$$\frac{(c+a-b)^4}{a(a+b-c)} + \frac{(a+b-c)^4}{b(b+c-a)} + \frac{(b+c-a)^4}{c(c+a-b)} \geq ab + bc + ca.$$

Analysis. In order to make the structure of the inequality simpler and clearer, we first give the "tangent length substitution":

$$c + a - b = x, \quad a + b - c = y, \quad b + c - a = z.$$

Proof. Let $c + a - b = x, a + b - c = y$, and $b + c - a = z$. Then

$$a = \frac{x+y}{2}, \quad b = \frac{y+z}{2}, \quad c = \frac{z+x}{2},$$

$$a + b + c = x + y + z.$$

Denote the left hand side of the inequality to be proved as S. Then

$$S = \frac{2x^4}{y(x+y)} + \frac{2y^4}{z(y+z)} + \frac{2z^4}{x(z+x)}.$$

The Cauchy inequality ensures

$$\frac{S}{2}[y(x+y) + z(y+z) + x(z+x)] \geq (x^2 + y^2 + z^2)^2.$$

It follows that

$$S \geq \frac{2(x^2 + y^2 + z^2)^2}{x^2 + y^2 + z^2 + xy + yz + zx}$$

$$\geq \frac{2(x^2 + y^2 + z^2)^2}{x^2 + y^2 + z^2 + (x^2 + y^2 + z^2)}$$

$$= x^2 + y^2 + z^2 \geq \frac{(x+y+z)^2}{3}$$

$$= \frac{(a+b+c)^2}{3} \geq ab + bc + ca. \qquad\square$$

Example 4. Let $a_1, a_2, \ldots, a_n$ $(n \geq 2)$ be real numbers. Prove: One may choose $\varepsilon_1, \varepsilon_2, \ldots, \varepsilon_n \in (-1, 1)$ such that

$$\left(\sum_{i=1}^{n} a_i \right)^2 + \left(\sum_{i=1}^{n} \varepsilon_i a_i \right)^2 \leq (n+1) \sum_{i=1}^{n} a_i^2.$$

Proof. We show that

$$\left(\sum_{i=1}^{n} a_i \right)^2 + \left(\sum_{i=1}^{\left[\frac{n}{2} \right]} a_i - \sum_{j=\left[\frac{n}{2} \right]+1}^{n} a_j \right)^2 \leq (n+1) \sum_{i=1}^{n} a_i^2. \qquad (21.3)$$

This means that, if we choose $\varepsilon_i = 1$ for $i = 1, \ldots, \left[\frac{n}{2} \right]$ and $\varepsilon_i = -1$ for $i = \left[\frac{n}{2} \right] + 1$, then the conclusion is true. (Here, $[x]$ represents the integral part of the real number x.)

In fact, the left hand side of (21.3) is

$$\left(\sum_{i=1}^{\left[\frac{n}{2} \right]} a_i + \sum_{j=\left[\frac{n}{2} \right]+1}^{n} a_j \right)^2 + \left(\sum_{i=1}^{\left[\frac{n}{2} \right]} a_i - \sum_{j=\left[\frac{n}{2} \right]+1}^{n} a_j \right)^2$$

$$= 2 \left(\sum_{i=1}^{\left[\frac{n}{2} \right]} a_i \right)^2 + 2 \left(\sum_{j=\left[\frac{n}{2} \right]+1}^{n} a_j \right)^2.$$

By the Cauchy inequality, and the facts that $n - \left[\frac{n}{2} \right] = \left[\frac{n+1}{2} \right]$ and $[x] \leq x$, we have

$$2 \left(\sum_{i=1}^{\left[\frac{n}{2} \right]} a_i \right)^2 + 2 \left(\sum_{j=\left[\frac{n}{2} \right]+1}^{n} a_j \right)^2 \leq 2 \left[\frac{n}{2} \right] \sum_{i=1}^{\left[\frac{n}{2} \right]} a_i^2 + 2 \left(n - \left[\frac{n}{2} \right] \right) \sum_{j=\left[\frac{n}{2} \right]+1}^{n} a_j^2$$

$$= 2 \left[\frac{n}{2} \right] \sum_{i=1}^{\left[\frac{n}{2} \right]} a_i^2 + 2 \left[\frac{n+1}{2} \right] \sum_{j=\left[\frac{n}{2} \right]+1}^{n} a_j^2$$

$$\leq n \sum_{i=1}^{\left[\frac{n}{2} \right]} a_i^2 + (n+1) \sum_{j=\left[\frac{n}{2} \right]+1}^{n} a_j^2$$

$$\leq (n+1) \sum_{i=1}^{n} a_i^2,$$

which gives (21.3), so the proposition is proved. $\qquad \square$

Example 5. We are given n $(n \geq 2)$ positive real numbers $a_1, a_2, \ldots, a_n$ that satisfy

$$\sum_{i=1}^{n} a_i \cdot \sum_{i=1}^{n} \frac{1}{a_i} \leq \left(n + \frac{1}{2} \right)^2.$$

Prove: $\max\{a_1, a_2, \ldots, a_n\} \leq 4 \min\{a_1, a_2, \ldots, a_n\}$.

Proof. Denote $m = \min\{a_1, a_2, \ldots, a_n\}$ and $M = \max\{a_1, a_2, \ldots, a_n\}$. From the symmetry, without loss of generality we may assume that

$$m = a_1 \leq a_2 \leq \cdots \leq a_n = M.$$

By the Cauchy inequality, we have

$$\left(n + \frac{1}{2} \right)^2 \geq (a_1 + a_2 + \cdots + a_n) \left(\frac{1}{a_1} + \frac{1}{a_2} + \cdots + \frac{1}{a_n} \right)$$

$$= (m + a_2 + \cdots + a_{n-1} + M) \left(\frac{1}{M} + \frac{1}{a_2} + \cdots + \frac{1}{a_{n-1}} + \frac{1}{m} \right)$$

$$\geq \left(\sqrt{\frac{m}{M}} + 1 + \cdots + 1 + \sqrt{\frac{M}{m}} \right)^2$$

$$= \left(\sqrt{\frac{m}{M}} + n - 2 + \sqrt{\frac{M}{m}} \right)^2,$$

namely

$$2(m + M) \leq 5\sqrt{Mm},$$

which is the same as

$$2M - 5\sqrt{Mm} + 2m \leq 0,$$

$$(2\sqrt{M} - \sqrt{m})(\sqrt{M} - 2\sqrt{m}) \leq 0.$$

Therefore, $\sqrt{M} \leq 2\sqrt{m}$, that is $M \leq 4m$. $\qquad \square$

Example 6. Let $a_1, a_2, \ldots, a_n$ be given real numbers not all of which are zero, and let $r_1, r_2, \ldots, r_n$ be real numbers. If an inequality $\sum_{k=1}^{n} r_k (x_k - a_k) \leq \left(\sum_{k=1}^{n} x_k^2 \right)^{\frac{1}{2}} - \left(\sum_{k=1}^{n} a_k^2 \right)^{\frac{1}{2}}$ is satisfied for any real numbers $x_1, x_2, \ldots, x_n$, find the values of $r_1, r_2, \ldots, r_n$.

Solution. From the given condition, if we choose $x_1 = x_2 = \cdots = x_n = 0$, then

$$\sum_{k=1}^{n} r_k a_k \geq \left(\sum_{k=1}^{n} a_k^2 \right)^{\frac{1}{2}}, \tag{21.4}$$

and if we choose $x_k = 2a_k$ $(k = 1, 2, \ldots, n)$, then

$$\sum_{k=1}^{n} r_k a_k \leq \left(\sum_{k=1}^{n} a_k^2 \right)^{\frac{1}{2}}. \tag{21.5}$$

So, (21.4) and (21.5) imply that

$$\sum_{k=1}^{n} r_k a_k = \left(\sum_{k=1}^{n} a_k^2 \right)^{\frac{1}{2}}. \tag{21.6}$$

By the Cauchy inequality,

$$\sum_{k=1}^{n} r_k a_k \leq \left(\sum_{k=1}^{n} r_k^2 \right)^{\frac{1}{2}} \left(\sum_{k=1}^{n} a_k^2 \right)^{\frac{1}{2}}. \tag{21.7}$$

Since $a_1, a_2, \ldots, a_n$ are not all zero, the formulas (21.6) and (21.7) ensure that

$$\left(\sum_{k=1}^{n} r_k^2 \right)^{\frac{1}{2}} \geq 1. \tag{21.8}$$

Choose $x_k = r_k$ $(k = 1, 2, \ldots, n)$. From the given condition and (21.6), we obtain

$$\sum_{k=1}^{n} r_k^2 \leq \left(\sum_{k=1}^{n} r_k^2 \right)^{\frac{1}{2}},$$

namely

$$\sum_{k=1}^{n} r_k^2 \leq 1. \tag{21.9}$$

Combining the formulas (21.8) and (21.9), we have

$$\sum_{k=1}^{n} r_k^2 = 1. \tag{21.10}$$

By multiplying the same number 1 and $\sum_{k=1}^{n} r_k^2$ to the left and right hand sides of (21.6), respectively, it follows that the equality in the Cauchy inequality (21.7) is valid, hence there is a real number λ such that

$$r_k = \lambda a_k, \quad k = 1, 2, \ldots, n. \tag{21.11}$$

Substituting (21.11) into (21.6) results in

$$\lambda = \left(\sum_{k=1}^{n} a_k^2 \right)^{-\frac{1}{2}}.$$

Substituting the value of λ back into (21.11), we finally get

$$r_j = a_j \left(\sum_{k=1}^{n} a_k^2 \right)^{-\frac{1}{2}} \quad (j = 1, 2, 3, \ldots, n).$$

$\square$

Example 7. Suppose that $x_1, x_2, \ldots, x_n$ $(n \geq 2)$ are all positive numbers and $\sum_{i=1}^{n} x_i = 1$. Prove:

$$\sum_{i=1}^{n} \frac{x_i}{\sqrt{1 - x_i}} \geq \frac{1}{\sqrt{n-1}} \sum_{i=1}^{n} \sqrt{x_i}.$$

Proof. The left hand side of the inequality can be written as

$$\sum_{i=1}^{n} \frac{x_i}{\sqrt{1 - x_i}} = \sum_{i=1}^{n} \frac{1}{\sqrt{1 - x_i}} - \sum_{i=1}^{n} \sqrt{1 - x_i}. \tag{21.12}$$

For positive numbers $y_1, y_2, \ldots, y_n$, the following inequality

$$\sum_{i=1}^{n} \frac{1}{y_i} \geq \frac{n^2}{\sum_{i=1}^{n} y_i}$$

is a simple corollary of the Cauchy inequality. Choosing $y_i = \sqrt{1 - x_i}$ in the above inequality gives

$$\sum_{i=1}^{n} \frac{1}{\sqrt{1 - x_i}} \geq \frac{n^2}{\sum_{i=1}^{n} \sqrt{1 - x_i}}. \tag{21.13}$$

Using the Cauchy inequality again, we have

$$\sum_{i=1}^{n} \sqrt{1 - x_i} \leq \left(\sum_{i=1}^{n} 1 \right)^{\frac{1}{2}} \left[\sum_{i=1}^{n} (1 - x_i) \right]^{\frac{1}{2}} = \sqrt{n(n-1)} \tag{21.14}$$

and

$$\sum_{i=1}^{n} \sqrt{x_i} \leq \sqrt{n}. \tag{21.15}$$

Combining the formulas (21.12), (21.13), (21.14), and (21.15) gives

$$\sum_{i=1}^{n} \frac{x_i}{\sqrt{1-x_i}} = \sum_{i=1}^{n} \frac{1}{\sqrt{1-x_i}} - \sum_{i=1}^{n} \sqrt{1-x_i}$$

$$\geq \frac{n^2}{\sum_{i=1}^{n} \sqrt{1-x_i}} - \sum_{i=1}^{n} \sqrt{1-x_i}$$

$$\geq \frac{n^2}{\sqrt{n(n-1)}} - \sqrt{n(n-1)} = \sqrt{\frac{n}{n-1}}$$

$$\geq \frac{1}{\sqrt{n-1}} \sum_{i=1}^{n} \sqrt{x_i}.$$

$\square$

Remark. The Cauchy inequality has a very good generalization, namely the famous Hölder inequality:

Assume that $a_i > 0, b_i > 0$ $(i = 1, 2, \ldots, n), p > 1, q > 1$, and $\frac{1}{p} + \frac{1}{q} = 1$. Then

$$\sum_{i=1}^{n} a_i b_i \leq \left(\sum_{i=1}^{n} a_i^p \right)^{\frac{1}{p}} \left(\sum_{i=1}^{n} b_i^q \right)^{\frac{1}{q}},$$

and the equality is valid if and only if $a_i^p = \lambda b_i^q$ $(i = 1, 2, \ldots, n;$ and $\lambda > 0)$.

Example 8. (1) Let three positive real numbers $a, b,$ and c satisfy the condition

$$(a^2 + b^2 + c^2)^2 > 2(a^4 + b^4 + c^4). \tag{21.16}$$

Prove: The numbers $a, b,$ and c must be the lengths of the three sides of a triangle.

(2) Let n $(n > 3)$ positive real numbers $a_1, a_2, \ldots, a_n$ satisfy the inequality

$$(a_1^2 + a_2^2 + \cdots + a_n^2)^2 > (n-1)(a_1^4 + a_2^4 + \cdots + a_1^4). \tag{21.17}$$

Prove: Any three numbers in $\{a_1, a_2, \ldots, a_n\}$ must be the lengths of the three sides of a triangle.

Proof. (1) From the given condition, we know that

$$2a^2b^2 + 2b^2c^2 + 2c^2a^2 - a^4 - b^4 - c^4 > 0.$$

Factorizing the left hand side of the above inequality gives

$$(a + b + c)(a + b - c)(b + c - a)(c + a - b) > 0.$$

By symmetry, we may assume $a \geq b \geq c$, so that three factors of the left hand side are positive, and the above inequality implies that the fourth one is also positive, that is $b + c > a$. This indicates that $a, b,$ and c can be the lengths of the three sides of some triangle.

(2) **Proof 1.** Since $n > 3$, by the Cauchy inequality,

$$(n - 1)(a_1^4 + a_2^4 + \cdots + a_n^4) < (a_1^2 + a_2^2 + \cdots + a_n^2)^2$$

$$= \left(\frac{a_1^2 + a_2^2 + a_3^2}{2} + \frac{a_1^2 + a_2^2 + a_3^2}{2} + a_4^2 + \cdots + a_n^2 \right)^2$$

$$\leq (n - 1) \left[\frac{(a_1^2 + a_2^2 + a_3^2)^2}{4} + \frac{(a_1^2 + a_2^2 + a_3^2)^2}{4} + a_4^4 + \cdots + a_n^4 \right].$$

Canceling the same factor $n - 1$ and eliminating the same terms from the both sides in the above, we have

$$2(a_1^4 + a_2^4 + a_3^4) < (a_1^2 + a_2^2 + a_3^2)^2.$$

This is exactly the same condition as in (1), so by (21.16), $a_1, a_2,$ and a_3 can be the lengths of the three sides of a triangle. From the symmetry of $a_1, a_2, \ldots, a_n$, we know that any three numbers from them can be the lengths of the three sides of a triangle. □

Proof 2. Rewriting the formula (21.17) and using the Cauchy inequality with parameters, we have

$$(n - 1)(a_1^4 + a_2^4 + \cdots + a_1^4) < (a_1^2 + a_2^2 + \cdots + a_n^2)^2$$

$$= \left[\lambda \left(a_1^2 + a_2^2 + a_3^2 \right) \cdot \frac{1}{\lambda} + a_4^2 + \cdots + a_n^2 \right]^2$$

$$\leq \left[\lambda^2 \left(a_1^2 + a_2^2 + a_3^2 \right)^2 + a_4^4 + \cdots + a_n^4 \right] \left(\frac{1}{\lambda^2} + n - 3 \right).$$

In order to eliminate $a_4, \ldots, a_n$ from the both sides of the above, let $\frac{1}{\lambda^2} + n - 3 = n - 1$. Then $\lambda^2 = \frac{1}{2}$. Substituting $\lambda^2 = \frac{1}{2}$ into the above inequality gives

$$2(a_1^4 + a_2^4 + a_3^4) < (a_1^2 + a_2^2 + a_3^2)^2,$$

and the remaining part of the proof is the same as in Proof 1. □

Remark. Many important inequalities can be written as a form with parameters. For example:

The mean value inequality

$$\sqrt[n]{a_1 a_2 \cdots a_n} \le \frac{1}{n} \sum_{i=1}^{n} \lambda_i a_i,$$

where $a_i > 0, \lambda_i > 0 \ (i = 1, 2, \ldots, n)$, and $\lambda_1 \lambda_2 \cdots \lambda_n = 1$.

The Cauchy inequality:

$$\left(\sum_{i=1}^{n} a_i b_i \right)^2 \le \sum_{i=1}^{n} \lambda_i a_i^2 \cdot \sum_{i=1}^{n} \frac{1}{\lambda_i} b_i^2,$$

where $\lambda_i > 0 \ (i = 1, 2, \ldots, n)$.

3. Exercises

Group A

I. Filling Problems

1. Three real numbers x, y, and z satisfy two equations

$$\begin{cases} 2x + 3y + z = 13, \\ 4x^2 + 9y^2 + z^2 - 2x + 15y + 3z = 82. \end{cases}$$

Then the respective values of x, y, and z are _____.

2. The maximum value of $y = \sqrt{x^2 - x^4} + \sqrt{2x^2 - x^4}$ is _____.

3. Suppose that a, b, and c are all real numbers and $a + 2b + 3c + 4d = \sqrt{10}$. Then the minimum value of $a^2 + b^2 + c^2 + d^2 + (a + b + c + d)^2$ is

_____.

4. Let a_1, a_2, a_3, and a_4 be different numbers among $1, 2, \ldots, 100$ that satisfy

$$(a_1^2 + a_2^2 + a_3^2)(a_2^2 + a_3^2 + a_4^2) = (a_1 a_2 + a_2 a_3 + a_3 a_4)^2.$$

The count of such ordered number arrays (a_1, a_2, a_3, a_4) is

_____.

II. Calculation Problems

5. Let a, b, and c be positive real numbers and $a \cos^2 \theta + b \sin^2 \theta < c$. Prove:

$$\sqrt{a} \cos^2 \theta + \sqrt{b} \sin^2 \theta < \sqrt{c}.$$

6. Find the minimal real number k such that for any $a, b, c, d \in \mathbf{R}$,

$$\sqrt{(a^2 + 1)(b^2 + 1)(c^2 + 1)} + \sqrt{(b^2 + 1)(c^2 + 1)(d^2 + 1)}$$
$$+ \sqrt{(c^2 + 1)(d^2 + 1)(a^2 + 1)} + \sqrt{(d^2 + 1)(a^2 + 1)(b^2 + 1)}$$
$$\geq 2(ab + bc + cd + da) - k.$$

7. Let $a, b, c, d > 0$ and $a + b + c + d = 3$. Find the minimum value of

$$\frac{a^3}{b + c + d} + \frac{b^3}{a + c + d} + \frac{c^3}{a + b + d} + \frac{d^3}{a + b + c}.$$

Group B

8. Let $a_1, a_2, \ldots, a_n$ be all positive numbers and $n \in \mathbf{N}_+$. Prove:

$$\sum_{k=1}^{n} \frac{k^2}{a_1 + a_2 + \cdots + a_k} < 4 \sum_{k=1}^{n} \frac{k}{a_k}.$$

9. Suppose that P is a point inside $\triangle ABC$, numbers r_1, r_2, and r_3 are the distances of P to the three sides whose lengths are a_1, a_2, and a_3, respectively, and R is the radius of the circumscribed circle of $\triangle ABC$. Prove:

$$\sqrt{r_1} + \sqrt{r_2} + \sqrt{r_3} \leq \frac{1}{\sqrt{2R}} (a_1^2 + a_2^2 + a_3^2)^{\frac{1}{2}}.$$

Also find a condition for the equality to be true.

10. Let $a_i \in \mathbf{R}_+$ with $1 \leq i \leq n$. Prove:

$$\frac{1}{\frac{1}{1+a_1} + \frac{1}{1+a_2} + \cdots + \frac{1}{1+a_n}} - \frac{1}{\frac{1}{a_1} + \frac{1}{a_2} + \cdots + \frac{1}{a_n}} \geq \frac{1}{n}.$$

11. Assume that $x \in \left(0, \frac{\pi}{2}\right)$ and $n \in \mathbf{N}_+$. Prove:

$$\frac{1 - \sin^{2n} x}{\sin^{2n} x} \cdot \frac{1 - \cos^{2n} x}{\cos^{2n} x} \geq (2^n - 1)^2.$$

12. Let n be a positive integer and $a_1, a_2, \ldots, a_n$ be real numbers. If for any $1 \leq m \leq n$,

$$a_m + a_{m+1} + \cdots + a_n \geq m + (m+1) + \cdots + n,$$

prove: $a_1^2 + a_2^2 + \cdots + a_n^2 \geq \frac{1}{6}n(n+1)(2n+1)$.

Chapter 22

Rearrangement Inequalities

1. Key Points of Knowledge and Basic Methods

Rearrangement inequalities. Let two groups of numbers $a_1, a_2, \ldots, a_n$ and $b_1, b_2, \ldots, b_n$ satisfy $a_1 \leq a_2 \leq \cdots \leq a_n$ and $b_1 \leq b_2 \leq \cdots \leq b_n$. Then

$$a_1 b_n + a_2 b_{n-1} + \cdots + a_n b_1 \qquad \text{(reverse order sum)}$$

$$\leq a_1 b_{t_1} + a_2 b_{t_2} + \cdots + a_n b_{t_n} \qquad \text{(random order sum)}$$

$$\leq a_1 b_1 + a_2 b_2 + \cdots + a_n b_n, \qquad \text{(same order sum)}$$

where $\{t_1, t_2, \ldots, t_n\} = \{1, 2, \ldots, n\}$. If and only if $a_1 = a_2 = \cdots = a_n$ or $b_1 = b_2 = \cdots = b_n$, the equalities are valid.

Proof. (1) Since

$$(a_n - a_k)(b_n - b_{t_n}) \geq 0 \ (t_n < n, k < n, \text{ and } k \in \mathbf{N}_+),$$

so

$$a_n b_n + a_k b_{t_n} \geq a_k b_n + a_n b_{t_n}.$$

If $b_{t_k} = b_n \ (t_k < n)$, then the above inequality implies that

$$a_1 b_{t_1} + a_2 b_{t_2} + \cdots + a_k b_{t_k} + \cdots + a_n b_{t_n}$$

$$\leq a_1 b_{t_1} + a_2 b_{t_2} + \cdots + a_k b_{t_k} + \cdots + a_n b_{t_n}.$$

If $b_{t_r} = b_{n-1} \ (t_r < n-1)$, then by the same inequality,

$$a_1 b_{t_1} + a_2 b_{t_2} + \cdots + a_r b_{t_r} + \cdots + a_{k-1} b_{t_{k-1}} + \cdots + a_n b_{t_n}$$

$$\leq a_1 b_{t_1} + a_2 b_{t_2} + \cdots + a_r b_{t_r} + \cdots + a_{n-1} b_{n-1} + a_n b_{t_n},$$

Continuing in this way, we obtain the inequalities on the random order sum and the same order sum.

(2) Since $b_1 \leq b_2 \leq \cdots \leq b_n$,

$$-b_1 \geq -b_2 \geq \cdots \geq -b_n.$$

From (1) we know that

$$(-b_n)a_1 + (-b_{n-1})a_2 + \cdots + (-b_1)a_n$$
$$\geq (-b_{t_1})a_1 + (-b_{t_2})a_2 + \cdots + (-b_{t_n})a_n.$$

Dividing (-1) from the both sides of the above proves the inequalities on the reverse order sum and the random order sum. □

The rearrangement inequalities have wide applications in solving competition problems. Their applications appear after reasonably constructing two groups of ordered numbers $a_1 \leq a_2 \leq \cdots \leq a_n$ and $b_1 \leq b_2 \leq \cdots \leq b_n$ according to the need.

2. Illustrative Examples

Example 1. Assume that $a > 0, b > 0$, and $c > 0$. Prove:

$$\frac{1}{a} + \frac{1}{b} + \frac{1}{c} \leq \frac{a^8 + b^8 + c^8}{a^3 b^3 c^3}.$$

Proof. Without loss of generality, assume that $a \geq b \geq c > 0$. Then

$$a^2 \geq b^2 \geq c^2, \quad \frac{1}{c^3} \geq \frac{1}{b^3} \geq \frac{1}{a^3}, \quad \frac{1}{bc} \geq \frac{1}{ca} \geq \frac{1}{ab}.$$

Hence, by the rearrangement inequalities,

$$\frac{1}{a} + \frac{1}{b} + \frac{1}{c} = a^2 \cdot \frac{1}{a^3} + b^2 \cdot \frac{1}{b^3} + c^3 \cdot \frac{1}{c^3}$$
$$\leq a^2 \cdot \frac{1}{c^3} + b^2 \cdot \frac{1}{a^3} + c^3 \cdot \frac{1}{b^3}$$
$$= a^5 \cdot \frac{1}{c^3 a^3} + b^5 \cdot \frac{1}{a^3 b^3} + c^5 \cdot \frac{1}{b^3 c^3}$$
$$\leq a^5 \cdot \frac{1}{b^3 c^3} + b^5 \cdot \frac{1}{c^3 a^3} + c^5 \cdot \frac{1}{a^3 b^3}$$
$$= \frac{a^8 + b^8 + c^8}{a^3 b^3 c^3}.$$

□

Example 2. There are 2007! different sequences from the permutations of the 2007 numbers of $1, 2, 3, \ldots, 2007$. Do there exist four sequences:

$$a_1, a_2, \ldots, a_{2007}; \ b_1, b_2, \ldots, b_{2007}; \ c_1, c_2, \ldots, c_{2007}; \ d_1, d_2, \ldots, d_{2007},$$

such that $a_1 b_1 + a_2 b_2 + \cdots + a_{2007} b_{2007} = 2(c_1 d_1 + c_2 d_2 + \cdots + c_{2007} d_{2007})$? Prove your conclusion.

Analysis. By the rearrangement inequalities, we know that the maximum value of $a_1 b_1 + a_2 b_2 + \cdots + a_{2007} b_{2007}$ is $\sum_{i=1}^{2007} i^2$, while the minimum value of $c_1 d_1 + c_2 d_2 + \cdots + c_{2007} d_{2007}$ is $\sum_{i=1}^{2007} i(2007 - i + 1)$. If

$$\sum_{i=1}^{2007} i^2 < 2 \sum_{i=1}^{2007} i(2007 - i + 1),$$

then the four sequences in the problem do not exist.

Solution. From the rearrangement inequalities,

$$a_1 b_1 + a_2 b_2 + \cdots + a_{2007} b_{2007} \leq 1^2 + 2^2 + \cdots + 2007^2$$

$$= \frac{2007 \cdot 2008 \cdot 4015}{6} = 2696779140,$$

$$c_1 d_1 + c_2 d_2 + \cdots + c_{2007} d_{2007} \geq 1 \cdot 2007 + 2 \cdot 2006 + \cdots + 2007 \cdot 1$$

$$= \sum_{k=1}^{2007} k(2008 - k) = 2008 \sum_{k=1}^{2007} k - \sum_{k=1}^{2007} k^2$$

$$= 2008 \cdot \frac{2007 \cdot 2008}{2} - 2696779140$$

$$= 1349397084.$$

Thus,

$$2(c_1 d_1 + c_2 d_2 + \cdots + c_{2007} d_{2007}) \geq 2698794168 > 2696779140$$

$$\geq a_1 b_1 + a_2 b_2 + \cdots + a_{2007} b_{2007}.$$

Therefore, there do not exist four sequences that satisfy the condition. $\square$

Example 3. Let $x_1, x_2, \ldots, x_n$ and $a_1, a_2, \ldots, a_n$ $(n \geq 2)$ be two groups of arbitrary real numbers that satisfy the conditions

(1) $x_1 + x_2 + \cdots + x_n = 0$;
(2) $|x_1| + |x_2| + \cdots + |x_n| = 1$;
(3) $a_1 \geq a_2 \geq \cdots \geq a_n$.

Find the minimum value for a real number A that makes the inequality

$$|a_1 x_1 + a_2 x_2 + \cdots + a_n x_n| \le A(a_1 - a_n)$$

satisfied.

Solution. Let $i_1, i_2, \ldots, i_s, j_1, j_2, \ldots, j_t$ be a permutation of $1, 2, \ldots, n$ such that

$$x_{i_1} \ge x_{i_2} \ge \cdots \ge x_{i_s} \ge 0 > x_{j_1} \ge x_{j_2} \ge \cdots \ge x_{j_t}.$$

Denote $\lambda = x_{i_1} + x_{i_2} + \cdots + x_{i_s}$ and $\mu = -(x_{j_1} + x_{j_2} + \cdots + x_{j_t})$. Then by the condition of the problem,

$$\lambda - \mu = 0, \quad \lambda + \mu = 1.$$

Hence, $\lambda = \mu = \frac{1}{2}$.

Without loss of generality, assume that $a_1 x_1 + a_2 x_2 + \cdots + a_n x_n \ge 0$ (otherwise, if $a_1 x_1 + a_2 x_2 + \cdots + a_n x_n < 0$, then for $x_i' = -x_i$ with $i = 1, 2, \ldots, n$, the numbers $x_1', x_2', \ldots, x_n'$ still satisfy (1) and (2), and

$$|a_1 x_1' + a_2 x_2' + \cdots + a_n x_n'| = |a_1 x_1 + a_2 x_2 + \cdots + a_n x_n|,$$

so (3) is also satisfied). By the rearrangement inequalities,

$$
\begin{aligned}
a_1 x_1 &+ a_2 x_2 + \cdots + a_n x_n \\
&\le a_1 x_{i_1} + a_2 x_{i_2} + \cdots + a_s x_{i_s} + a_{s+1} x_{j_1} + \cdots + a_n x_{j_t} \\
&\le a_1 (x_{i_1} + x_{i_2} + \cdots + x_{i_s}) + a_n (x_{j_1} + x_{j_2} + \cdots + x_{j_t}) \\
&= \frac{1}{2}(a_1 - a_n).
\end{aligned}
$$

On the other hand, if one chooses $x_1 = \frac{1}{2}, x_2 = x_3 = \cdots = x_{n-1} = 0$, and $x_n = -\frac{1}{2}$, then $x_1, x_2, \ldots, x_n$ satisfy (1) and (2), and

$$|a_1 x_1 + a_2 x_2 + \cdots + a_n x_n| = \frac{1}{2}(a_1 - a_n).$$

Therefore, the minimum value of A that satisfies the inequality in the problem is $\frac{1}{2}$. $\qquad\square$

Remark. In this problem, if we choose $a_i = \frac{1}{i}$ with $i = 1, 2, \ldots, n$, then

$$\left| x_1 + \frac{x_2}{2} + \cdots + \frac{x_n}{n} \right| \le \frac{1}{2} - \frac{1}{2n}.$$

This is the second problem in the 1989 National Mathematics Competition for High Schools.

Example 4. Suppose that $x \geq y \geq z > 0$. Prove:

$$\frac{x^2y}{z} + \frac{y^2z}{x} + \frac{z^2x}{y} \geq x^2 + y^2 + z^2.$$

Proof. Since $x^2 \geq yx \geq y^2$ and $z^2 \leq yz \leq y^2$,

$$x^2 \cdot z^2 + yx \cdot y^2 + y^2 \cdot yz \leq x^2 \cdot y^2 + yx \cdot yz + y^2 \cdot z^2,$$

that is

$$x^2z^2 + y^3x + y^3z \leq x^2y^2 + y^2xz + y^2z^2,$$

and so

$$y^3(x + z) \leq y^2(x^2 + xz + z^2) - x^2z^2.$$

Since $\frac{x-z}{y} \geq 0$,

$$\frac{x-z}{y} \cdot y^3(x + z) \leq \frac{x-z}{y} \cdot y^2(x^2 + xz + z^2) - \frac{x-z}{y} \cdot x^2z^2,$$

namely

$$y^2(x^2 - z^2) \leq y(x^3 - z^3) - \frac{x-z}{y} \cdot x^2z^2.$$

It follows that

$$\frac{y^2x}{z} + \frac{x^2z}{y} + \frac{z^2y}{x} \leq \frac{x^2y}{z} + \frac{z^2x}{y} + \frac{y^2z}{x}.$$

Hence,

$$\left(\frac{x^2y}{z} + \frac{z^2x}{y} + \frac{y^2z}{x}\right)^2$$

$$\geq \left(x^2 \cdot \frac{y}{z} + z^2 \cdot \frac{x}{y} + y^2 \cdot \frac{z}{x}\right)\left(x^2 \cdot \frac{z}{y} + z^2 \cdot \frac{y}{x} + y^2 \cdot \frac{x}{z}\right)$$

$$\geq (x^2 + y^2 + z^2)^2. \tag{22.1}$$

In other words,

$$\frac{x^2y}{z} + \frac{z^2x}{y} + \frac{y^2z}{x} \geq x^2 + y^2 + z^2,$$

and the necessary and sufficient condition for the equality to be satisfied is $x = y = z$. □

Remark. The second inequality in the formula (22.1) was from using the Cauchy inequality. In the process of solving many mathematical problems, one needs to use flexibly the tools of mean value inequalities, the Cauchy inequality, the rearrangement inequalities, and etc.

Example 5. Assume that $a_1 \geq a_2 \geq \cdots \geq a_n > 0$ and $b_1 \geq b_2 \geq \cdots \geq b_n > 0$. Let $i_1, i_2, \ldots, i_n$ and $j_1, j_2, \ldots, j_n$ be two arbitrary permutations of $1, 2, \ldots, n$. Prove:

$$\sum_{r=1}^{n} \sum_{s=1}^{n} \frac{a_{i_r} b_{j_s}}{r+s} \leq \sum_{r=1}^{n} \sum_{s=1}^{n} \frac{a_r b_s}{r+s}.$$

Proof. By the rearrangement inequalities, we obtain that

$$\sum_{s=1}^{n} \frac{b_{j_s}}{r+s} \leq \sum_{s=1}^{n} \frac{b_s}{r+s}, \quad \sum_{r=1}^{n} \frac{a_{i_r}}{r+s} \leq \sum_{r=1}^{n} \frac{a_r}{r+s}.$$

Consequently,

$$\sum_{r=1}^{n} \sum_{s=1}^{n} \frac{a_{i_r} b_{j_s}}{r+s} \leq \sum_{r=1}^{n} a_{i_r} \sum_{s=1}^{n} \frac{b_{j_s}}{r+s} = \sum_{s=1}^{n} b_s \sum_{r=1}^{n} \frac{a_{i_r}}{r+s}$$

$$\leq \sum_{s=1}^{n} b_s \sum_{r=1}^{n} \frac{a_r}{r+s} = \sum_{r=1}^{n} \sum_{s=1}^{n} \frac{a_r b_s}{r+s}. \qquad \square$$

Example 6. Suppose that $0 < p \leq a_i \leq q$ $(i = 1, 2, \ldots, n)$, and $b_1, b_2, \ldots, b_n$ is a permutation of $a_1, a_2, \ldots, a_n$. Prove:

$$n \leq \frac{a_1}{b_1} + \frac{a_2}{b_2} + \cdots + \frac{a_n}{b_n} \leq n + \left[\frac{n}{2}\right] \left(\sqrt{\frac{p}{q}} - \sqrt{\frac{q}{p}}\right)^2.$$

Proof. By the arithmetic mean-geometric mean inequality,

$$\frac{a_1}{b_1} + \frac{a_2}{b_2} + \cdots + \frac{a_n}{b_n} \geq n.$$

So, we only need to show the left inequality in the problem. Without loss of generality, assume that $a_1 \leq a_2 \leq \cdots \leq a_n$. The rearrangement inequalities imply that

$$\frac{a_1}{b_1} + \frac{a_2}{b_2} + \cdots + \frac{a_n}{b_n} \leq \frac{a_1}{a_n} + \frac{a_2}{a_{n-1}} + \cdots + \frac{a_n}{a_1}.$$

When $n = 2k$ is an even integer, the above inequality can be rewritten as

$$\frac{a_1}{b_1} + \frac{a_2}{b_2} + \cdots + \frac{a_n}{b_n} \leq \left(\frac{a_1}{a_n} + \frac{a_n}{a_1}\right) + \cdots + \left(\frac{a_k}{a_{k+1}} + \frac{a_{k+1}}{a_k}\right).$$

Now, for $x \in \left[\frac{p}{q}, \frac{q}{p}\right]$ it holds that

$$x + \frac{1}{x} = \left(\sqrt{x} - \frac{1}{\sqrt{x}}\right)^2 + 2 \leq \left(\sqrt{\frac{p}{q}} - \sqrt{\frac{q}{p}}\right)^2 + 2,$$

and thus

$$\frac{a_1}{b_1} + \frac{a_2}{b_2} + \cdots + \frac{a_n}{b_n} \leq 2k + k\left(\sqrt{\frac{p}{q}} - \sqrt{\frac{q}{p}}\right)^2$$

$$= n + \frac{n}{2}\left(\sqrt{\frac{p}{q}} - \sqrt{\frac{q}{p}}\right)^2.$$

When $n = 2k + 1$ is an odd integer, we have

$$\frac{a_1}{b_1} + \frac{a_2}{b_2} + \cdots + \frac{a_n}{b_n} \leq \left(\frac{a_1}{a_n} + \frac{a_n}{a_1}\right) + \cdots + \left(\frac{a_{k-1}}{a_{k+1}} + \frac{a_{k+1}}{a_{k-1}}\right) + \frac{a_k}{a_k}$$

$$\leq 2(k-1) + (k-1)\left(\sqrt{\frac{p}{q}} - \sqrt{\frac{q}{p}}\right)^2 + 1$$

$$= n + \frac{n-1}{2}\left(\sqrt{\frac{p}{q}} - \sqrt{\frac{q}{p}}\right)^2.$$

To summarize the above two cases, we have proved the inequality. $\square$

Example 7. Let two groups of ordered numbers $a_1 \leq a_2 \leq \cdots \leq a_n$ and $b_1 \leq b_2 \leq \cdots \leq b_n$ be given. Prove that

$$\frac{1}{n}(a_1 b_n + a_2 b_{n-1} + \cdots + a_n b_1) \leq \frac{a_1 + a_2 + \cdots + a_n}{n} \cdot \frac{b_1 + b_2 + \cdots + b_n}{n}$$

$$\leq \frac{1}{n}(a_1 b_1 + a_2 b_2 + \cdots + a_n b_n).$$

If and only if $a_1 = a_2 = \cdots = a_n$ or $b_1 = b_2 = \cdots = b_n$, the equalities are valid.

Proof. The rearrangement inequalities ensure that

$$a_1 b_n + a_2 b_{n-1} + \cdots + a_n b_1 \leq a_1 b_1 + a_2 b_2 + \cdots + a_n b_n,$$

$$a_1 b_n + a_2 b_{n-1} + \cdots + a_n b_1 \leq a_1 b_2 + a_2 b_3 + \cdots + a_{n-1} b_n + a_n b_1,$$

$$\cdots$$

$$a_1 b_n + a_2 b_{n-1} + \cdots + a_n b_1 \leq a_1 b_{n-1} + a_2 b_n + a_3 b_1 + \cdots + a_n b_{n-2},$$

$$a_1 b_n + a_2 b_{n-1} + \cdots + a_n b_1 \leq a_1 b_n + a_2 b_1 + \cdots + a_n b_{n-1}.$$

Adding up the above n inequalities and then dividing by n^2 give

$$\frac{a_1 b_n + a_2 b_{n-1} + \cdots + a_n b_1}{n} \leq \frac{a_1 + a_2 + \cdots + a_n}{n} \cdot \frac{b_1 + b_2 + \cdots + b_n}{n}.$$

By the same token, we can show that

$$\frac{a_1 + a_2 + \cdots + a_n}{n} \cdot \frac{b_1 + b_2 + \cdots + b_n}{n} \leq \frac{a_1 b_1 + a_2 b_2 + \cdots + a_n b_n}{n}. \qquad \square$$

Remark. The above two inequalities are called Chebyshev's inequalities.

In the following, we look at an example of applying Chebyshev's inequalities.

Example 8. Let $x_i > 0$ $(i = 1, 2, \ldots, n)$ and $k \geq 1$. Prove:

$$\sum_{i=1}^{n} \frac{1}{1 + x_i} \cdot \sum_{i=1}^{n} x_i \leq \sum_{i=1}^{n} \frac{1}{x_i^k} \cdot \sum_{i=1}^{n} \frac{x_i^{k+1}}{1 + x_i}.$$

Proof. By symmetry, assume that $x_1 \geq x_2 \geq \cdots \geq x_n > 0$ without loss of generality. Then

$$\frac{1}{x_1^k} \leq \frac{1}{x_2^k} \leq \cdots \leq \frac{1}{x_n^k},$$

$$\frac{x_1^k}{1 + x_1} \geq \frac{x_2^k}{1 + x_2} \geq \cdots \geq \frac{x_n^k}{1 + x_n}.$$

Using Chebyshev's inequalities twice, we obtain that

$$\sum_{i=1}^{n} \frac{1}{1 + x_i} \cdot \sum_{i=1}^{n} x_i = \sum_{i=1}^{n} x_i \cdot \sum_{i=1}^{n} \frac{1}{x_i^k} \frac{x_i^k}{1 + x_i}$$

$$\leq \frac{1}{n} \sum_{i=1}^{n} x_i \cdot \sum_{i=1}^{n} \frac{1}{x_i^k} \cdot \sum_{i=1}^{n} \frac{x_i^k}{1 + x_i}$$

$$\leq \frac{1}{n} \sum_{i=1}^{n} \frac{1}{x_i^k} \cdot \sum_{i=1}^{n} \left(n x_i \frac{x_i^k}{1 + x_i} \right)$$

$$= \sum_{i=1}^{n} \frac{1}{x_i^k} \sum_{i=1}^{n} \frac{x_i^{k+1}}{1 + x_i}.$$

$$\square$$

3. Exercises

Group A

I. Filling Problems

1. Let $b_1, b_2, \ldots, b_n$ be a permutation of positive numbers $a_1, a_2, \ldots, a_n$. Then the minimum value of $\frac{a_1}{b_1} + \frac{a_2}{b_2} + \cdots + \frac{a_n}{b_n}$ is _____.
2. Let the radius of a circle O be R, let $\triangle ABC$ be the inscribed triangle of the circle O, and let h_a, h_b, and h_c be the altitudes on the sides BC, AC, and AB of $\triangle ABC$, respectively. Then the minimum value of $\frac{\sin A}{h_a} + \frac{\sin B}{h_b} + \frac{\sin C}{h_c}$ is _____.

II. Calculation Problems

3. Let x, y, and z be positive numbers. Prove:
$$\frac{z^2 - x^2}{x + y} + \frac{x^2 - y^2}{y + z} + \frac{y^2 - z^2}{z + x} \geq 0.$$

4. Suppose that x_i and y_i $(i = 1, 2, \ldots, n)$ are real numbers satisfying $x_1 \geq x_2 \geq \cdots \geq x_n$ and $y_1 \geq y_2 \geq \cdots \geq y_n$. If $z_1, z_2, \ldots, z_n$ is any permutation of $y_1, y_2, \ldots, y_n$, prove:
$$\sum_{i=1}^{n}(x_i - y_i)^2 \leq \sum_{i=1}^{n}(x_i - z_i)^2.$$

5. Assume that $a_1 \geq a_2 \geq \cdots \geq a_n > 0$ and $b_1 \geq b_2 \geq \cdots \geq b_n > 0$. Prove: $a_1^{b_1} a_2^{b_2} \cdots a_n^{b_n} \leq a_1^{b_{i_1}} a_2^{b_{i_2}} \cdots a_n^{b_{i_n}} \leq a_1^{b_n} a_2^{b_{n-1}} \cdots a_n^{b_1}$, where $b_{i_1}, b_{i_2}, \ldots, b_{i_n}$ is any permutation of $b_1, b_2, \ldots, b_n$.
6. Let a set $\{a_1, a_2, \ldots, a_n\} = \{1, 2, \ldots, n\}$. Prove:
$$\frac{1}{2} + \frac{2}{3} + \cdots + \frac{n-1}{n} \leq \frac{a_1}{a_2} + \frac{a_2}{a_3} + \cdots + \frac{a_{n-1}}{a_n}.$$

7. Suppose that $a_1, a_2, \ldots, a_n \in \mathbf{N}$ are mutually different. Prove:
$$1 + \frac{1}{2} + \cdots + \frac{1}{n} \leq a_1 + \frac{a_2}{2^2} + \cdots + \frac{a_n}{n^2}.$$

Group B

8. Let $\frac{1}{2} \leq p \leq 1$. Suppose that $a_i \geq 0$ and $0 \leq b_i \leq p$ $(i = 1, 2, \ldots, n$ with $n \geq 2)$. If $\sum_{i=1}^{n} a_i = \sum_{i=1}^{n} b_i = 1$, prove:
$$\sum_{i=1}^{n}\left(b_i \prod_{1 \leq j \leq n, j \neq i} a_j\right) \leq \frac{p}{(n-1)^{n-1}}.$$

9. Assume that $a_1 \geq a_2 \geq \cdots \geq a_n > 0$ and $0 < b_1 \leq b_2 \leq \cdots \leq b_n$, or $0 < a_1 \leq a_2 \leq \cdots \leq a_n$ and $b_1 \geq b_2 \geq \cdots \geq b_n > 0$. Let $r, s \geq 1$. Prove:

$$\sum_{i=1}^{n} \frac{a_i^r}{b_i^s} \geq n^{1+s-r} \left(\sum_{i=1}^{n} a_i \right)^r \bigg/ \left(\sum_{i=1}^{n} b_i \right)^s.$$

10. Let $x_1, x_2, \ldots, x_n \in \mathbf{R}_+$ satisfy $\frac{x_1}{1+x_1} + \frac{x_2}{1+x_2} + \cdots + \frac{x_n}{1+x_n} = 1$. Prove:

$$x_1 + x_2 + \cdots + x_n \geq \frac{n}{n-1}.$$

11. Let $a, b,$ and c be the lengths of the three sides of a triangle such that $a + b + c = 2S$. Prove: For any positive integer n,

$$\frac{a^n}{b+c} + \frac{b^n}{c+a} + \frac{c^n}{a+b} \geq \left(\frac{2}{3} \right)^{n-2} \cdot S^{n-1}.$$

12. Let $a_1, a_2, \ldots, a_n$ be all positive numbers and let k be a positive integer. Denote $a_{n+1} = a_1$. Prove:

$$\sum_{i=1}^{n} \frac{a_i^{k+1}}{a_i^k + a_i^{k-1} a_{i+1} + \cdots + a_i a_{i+1}^{k-1} + a_{i+1}^k} \geq \frac{1}{k+1} \sum_{i=1}^{n} a_i.$$

Chapter 23

Convex Functions and Jensen's Inequality

1. Key Points of Knowledge and Basic Methods

A. Definition of convex functions

Let a function $f(x)$ be continuous on an interval I. If for any two numbers x_1 and x_2 in I, the inequality

$$f\left(\frac{x_1 + x_2}{2}\right) \le \frac{f(x_1) + f(x_2)}{2} \tag{23.1}$$

is always satisfied, then $f(x)$ is called a convex function on I.

If the inequality sign in (23.1) is reversed, then such $f(x)$ is called a concave function on I.

If the inequality sign in the above becomes the strict one for all distinct points x_1 and x_2 in I, then the convex (concave) function is called strictly convex (concave) on I.

The geometric meaning of a convex function is: For any chord passing through two points on the graph of $y = f(x)$, the middle point of the chord must be above the curve or on the curve. The situation is just opposite for concave functions. For strictly convex (concave) functions, for any chord passing through two distinct points on the graph of $y = f(x)$, the middle point of the chord must be above (below) the curve.

B. Determination of convex functions

One can use the definition of convex functions to determine the convexity of a function, and the following theorem can also be used to determine quickly the convexity of certain functions.

Theorem. *Assume that $f(x)$ is twice differentiable in an open interval (a, b). Then a necessary and sufficient condition for $f(x)$ to be a convex (concave) function on (a, b) is: For all $x \in (a, b)$, it is always true that $f''(x) \geq 0 (\leq 0)$. If $f''(x) > 0 (< 0)$ for all $x \in (a, b)$, then the function $f(x)$ is strictly convex (concave) on (a, b).*

C. Jensen's inequality

Suppose that $f(x)$ is a convex function on an interval I. Then for any $x_1, x_2, \ldots, x_n \in I$, there holds that

$$f\left(\frac{x_1 + x_2 + \cdots + x_n}{n}\right) \leq \frac{1}{n}[f(x_1) + f(x_2) + \cdots + f(x_n)],$$

and the equality is valid when $x_1 = x_2 = \cdots = x_n$. If in addition $f(x)$ is strictly convex, then the above inequality is strict for all mutually distinct $x_1, x_2, \ldots, x_n \in I$, and the equality is valid if and only if $x_1 = x_2 = \cdots = x_n$.

  Suppose that $f(x)$ is a concave function on an interval I. Then for any $x_1, x_2, \ldots, x_n \in I$, there holds that

$$f\left(\frac{x_1 + x_2 + \cdots + x_n}{n}\right) \geq \frac{1}{n}[f(x_1) + f(x_2) + \cdots + f(x_n)],$$

and the equality is valid when $x_1 = x_2 = \cdots = x_n$. If in addition $f(x)$ is strictly concave, then the above inequality is strict for all mutually distinct $x_1, x_2, \ldots, x_n \in I$, and the equality is valid if and only if $x_1 = x_2 = \cdots = x_n$.

Proof. We only prove it for the case of convex functions. When $n = 1$, the proposition is obviously true. Assume that the proposition is satisfied for $n = k$. When $n = k + 1$, denote

$$A = \frac{1}{k + 1}(x_1 + x_2 + \cdots + x_k + x_{k+1}).$$

Then

$$A = \frac{(k + 1)A + (k - 1)A}{2k} = \frac{x_1 + x_2 + \cdots + x_k + x_{k+1} + (k - 1)A}{2k}.$$

Also denote

$$B = \frac{x_1 + x_2 + \cdots + x_k}{k},$$

$$C = \frac{x_{k+1} + (k - 1)A}{k}.$$

Then

$$f(A) = f\left(\frac{B+C}{2}\right) \le \frac{1}{2}[f(B) + f(C)]$$

$$= \frac{1}{2}\left[f\left(\frac{x_1 + x_2 + \cdots + x_k}{k}\right) + f\left(\frac{x_{k+1} + (k-1)A}{k}\right)\right]$$

$$\le \frac{1}{2}\left\{\frac{1}{k}[f(x_1) + f(x_2) + \cdots + f(x_k)]\right.$$

$$\left. + \frac{1}{k}[f(x_{k+1}) + f(A) + \cdots + f(A)]\right\}$$

$$= \frac{1}{2k}[f(x_1) + f(x_2) + \cdots + f(x_{k+1}) + (k-1)f(A)].$$

Therefore,

$$f(A) \le \frac{1}{k+1}[f(x_1) + f(x_2) + \cdots + f(x_k) + f(x_{k+1})],$$

and the equality is valid when $x_1 = x_2 = \cdots = x_k = x_{k+1}$.

In summary, the proposition is true for all natural numbers n. □

D. Corollary of Jensen's inequality

Suppose that $f(x)$ is a convex function on an interval I. Then for any $x_1, x_2, \ldots, x_n \in I$ and any nonnegative numbers $\lambda_1, \lambda_2, \ldots, \lambda_n$ satisfying $\lambda_1 + \lambda_2 + \cdots + \lambda_n = 1$, there holds that

$$f(\lambda_1 x_1 + \lambda_2 x_2 + \cdots + \lambda_n x_n) \le \lambda_1 f(x_1) + \lambda_2 f(x_2) + \cdots + \lambda_n f(x_n),$$

and the equality is valid when $x_1 = x_2 = \cdots = x_n$. If in addition $f(x)$ is strictly convex, then the above inequality is strict for all mutually distinct $x_1, x_2, \ldots, x_n \in I$ and any positive numbers $\lambda_1, \lambda_2, \ldots, \lambda_n$ satisfying $\lambda_1 + \lambda_2 + \cdots + \lambda_n = 1$, and the equality is valid if and only if $x_1 = x_2 = \cdots = x_n$.

For a concave function $f(x)$, there is a similar corollary with the inequality sign just reversed.

2. Illustrative Examples

Example 1. Prove: (1) $f(x) = \sin x$ is a concave function on $[0, \pi]$.

(2) $g(x) = \log x$ is a concave function on $(0, +\infty)$.

(3) $h(x) = \tan x$ is a convex function on $\left[0, \frac{\pi}{2}\right)$.

Proof. (1) For any $x_1, x_2 \in [0, \pi]$, we have

$$\frac{f(x_1) + f(x_2)}{2} = \frac{1}{2}(\sin x_1 + \sin x_2)$$

$$= \sin \frac{x_1 + x_2}{2} \cos \frac{x_1 - x_2}{2} \leq \sin \frac{x_1 + x_2}{2},$$

namely

$$\frac{f(x_1) + f(x_2)}{2} \leq f\left(\frac{x_1 + x_2}{2}\right).$$

Hence, $f(x) = \sin x$ is a concave function on $[0, \pi]$.

(2) Let $x_1, x_2 \in (0, +\infty)$. Then

$$\frac{\log x_1 + \log x_2}{2} = \log \sqrt{x_1 x_2} \leq \log \frac{x_1 + x_2}{2},$$

and so

$$\frac{g(x_1) + g(x_2)}{2} \leq g\left(\frac{x_1 + x_2}{2}\right).$$

Thus, $g(x) = \log x$ is a concave function on $(0, +\infty)$.

(3) When $0 \leq x_1, x_2 < \frac{\pi}{2}$,

$$\tan x_1 + \tan x_2 = \frac{\sin x_1}{\cos x_1} + \frac{\sin x_2}{\cos x_2} = \frac{\sin(x_1 + x_2)}{\cos x_1 \cos x_2}$$

$$= \frac{2 \sin(x_1 + x_2)}{\cos(x_1 + x_2) + \cos(x_1 - x_2)}$$

$$\geq \frac{2 \sin(x_1 + x_2)}{\cos(x_1 + x_2) + 1}$$

$$= 2 \tan \frac{x_1 + x_2}{2},$$

that is

$$\frac{h(x_1) + h(x_2)}{2} \geq h\left(\frac{x_1 + x_2}{2}\right).$$

Therefore, $h(x) = \tan x$ is a convex function on $[0, \frac{\pi}{2})$. $\square$

Remark. In this example, one can use the second derivative to prove the convexity or concavity. The convexity or concavity of these special functions have wide applications.

Example 2. Suppose that $\frac{3}{2} \leq x \leq 5$. Prove:

$$2\sqrt{x+1} + \sqrt{2x-3} + \sqrt{15-3x} < 2\sqrt{19}.$$

Proof. It is easy to show that $y = \sqrt{x}$ is a strictly concave function on $[0, +\infty)$. By Jensen's inequality,

$$2\sqrt{x+1} + \sqrt{2x-3} + \sqrt{15-3x} = \sqrt{x+1} + \sqrt{x+1} + \sqrt{2x-3} + \sqrt{15-3x}$$

$$\leq 4\sqrt{\frac{x+1+x+1+2x-3+15-3x}{4}}$$

$$= 2\sqrt{x+14}.$$

Since $x+1, 2x-3$, and $15-3x$ cannot be the same number, so

$$2\sqrt{x+1} + \sqrt{2x-3} + \sqrt{15-3x} < 2\sqrt{x+14} \leq 2\sqrt{19}. \qquad \square$$

Example 3. Assume that $x, y, z > 0$ and $x + y + z = 1$. Prove:

$$\sqrt{\frac{yz}{x+yz}} + \sqrt{\frac{zx}{y+zx}} + \sqrt{\frac{xy}{z+xy}} \leq \frac{3}{2}.$$

Proof. Construct $\triangle ABC$ with $y+z, z+x$, and $x+y$ as the three sides. By the law of cosines,

$$\cos A = \frac{(x+y)^2 + (x+z)^2 - (y+z)^2}{2(x+y)(x+z)} = \frac{x^2 + xy + xz - yz}{(x+y)(x+z)},$$

so

$$\sin \frac{A}{2} = \sqrt{\frac{yz}{x+yz}}.$$

By the same token,

$$\sin \frac{B}{2} = \sqrt{\frac{zx}{y+zx}}, \quad \sin \frac{C}{2} = \sqrt{\frac{xy}{z+xy}}.$$

Since $y = \sin x$ is a concave function for $x \in \left(0, \frac{\pi}{2}\right)$,

$$\frac{\sin \frac{A}{2} + \sin \frac{B}{2} + \sin \frac{C}{2}}{3} \leq \sin \frac{A+B+C}{3} = \sin \frac{\pi}{6},$$

in other words,

$$\sqrt{\frac{yz}{x+yz}} + \sqrt{\frac{zx}{y+zx}} + \sqrt{\frac{xy}{z+xy}} = \sin \frac{A}{2} + \sin \frac{B}{2} + \sin \frac{C}{2}$$

$$\leq 3 \sin \frac{\pi}{6} = \frac{3}{2},$$

and the equality is true if and only if $x = y = z = \frac{1}{3}$. $\qquad \square$

Remark. Since x, y, and z are positive numbers, using $x + y, y + z$, and $z + x$ to represent the sides of a triangle can reduce an algebraic problem to a geometric problem or trigonometric problem.

Example 4. Suppose that a, b, and c are positive real numbers and $a^4 + b^4 + c^4 = 3$. Prove:

$$\frac{1}{4 - ab} + \frac{1}{4 - bc} + \frac{1}{4 - ca} \leq 1.$$

Proof. Let $f(x) = \frac{1}{4 - \sqrt{x}}$ with $x \in \left[0, \frac{16}{9}\right]$. From

$$f'(x) = \frac{1}{2\sqrt{x} \cdot (4 - \sqrt{x})^2},$$

$$f''(x) = -\frac{(3\sqrt{x} - 4)(\sqrt{x} - 4)}{4x\sqrt{x}(4 - \sqrt{x})^4} < 0,$$

we know that $f(x)$ is a strictly concave function on $\left[0, \frac{16}{9}\right]$. By Jensen's inequality, we get

$$\frac{1}{4 - ab} + \frac{1}{4 - bc} + \frac{1}{4 - ca} \leq \frac{3}{4 - \sqrt{\frac{a^2b^2 + b^2c^2 + c^2a^2}{3}}}$$

$$\leq \frac{3}{4 - \sqrt{\frac{a^4 + b^4 + c^4}{3}}} = 1.$$

$\square$

Example 5. Find the maximal possible constant k such that for all $x, y, z \in \mathbf{R}_+$, the following inequality

$$\frac{x}{\sqrt{y + z}} + \frac{y}{\sqrt{z + x}} + \frac{z}{\sqrt{x + y}} \geq k\sqrt{x + y + z}$$

is always satisfied.

Solution. By the homogeneity, assume that $x + y + z = 1$ without loss of generality. Since

$$x(1 - x)^2 = \frac{1}{2} \cdot 2x(1 - x)(1 - x) \leq \frac{1}{2}\left(\frac{2}{3}\right)^2 = \frac{4}{27},$$

and so

$$\frac{x}{\sqrt{y+z}} + \frac{y}{\sqrt{z+x}} + \frac{z}{\sqrt{x+y}} = \frac{x}{\sqrt{1-x}} + \frac{y}{\sqrt{1-y}} + \frac{z}{\sqrt{1-z}}$$

$$= \frac{x^{\frac{5}{4}}}{\sqrt[4]{x(1-x)^2}} + \frac{y^{\frac{5}{4}}}{\sqrt[4]{y(1-y)^2}} + \frac{z^{\frac{5}{4}}}{\sqrt[4]{z(1-z)^2}}$$

$$\geq \frac{x^{\frac{5}{4}}}{\sqrt[4]{\frac{4}{27}}} + \frac{y^{\frac{5}{4}}}{\sqrt[4]{\frac{4}{27}}} + \frac{z^{\frac{5}{4}}}{\sqrt[4]{\frac{4}{27}}}$$

$$= \sqrt[4]{\frac{27}{4}} \left(x^{\frac{5}{4}} + y^{\frac{5}{4}} + z^{\frac{5}{4}} \right).$$

On the other hand, since $y = x^{\frac{5}{4}}$ is a convex function on the interval $(0, +\infty)$, Jensen's inequality implies that

$$\sqrt[4]{\frac{27}{4}} \left(x^{\frac{5}{4}} + y^{\frac{5}{4}} + z^{\frac{5}{4}} \right) \geq 3 \sqrt[4]{\frac{27}{4}} \left(\frac{x+y+z}{3} \right)^{\frac{5}{4}} = \frac{\sqrt{6}}{2},$$

and the equality in the above inequality is satisfied if and only if $x = y = z = \frac{1}{3}$. Hence, the maximum value of k is $\frac{\sqrt{6}}{2}$. $\square$

Remark. Assuming that $x + y + z = 1$ thanks to the homogeneity may give the benefit of simplifying the inequality.

Example 6. Let $a_{ij} > 0$ and $x_i > 0$ with $a_{ij} = a_{ji}$ for $i, j = 1, 2, \ldots, n$, and assume that

$$\sum_{k=1}^{n} a_{ik} = 1, \quad i = 1, 2, \ldots, n.$$

Denote $y_i = \sum_{k=1}^{n} a_{ik} x_k$ for $i = 1, 2, \ldots, n$. Prove: $y_1 y_2 \cdots y_n \geq x_1 x_2 \cdots x_n$.

Proof. Applying Jensen's inequality to the concave function $f(x) = \ln x$ implies that

$$\ln y_i = \ln \left(\sum_{k=1}^{n} a_{ik} x_k \right) \geq \sum_{k=1}^{n} a_{ik} \ln x_k \quad (i = 1, 2, \ldots, n).$$

Since $\ln x$ is an increasing function, we see that

$$y_i \geq x_1^{a_{i1}} x_2^{a_{i2}} \cdots x_n^{a_{in}} \quad (i = 1, 2, \ldots, n).$$

It follows that

$$y_1 y_2 \cdots y_n \geq x_1^{\sum_{i=1}^n a_{i1}} \cdot x_2^{\sum_{i=1}^n a_{i2}} \cdots \cdots x_n^{\sum_{i=1}^n a_{in}} = x_1 x_2 \cdots x_n,$$

namely the original inequality is satisfied. □

Example 7. Let $n \in \mathbf{N}$ satisfy $n \geq 2$. Prove: When $x \in \left(0, \frac{\pi}{2^n}\right)$,

$$\cot 2^n x + \frac{n}{\sin \frac{2^{n+1}-2}{n} x} < \cot x.$$

Proof. Repeatedly using the obvious trigonometric identity

$$\cot 2x + \frac{1}{\sin 2x} = \cot x$$

gives

$$\frac{1}{\sin 2x} + \frac{1}{\sin 4x} + \cdots + \frac{1}{\sin 2^n x} = \cot x - \cot 2^n x.$$

Let $f(x) = \frac{1}{\sin x}$. Then $f'(x) = -\frac{\cos x}{\sin^2 x}$ and $f''(x) = \frac{\sin^3 x + 2\cos^2 x}{\sin^3 x} > 0$, so $f(x)$ is a strictly convex function on $\left(0, \frac{\pi}{2^n}\right)$.

Jensen's inequality now ensures that

$$f(2x) + f(4x) + \cdots + f(2^n x) \geq n \cdot f\left(\frac{2x + 4x + \cdots + 2^n x}{n}\right)$$

$$= n \cdot \frac{1}{\sin \frac{2^{n+1}-2}{n} x},$$

and consequently

$$\cot x - \cot 2^n x \geq \frac{n}{\sin \frac{2^{n+1}-2}{n} x},$$

namely

$$\cot 2^n x + \frac{n}{\sin \frac{2^{n+1}-2}{n} x} \leq \cot x.$$

By the condition for the equality of Jensen's inequality, we know that the equality cannot hold in the above, thus the original proposition is true. □

Example 8. Let $a_i \in \mathbf{R}_+, 0 \le x_i \le 1$ $(i = 1, 2, \ldots, n)$, and $\sum_{i=1}^{n} a_i = 1$. Prove:

$$\sum_{i=1}^{n} \frac{a_i}{1 + x_i} \le \frac{1}{1 + x_1^{a_1} x_2^{a_2} \cdots x_n^{a_n}},$$

and the equality is valid if and only if $x_1 = x_2 = \cdots = x_n$.

Proof. If some x_i equals 0, then the inequality is clearly correct. So, in the following we assume that $0 < x_i \le 1$ for $i = 1, 2, \ldots, n$. Let $y_i = \ln x_i$. Then $-\infty < y_i \le 0$ and $x_i = e^{y_i}$ $(i = 1, 2, \ldots, n)$. Consider the function

$$f(t) = \frac{1}{1 + e^t} \quad (-\infty < t \le 0).$$

Since $f'(t) = -\frac{e^t}{(1+e^t)^2}$ and $f''(t) = e^t(e^t - 1)\frac{1}{(1+e^t)^3} < 0$, so $f(x)$ is a strictly concave function on $(-\infty, 0)$. By Jensen's inequality,

$$\sum_{i=1}^{n} \frac{a_i}{1 + e^{y_i}} \le \left(1 + e^{\sum_{i=1}^{n} a_i y_i} \right)^{-1},$$

that is

$$\sum_{i=1}^{n} \frac{a_i}{1 + x_i} \le (1 + x_1^{a_1} x_2^{a_2} \cdots x_n^{a_n})^{-1}.$$

It is easy to see that the equality holds if and only if $y_1 = y_2 = \cdots = x_n$, namely $x_1 = x_2 = \cdots = x_n$. $\qquad\square$

3. Exercises

Group A

I. Filling Problems

1. In $\triangle ABC$, the maximum value of $\sin A + \sin B + \sin C$ is _____.

2. Let $A = \sqrt[3]{3 - \sqrt[3]{3}} + \sqrt[3]{3 + \sqrt[3]{3}}$ and $B = 2\sqrt[3]{3}$. Then the size relationship of A and B is _____.

II. Calculation Problems

3. Suppose that positive numbers a_i $(i = 1, 2, \ldots, n)$ satisfy $\sum_{i=1}^{n} a_i = 1$. Prove:

$$\prod_{i=1}^{n} \left(a_i + \frac{1}{a_i} \right) \geq \left(n + \frac{1}{n} \right)^n.$$

4. Let p be a natural number, a a positive constant, and $x_1, x_2, \ldots, x_n$ positive numbers such that $x_1 + x_2 + \cdots + x_n = 1$. Prove:

$$\frac{1}{n} \left[\left(x_1 + \frac{a}{x_1} \right)^p + \left(x_2 + \frac{a}{x_2} \right)^p + \cdots + \left(x_n + \frac{a}{x_n} \right)^p \right] \geq \left(\frac{n^2 a + 1}{n} \right)^p.$$

5. Given a positive integer $n \geq 3$. Prove:

$$\sqrt{9n + 8} < \sqrt{n} + \sqrt{n+1} + \sqrt{n+2} < 3\sqrt{n+1}.$$

6. With $n \geq 2$, let $x_1, x_2, \ldots, x_n$ be positive numbers satisfying $x_1 + x_2 + \cdots + x_n = 1$. Prove:

$$\frac{x_1}{\sqrt{1 - x_1}} + \frac{x_2}{\sqrt{1 - x_2}} + \cdots + \frac{x_n}{\sqrt{1 - x_n}} \geq \frac{\sqrt{x_1} + \sqrt{x_2} + \cdots + \sqrt{x_n}}{\sqrt{n - 1}}.$$

7. Let P be any point inside $\triangle ABC$. Prove: There is at least one among $\angle PAB, \angle PBC$, and $\angle PCA$ that is less than or equal to $30°$.

8. Let $m \geq 2$. Then in $\triangle ABC$, there holds that

$$\tan \frac{A}{m} + \tan \frac{B}{m} + \tan \frac{C}{m} \geq 3 \tan \frac{\pi}{3m}.$$

Group B

9. Suppose that $x, y, z > 0$ and $x + y + z = 1$. Prove:

$$\left(\frac{1}{x^2} + x \right) \left(\frac{1}{y^2} + y \right) \left(\frac{1}{z^2} + z \right) \geq \left(\frac{28}{3} \right)^3.$$

10. Suppose that $x_i > 0$ $(i = 1, 2, \ldots, n)$ satisfy $x_1 + x_2 + \cdots + x_n \geq x_1 x_2 \cdots x_n$ $(n \geq 2)$. Let $1 \leq \alpha \leq n$. Prove:

$$\frac{x_1^\alpha + x_2^\alpha + \cdots + x_n^\alpha}{x_1 x_2 \cdots x_n} \geq n^{\frac{\alpha - 1}{n - 1}}.$$

11. Let x, y, and z be positive numbers such that $xyz = 1$. Prove:

$$\frac{x^3}{(1+y)(1+z)} + \frac{y^3}{(1+z)(1+x)} + \frac{z^3}{(1+x)(1+y)} \geq \frac{3}{4}.$$

12. Assume that $\alpha > \beta > 0$ and $a_1, a_2, \ldots, a_n \in \mathbf{R}_+$. Prove:

$$\left(\frac{a_1^\alpha + a_2^\alpha + \cdots + a_n^\alpha}{n}\right)^{\frac{1}{\alpha}} \geq \left(\frac{a_1^\beta + a_2^\beta + \cdots + a_n^\beta}{n}\right)^{\frac{1}{\beta}}.$$

Chapter 24

Recursive Sequences

1. Key Points of Knowledge and Basic Methods

A. *Recursive sequences*

A sequence $\{a_n\}$ determined by initial values and an equation of the form

$$a_{n+k} = F(a_{n+k-1}, \ldots, a_n) \tag{24.1}$$

is called a kth order recursive sequence.

In particular, when the form of (24.1) is

$$a_{n+k} = c_1 a_{n+k-1} + c_2 a_{n+k-2} + \cdots + c_k a_n + f(n), \tag{24.2}$$

the sequence $\{a_n\}$ is called a kth order linear recursive sequence with constant coefficients. Here, $c_1, c_2, \ldots, c_k$ are constants and $c_k \neq 0$. If $f(n) = 0$, then the sequence $\{a_n\}$ determined by (24.2) is called a kth order homogeneous linear recursive sequence with constant coefficients.

Arithmetic sequences satisfy the recursive formula $a_{n+2} = 2a_{n+1} - a_n$, and geometric sequences satisfy $a_{n+1} = qa_n$ in which q is a nonzero constant. They are the simplest homogeneous linear recursive sequences with constant coefficients.

B. *General terms of homogeneous linear recursive sequences with constant coefficients*

The formula for general terms of a homogeneous linear recursive sequence with constant coefficients can be given by its characteristic roots.

Definition 1. The equation

$$x^k = c_1 x^{k-1} + \cdots + c_k$$

is called the characteristic equation of (24.2), and the roots of this equation are called the characteristic roots of the sequence $\{a_n\}$, which are denoted as $\lambda_1, \lambda_2, \ldots, \lambda_k$.

Theorem 1. *In* (24.2), *if* $f(n) = 0$ *and* $\lambda_1, \lambda_2, \ldots, \lambda_k$ *are mutually distinct, then the general term formula for the sequence* $\{a_n\}$ *is*

$$a_n = A_1 \lambda_1^n + A_2 \lambda_2^n + \cdots + A_k \lambda_k^n,$$

where $A_1, A_2, \ldots, A_k$ *are constants and are determined by the initial values* $a_1, a_2, \ldots, a_k$.

Theorem 2. *Under the same conditions of Theorem 1 except that there are multiple roots among* $\lambda_1, \lambda_2, \ldots, \lambda_k$, *if the distinct characteristic roots are* $\lambda_1, \lambda_2, \ldots, \lambda_s$ $(s < k)$, *and the multiplicity of* λ_i *is* t_i *for* $1 \le i \le s$, *then the general term formula for the sequence* $\{a_n\}$ *is*

$$a_n = A_1(n)\lambda_1^n + A_2(n)\lambda_2^n + \cdots + A_s(n)\lambda_s^n,$$

where $A_i(n) = B_1^{(i)} + B_2^{(i)} n + \cdots + B_{t_i}^{(i)} n^{t_i - 1}$ *with* $1 \le i \le s$. *Here,* $B_j^{(i)}$ *are constants, which can be determined by the initial values.*

When $f(n)$ are of some special nonzero forms, the general term of the sequence (24.2) can be solved out by transforming to the homogeneous form. For example, when $f(n) = c$ (c is a nonzero constant), the transformation $b_n = a_n + \frac{c}{c_1 + c_2 + \cdots + c_k - 1}$ can change the nonhomogeneous problem for $\{a_n\}$ to a homogeneous problem for the linear recursive sequence $\{b_n\}$ with constant coefficients.

C. Ideas and methods for problems of recursive sequences

The problems related to recursive sequences are mainly of two types.

(1) Find the general term of a recursive sequence

There is no universal method for general terms of arbitrary recursive sequences, but for some concrete and special recursive sequences, there are quite a few frequently used methods. For example, ideas of changing variables (a reduction to homogeneous linear recursive sequences with constant coefficients, including arithmetic or geometric sequences, etc.), the fixed point method, the guess first-proof second method (combined with mathematical induction), and so on.

(2) Investigate properties of recursive sequences

It may not (or may) be required to find the general term, but it is often needed to make a flexible deformation to the recursive sequence, and combine the knowledge of algebra, number theory, and combinatorics to solve the problem.

D. *Recursive method*

Some problems in algebra, number theory, and combinatorics may be solved via establishing a model of recursive sequences, and especially, recursion is an extremely useful computational method. The related concrete materials will be introduced in the third book of this series for high school students, so this part will not be touched in this chapter.

2. **Illustrative Examples**

Example 1. Let a sequence $\{a_n\}$ satisfy $a_1 = \frac{\pi}{6}$ and $a_{n+1} = \arctan(\sec a_n)$ with $n \in \mathbf{N}_+$. Find a positive integer m such that

$$\sin a_1 \cdot \sin a_2 \cdots \cdot \sin a_m = \frac{1}{100}.$$

Analysis. A key point of solving this problem is to find the general term of the recursive sequence. From the condition it is easy to see that $\sec a_n > 0$, so $\tan^2 a_{n+1} = \sec^2 a_n$, namely $\tan^2 a_{n+1} - \tan^2 a_n = 1$.

Solution. From the given condition we see that $a_{n+1} \in \left(-\frac{\pi}{2}, \frac{\pi}{2}\right)$ for all positive integers n, and

$$\tan a_{n+1} = \sec a_n. \tag{24.3}$$

Since $\sec a_n > 0$, actually $a_{n+1} \in \left(0, \frac{\pi}{2}\right)$. By (24.3), $\tan^2 a_{n+1} = \sec^2 a_n = 1 + \tan^2 a_n$, and so

$$\tan^2 a_n = n - 1 + \tan^2 a_1 = n - 1 + \frac{1}{3} = \frac{3n - 2}{3},$$

that is,

$$\tan a_n = \sqrt{\frac{3n-2}{3}}.$$

Therefore,

$$\sin a_1 \sin a_2 \cdots \sin a_m = \frac{\tan a_1}{\sec a_1} \cdot \frac{\tan a_2}{\sec a_2} \cdots \cdots \frac{\tan a_m}{\sec a_m}$$

$$= \frac{\tan a_1}{\tan a_2} \cdot \frac{\tan a_2}{\tan a_3} \cdots \cdots \frac{\tan a_m}{\tan a_{m+1}} \quad \text{(by (24.3))}$$

$$= \frac{\tan a_1}{\tan a_{m+1}} = \sqrt{\frac{1}{3m+1}}.$$

Solving the equation $\sqrt{\frac{1}{3m+1}} = \frac{1}{100}$, we obtain $m = 3333$. $\qquad\square$

Example 2. Assume that a sequence $\{a_n\}$ satisfies $a_1 = a_2 = 1$ and $a_n = \frac{a_{n-1}^2+2}{a_{n-2}}$ $(n \geq 3)$. Find a formula for the general term.

Solution. The provided condition gives $a_3 = 3$ and

$$a_n a_{n-2} = a_{n-1}^2 + 2, \quad a_{n+1} a_{n-1} = a_n^2 + 2.$$

Subtracting the first equality from the second one in the above to eliminate the constant 2, we have $a_{n+1}a_{n-1} - a_n a_{n-2} = a_n^2 - a_{n-1}^2$, namely

$$a_{n-1}(a_{n+1} + a_{n-1}) = a_n(a_n + a_{n-2}).$$

Thus, $\frac{a_{n+1}+a_{n-1}}{a_n+a_{n-2}} = \frac{a_n}{a_{n-1}}$. Consequently,

$$\frac{a_n + a_{n-2}}{a_{n-1} + a_{n-3}} = \frac{a_{n-1}}{a_{n-2}}, \ldots, \frac{a_4 + a_2}{a_3 + a_1} = \frac{a_3}{a_2}.$$

Multiplying all the above equalities results in $\frac{a_{n+1}+a_{n-1}}{a_3+a_1} = \frac{a_n}{a_2}$, in other words

$$a_{n+1} + a_{n-1} = 4a_n. \tag{24.4}$$

The characteristic equation of (24.4) is $x^2 - 4x + 1 = 0$, and its characteristic roots are $x_1 = 2 + \sqrt{3}$ and $x_2 = 2 - \sqrt{3}$. Hence,

$$a_n = A_1(2 + \sqrt{3})^n + A_2(2 - \sqrt{3})^n.$$

The initial values $a_1 = a_2 = 1$ give

$$\begin{cases} (2 + \sqrt{3})A_1 + (2 - \sqrt{3})A_2 = 1, \\ (7 + 4\sqrt{3})A_1 + (7 - 4\sqrt{3})A_2 = 1, \end{cases}$$

the solutions of which are $A_1 = \frac{3\sqrt{3}-5}{2\sqrt{3}}$ and $A_2 = \frac{5+3\sqrt{3}}{2\sqrt{3}}$. Therefore, the general term of the sequence $\{a_n\}$ is

$$a_n = \frac{1}{2\sqrt{3}}\left[(3\sqrt{3}-5)(2+\sqrt{3})^n + (5+3\sqrt{3})(2-\sqrt{3})^n\right].$$ □

Remark. Some recursive sequences are nonlinear in format, but are essentially linear recursions. This requires us to "discard the false and retain the true."

Example 3. Let n be a given positive integer. A sequence $a_0, a_1, \ldots, a_n$ is defined by: $a_0 = \frac{1}{2}$ and $a_k = a_{k-1} + \frac{a_{k-1}^2}{n}$ for $k = 1, 2, \ldots, n$. Prove: $1 - \frac{1}{n} < a_n < 1$.

Proof. The recursive relation gives that

$$\frac{1}{a_k} = \frac{n}{a_{k-1}(a_{k-1}+n)} = \frac{1}{a_{k-1}} - \frac{1}{a_{k-1}+n}, \quad k = 1, 2, \ldots, n.$$

Consequently, we have

$$\frac{1}{a_{k-1}+n} = \frac{1}{a_{k-1}} - \frac{1}{a_k}.$$

Summing up the above equalities with $k = 1, 2, \ldots, n$, we find that

$$\frac{1}{a_0} - \frac{1}{a_n} = \sum_{k=1}^{n} \frac{1}{a_{k-1}+n} < \sum_{k=1}^{n} \frac{1}{n} = 1,$$

where we have used the property that $a_{k-1} > 0$ for $k = 1, 2, \ldots, n$, which is easy to see from the recursive formula. Therefore, we have $a_n < 1$ since $a_0 = \frac{1}{2}$.

On the other hand, the recursive formula implies that $a_k > a_{k-1}$ for $k = 1, 2, \ldots, n$, so for any $k \in \{1, 2, \ldots, n\}$, we have $a_{k-1} < a_n < 1$. Thus, we obtain

$$\frac{1}{a_0} - \frac{1}{a_n} = \sum_{k=1}^{n} \frac{1}{a_{k-1}+n} > \sum_{k=1}^{n} \frac{1}{n+1} = \frac{n}{n+1}.$$

It follows that

$$a_n > \frac{n+1}{n+2} = 1 - \frac{1}{n+2} > 1 - \frac{1}{n}.$$

In summary, the proposition is true. □

Remark. For many problems on recursive sequences, it is not necessary to find the general term, but to show some properties of the general term. Then one gives a concrete analysis to concrete problems according to the feature of the recursive formula. In this problem, we considered to choose the reciprocals and then split the terms to sum up, starting from the feature of the recursive formula to go forward.

Example 4. Let a sequence $\{a_n\}$ be defined as follows: $a_1 = 1$ and

$$a_{n+1} = \frac{a_n}{2} + \frac{1}{4a_n}, \quad n = 1, 2, \ldots.$$

Prove: When $n \geq 2$, the numbers $\sqrt{\frac{2}{2a_n^2-1}}$ are all positive integers.

Analysis. Since the conclusion involves the radical sign and the term a_n^2, we let $b_n = \sqrt{\frac{2}{2a_n^2-1}}$, and square the both sides of the given recursive relation, from which one can find ideas easily for solving the problem.

Proof. Denote $b_n = \sqrt{\frac{2}{2a_n^2-1}}$. Then $b_n^2 = \frac{2}{2a_n^2-1}$, and so we have $a_n = \frac{1}{b_n^2} + \frac{1}{2}$. Since $a_{n+1}^2 = \frac{a_n^2}{4} + \frac{1}{16a_n^2} + \frac{1}{4}$, it follows that

$$\frac{1}{b_{n+1}^2} + \frac{1}{2} = \frac{1}{4}\left(\frac{1}{b_n^2} + \frac{1}{2}\right) + \frac{1}{16\left(\frac{1}{b_n^2} + \frac{1}{2}\right)} + \frac{1}{4},$$

that is

$$b_{n+1}^2 = 2b_n^2(b_n^2 + 2).$$

Consequently,

$$b_{n+1}^2 = 2b_n^2\left[2b_{n-1}^2(b_{n-1}^2 + 2) + 2\right] = 4b_n^2(b_{n-1}^2 + 1)^2. \tag{24.5}$$

Since $b_2 = \sqrt{\frac{2}{2a_2^2-1}} = 4$ and $b_3 = \sqrt{\frac{2}{2a_3^2-1}} = 24$, by (24.5) and the fact that $b_2, b_3 \in \mathbf{N}_+$, the numbers $b_n \in \mathbf{N}_+$ when $n > 1$. $\qquad\square$

Remark. Transforming the problem about a_n to that about b_n via changing the variable may solve the problem smoothly. We may have used the fixed point method to find the general term of the recursive sequence, but it would be difficult to show that the radial expressions are all positive integers.

Example 5. Suppose that a is an irrational number greater than 1 and n is an integer greater than 1. Prove: The number $(a + \sqrt{a^2 - 1})^{\frac{1}{n}} + (a - \sqrt{a^2 - 1})^{\frac{1}{n}}$ is an irrational number.

Proof. Denote $x = (a + \sqrt{a^2 - 1})^{\frac{1}{n}}$. Then $(a - \sqrt{a^2 - 1})^{\frac{1}{n}} = \frac{1}{x}$. Let $a_k = x^k + \frac{1}{x^k}$ with $k = 0, 1, 2, \ldots$. Then the sequence $\{a_k\}$ satisfies the following recursive formula

$$a_{k+2} = \left(x + \frac{1}{x} \right) a_{k+1} - a_k, \quad k = 0, 1, 2, \ldots.$$

If $x + \frac{1}{x}$ is a rational number, then combining $a_0 = 2$ and $a_1 = x + \frac{1}{x}$, we know that for any positive integer k the number a_k is always a rational number. In particular, $a_n = x^n + \frac{1}{x^n} = 2a$ must be a rational number, which contradicts the assumption that a is an irrational number. Therefore, $x + \frac{1}{x}$ is an irrational number, thus proving the proposition. $\quad\square$

Example 6. A sequence $a_1, a_2, \ldots, a_{201}$ of positive integers satisfies the properties that $a_1 = a_{201} = 19999$ and $\frac{a_{i-1} + a_{i+1}}{2} - a_i$ is always the same positive integer for each $i = 2, 3, \ldots, 200$. Find a_{200}.

Solution. Let $\frac{a_{i-1} + a_{i+1}}{2} - a_i = c$. Then $c \in \mathbf{Z}_+$, and

$$a_{i+1} - a_i = a_i - a_{i-1} + 2c.$$

Denote $d_i = a_{i+1} - a_i$ $(i = 1, 2, \ldots, 200)$. Then

$$d_i = d_{i-1} + 2c \ (i = 2, 3, \ldots, 200).$$

So, $d_i = d_1 + 2(i - 1)c \ (i = 1, 2, \ldots, 200)$.
When $i \geq 2$,

$$a_i = \sum_{j=1}^{i-1} (a_{j+1} - a_j) + a_1$$

$$= \sum_{j=1}^{i-1} d_j + a_1 = a_1 + \sum_{j=1}^{i-1} [d_1 + 2(j-1)c]$$

$$= a_1 + (i-1)[d_1 + 2(i-1)c].$$

Clearly, the above equality is also satisfied by a_1.

From $a_{201} = a_1 + 200(d_1 + 199c) = a_1$ we know that $d_1 = -199c$.

Thus, $a_i = a_1 + (i - 1)(i - 201)c$.

Also from $a_{101} = a_1 + 100(-100)c = a_1 - 100^2 c = 19999 - 100^2 c \in \mathbf{Z}_+$,

$$c \leq \frac{19999}{10000} < 2.$$

That is, $c = 1$.

Therefore, $a_i = 19999 + (i - 1)(i - 201)$, and in particular

$$a_{200} = 19999 + 199(-1) = 19800. \qquad \qquad \square$$

Example 7. A positive integer sequence $\{a_n\}$ is defined as follows: $a_1 = 2, a_2 = 7$, and $-\frac{1}{2} < a_{n+1} - \frac{a_n^2}{a_{n-1}} \leq \frac{1}{2}$ for $n = 2, 3, \dots$. Find the general term of the sequence.

Solution. It is difficult to determine a_n from the given recursive formula, but can we obtain a linear recursive sequence with constant coefficients that is familiar to us from the given condition? A bold conjecture: $a_{n+1} = pa_n + qa_{n-1}$, where p and q are constants to be determined.

By trying the first several terms of the sequence, we find that $a_1 = 2, a_2 = 7, a_3 = 25, a_4 = 89, \dots$ to guess the values of p and q. Conjecture: $a_{n+1} = 3a_n + 2a_{n-1}$ for $n \geq 2$.

We prove the above conjecture with mathematical induction in the following.

When $n = 2$ and 3, the above conjecture is correct.

Suppose that $a_{k+1} = 3a_k + 2a_{k-1}$ for all $k \leq n$. Then for $k = n + 1$,

$$\frac{a_{n+1}^2}{a_n} = \frac{a_{n+1}(3a_n + 2a_{n-1})}{a_n} = 3a_{n+1} + 2a_n + 2 \cdot \frac{a_{n+1}a_{n-1} - a_n^2}{a_n}.$$

Note that

$$\left| 2 \cdot \frac{a_{n+1}a_{n-1} - a_n^2}{a_n} \right| = \left| \frac{2a_{n-1}}{a_n} \right| \cdot \left| a_{n+1} - \frac{a_n^2}{a_{n-1}} \right| \leq \frac{1}{2} \left| \frac{2a_{n-1}}{a_n} \right|.$$

By the inductive hypothesis, we see easily that $a_n > 2a_{n-1}$, hence

$$\left| 3a_{n+1} + 2a_n - \frac{a_{n+1}^2}{a_n} \right| < \frac{1}{2}.$$

Since a_{n+2} is an integer and $\left| a_{n+2} - \frac{a_{n+1}^2}{a_n} \right| \le \frac{1}{2}$, we find that

$$\left| a_{n+2} - (3a_{n+1} + 2a_n) \right| \le \left| a_{n+2} - \frac{a_{n+1}^2}{a_n} \right| + \left| \frac{a_{n+1}^2}{a_n} - (3a_{n+1} + 2a_n) \right|$$

$$< \frac{1}{2} + \frac{1}{2} = 1.$$

Thus, $a_{n+2} = 3a_{n+1} + 2a_n$, and so the conjecture is true for $k = n + 1$.

To summarize, the sequence $\{a_n\}$ satisfies $a_1 = 2, a_2 = 7$ and $a_n = 3a_{n-1} + 2a_{n-2}$ for $n = 2, 3, \ldots$. Using the characteristic equation to solve the homogeneous linear recursive equation with constant coefficients, we find the general term

$$a_n = \frac{17 + 5\sqrt{17}}{68} \cdot \left(\frac{3 + \sqrt{17}}{2} \right)^n + \frac{17 - 5\sqrt{17}}{68} \cdot \left(\frac{3 - \sqrt{17}}{2} \right)^n. \qquad \Box$$

Remark. If a mathematical problem already has a conclusion, it is usually much easier to solve it, but obtaining a conclusion frequently needs a conjecture. Problems related to sequences are more so. After accumulating some experience, making a conjecture is often successful. That is also a kind of mathematical intuition.

Example 8. Prove: There exists a unique infinite sequence $\{a_n\}$ consisting of positive integers such that $a_1 = 1, a_2 > 1$, and $a_{n+1}^3 + 1 = a_n a_{n+2}$ for $n = 1, 2, \ldots$.

Proof. If $\{a_n\}$ is a sequence of positive integers satisfying the condition, then $a_n \in \mathbf{N}_+$ for any $n \in \mathbf{N}_+$, and from $a_4 = \frac{a_3^3 + 1}{a_2} = \frac{(a_2^3 + 1)^3 + 1}{a_2} \in \mathbf{N}_+$ we deduce that $a_2 \mid 2$. Since $a_2 > 1$, we must have $a_2 = 2$. Since $\{a_n\}$ is a second order recursive sequence, every term of the sequence $\{a_n\}$ is uniquely determined by the recursive formula.

We still need to prove in the following: Each term of the sequence $\{a_n\}$ defined by $a_1 = 1, a_2 = 2$, and $a_{n+1}^3 + 1 = a_n a_{n+2}$ for $n = 1, 2, \ldots$ is indeed a positive integer.

In fact, when $n \le 4$, it is easy to see that $a_n \in \mathbf{N}_+$. Suppose that $a_n \in \mathbf{N}_+$ for $n \le k$ $(k \ge 4)$. We show below: $a_{k+1} \in \mathbf{N}_+$. First,

$$a_{k+1} = \frac{a_k^3 + 1}{a_{k-1}} = \frac{1}{a_{k-1}} \left[\left(\frac{a_{k-1}^3 + 1}{a_{k-2}} \right)^3 + 1 \right]$$

$$= \frac{1}{a_{k-1} a_{k-2}^3} \left(a_{k-1}^9 + 3a_{k-1}^6 + 3a_{k-1}^3 + 1 + a_{k-2}^3 \right).$$

Note that

$$a_k^3 + 1 = \frac{1}{a_{k-2}^3} \left(a_{k-1}^9 + 3a_{k-1}^6 + 3a_{k-1}^3 + 1 + a_{k-2}^3 \right) \in \mathbf{N}_+,$$

and

$$\frac{1}{a_{k-1}} \left(a_{k-1}^9 + 3a_{k-1}^6 + 3a_{k-1}^3 + 1 + a_{k-2}^3 \right)$$
$$= a_{k-1}^8 + 3a_{k-1}^5 + 3a_{k-1}^2 + a_{k-3} \in \mathbf{N}_+.$$

So, to show that $a_{k+1} \in \mathbf{N}_+$, it is sufficient to show:

$$(a_{k-1}, a_{k-2}) = 1.$$

Since $(a_{k-1}, a_{k-2}) = \left(\frac{a_{k-2}^3 + 1}{a_{k-3}}, a_{k-2} \right) \le (a_{k-2}^3 + 1, a_{k-2}) = 1$, it follows that $a_{k+1} \in \mathbf{N}_+$. ☐

3. Exercises

Group A

I. Filling Problems

1. If a sequence $\{a_n\}$ satisfies $a_1 = \frac{1}{3}$ and $a_{n+1} = a_n^2 + a_n$ for any $n \in \mathbf{N}_+$, then the integral part of $\sum_{n=1}^{2016} \frac{1}{a_n + 1}$ is _____.
2. Suppose that a sequence $\{a_n\}$ satisfies $a_1 = 1$ and $a_{n+1}a_n - 2n^2(a_{n+1} - a_n) + 1 = 0$ for $n \ge 1$. Then the general term of the sequence is $a_n = $ _____.

II. Calculation Problems

3. Let $a_0 = 2, a_1 = 3, a_2 = 6$, and for $n \ge 3$,

$$a_n = (n+4)a_{n-1} - 4na_{n-2} + (4n - 8)a_{n-3}.$$

Find the general term a_n.
4. A sequence $\{a_n\}$ satisfies $a_1 = 2$ and $a_{n+1} = \frac{a_n}{2} + \frac{1}{a_n}$ for $n \ge 1$. Find the general term a_n of this sequence.
5. Find the minimal positive integer k such that there exist at least two sequences $\{a_n\}$ satisfying the following conditions:

   (1) For any positive integer n, there holds that $a_n \le a_{n+1}$.
   (2) For any positive integer n, it is true that $a_{n+2} = a_{n+1} + a_n$.
   (3) $a_9 = k$.

6. Given a sequence of positive integers $a_1, a_2, \ldots$ with the property that every positive integer appears just one time, prove: There exist integers m and n such that $1 < n < m$ and $a_1 + a_m = 2a_n$.

7. A sequence $\{x_n\}$ is defined as follows: $x_1 = 2$ and $nx_n = 2(2n-1)x_{n-1}$ for $n = 2, 3, \ldots$. Prove: For every positive integer n, the number x_n is a positive integer.

8. A sequence $\{x_n\}$ is defined as follows: $x_1 = 2$ and $x_{n+1} = \frac{x_n^4+1}{5x_n}$ for $n = 1, 2, \ldots$. Prove: When $n \geq 2$, it is true that $\frac{1}{5} < x_n < 2$.

9. The definitions of sequences $\{a_n\}$ and $\{b_n\}$ are:

$$a_1 = 1, \quad b_1 = 2, \quad a_{n+1} = \frac{1 + a_n + a_n b_n}{b_n}, \quad b_{n+1} = \frac{1 + b_n + a_n b_n}{a_n}.$$

Prove: $a_{2015} < 5$.

10. Let two sequences $\{a_n\}$ and $\{b_n\}$ satisfy $a_0 = 1, b_0 = 0$, and

$$\begin{cases} a_{n+1} = 7a_n + 6b_n - 3, \\ b_{n+1} = 8a_n + 7b_n - 4, \end{cases} \quad (n = 0, 1, 2, \ldots).$$

Prove: $a_n \ (n = 0, 1, 2, \ldots)$ is a perfect square.

Group B

11. A sequence $\{a_n\}$ satisfies: $a_0 = 1$ and $a_{n+1} = \frac{7a_n + \sqrt{45a_n^2 - 36}}{2}$ for $n \in \mathbf{N}$. Prove:

   (1) For any $n \in \mathbf{N}$, the number a_n is a positive integer.
   (2) For any $n \in \mathbf{N}$, the number $a_n a_{n+1} - 1$ is a perfect square.

12. An infinite sequence $a_0, a_1, \ldots$ consists of real numbers, in which a_0 and a_1 are two different positive real numbers, and $a_n = |a_{n+1} - a_{n+2}|$ for $n = 0, 1, 2, \ldots$. Question: Is it possible that this sequence can be a bounded one? Prove your conclusion.

Chapter 25

Periodic Sequences

1. Key Points of Knowledge and Basic Methods

A. Concept of periodic sequences

(1) For a sequence $\{a_n\}$, if there exist definite positive integers T and n_0 such that for all positive integers $n \geq n_0$ the equality $a_{n+T} = a_n$ is always satisfied, then we call $\{a_n\}$ a periodic sequence with period T starting from the n_0th term. When $n_0 = 1$, we say that $\{a_n\}$ is a purely periodic sequence, and when $n_0 \geq 2$, we say that $\{a_n\}$ is a mixed periodic sequence.

(2) Let $\{a_n\}$ be a sequence of integers and let m be some chosen positive integer greater than 1. If b_n is the remainder of a_n after divided by m, that is $b_n \equiv a_n \pmod{m}$ and $b_n \in \{0, 1, 2, \ldots, m-1\}$, then we call the sequence $\{b_n\}$ the modular sequence of $\{a_n\}$ with respect to m, which is denoted as $\{a_n \pmod{m}\}$.

  If a modular sequence $\{a_n \pmod{m}\}$ is periodic, then we say that $\{a_n\}$ is a periodic sequence with respect to the module m.

B. Important properties and conclusions of periodic sequences

(1) The range of a periodic sequence is a finite set.
(2) If T is a period of $\{a_n\}$, then kT is also a period of $\{a_n\}$ for any $k \in \mathbf{N}_+$.
(3) A periodic sequence must have the minimal period.
(4) If T is the minimal period of a periodic sequence $\{a_n\}$ and T' is any period of $\{a_n\}$, then $T \mid T'$.

(5) Any kth order linear recursive sequence satisfying a homogeneous linear recursive relation

$$a_{n+k} = c_1 a_{n+k-1} + c_2 a_{n+k-2} + \cdots + c_k a_n \ (n \in \mathbf{N}_+ \text{ and } c_k \neq 0)$$

with constant coefficients is a modular periodic sequence with respect to any positive integer $m \ (\geq 2)$.

C. *Problems and methods related to periodic sequences*

(1) Given a general term expression, find the sum of the first n terms of the sequence: First calculate several terms of the sequence to make an observation on periodicity of the sequence, and if so then reduce the original problem to the sum of several terms within one period; or take some algebraic transformation according to the general term formula to find its period, and then solve the problem by means of periodicity of the sequence.

(2) Given a recursive formula, find some particular term, the sum of several terms, or determine some features of the general term (such as divisibility and rationality and etc.): From the recursive formula, make an appropriate transformation to deduce the period, or by calculating several terms to make observations, conjectures, and give an inductive proof on periodicity. Then solve the problem by using periodicity.

(3) The concerned problems of modular periodic sequences are often about judging divisibility of terms, determining some digits of the terms, or testing whether a term is a composite number or a prime number, etc. The "module" in a modular periodic sequence is a key of solving the problem. Finding a proper module means that the problem has been simplified, and so it is much easier to solve. There is no general method for finding the module, and an analysis should be performed according to the condition of the problem. Trying, and sometimes several times of trying, may lead to the possibility of success.

2. Illustrative Examples

Example 1. Define a sequence $\{a_n\}$: $a_n = \frac{1+a_{n-1}}{1-a_{n-1}}$ $(n \geq 2, a_1 \neq 0$ and $a_1 \neq \pm 1)$. Judge periodicity of the sequence $\{a_n\}$.

Solution. Make a trigonometric substitution: Let $a_n = \tan \theta_n$ for $n = 1, 2, \ldots, n$. Then

$$a_{n+1} = \frac{\tan \frac{\pi}{4} + \tan \theta_n}{1 - \tan \frac{\pi}{4} \tan \theta_n} = \tan \left(\frac{\pi}{4} + \theta_n \right).$$

So,

$$\tan \theta_{n+1} = \tan \left(\frac{\pi}{4} + \theta_n \right) = \tan \left(\frac{\pi}{2} + \theta_{n-1} \right)$$

$$= \tan \left(\frac{3\pi}{4} + \theta_{n-2} \right) = \cdots = \tan \left(\frac{\pi}{4} n + \theta_1 \right).$$

Hence,

$$\tan \theta_n = \tan \left[\frac{\pi}{4}(n-1) + \theta_1 \right].$$

Since $\tan x$ is a periodic function of period π, thus $\{a_n\}$ is a purely periodic sequence of period 4. □

Example 2. Let $\{a_n\}$ be a sequence of real numbers that satisfies $a_{n+2} = |a_{n+1}| - a_n$. Prove: There exists $N_0 \in \mathbf{N}_+$ such that when $n \geq N_0$, it is always true that $a_{n+9} = a_n$.

Proof. When a_n is always 0, clearly the conclusion is true. So, in the following we assume that a_n is not always 0. Then there must exist $N_0 \in \mathbf{N}_+$ such that a_n are not always positive or always negative for all $n \geq N_0$. Otherwise, if a_n are always positive, then $a_{n+3} = -a_n < 0$, a contradiction; if a_n are always negative, then $a_{n+2} = |a_{n+1}| - a_n > 0$, also a contradiction. Thus, we may assume that $a_{N_0} = -a$ and $a_{N_0+1} = b$ ($a, b \geq 0$, and they are not both zero). It follows that

$$a_{N_0+2} = |a_{N_0+1}| - a_{N_0} = b + a,$$
$$a_{N_0+3} = |a_{N_0+2}| - a_{N_0+1} = a,$$
$$a_{N_0+4} = |a_{N_0+3}| - a_{N_0+2} = -b,$$
$$a_{N_0+5} = |a_{N_0+4}| - a_{N_0+3} = b - a.$$

(1) If $b \geq a$, then

$$a_{N_0+6} = |a_{N_0+5}| - a_{N_0+4} = 2b - a,$$
$$a_{N_0+7} = |a_{N_0+6}| - a_{N_0+5} = (2b - a) - (b - a) = b,$$
$$a_{N_0+8} = |a_{N_0+7}| - a_{N_0+6} = b - (2b - a) = a - b \leq 0,$$
$$a_{N_0+9} = |a_{N_0+8}| - a_{N_0+7} = b - a - b = -a,$$
$$a_{N_0+10} = a - (a - b) = b.$$

(2) If $b < a$, then we find that

$$a_{N_0+6} = a, \ a_{N_0+7} = 2a - b > 0, \ a_{N_0+8} = a - b > 0,$$
$$a_{N_0+9} = -a, \ a_{N_0+10} = b.$$

Hence, we always have $a_{N_0+9} = a_{N_0}$ and $a_{N_0+10} = a_{N_0+1}$. Then by using the recursive formula $a_{n+2} = |a_{n+1}| - a_n$ and mathematical induction, we can show that $a_{n+9} = a_n$ for all $n \geq N_0$. □

Example 3. A sequence $a_1, a_2, \ldots$ of integers is defined as follows: $a_1 = 1, a_2 = 2$, and for $n \geq 1$,

$$a_{n+2} = \begin{cases} 5a_{n+1} - 3a_n, & \text{if } a_n a_{n+1} \text{ is an even number,} \\ a_{n+1} - a_n, & \text{if } a_n a_{n+1} \text{ is an odd number.} \end{cases}$$

Prove: $a_n \neq 0$ for any positive integer n.

Proof. Calculate the first couple of terms of the sequence:

$$a_1 = 1, \ a_2 = 2, \ a_3 = 7, \ a_4 = 29, \ a_5 = 22,$$
$$a_6 = 23, \ a_7 = 49, \ a_8 = 26, \ a_9 = -17.$$

An observation finds: For $k \in \mathbf{N}$,

$$a_{3k+1} \equiv 1 \ (\text{mod } 4), \quad a_{3k+2} \equiv 2 \ (\text{mod } 4), \quad a_{3k+1} \equiv 3 \ (\text{mod } 4).$$

In the following, we use mathematical induction to prove the above claim.

The previous calculation shows that the proposition is true for $k = 0$.

Assume that the proposition is true for $k \in \mathbf{N}$. Then $a_{3k+2} \cdot a_{3k+3}$ is an even number, so

$$a_{3(k+1)+1} = 5a_{3k+3} - 3a_{3k+2} \equiv 3 - 2 = 1 \ (\text{mod } 4).$$

From this and the inductive hypothesis, we know that $a_{3k+3} \cdot a_{3(k+1)+1}$ is an odd number, hence

$$a_{3(k+1)+2} = a_{3(k+1)+1} - a_{3k+3} \equiv 1 - 3 = 2 \ (\text{mod } 4).$$

Also from what has been proved, $a_{3(k+1)+1} \cdot a_{3(k+1)+2}$ is an even number, therefore

$$a_{3(k+1)+3} = 5a_{3(k+1)+2} - 3a_{3(k+1)+1} \equiv 2 - 3 = 3 \ (\text{mod } 4).$$

In summary, the proposition is true for $k + 1$.

This means that $a_n \not\equiv 0 \pmod 4$ for any positive integer n, so $a_n \neq 0$. □

Remark. We actually proved that the modular sequence of the sequence in the problem with respect to 4 is a periodic sequence of period 3.

Example 4. Suppose that a sequence $\{a_n\}$ of positive integers satisfies that a_{n+1} is the minimal prime factor of $a_n + a_{n-1}$ for all $n \geq 2$. Prove: $\{a_n\}$ is a periodic sequence from some term on.

Proof. When a_n and a_{n-1} are both odd numbers, clearly $a_{n+1} = 2$. Thus, from the odd-even property we know that there must be one term to be 2 among $a_1, a_2, \ldots, a_5$.

In the sequence $\{a_n\}$, if there exists $i \in \mathbf{Z}_+$ such that $a_i = a_{i+1} = 2$, then $a_n = 2 \ (n \geq i)$.

If two nearby numbers in the sequence are 2 and 3 (or 3 and 2), then we have the periodic sequence

$$2, \ 3, \ 5, \ 2, \ 7, \ 3, \ 2, \ 5, \ 7, \ 2, \ 3, \ \ldots.$$

If two nearby numbers in the sequence are 2 and $6k+1$ (or $6k+1$ and 2), then we get the sequence

$$2, \ 6k + 1, \ 3, \ 2, \ \ldots \quad \text{or} \quad 6k + 1, \ 2, \ 3, \ \ldots,$$

namely 2 and 3 (or 3 and 2) appear as nearby numbers, resulting in a period sequence.

Suppose that the above cases do not happen. Then for any $a_i = 2$ $(i \geq 2)$, we always have

$$a_{i-1} \equiv a_{i+1} \equiv -1 \pmod 6.$$

Let

$$a_{i-1} = 6k - 1, \quad a_{i+1} = 6l - 1 \quad (k, l \in \mathbf{Z}_+).$$

Now, a_{i+2} is an odd number and $a_{i+3} = 2$. So, $a_{i+2} \equiv -1 \pmod 6$, and can be written as

$$a_{i+2} = 6m - 2 \quad (m \in \mathbf{Z}_+).$$

From the given condition,

$$a_{i+1} \mid (a_i + a_{i-1}), \quad a_{i+2} \mid (a_{i+1} + a_i),$$

so $(6l - 1) \mid (6k + 1)$ and $(6m - 1) \mid (6l + 1)$.

It follows that $m < l < k$, from which $a_{i+1} < a_{i-1}$.

Continuing doing the above in this way, we find that $a_{i-1}, a_{i+2}, a_{i+5}, \ldots$ is a strictly decreasing sequence of prime numbers, which is a contradiction.

In summary, $\{a_n\}$ must be a periodic sequence from some term on. $\square$

Example 5. For nonnegative integers m and n, we use $a_{m,n}$ to denote the coefficient of the x^n term in the expansion of the polynomial $(1 + x + x^2)^m$. Prove: For any nonnegative integer k, it is true that

$$0 \le \sum_{i=0}^{\left[\frac{2k}{3}\right]} (-1)^i a_{k-i,i} \le 1.$$

Proof. Note that

$$(1 + x + x^2)^{m+1} = (1 + x + x^2)(a_{m,0} + a_{m,1}x + \cdots + a_{m,n}x^n + \cdots),$$

so

$$a_{m+1,n} = a_{m,n} + a_{m,n-1} + a_{m,n-2}.$$

Now, let $S_k = \sum_{i=0}^{\left[\frac{2k}{3}\right]} (-1)^i a_{k-i,i}$. Then from the above recursive formula, we see that

$$S_k - S_{k+1} - S_{k+2} = \sum_{i=0}^{\left[\frac{2k}{3}\right]} (-1)^i (a_{k-i,i} + a_{k-i,i+1} + a_{k-i,i+2})$$

$$= \sum_{i=0}^{\left[\frac{2k}{3}\right]} (-1)^i a_{k+1-i,i+2}$$

$$= \sum_{i=0}^{\left[\frac{2(k+3)}{3}\right]} (-1)^i a_{k+3-i,i} = S_{k+3}.$$

In the above calculation, we assigned $a_{m,j} = 0$ for $j < 0$ or $j > 2(m+1)$.

Consequently, we have

$$S_{k+4} = S_{k+1} - S_{k+2} + S_{k+3}$$

$$= (S_{k+1} - S_{k+2}) + (S_k - S_{k+1} + S_{k+2})$$

$$= S_k.$$

That is, $\{S_k\}$ is a purely periodic sequence of period 4.

Calculating the initial values of the sequence $\{S_k\}$ directly gives $S_0 = S_1 = 1$ and $S_2 = S_3 = 0$, and combining it with periodicity of $\{S_n\}$, we know that the proposition is true. □

Example 6. Let a sequence $\{a_n\}$ satisfy $a_0 = 1, a_1 = 3$, and $a_{n+2} = 4a_{n+1} - a_n$ for $n \geq 0$. Try to determine the last two digits of $2a_n^3 - a_n$ for any nonnegative integer n.

Solution. Since $a_0 = 1, a_1 = 3$, and $a_{n+2} \equiv -a_n \pmod 4$, we have

$$a_n \equiv \begin{cases} 1 \pmod 4, & n \equiv 0 \text{ or } 3 \pmod 4, \\ 3 \pmod 4, & n \equiv 1 \text{ or } 2 \pmod 4. \end{cases}$$

Hence, $2a_n^3 - a_n \equiv 1$ or $3 \pmod 4$.

On the other hand, the remainders of a_n divided by 25 form a periodic sequence of period 15:

$$1, 3, 11, 16, 3, 21, 6, 3, 6, 21, 3, 16, 11, 3, 1; 1, 3, 11, 16, 3, \ldots,$$

thus $a_n \equiv 1, 3, 6, 11, 16$ or $21 \pmod 4$. A direct calculation finds that $2a_n^3 - a_n \equiv 1 \pmod 4$. Consequently,

$$2a_n^3 - a_n \equiv \begin{cases} 1 \pmod{100}, & n \equiv 0 \text{ or } 3 \pmod 4, \\ 5 \pmod{100}, & n \equiv 1 \text{ or } 2 \pmod 4. \end{cases}$$

Therefore, when $n = 4k$ or $4k + 3$, the last two digits of $2a_n^3 - a_n$ are 01; when $n = 4k + 1$ or $4k + 2$, the last two digits of $2a_n^3 - a_n$ are 51. □

Example 7. Let m be a given positive integer. Suppose that a sequence $\{x_n\}$ satisfies:

$$x_k = k, \ k = 1, 2, \ldots, m+1; \quad x_{k+1} = x_k + x_{k-m}, \ k = m+1, \ldots.$$

Prove: there exist m consecutive terms in $\{x_n\}$, all of which are multiples of $m + 1$.

Proof. We study the sequence $\{x_n \pmod{(m+1)}\}_{n=1}^{+\infty}$, in which $x_n \pmod{(m+1)}$ denotes the remainder of x_n divided by $m + 1$, which is written as y_n. Since there are only finitely many values (at most $m + 1$ different values) for each number of $y_1, y_2, \ldots$, there must exit $k < n$ with $k, n \in \mathbf{N}_+$ such that

$$(y_k, y_{k+1}, \ldots, y_{k+m}) = (y_n, y_{n+1}, \ldots, y_{n+m}). \tag{25.1}$$

The above conclusion is from the drawer principle.

By the recursive formula, we know that $y_{k-1} = y_{k+m} - y_{k+m-1}$ and $y_{n-1} = y_{n+m} - y_{n+m-1}$. Combined with (25.1) it follows that $y_{k-1} = y_{n-1}$. Deducing backward with the same idea gives that $y_l = y_{l+(n-k)}$ for any $l \in \{1, 2, \ldots, m\}$. Furthermore, by deducing backward to negative integers according to the recursive formula, one sees that $y_l = y_{l+(n-k)}$ for any $l \leq m$ with $l \in \mathbf{Z}$.

Using the recursive formula and the initial value conditions to deduce backward, we obtain that $x_0 = x_{-1} = \cdots = x_{-(m-1)} = 1$ and $x_m = x_{-(m+1)} = \cdots = x_{-(2m-1)} = 0$. Consequently,

$$(y_{n-k-(2m-1)}, \ldots, y_{n-k-m}) = (y_{-(2m-1)}, \ldots, y_{-m}) = (0, 0, \ldots, 0).$$

Also from $y_{-(m-1)} = \cdots = y_0 = 1$, the subscript $n - k - (2m - 1) \geq 1$.

To summarize, there are m consecutive terms in $\{x_n\}$ that are all multiples of $m + 1$. $\qquad\square$

Example 8. Let m be a given positive integer. For any positive integer n let $S_m(n)$ denote the sum of the mth powers of all digits of n in the decimal number system. For example, $S_3(172) = 1^3 + 7^3 + 2^3 = 352$. Define the sequence $\{n_k\}_{k=0}^{+\infty}$ as follows: $n_0 \in \mathbf{N}_+$ and $n_k = S_m(n_{k-1})$ for $k = 1, 2, \ldots$. Prove:

(1) For any $n_0 \in \mathbf{N}_+$, the sequence $\{n_k\}_{k=0}^{+\infty}$ is a periodic one.
(2) When n_0 varies, the set consisting of the minimal periods of the corresponding sequences in (1) is a finite set.

Proof. If $n \geq 10^{m+1}$, then there exists a positive integer $p \geq m + 1$ such that $10^p \leq n < 10^{p+1}$. Then

$$S_m(n) \leq (p + 1) \cdot 9^m < 9^p + C_p^1 \cdot 9^{p-1} < (9 + 1)^p \leq n.$$

Thus, $n_{k+1} < n_k$ when $n_k \geq 10^{m+1}$.

If $n < 10^{m+1}$, then

$$S_m(n) \leq (m + 1) \cdot 9^m < (9 + 1)^{m+1} = 10^{m+1}.$$

The above discussion indicates that for $n_0 \in \mathbf{N}_+$, when k is sufficiently large, it is always valid that $n_k < 10^{m+1}$. So, from some term on, every term of the sequence belongs to $\{1, 2, \ldots, 10^{m+1} - 1\}$.

Combining the above with the definition of $\{n_k\}$, we deduce that for any $n_0 \in \mathbf{N}_+$, the sequence $\{n_k\}_{k=0}^{+\infty}$ is a periodic one, thus proving (1).

Furthermore, we know that when n_0 varies, as n_k eventually enters the set $\{1, 2, \ldots, 10^{m+1} - 1\}$, by the drawer principle, the next 10^{m+1} numbers must have two among them that are equal to each other. Therefore, the minimal periods of all $\{n_k\}$ are always less than 10^{m+1}, hence (2) is satisfied. $\qquad\qquad\qquad\qquad\qquad\qquad\qquad\qquad\qquad\qquad\square$

3. Exercises

Group A

I. Filling Problems

1. Suppose that the sum of any three consecutive terms of a sequence $\{x_n\}$ is 20. Let $x_4 = 9$ and $x_{11} = 7$. Then the value of x_{2014} is _____.

2. A sequence $\{x_n\}$ of real numbers is defined as follows: x_0 and x_1 are any given positive real numbers, and $x_{n+2} = \frac{1+x_{n+1}}{x_n}$ for $n = 0, 1, 2, \ldots$. Then $x_{2010} =$ _____.

3. A sequence $\{x_n\}$ satisfies: $x_1 = 1, x_2 = -2$, and $x_{n+2} = \frac{x_{n+1}^2 - 5}{x_n}$ for $n = 1, 2, \ldots$. Then the value of $\sum_{k=1}^{2014} x_k$ is _____.

4. Let a sequence $\{x_n\}$ satisfy: $x_1 = 2014$ and $x_{n+1} = \frac{(\sqrt{2}+1)x_n - 1}{(\sqrt{2}+1) + x_n}$ for $n \geq 1$. Then the value of x_{2015} is _____.

II. Calculation Problems

5. A sequence $\{a_n\}$ is given in the following way: When $n \geq 1$, let $a_{2n} = a_n$, and when $n \geq 0$, let $a_{4n+1} = 1$ and $a_{4n+3} = 0$. Prove: This sequence is not a periodic one.

6. Let four periodic sequences $\{a_n\}, \{b_n\}, \{c_n\}$, and $\{d_n\}$ satisfy: $a_{n+1} = a_n + b_n, b_{n+1} = b_n + c_n, c_{n+1} = c_n + d_n$, and $d_{n+1} = d_n + a_n$ for $n = 1, 2, \ldots$. Prove: $a_2 = b_2 = c_2 = d_2 = 0$.

7. For any positive integer k, let $f_1(k)$ be the square of the sum of each digit of k. For $n \geq 2$, let $f_n(k) = f_1(f_{n-1}(k))$. Find $f_{100}(11)$.

8. In a sequence $a_0, a_1, \ldots$ of positive integers, a_0 is an arbitrary integer, and for each nonnegative integer n,

$$a_{n+1} = \begin{cases} \dfrac{1}{2}a_n, & \text{if } a_n \text{ is even,} \\[2mm] a + a_n, & \text{if } a_n \text{ is odd.} \end{cases}$$

Here, a is a fixed odd number. Prove: This sequence is a periodic one.

Group B

9. Suppose that $U_0 = 0, U_1 = 1$, and $U_{n+1} = 8U_n - U_{n-1}$ with $n = 1, 2, \ldots$. Prove: There are no terms of the form $3^\alpha \cdot 5^\beta$ in the sequence, where $\alpha, \beta \in \mathbf{N}_+$.

10. Let S_0 be a set of finitely many positive integers. Define a sequence of sets $S_0, S_1, S_2, \ldots$ as follows: For $n = 0, 1, 2, \ldots$, if and only if exactly one of the conditions $a - 1 \in S_n$ and $a \in S_n$ is satisfied, $a \in S_{n+1}$. Prove: There exist infinitely many positive integers N such that $S_N = S_0 \cup \{N + a \mid a \in S_0\}$.

11. Let X be a subset of the set $\mathbf{Z}$ of all integers. Denote $X + a = \{x + a \mid x \in X\}$. Prove: If there exist integers $a_1, a_2, \ldots, a_n$ such that $X + a_1, X + a_2, \ldots, X + a_n$ constitute a partition of $\mathbf{Z}$, then there exists an integer N such that $X + N = X$.

12. Suppose that a sequence $\{x_n\}$ satisfies: x_1 and x_2 are two relatively prime positive integers, and $x_{n+1} = x_n x_{n-1} + 1$ for $n \geq 2$.

   (1) Prove: For any integer i, there exists an integer $j > i$ such that $x_i^i \mid x_j^j$.

   (2) Does there have to exist $j > 1$ such that $x_1 \mid x_j^j$?

Chapter 26

Polar Coordinates

1. Key Points of Knowledge and Basic Methods

A. *Polar coordinates system*

As in Figure 26.1, choose one point O in a plane, called the pole. Draw a ray Ox from the pole, called the polar axis. Then provide a unit length and a positive direction (often the counter clockwise direction) of an angle. Thus, a polar coordinates system is established, so that any point P in the plane may be represented by a pair of ordered numbers (ρ, θ), in which ρ represents the length of the line segment OP, called the polar radius, and θ represents the amount of the rotational angle from Ox to OP, called the polar angle of the point P.

Sometimes for the purpose of convenience, we may also allow ρ to take negative values, and in this case, the coordinates (ρ, θ) represent the point on the opposite direction extended line of the ray with the polar angle θ and the distance $-\rho$ to the pole.

Obviously, any pair of polar coordinates represents a unique point of the plane, but the polar coordinates of a point are not unique. However, if one makes the restriction of $\rho > 0$ and $0 \leq \theta < 2\pi$ (or $-\pi < \theta \leq \pi$), then except for the pole, all points of the plane and polar coordinates have a one-to-one correspondence.

B. *Mutual transformations of polar coordinates and rectangular coordinates*

If the pole and the origin coincide, and the polar axis and the positive direction of the x-axis coincide, then under the two coordinates systems,

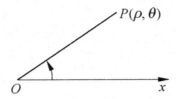

Figure 26.1 Polar coordinates.

the coordinates of a point can be mutually transformed:

$$(1) \begin{cases} x = \rho \cos \theta, \\ y = \rho \sin \theta; \end{cases} \quad (2) \begin{cases} \rho^2 = x^2 + y^2, \\ \tan \theta = \dfrac{y}{x} \ (x \neq 0). \end{cases}$$

C. *Equations of curves under the polar coordinates system*

(1) Equations of straight lines:

    (i) The equation $\theta = \theta_0$ ($\rho \in \mathbf{R}$) represents the straight line passing through the pole with the polar angle θ_0.

    (ii) The equation $\rho \cdot \cos(\theta - \theta_0) = a$ represents the straight line that passes through the point $A(a, \theta_0)$ and is perpendicular to OA, where O is the pole.

(2) Equations of circles:

    (i) The equation $\rho = r$ represents the circle centered at the pole of radius r.

    (ii) The equation $\rho^2 + \rho_0^2 - 2\rho\rho_0 \cos(\theta - \theta_0) = r^2$ represents the circle centered at (ρ_0, θ_0) of radius r.

(3) Equations of conic sections:

    (i) The equation $\rho = \dfrac{ep}{1 - e \cos \theta}$ represents a conic section with e the eccentricity and p the distance of a focus to the corresponding directrix.

        When $e = 1$, the pole is just the focus of the parabola.

        When $e > 1$, the pole is the right focus of the hyperbola. For $\rho \in \mathbf{R}$, the equation represents the whole hyperbola, and if ρ is restricted to $\mathbf{R}_+$, then the equation represents the right branch of the hyperbola.

        When $0 < e < 1$, the pole is the left focus of the ellipse.

(ii) When the pole and the origin of the rectangular coordinates coincide and the polar axis coincides with the positive direction of the x-axis, an equation of a conic section can be obtained via the coordinates transformation formulas.

D. Basic methods

An advantage of polar coordinates is to treat the distance and angle as the investigation objects, while the rectangular coordinates system is more systematic. These two systems have their merits and shortcomings, so the mutual transformations of the two coordinates systems appear to be extremely important.

2. Illustrative Examples

Example 1. Given two points A and B on an ellipse $\rho = \frac{3}{2-\cos\theta}$ that satisfy $\overrightarrow{AO} = 2\overrightarrow{OB}$, where O is the pole. Find the equation of the straight line AB and the length $|AB|$ of the line segment AB.

Analysis. From the condition of the problem, the three points A, O, and B are co-linear, so the polar angles of A and B differ by π. Thus, it is enough to find the coordinates of A and B to obtain the solution.

Solution. Since $\overrightarrow{AO} = 2\overrightarrow{OB}$, the three points A, O, and B are co-linear. Write the polar coordinates of A and B as $A(\rho_1, \theta)$ and $B(\rho_2, \theta + \pi)$ with $\rho_1 = 2\rho_2$. Then

$$\frac{3}{2-\cos\theta} = 2 \cdot \frac{3}{2 - \cos(\theta + \pi)} = 2 \cdot \frac{3}{2 + \cos\theta},$$

the solutions of which are

$$\cos\theta = \frac{2}{3}, \quad \rho_1 = \frac{9}{4}, \quad \rho_2 = \frac{9}{8}.$$

Therefore, the desired equation of the straight line passing though A and B is $\theta = \pm\arccos\frac{2}{3}$ $(\rho \in \mathbf{R})$, and

$$|AB| = \rho_1 + \rho_2 = \frac{27}{8}. \qquad \square$$

Remark. By means of polar coordinates, it is quite convenient to find the length of a line segment passing through the pole and solve the angle problem with the pole as the vertex of the angle.

Example 2. There are an ellipse E_1, a parabola E_2, and a hyperbola E_3 with a common focus F and directrix l_1 corresponding to F. Let a straight line l_2 pass through F and let the chords obtained from its intersections with E_1, E_2, and E_3 be AB, PQ, and MN, respectively (as Figure 26.2 shows). Assume that $\left(\frac{|PQ|}{|AB|}\right)_{\min} = m$ and $\left(\frac{|MN|}{|PQ|}\right)_{\min} = n$. If $m = n$, find the relation between the eccentricity e_1 of the ellipse and the eccentricity e_2 of the hyperbola.

Solution. Take F as the pole and the ray Fx that is perpendicular to l_1 as the polar axis. Let the distance between F and l_1 be p. Then the equations of E_1, E_2, and E_3 are respectively

$$\rho = \frac{e_1 p}{1 - e_1 \cos\theta}, \quad \rho = \frac{p}{1 - \cos\theta}, \quad \rho = \frac{e_2 p}{1 - e_2 \cos\theta}.$$

Suppose that the inclination angle of the straight line l_2 is α. Then

$$|FA| = \frac{e_1 p}{1 + e_1 \cos\alpha}, \quad |FB| = \frac{e_1 p}{1 - e_1 \cos\alpha},$$

$$|AB| = |FA| + |FB| = \frac{e_1 p}{1 + e_1 \cos\alpha} + \frac{e_1 p}{1 - e_1 \cos\alpha} = \frac{2e_1 p}{1 - e_1^2 \cos^2\alpha}.$$

By the same token, $|PQ| = \frac{2p}{\sin^2\alpha}$ and $|MN| = \frac{2e_2 p}{1 - e_2^2 \cos^2\alpha}$. Consequently,

$$\frac{|PQ|}{|AB|} = \frac{2p}{\sin^2\alpha} \cdot \frac{1 - e_2^2 \cos^2\alpha}{2e_1 p} = \frac{1 - e_1^2 + e_1^2 \sin^2\alpha}{e_1 \sin^2\alpha}$$

$$= e_1 + \frac{1 - e_1^2}{e_1 \sin^2\alpha} \geq e_1 + \frac{1 - e_1^2}{e_1} = \frac{1}{e_1}.$$

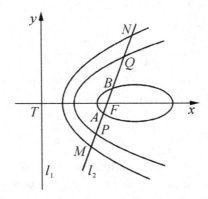

Figure 26.2   Figure for Example 2.

Therefore, $m = \frac{1}{e_1}$. The same argument implies that $\frac{|MN|}{|PQ|} \geq e_2$, so $n = e_2$. The condition that $m = n$ means that $\frac{1}{e_1} = e_2$, that is, $e_1 e_2 = 1$. $\qquad\square$

Example 3. Given the ellipse $\frac{x^2}{2} + y^2 = 1$ with F as its left focus, at F draw two straight lines l_1 and l_2 that intersect the ellipse at P, Q and M, N, respectively. Assume that $l_1 \perp l_2$. Find the maximum value and the minimum value of the area of the quadrilateral $PMQN$.

Analysis. The area of the quadrilateral $PMQN$ is $S = \frac{1}{2}|PQ| \cdot |MN|$, and PQ and MN are the chords that pass through the focus F, so the ellipse problem can be transformed to the unified equation of conic sections to deal with.

Solution. From $\frac{x^2}{2} + y^2 = 1$, we get $a = \sqrt{2}, b = 1, c = 1, e = \frac{\sqrt{2}}{2}$, and $p = 1$.

Establish a polar coordinates system with F as the pole and Fx as the polar axis. Then the polar equation of the ellipse is

$$\rho = \frac{\frac{\sqrt{2}}{2}}{1 - \frac{\sqrt{2}}{2}\cos\theta} = \frac{1}{\sqrt{2} - \cos\theta}.$$

We can write $P(\rho_1, \theta), Q(\rho_2, \theta + \pi), M\left(\rho_3, \theta + \frac{\pi}{2}\right)$, and $N\left(\rho_4, \theta + \frac{3\pi}{4}\right)$ by the condition of the problem, where $\theta \in [0, 2\pi)$. Then

$$|PQ| = \rho_1 + \rho_2 = \frac{1}{\sqrt{2} - \cos\theta} + \frac{1}{\sqrt{2} + \cos\theta} = \frac{2\sqrt{2}}{2 - \cos^2\theta},$$

$$|MN| = \rho_3 + \rho_4 = \frac{1}{\sqrt{2} - \cos\left(\theta + \frac{\pi}{2}\right)} + \frac{1}{\sqrt{2} + \cos\left(\theta + \frac{\pi}{2}\right)} = \frac{2\sqrt{2}}{2 - \sin^2\theta},$$

and so

$$S = \frac{1}{2}|PQ||MN| = \frac{4}{2 + \cos^2\theta \sin^2\theta} = \frac{16}{8 + \sin^2 2\theta}.$$

Since $0 \leq \sin^2 2\theta \leq 1$, we obtain $\frac{16}{9} \leq S \leq 2$.

When $\sin^2 2\theta = 0$, the area S achieves the maximum value 2.

When $\sin^2 2\theta = 1$, the area S achieves the minimum value $\frac{16}{9}$. $\qquad\square$

Remark. When exploring a problem on the focal radius or focal chord of a conic section, it is often more convenient to use the unified polar coordinates equation of conic sections.

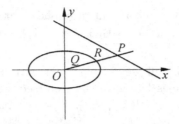

Figure 26.3　Figure for Example 4.

Example 4. As shown in Figure 26.3, the ellipse $\frac{x^2}{24} + \frac{y^2}{16} = 1$ and the straight line $l : \frac{x}{12} + \frac{y}{8} = 1$ are given, P is a point on l, a ray OP intersects the ellipse at a point R, and a point Q is on OP that satisfies $|OP||OQ| = |OR|^2$. When the point P moves on l, find the equation of the trajectory of Q, and state what curve the trajectory is.

Analysis. The three points $P, Q,$ and R are on the same ray, and $OP, OQ,$ and OR are the line segments with the coordinates origin as their initial point. So we can consider to transform the origin to the pole and the x-axis to the polar axis, resulting in a polar coordinates system. In this way we can obtain the relation of the polar radii of $P, Q,$ and R that have the same polar angle, solving the problem conveniently.

Solution. Set up a polar coordinates system with the origin O as the pole and the positive direction x-axis as the polar axis. Then the polar coordinates equations of the ellipse and the straight line l are respectively

$$\rho^2 = \frac{48}{2 + \sin^2 \theta} \quad \text{and} \quad \rho = \frac{24}{2 \cos \theta + 3 \sin \theta}.$$

Write $P(\rho_1, \theta), Q(\rho, \theta),$ and $R(\rho_2, \theta)$. Then $|OP||OQ| = |OR|^2$ implies that $\rho_1 \rho = \rho_2^2$, in other words,

$$\frac{24\rho}{2 \cos \theta + 3 \sin \theta} = \frac{48}{2 + \sin^2 \theta}.$$

Rewriting the equality as

$$24\rho^2(2 + \sin^2 \theta) = 48\rho(2 \cos \theta + 3 \sin \theta).$$

Transforming the above polar coordinates equation to the equation under the rectangular coordinates, we have $2x^2 + 3y^2 - 4x - 6y = 0$.

Therefore, the sought trajectory of the point is the ellipse with center $(1, 2)$ (the origin is excluded). The length of its long axis is $\sqrt{10}$ and the length of its short axis is $\frac{2\sqrt{15}}{3}$. $\qquad\square$

Example 5. Suppose that AB and CD are two moving chords passing through a focus F of a conic section, and the line segments AC and BD that connect two end points of the chords intersect at a point M. Prove: The trajectory of the moving point M is the corresponding directrix of the conic section.

Proof. As seen from Figure 26.4, let the polar equation of the conic section be $\rho = \frac{ep}{1 - e\cos\theta}$. Write $A(\rho_1, \theta_1), B(\rho_1', \theta_1 + \pi), C(\rho_2, \theta_2)$, and $D(\rho_2', \theta_2 + \pi)$.

Denote the moving point on the straight line AC as $P(\rho, \theta)$. Then by the angle formula in plane geometry, $\frac{\sin(\theta_1 - \theta_2)}{\rho} = \frac{\sin(\theta_1 - \theta)}{\rho_2} + \frac{\sin(\theta - \theta_2)}{\rho_1}$. After a manipulation we obtain the equation of the straight line AC:

$$\rho\left[\cos\left(\frac{\theta_1 + \theta_2}{2} - \theta\right) - e\cos\frac{\theta_1 - \theta_2}{2}\cos\theta\right] = ep\cos\frac{\theta_1 - \theta_2}{2}. \qquad (26.1)$$

By the same way, we have the equation of the straight line BD:

$$\rho\left[\cos\left(\frac{\theta_1 + \pi + \theta_2 + \pi}{2} - \theta\right) - e\cos\frac{\theta_1 - \theta_2}{2}\cos\theta\right] = ep\cos\frac{\theta_1 - \theta_2}{2}. \qquad (26.2)$$

$(26.1) + (26.2)$ and simplifying the resulting expression give $\rho\cos\theta = -p$, which is just the equation for the corresponding directrix of the conic section. Hence, the trajectory of the point M is the directrix of the conic section. $\qquad\square$

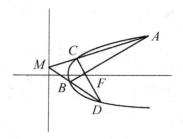

Figure 26.4 Figure for Example 5.

Example 6. Given the point $A(1,0)$ and the straight line $l : x = 3$, the distance from a moving point M to A is m and to l is n. Assume that $m + n = 4$.

(1) Find the trajectory of the point M.
(2) Draw a straight line at A with an inclination angle α, and it intersects the trajectory of M at two points P and Q. Let $d = |PQ|$. Find an analytic expression of $d = f(\alpha)$.
(3) At what values of α, does $f(\alpha)$ have the maximum value? What is the maximum value?

Solution. (1) Denote $M(x,y)$. Then $\sqrt{(x-1)^2 + y^2} + |x - 3| = 4$, from which

$$y^2 = 4x \quad (0 \le x \le 3)$$

or

$$y^2 = -12(x - 4) \quad (3 < x \le 4).$$

Hence, the trajectory of the point M is the union of the two parabolas, which is a closed curve.

(2) The two parabolas intersect at $B(3, 2\sqrt{3})$ and $C(3, -2\sqrt{3})$. From (1), we know that the two parabolas share the same focus A. Set up a polar coordinates system with A as the pole and the positive direction of the x-axis as the polar axis. Then the polar coordinates of the point B are $\left(4, \frac{\pi}{3}\right)$ and the polar coordinates of the point C are $\left(4, \frac{5}{3}\pi\right)$. Since the inclination angle of PQ is α, we write $P(\rho_1, \alpha)$ and $Q(\rho_2, \alpha + \pi)$. The locations of P and Q are different and vary along with the changes of α. See Figure 26.5.

The polar coordinates equation of $y^2 = 4x$ $(0 \le x \le 3)$ is $\rho = \frac{2}{1 - \cos\theta}$.
The polar coordinates equation of $y^2 = -12(x - 4)$ $(3 < x \le 4)$ is $\rho = \frac{6}{1 + \cos\theta}$.

(i) When $0 \le \alpha < \frac{\pi}{3}$,

$$|PQ| = \rho_1 + \rho_2 = \frac{6}{1 + \cos\theta} + \frac{2}{1 - \cos(\theta + \pi)} = \frac{8}{1 + \cos\theta}.$$

(ii) When $\frac{\pi}{3} \le \alpha < \frac{2\pi}{3}$,

$$|PQ| = \rho_1 + \rho_2 = \frac{2}{1 - \cos\theta} + \frac{2}{1 - \cos(\theta + \pi)} = \frac{4}{\sin^2\alpha}.$$

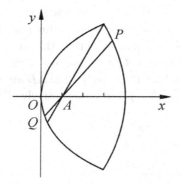

Figure 26.5 Figure for Example 6.

(iii) When $\frac{2\pi}{3} \le \alpha < \pi$,

$$|PQ| = \rho_1 + \rho_2 = \frac{2}{1 - \cos\theta} + \frac{6}{1 + \cos(\theta + \pi)} = \frac{8}{1 - \cos\theta}.$$

Therefore,

$$f(\alpha) = \begin{cases} \dfrac{8}{1 + \cos\theta}, & 0 \le \alpha < \dfrac{\pi}{3}, \\[2mm] \dfrac{4}{\sin^2\alpha}, & \dfrac{\pi}{3} \le \alpha < \dfrac{2\pi}{3}, \\[2mm] \dfrac{8}{1 - \cos\theta}, & \dfrac{2\pi}{3} \le \alpha < \pi. \end{cases}$$

(3) Clearly, when $\alpha = \frac{\pi}{3}$ or $\frac{2\pi}{3}$, the function $f(\alpha)$ has the maximum value $\frac{16}{3}$. In other words, the maximum value of $|PQ|$ is $\frac{16}{3}$. ☐

Example 7. An arbitrary straight line passing through a fixed point O intersects two fixed straight lines at A and B, respectively. A point P is on the straight line AB. Prove: The point P satisfying $\frac{2}{|OP|} = \frac{1}{|OA|} + \frac{1}{|OB|}$ must be on another straight line.

Analysis. Since the directed line segments OA, OB, and OP all lie in the same straight line, so it is rather convenient to solve the problem by using the analytic method after setting up a polar coordinates system with O as its pole.

Proof. Set up a polar coordinates system with O as its pole. Suppose that the equations of the two fixed straight lines are $a_i = \rho \cos(\theta - \alpha_i)$ for $i = 1$ and 2, respectively. Also assume that the angle between an arbitrary straight line passing through O and the polar axis is θ, and the polar coordinates of the point P on the straight line AB are (ρ, θ). Then $|OA|$ and $|OB|$ equal $\frac{a_i}{\cos(\theta - \alpha_i)}$ with $i = 1$ and 2 respectively. By the given condition,

$$\frac{2}{\rho} = \frac{\cos(\theta - \alpha_1)}{a_1} + \frac{\cos(\theta - \alpha_2)}{a_2}$$

$$= \frac{(a_2 \cos \alpha_1 + a_1 \cos \alpha_2) \cos \theta + (a_2 \sin \alpha_1 + a_1 \sin \alpha_2) \sin \theta}{a_1 a_2},$$

namely $a = \rho \cos(\theta - \theta_0)$, where the constants a and θ_0 are given by

$$a = \frac{2 a_1 a_2}{\sqrt{a_1^2 + a_2^2 + 2 a_1 a_2 \cos(\alpha_1 - \alpha_2)}}, \quad \tan \theta_0 = \frac{a_2 \sin \alpha_1 + a_1 \sin \alpha_2}{a_2 \cos \alpha_1 + a_1 \cos \alpha_2},$$

with the terminal side of the angle θ_0 containing the point of the xy-coordinates

$$(a_2 \cos \alpha_1 + a_1 \cos \alpha_2, a_2 \sin \alpha_1 + a_1 \sin \alpha_2).$$

The graph of the polar equation $a = \rho \cos(\theta - \theta_0)$ is a straight line whose distance to the pole is a, which is perpendicular to the straight line $\theta = \theta_0$. $\qquad \square$

Example 8. At the center O of an ellipse $\frac{x^2}{a^2} + \frac{y^2}{b^2} = 1$, draw in succession n radii $r_1, r_2, \ldots, r_n$, with nearby two radii making the angle of $\frac{2\pi}{n}$. Prove: $\sum_{i=1}^{n} \frac{1}{r_i^2} = \frac{n}{2} \left(\frac{1}{a^2} + \frac{1}{b^2} \right)$.

Analysis. In a polar coordinates system, it is quite convenient to deal with problems related to the length of a line segment with the pole as one endpoint. So we set up a polar coordinates system with the center O of the ellipse as the pole, and transform the equation in the rectangular coordinates to that in the polar coordinates to solve the problem.

Proof. Set up a polar coordinates system with the center O of the ellipse as the pole and the positive direction of the x-axis as the polar axis. Then the polar coordinates equation of the ellipse is $\frac{\rho^2 \cos^2 \theta}{a^2} + \frac{\rho^2 \sin^2 \theta}{b^2} = 1$, that is, $\frac{1}{\rho^2} = \frac{\cos^2 \theta}{a^2} + \frac{\sin^2 \theta}{b^2}$.

Suppose that the intersection points of the n radii with the ellipse are $P_1, P_2, \ldots, P_n$ with their polar coordinates

$$P_1(\rho_1, \theta), \ P_2\left(\rho_2, \theta + \frac{2\pi}{n}\right), \ \ldots, \ P_n\left(\rho_n, \theta + \frac{n-1}{n}2\pi\right).$$

Hence,

$$\sum_{i=1}^{n} \frac{1}{r_i^2} = \sum_{i=1}^{n} \frac{1}{\rho_i^2}$$

$$= \frac{1}{a^2}\left\{\cos^2\theta + \cos^2\left(\theta + \frac{2\pi}{n}\right) + \cdots + \cos^2\left[\theta + \frac{2(n-1)\pi}{n}\right]\right\}$$

$$+ \frac{1}{b^2}\left\{\sin^2\theta + \sin^2\left(\theta + \frac{2\pi}{n}\right) + \cdots + \sin^2\left[\theta + \frac{2(n-1)\pi}{n}\right]\right\}$$

$$= \frac{n}{2}\left(\frac{1}{a^2} + \frac{1}{b^2}\right) + \frac{1}{2}\left(\frac{1}{a^2} - \frac{1}{b^2}\right)\left\{\cos 2\theta + \cos\left(2\theta + \frac{4\pi}{n}\right)\right.$$

$$\left. + \cdots + \cos\left[2\theta + \frac{4(n-1)\pi}{n}\right]\right\}$$

$$= \frac{n}{2}\left(\frac{1}{a^2} + \frac{1}{b^2}\right) + \frac{1}{2}\left(\frac{1}{a^2} - \frac{1}{b^2}\right)\frac{1}{2\sin\frac{2\pi}{n}}\left\{\sin\left(2\theta + \frac{2\pi}{n}\right)\right.$$

$$- \sin\left(2\theta - \frac{2\pi}{n}\right) + \sin\left(2\theta + \frac{6\pi}{n}\right) - \sin\left(2\theta + \frac{2\pi}{n}\right) + \cdots$$

$$\left. + \sin\left[2\theta + \frac{2(2n-1)\pi}{n}\right] - \sin\left[2\theta + \frac{2(2n-3)\pi}{n}\right]\right\}$$

$$= \frac{n}{2}\left(\frac{1}{a^2} + \frac{1}{b^2}\right) + \left(\frac{1}{a^2} - \frac{1}{b^2}\right)\frac{1}{4\sin\frac{2\pi}{n}}\left\{\sin\left[2\theta + \frac{2(2n-1)\pi}{n}\right]\right.$$

$$\left. - \sin\left(2\theta - \frac{2\pi}{n}\right)\right\}$$

$$= \frac{n}{2}\left(\frac{1}{a^2} + \frac{1}{b^2}\right) + \left(\frac{1}{a^2} - \frac{1}{b^2}\right)\frac{1}{2\sin\frac{2\pi}{n}}\sin 2\pi \cos\left[2\theta + \frac{2(n-1)\pi}{n}\right]$$

$$= \frac{n}{2}\left(\frac{1}{a^2} + \frac{1}{b^2}\right).$$

This proves the equality. $\qquad\qquad\square$

3. Exercises

Group A

I. Filling Problems

1. For all integers m and n satisfying $1 \leq n \leq m \leq 5$, the number of different hyperbolas represented by the equation $\rho = \frac{1}{1 - C_m^n \cos \theta}$ of polar coordinates is _____.

2. The curve represented by the equation $\rho = \frac{5}{3 - 4\cos \theta + 4\sin \theta}$ of polar coordinates is _____.

3. At the focus of the parabola $y^2 = 8x$, draw a chord with an inclination angle θ. If the length of the chord does not exceed 16, then the range of the values for θ is _____.

II. Calculation Problems

4. A Straight line passing through the focus of a parabola intersects the parabola at two points P and Q. If the perpendicular bisector of PQ intersects the symmetric axis at R. Prove: $|FR| = \frac{1}{2}|PQ|$.

5. At the focus of a parabola $y^2 = 2px$ $(p > 0)$, draw a chord $P_1 P_2$. Find the value of $\frac{1}{|FP_1|} + \frac{1}{|FP_2|}$.

6. Let A, B, and C be three points on the ellipse $\frac{x^2}{16} + \frac{y^2}{7} = 1$, F be the left focus, and $\angle AFB = \angle BFC = \angle CFA$. Find $\frac{1}{|AF|} + \frac{1}{|BF|} + \frac{1}{|CF|}$.

7. Let the center of a circle A be $A(0,1)$, the radius be 1, and Q be any point on the circle A. Choose a point on the ray OQ such that the distance from P to Q equals the distance from P to the straight line $y = 2$. Find the equation of the trajectory for the point P.

Group B

8. Let F be a fixed point, l be a fixed straight line, the distance from F to l be p $(p > 0)$, and a point M be on l. A moving point N is on the extended line of MF and satisfies $\frac{|FN|}{|MN|} = \frac{1}{|MF|}$. Find the minimum value of $|MN|$.

9. An ellipse $\frac{x^2}{a^2} + \frac{y^2}{b^2} = 1$ $(a > b > 0)$ has an inscribed parallelogram. If one pair of the opposite sides of the parallelogram pass through the two foci respectively, find the maximum value for the area of the parallelogram.

10. For a given parabola $y^2 = 2px$ $(p > 0)$, AB is a chord passing through the focus F. If the angle formed by AB and the x-axis is θ $(0 < \theta \le \frac{\pi}{2})$, find $\angle AOB$.

11. At one focus F of an ellipse $b^2x^2 + a^2y^2 = a^2b^2$ $(a > b > 0)$, draw two mutually perpendicular chords AB and CD.

 (1) Prove: $\frac{1}{|FA|} + \frac{1}{|FB|}$ keeps the same value, and $\frac{1}{|AB|} + \frac{1}{|CD|}$ keeps the same value.

 (2) Find the minimum values of $|AB| + |CD|$ and $|AB||CD|$.

 (3) If M is the end point of the long axis near F, find the maximum value for the area of $\triangle ABM$.

12. As shown in Figure 26.6, suppose that A and B are two points on an ellipse $\frac{x^2}{a^2} + \frac{y^2}{b^2} = 1$ $(a > b > 0)$, O is the center of the ellipse, and $OA \perp OB$.

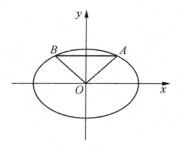

Figure 26.6   Figure for Exercise 12.

 (1) Prove: $\frac{1}{|OA|^2} + \frac{1}{|OB|^2}$ is of a fixed value.

 (2) Find the maximum value and minimum value of $\frac{1}{|OA|} + \frac{1}{|OB|}$.

 (3) Find the maximum value and minimum value for the area of $\triangle AOB$.

Chapter 27

Analytic Method for Plane Geometry

1. Key Points of Knowledge and Basic Methods

The analytic method is a general method that transforms geometric problems to algebraic ones, and it perfectly reflects the important thought of the number-figure combination. By means of a rectangular coordinates system, we can easily combine organically the information about numbers and that of figures. This method opens a new road when a purely plane geometry method is found difficult to solve problems.

The analytic method does not always represent tedious computations, but rather if it is possible to use techniques appropriately, not only it can make the objective clear, but also realize the goal of simplifying the computation. In the following we introduce some commonly used techniques.

A. *Set up coordinates systems cleverly*

Properly establishing a coordinates system can reduce much trouble for the computation, realizing the purpose of making the tedious to the simple. Generally speaking, the choice of the origin is a very important step. We often choose those extremely important and unique "special points" as the origin (for example the center of a circle or the vertex of a right angle). After the origin is determined, in general we set up a rectangular coordinates system by selecting a given straight line as the x-axis.

B. Designation without calculation

In analytic geometry, many quantities do not need to be calculated out completely. These quantities may only play an "intermediate" role. so we only need to design such quantities. For example, if the horizontal coordinates of two points A and B satisfy a quadratic equation with $x_A \neq x_B$, then in many cases we do not have to calculate x_A and x_B, but it is enough only to know what the values of $x_A + x_B$ and $x_A x_B$ are.

The essence of "designation without calculation" lies in "catching a key quantity."

2. Illustrative Examples

Example 1. Extend the two sides AB and AC of $\triangle ABC$ to D and E such that $|BD| = |CE|$. Suppose that the middle points of BC and DE are M and N, respectively, and AT is the angular bisector of $\angle A$. Prove: $MN \parallel AT$.

Analysis. Without doubt A is the most important point, and the straight lines AB and AC are symmetric with respect to the straight line AT. Hence, it is not difficult for us to think of setting up a rectangular coordinates system with A as the origin and AT as the x-axis.

Proof. As Figure 27.1 shows, set up a rectangular coordinates system with A as the origin and the straight line containing AT as the x-axis. Assume that $\angle DAT = \theta$, and write $B(c\cos\theta, c\sin\theta)$ and $C(b\cos\theta, -b\sin\theta)$. If we let $|BD| = t$, then the vertical coordinates of the two points D and E are

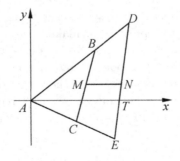

Figure 27.1 Figure for Example 1.

$(c + t) \sin \theta$ and $-(b + t) \sin \theta$, respectively. Thus,

$$y_M = \frac{y_B + y_C}{2} = \frac{c - b}{2} \sin \theta,$$

$$y_N = \frac{y_D + y_E}{2} = \frac{(c + t) \sin \theta + (-b - t) \sin \theta}{2} \sin \theta = \frac{c - b}{2}.$$

So, $y_M = y_N$. Therefore, $MN \parallel$ the x-axis, in other words, $MN \parallel AT$. $\square$

Example 2. In Figure 27.2, The tangent points of the inscribed circle of an equilateral triangle ABC with the three sides are D, E, and F, respectively. If the distances of any point P on the arc EF to the three sides are p, q, and r, respectively, prove: $p^{\frac{1}{2}} + r^{\frac{1}{2}} = q^{\frac{1}{2}}$.

Proof. Set up a rectangular coordinates system with the center of the triangle ABC as the origin so that the vertex A is on the y-axis. Assign the radius of the inscribed circle to $\triangle ABC$ as the unit length. Then the equations of the three sides AC, AB, and BC are respectively

$$\begin{cases} x \cos \dfrac{\pi}{6} + y \sin \dfrac{\pi}{6} = 1, \\[2mm] x \cos \dfrac{5\pi}{6} + y \sin \dfrac{5\pi}{6} = 1, \\[2mm] x \cos \dfrac{3\pi}{2} + y \sin \dfrac{3\pi}{2} = 1. \end{cases}$$

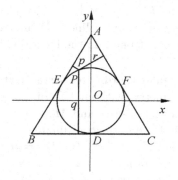

Figure 27.2 Figure for Example 2.

Let the coordinates of the point P be $(\cos\theta, \sin\theta)$ with $\theta \in \left(\frac{\pi}{6}, \frac{5\pi}{6}\right)$. Then

$$r^{\frac{1}{2}} = \left|\cos\theta\cos\frac{\pi}{6} + \sin\theta\sin\frac{\pi}{6} - 1\right|^{\frac{1}{2}}$$

$$= \left[1 - \cos\left(\theta - \frac{\pi}{6}\right)\right]^{\frac{1}{2}} = \sqrt{2}\sin\left(\frac{\theta}{2} - \frac{\pi}{12}\right).$$

By the same token, $p^{\frac{1}{2}} = \sqrt{2}\sin\left(\frac{5}{12}\pi - \frac{\theta}{2}\right)$ and $q^{\frac{1}{2}} = \sqrt{2}\sin\left(\frac{3}{4}\pi - \frac{\theta}{2}\right)$. Therefore,

$$p^{\frac{1}{2}} + r^{\frac{1}{2}} = \sqrt{2}\left[\sin\left(\frac{\theta}{2} - \frac{\pi}{12}\right) + \sin\left(\frac{5}{12}\pi - \frac{\theta}{2}\right)\right]$$

$$= 2\sqrt{2}\sin\frac{\pi}{6}\cos\left(\frac{\theta}{2} - \frac{\pi}{4}\right)$$

$$= \sqrt{2}\sin\left(\frac{3}{4}\pi - \frac{\theta}{2}\right) = q^{\frac{1}{2}}. \qquad \square$$

Remark. In the current example, the geometric meaning of the parameter in the parametric equations of the circle was used, so that the forms of $p^{\frac{1}{2}}, q^{\frac{1}{2}}$, and $r^{\frac{1}{2}}$ had a common property, transforming the proof to the simple trigonometric operations.

Example 3. Suppose that a straight line l and a circle centered at O do not intersect, E is a point on l, $OE \perp l$, and M is any point on l that is different from E. From M draw two tangent lines to the circle O with two tangent points A and B, and from E draw two lines EC and ED that are perpendicular to MA and MB, respectively, with C and D their respective intersection points. Let the straight line CD intersect OE at F. Prove: The location of F does not depend on the location of M.

Proof. Set up a rectangular coordinates system with E as the origin and l as the x-axis. By symmetry, without loss of generality we may assume that O is on the positive half y-axis and M is on the negative half x-axis. Let the radius of the circle be r, the coordinates of O be $(0, a)$ $(a > 0)$, the coordinates of M be $(-d, 0)$ $(d > 0)$, $\angle OME = \alpha$, and $\angle OMA = \theta$. As Figure 27.3 shows, $k_{AM} = \tan(\alpha - \theta)$ and $k_{EC} = -\cot(\alpha - \theta)$. Since the coordinates of M are $(-d, 0)$, we find the coordinates of the point C to be $\left(-d\sin^2(\alpha - \theta), \frac{1}{2}d\sin(2\alpha - 2\theta)\right)$. Similarly, the coordinates of the point D

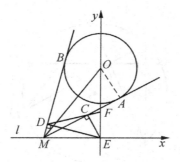

Figure 27.3 Figure for Example 3.

are $\left(-d\sin^2(\alpha+\theta), \frac{1}{2}d\sin(2\alpha+2\theta)\right)$. So, the equation of CD is

$$y-\frac{1}{2}d\sin(2\alpha+2\theta) = \frac{\frac{1}{2}d\sin(2\alpha+2\theta)-\frac{1}{2}d\sin(2\alpha-2\theta)}{-d\left[\sin^2(\alpha+\theta)-\sin^2(\alpha-\theta)\right]}\left[x+d\sin^2(\alpha+\theta)\right].$$

Letting $x=0$ in the above equation and simplifying the expression, we obtain that $y_F = d\sin(\alpha+\theta)\cos(\alpha+\theta)-d\sin^2(\alpha+\theta)\cot 2\alpha$. Also $d = |ME| = \frac{a}{\tan\alpha}$, and consequently

$$y_F = \frac{a}{\tan\alpha}\sin(\alpha+\theta)[\cos(\alpha+\theta)-\sin(\alpha+\theta)\cot 2\alpha]$$

$$= a\cot\alpha\sin(\alpha+\theta)\sin(\alpha-\theta)\frac{1}{\sin 2\alpha}$$

$$= a\frac{\cos\alpha}{\sin\alpha}\frac{\sin^2\alpha\cos^2\theta-\cos^2\alpha\sin^2\theta}{2\sin\alpha\cos\alpha}$$

$$= \frac{a}{2}\frac{1}{\sin^2\alpha}\left[\sin^2\alpha(1-\sin^2\theta)-(1-\sin^2\alpha)\sin^2\theta\right].$$

In $\triangle OMA$ and $\triangle OME$, we have $\frac{a}{\sin\alpha}=\frac{r}{\sin\theta}=|OM|$. Hence,

$$y_F = \frac{a}{2}\left(1-\sin^2\theta-\frac{\sin^2\theta}{\sin^2\alpha}+\sin^2\theta\right) = \frac{a}{2}\left(1-\frac{r^2}{a^2}\right) = \frac{a^2-r^2}{2a},$$

which is independent of the location of M. $\qquad\square$

Remark. By symmetry, with the point E as the origin and the straight line l as the x-axis, establishing the rectangular coordinates system may write relatively easily the coordinates of the points and the equations of the curves involved in the problem.

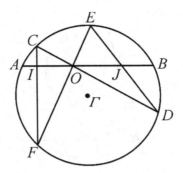

Figure 27.4   Figure for Example 4.

Example 4. Let AB be a fixed chord of a circle Γ and O be a fixed point on the chord AB. At O draw two chords CD and EF of the circle Γ (Figure 27.4). Suppose that the chord CF intersects OA at I and the chord DE intersects OB at J. Prove: $\frac{1}{|OA|} - \frac{1}{|OB|} = \frac{1}{|OI|} - \frac{1}{|OJ|}$.

Analysis. Since the relation formula to be shown is complicated and the known condition "seems" not sufficient enough, it is not easy to do this problem from the purely geometric knowledge, so we may try to use the analytic method. Of course, if we calculate the coordinates of every point, then we can solve the problem, but we do not wish and do not need to do this way. What we only need is "designation without calculation."

Proof. Create a rectangular coordinates system with O as the origin and the straight line containing the chord AB as the x-axis. Assume that the equation of the circle is $x^2 + y^2 + dx + ey + f = 0$, the equation of the straight line CD is $x - t_1 y = 0$, and the equation of the straight line EF is $x - t_2 y = 0$. If we view the union of the two straight lines CD and EF as a degenerated conic section, then its equation is $(x - t_1 y)(x - t_2 y) = 0$.

In addition, the circle Γ is also a conic section that passes through the four points C, D, E, and F, so all conic sections that pass through the four points C, D, E, and F (except for the union of the two straight lines CD and EF) can be expressed as

$$x^2 + y^2 + dx + ey + f + \lambda(x - t_1 y)(x - t_2 y) = 0, \qquad (27.1)$$

where $\lambda \in \mathbf{R}$ is a parameter to be determined.

On the other hand, the union of the two straight lines CF and DE is also a conic section passing through the four points C, D, E, and F, so there exists λ such that the corresponding (27.1) is the equation of the union of the two straight lines CF and DE. Let the coordinates of I and J be $(x_1, 0)$ and $(x_2, 0)$, respectively. By letting $y = 0$ in (27.1), we know that x_1 and x_2 are the two roots of the equation $x^2 + dx + f + \lambda x^2 = 0$. Thus,

$$x_1 + x_2 = -\frac{d}{1+\lambda}, \tag{27.2}$$

$$x_1 x_2 = \frac{f}{1+\lambda}. \tag{27.3}$$

Dividing (27.2) by (27.3) gives $\frac{1}{x_1} + \frac{1}{x_2} = -\frac{d}{f}$.

Since O is the origin, the fact that $x_1 < 0 < x_2$ implies that

$$\frac{1}{x_1} + \frac{1}{x_2} == -\frac{1}{|OI|} + \frac{1}{|OJ|},$$

namely

$$\frac{1}{|OI|} - \frac{1}{|OJ|} = \frac{d}{f}.$$

Let $\lambda = 0$ and $y = 0$ in (27.1). Then we obtain the equation

$$x^2 + dx + f = 0$$

that x_A and x_B satisfy. Hence,

$$\frac{1}{x_A} + \frac{1}{x_B} = \frac{x_A + x_B}{x_A x_B} = -\frac{d}{f}.$$

Since $x_A < 0 < x_B$, it follows that

$$\frac{1}{|OA|} - \frac{1}{|OB|} = \frac{d}{f}.$$

Therefore,

$$\frac{1}{|OI|} - \frac{1}{|OJ|} = \frac{1}{|OA|} - \frac{1}{|OB|}. \qquad \square$$

Remark. The solution process of this problem represents the merit of "designation without calculation": Only design some equations and coordinates instead of calculating any coordinates, and then mutually transform the relations of the problem to find the nature that was "hidden in the darkness."

Example 5. In a circle O, a chord CD is parallel to another chord EF and intersects a diameter AB in the angle of 45°. If CD and EF intersect the diameter AB at P and Q, respectively, and the radius of the circle O is 1, prove: $|PC||QE| + |PD||QF| < 2$.

Analysis. Set up a rectangular coordinates system according to the symmetry of the circle. So, using the analytic method to solve the problem is rather appropriate.

Proof. With O as its origin and a straight line parallel to CD as its x-axis, set up a rectangular coordinates system (as in Figure 27.5). Then the equation of the circle is

$$x^2 + y^2 = 1.$$

Let the distance between CD and the x-axis be a $(-1 < a < 1)$ and the distance between EF and the x-axis be b $(-1 < b < 1)$. Then the equation of the straight line CD is $y = a$ and the equation of the straight line EF is $y = -b$, and the coordinates of P and Q are $P(a, a)$ and $Q(-b, -b)$.

Further, we can obtain the coordinates of C, D, E, and F as

$$C\left(-\sqrt{1-a^2}, a\right), \ D\left(\sqrt{1-a^2}, a\right), \ E\left(-\sqrt{1-b^2}, -b\right), \ F\left(\sqrt{1-b^2}, -b\right).$$

It follows that

$$|PC| = \sqrt{1-a^2} + a, \quad |PD| = \sqrt{1-a^2} - a,$$
$$|QE| = \sqrt{1-b^2} - b, \quad |QF| = \sqrt{1-b^2} + b.$$

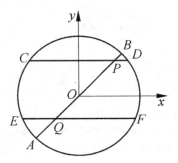

Figure 27.5   Figure for Example 5.

Consequently,

$$|PC||QE| + |PD||QF| = 2\left(\sqrt{(1-a^2)(1-b^2)} - ab\right)$$

$$\leq 2\left(\frac{1-a^2+1-b^2}{2} - ab\right)$$

$$= 2\left[\frac{2-(a+b)^2}{2}\right] \leq 2.$$

It is impossible that $a = b = 0$, hence the equality in the above cannot be satisfied. Thus,

$$|PC||QE| + |PD||QF| < 2.$$ □

Example 6. As Figure 27.6 shows, the inscribed circle O of a rhombus $ABCD$ is tangent to each side at E, F, G, and H, respectively. Draw a tangent line on the arc EF that intersects AB and BC at M and N, respectively, and draw a tangent line on the arc GH that intersects CD and DA at P and Q respectively. Prove: $MQ \parallel NP$.

Proof. It is easy to think of setting up a rectangular coordinates system with BD as the x-axis and AC as the y-axis. Without loss of generality, assume that the circle O is a unit one. Let the inclination angle of OH be θ. Then the coordinates of H are $H(\cos\theta, \sin\theta)$. Clearly, the coordinates of D and A are $D\left(\frac{1}{\cos\theta}, 0\right)$ and $A\left(0, \frac{1}{\sin\theta}\right)$. Thus, the equation of the straight line DA is $l_{DA} : x\cos\theta + y\sin\theta = 1$.

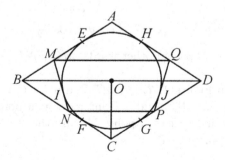

Figure 27.6   Figure for Example 6.

According to the symmetry, we obtain the equations of the other three straight lines:

$$l_{AB}: \quad -x\cos\theta + y\sin\theta = 1,$$
$$l_{BC}: \quad -x\cos\theta - y\sin\theta = 1,$$
$$l_{CD}: \quad x\cos\theta + y\sin\theta = 1.$$

Assume that PQ is tangent to the circle O at the point $(\cos\alpha, \sin\alpha)$. Then the equation of the straight line PQ is $l_{PQ}: x\cos\alpha + y\sin\alpha = 1$.

Assume that MN is tangent to the circle O at the point $(\cos\beta, \sin\beta)$. Then the equation of the straight line MN is $l_{MN}: x\cos\beta + y\sin\beta = 1$.

Solving the equations of DA and PQ together gives the coordinates of the point Q:

$$Q\left(\frac{\cos\frac{\alpha+\theta}{2}}{\cos\frac{\alpha-\theta}{2}}, \frac{\sin\frac{\alpha+\theta}{2}}{\sin\frac{\alpha-\theta}{2}}\right).$$

Solving the equations of AB and MN together gives the coordinates of the point M:

$$M\left(-\frac{\cos\frac{\theta-\beta}{2}}{\cos\frac{\theta+\beta}{2}}, \frac{\sin\frac{\theta-\beta}{2}}{\sin\frac{\theta+\beta}{2}}\right).$$

Hence,

$$k_{MQ} = \frac{y_M - y_Q}{x_M - x_Q} = \frac{\sin\frac{\alpha+\beta}{2}\cos\theta}{\cos\frac{\alpha+\beta}{2}\cos\theta + \cos\frac{\alpha-\beta}{2}}.$$

Similarly,

$$k_{PN} = \frac{\sin\frac{\alpha+\beta}{2}\cos\theta}{\cos\frac{\alpha+\beta}{2}\cos\theta + \cos\frac{\alpha-\beta}{2}}.$$

Therefore, $k_{MQ} = k_{PN}$, in other words, $MQ \parallel NP$. $\square$

Example 7. As in Figure 27.7, in $\triangle ABC$, a point O is the circumcenter, the three altitudes AD, BE, and CF intersect at a point H, the straight lines ED and AB intersect at a point M, and the straight lines FD and AC intersect at a point N. Prove:

(1) $OB \perp DF$ and $OC \perp DE$.
(2) $OH \perp MN$.

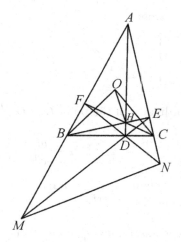

Figure 27.7 Figure for Example 7.

Analysis. Since there are too many perpendicular relations, we think of establishing a rectangular coordinates system. Then the problem of showing that the two straight lines are perpendicular is transformed to showing that the product of the slopes of the two straight lines is -1.

Proof. (1) With the straight line containing BC as the x-axis and D as the origin, we set up a rectangular coordinates system. Write $A(0, a), B(b, 0)$, and $C(c, 0)$. Then $k_{AC} = -\frac{a}{c}$ and $k_{AB} = -\frac{a}{b}$. Furthermore, the equation of the straight line AC is $y = -\frac{a}{c}(x - c)$, and the equation of the straight line BE is $y = \frac{c}{a}(x - b)$.

From the joint equations

$$\begin{cases} y = \dfrac{c}{a}(x - b), \\ y = -\dfrac{a}{c}(x - c), \end{cases}$$

we obtain that the coordinates of the point E is $E\left(\frac{a^2c+bc^2}{a^2+c^2}, \frac{ac^2-abc}{a^2+c^2}\right)$.

By the same token, we get $F\left(\frac{a^2b+b^2c}{a^2+b^2}, \frac{ab^2-abc}{a^2+b^2}\right)$.

On the other hand, the equation of the perpendicular bisector of the straight line AC is $y - \frac{a}{2} = \frac{c}{a}\left(x - \frac{c}{2}\right)$, and the equation of the perpendicular bisector of the straight line BC is $x = \frac{b+c}{2}$.

Similarly, from the system

$$\begin{cases} y - \dfrac{a}{2} = \dfrac{c}{a}\left(x - \dfrac{c}{2}\right), \\ x = \dfrac{b+c}{2}, \end{cases}$$

we obtain that the coordinates of the point O is $O\left(\dfrac{b+c}{2}, \dfrac{bc+a^2}{2a}\right)$. Then

$$k_{OB} = \dfrac{\frac{bc+a^2}{2a}}{\frac{b+c}{2} - b} = \dfrac{bc+a^2}{a(c-b)}, \quad k_{DF} = \dfrac{ab^2 - abc}{a^2b + b^2c} = \dfrac{a(b-c)}{bc+a^2}.$$

Hence, $k_{OB}k_{DF} = -1$, namely $OB \perp DF$.

By the same argument, $OC \perp DE$.

(2) Letting $x = 0$ in the equation $y = \frac{c}{a}(x - b)$ of the straight line BE gives the coordinates $\left(0, -\frac{bc}{a}\right)$ of the point H. Then

$$k_{OH} = \dfrac{\frac{bc+a^2}{2a} + \frac{bc}{a}}{\frac{b+c}{2} - 0} = \dfrac{a^2 + 3bc}{a(b+c)}.$$

While the equation of the straight line DF is $y = \frac{ab-ac}{a^2+bc}x$, from

$$\begin{cases} y = \dfrac{ab - ac}{a^2 + bc}x, \\ y = -\dfrac{a}{c}(x - c), \end{cases}$$

we obtain that the coordinates of the point N is

$$N\left(\dfrac{a^2c + bc^2}{a^2 + 2bc - c^2}, \dfrac{abc - ac^2}{a^2 + 2bc - c^2}\right).$$

The same method gives

$$M\left(\dfrac{a^2b + b^2c}{a^2 + 2bc - b^2}, \dfrac{abc - ab^2}{a^2 + 2bc - b^2}\right).$$

Therefore,

$$k_{MN} = \dfrac{a(b^2 - c^2)(a^2 + bc)}{(c - b)(a^2 + bc)(a^2 + 3bc)} = -\dfrac{a(b+c)}{a^2 + 3bc}.$$

Since $k_{MN}k_{OH} = -1$, we have $MN \perp OH$.   □

Example 8. A quadrilateral $ABCD$ is circumscribed to a circle O. Let the projection of O to the diagonal BD be M. Prove: BD bisects $\angle AMC$.

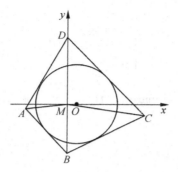

Figure 27.8 Figure for Example 8.

Proof. As in Figure 27.8, set up a rectangular coordinates system with BD as the y-axis and N as the origin.

Suppose that the coordinates of the point O are $(t, 0)$. Without loss of generality, assume that the radius of the circle O is 1. Then $|t| < 1$.

Write $D(0, d)$ and $B(0, -b)$ ($b > 0$ and $d > 0$). By the condition of the problem, to show that BD bisects $\angle AMC$, it is enough to show that $k_{MA} = -k_{MC}$.

Let the slope of the straight line that passes through D and is tangent to the circle O be k. Then the equation of the straight line is $y = kx + d$. Since the distance of O to this straight line is 1, we have that $1 = \frac{|tk+d|}{\sqrt{1+k^2}}$, that is $k^2 + 1 = (tk + d)^2$. In other words, k satisfies the equation $(1 - t^2)k^2 - 2tdk + 1 - d^2 = 0$.

Since the straight lines DA and DC are tangent to the circle O, the two slopes k_{DA} and k_{DC} are the two roots of the equation

$$(1 - t^2)k^2 - 2tdk + 1 - d^2 = 0. \tag{27.4}$$

Substituting $-b$ for d in (27.4), we see that k_{BA} and k_{BC} are the two roots of the equation

$$(1 - t^2)k^2 + 2tdk + 1 - b^2 = 0. \tag{27.5}$$

Vieta's theorem together with (27.4) and (27.5) imply that

$$k_{DA} + k_{DC} = \frac{2td}{1 - t^2}, \quad k_{BA} + k_{BC} = -\frac{2tb}{1 - t^2}. \tag{27.6}$$

Also via solving the system

$$\begin{cases} y = k_{DA}x + d, \\ y = k_{BA}x - b, \end{cases}$$

we obtain that the coordinates of the point A is $\left(\frac{b+d}{k_{BA}-k_{DA}}, \frac{bk_{DA}+dk_{BA}}{k_{BA}-k_{DA}} \right)$.
It follows that

$$k_{MA} = \frac{y_A}{x_A} = \frac{bk_{DA} + dk_{BA}}{b + d}.$$

By the same token,

$$k_{MC} = \frac{bk_{DC} + dk_{BC}}{b + d}.$$

Therefore, thanks to (27.6) we obtain that

$$k_{MA} + k_{MC} = \frac{b(k_{DA} + k_{DC}) + d(k_{BC} + k_{BA})}{b + d} = 0.$$

In other words, BD bisects $\angle AMC$ □

Remark. The solution method for this problem is clever. It has two aspects in "cleverness": (i) The choice of the coordinates system: The point M is the origin of the coordinates system, so the proof of the original proposition is reduced to proving $k_{MA} + k_{MC} = 0$, which greatly reduces the difficulty. (ii) Many "intermediate quantities" are not calculated out, so the "designation without calculation" largely leaves out tedious calculations.

3. Exercises

Group A

I. Filling Problems

1. Let A be a fixed point on a circle O with radius 3. Pick two other points B and C on the circle such that $\angle BAC = \frac{\pi}{3}$. Then the trajectory of the center of gravity G of $\triangle ABC$ is _____.
2. Assume that O is a point inside $\triangle ABC$. Then $S_{\triangle OAC}\overrightarrow{OB} + S_{\triangle OAB}\overrightarrow{OC} + S_{\triangle OBC}\overrightarrow{OA} = $ _____.

3. Two flag poles of heights h and k respectively stand on a horizontal plane and are $2a$ units away. Then the figure corresponding to the set consisting of all the points with the same elevations to the top of the poles is _____.

II. Calculation Problems

4. Let AM be the median line of the side BC of $\triangle ABC$. Draw an arbitrary straight line that intersects AB, AC, and AM at P, Q, and N in succession. Prove: The three ratios $\frac{|AB|}{|AP|}, \frac{|AM|}{|AN|}$, and $\frac{|AC|}{|AQ|}$ form an arithmetic sequence.

5. Suppose that Q is a point on a circle with AB as a diameter ($Q \neq A$ or B), and the projection of Q onto AB is H. A circle centered at Q with QH as radius intersects the circle with AB as diameter at two points C and D. Prove: CD bisects the line segment QH.

6. Draw a semicircle with the side BC of $\triangle ABC$ as diameter, which intersects AB and AC at points D and E, respectively. Draw the perpendicular lines of BC at D and E, with F and G the feet of the perpendicular respectively. Let the line segments DG and EF intersect at a point M. Prove: $AM \perp BC$.

Group B

7. In a "kite-shaped figure" $ABCD$, let $|AB| = |AD|$ and $|BC| = |CD|$. Draw any two straight lines at the intersection point O of AC and BD, which intersect AD, BC, AB, and CD at E, F, G, and H, respectively. If GF and EH intersect BD at I and J respectively, prove: $|IO| = |OJ|$.

8. Let the center of gravity of $\triangle ABC$ be G and form the line segments AG, BG, and CG whose extensions intersect the circumscribed circle of the triangle at P, M, and N, respectively. Prove:

$$\frac{|AG|}{|GP|} + \frac{|BG|}{|GM|} + \frac{|CG|}{|GN|} = 3.$$

9. As in Figure 27.9, circles O_1 and O_2 are tangent to the straight lines containing the three sides of $\triangle ABC$, with E, F, G, and H the tangent points. Suppose that the extensions of EG and FH intersect at a point P. Prove: The straight lines PA and BC are perpendicular.

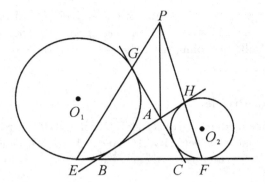

Figure 27.9 Figure for Exercise 9.

10. Let the three sides of $\triangle ABC$ be a, b, and c, the intersection of the three altitudes be H, the center of the circumscribed circle be O, the radius be R, and $|OH| = d$. Prove: $a^2 + b^2 + c^2 = 9R^2 - d^2$.

11. Draw a circle O with the base side BC of an acute triangle A_1BC that is not an isosceles one. The altitude A_1F on the side BC is a fixed straight line. At F draw the perpendicular lines to the two tangent lines to the circle from A_1, with D_1 and E_1 the respective feet of the perpendicular. Prove: all D_1E_1 pass through a fixed point.

12. Let the radii of two concentric circles in a plane be R and r $(R > r)$ respectively, P be a fixed point on the small circle, B be a moving point on the large circle, the straight line BP and the large circle intersect at another point C, the line l that passes through the point P and is perpendicular to BP intersects the small circle at another point A (if l and the small circle are tangent at the point P, then the points A and P coincide) Find:

    (1) The set of all possible values of the expression $|BC|^2 + |CA|^2 + |AB|^2$.

    (2) The trajectory of the middle point of the line segment AB.

Chapter 28

Synthetic Problems for Complex Numbers

1. Key Points of Knowledge and Basic Methods

A. Thanks to the introduction of complex numbers, there appear new contents on some problems related to algebraic equations: An nth order polynomial equation of one unknown has n roots; Vieta's theorem is still valid for equations of complex coefficients; the imaginary roots of equations with real coefficients appear in pairs of the conjugate form; and so on. Such materials enable us to investigate solutions of algebraic equations and related problems more deeply.

B. There are quite a few types of complex equations, and there are many methods for solving equations of complex numbers. For lower order equations, a usual method is to use the necessary and sufficient condition for the equality of complex numbers, resulting in a real number problem to solve. Sometimes, one may also use the modulus as a breaking point: First find the modulus $|z|$, and then find the complex number z.

C. Complex numbers correspond to points or vectors in the complex plane. Addition and subtraction of complex numbers correspond to the parallelogram rule of plane vectors; multiplication and division correspond to expansion or contraction and rotation of plane vectors. The geometric meaning of complex numbers and their operations pour great vigor further to complex numbers, bestow them with numerous functions, and provide a basis for us to solve related geometric problems with the knowledge of complex numbers.

D. By the root formula of complex numbers, one finds the n roots of the equation $x^n - 1 = 0$ $(n \in \mathbf{N}$ and $n \geq 2)$: $\epsilon_k = \cos \frac{2k}{n}\pi + i\sin \frac{2k}{n}\pi$ $(k = 0, 1, 2, \ldots, n - 1)$. They are all the nth roots of 1, called the nth order roots of unity. From the power formula of complex numbers, we have $\epsilon_k = \left(\cos \frac{2}{n}\pi + i\sin \frac{2}{n}\pi\right)^k = \epsilon_1^k$, which means that all nth order roots of unity can be represented as

$$1, \ \epsilon_1, \ \epsilon_1^2, \ \ldots, \ \epsilon_1^{n-1}.$$

There are following properties for the nth order roots of unity:

(1) $|\epsilon_k| = 1$ $(k = 0, 1, \ldots, n - 1)$.
(2) $\epsilon_j \epsilon_k = \epsilon_{\widetilde{j+k}}$ $(j, k = 0, 1, \ldots, n - 1$ and $j + k \equiv \widetilde{j+k} \pmod{n})$.
(3) $1 + \epsilon_1 + \epsilon_1^2 + \cdots + \epsilon_1^{n-1} = 0$ $(n \geq 2)$.
(4) Let m be an integer. Then

$$1 + \epsilon_1^m + \epsilon_2^m + \cdots + \epsilon_{n-1}^m = \begin{cases} n, & \text{if } m \text{ is a multiple of } n, \\ 0, & \text{if } m \text{ is not a multiple of } n. \end{cases}$$

2. Illustrative Examples

A. *Equations over sets of complex numbers*

Example 1. Given $z \in \mathbf{C}$. If the equation $x^2 - 2zx + \frac{3}{4} + i = 0$ with the unknown x has a real root, find the minimum value for the modulus $|z|$ of the complex number z.

Solution. Let $z = a + bi$ $(a, b \in \mathbf{R})$, and let $x = x_0$ be a real root of the equation $x^2 - 2zx + \frac{3}{4} + i = 0$. Then

$$x_0^2 - 2zx_0 + \frac{3}{4} + i = 0.$$

It follows that

$$\begin{cases} x_0^2 - 2ax_0 + \dfrac{3}{4} = 0, \\ -2bx_0 + 1 = 0. \end{cases}$$

The second equality above implies $x_0 = \frac{1}{2b}$. Substituting it into the first equality gives $\frac{1}{4b^2} - 2a\frac{1}{2b} + \frac{3}{4} = 0$, namely $3b^2 - 4ab + 1 = 0$. So, $a = \frac{3b^2 + 1}{4b}$.

Thus,

$$|z|^2 = a^2 + b^2 = \left(\frac{3b^2 + 1}{4b}\right)^2 + b^2$$

$$= \frac{25}{16}b^2 + \frac{1}{16b^2} + \frac{3}{8}$$

$$\geq \frac{5}{8} + \frac{3}{8} = 1,$$

and the equality holds if and only if $b = \pm\frac{\sqrt{5}}{5}$.

Hence, the minimum value of $|z|$ is 1, achieved when $a = \frac{2\sqrt{5}}{5}$ and $b = \frac{\sqrt{5}}{5}$, or $a = -\frac{2\sqrt{5}}{5}$ and $b = -\frac{\sqrt{5}}{5}$. That is,

$$z = \pm\left(\frac{2\sqrt{5}}{5} + \frac{\sqrt{5}}{5}i\right). \qquad \square$$

Example 2. For given angles $\alpha_1, \alpha_2, \ldots, \alpha_n$, discuss the equation

$$x^n + x^{n-1}\sin\alpha_1 + x^{n-2}\sin\alpha_2 + \cdots + x\sin\alpha_{n-1} + \sin\alpha_n = 0.$$

Does there exist a complex root with its modulus greater than 2?

Solution. The answer is "No." The proof is by reduction to absurdity.

Suppose that there exists an x_0 that is a complex solution of the original equation such that $|x_0| > 2$. Then

$$x_0^n = -x_0^{n-1}\sin\alpha_1 - \cdots - x_0\sin\alpha_{n-1} - \sin\alpha_n.$$

Taking the modulus for each side and using the modulus inequality, we obtain that

$$|x_0|^n \leq |x_0|^{n-1}|\sin\alpha_1| + \cdots + |x_0||\sin\alpha_{n-1}| + |\sin\alpha_n|$$

$$\leq |x_0|^{n-1} + \cdots + |x_0| + 1$$

$$= \frac{|x_0|^n - 1}{|x_0| - 1} < \frac{|x_0|^n}{|x_0| - 1} < \frac{|x_0|^n}{2 - 1} = |x_0|^n.$$

This clearly gives a contradiction, and it means that the original equation has no complex roots whose modulus is greater than 2. $\qquad \square$

Example 3. Prove: A necessary and sufficient condition for a polynomial $f(x) = x^3 + a_1 x^2 + a_2 x + a_3$ with real coefficients to have all three roots in the left half complex plane (namely the real part is less than 0) is $a_i > 0$ with $i = 1, 2,$ and 3, and $a_3 < a_1 a_2$.

Analysis. Note that an nth order polynomial of one unknown with real coefficients must have n roots in $\mathbf{C}$, in which imaginary roots appear in pairs. This means that the equation of the problem must have a real root, and so the other two roots are either both real or mutually conjugate complex numbers.

Proof. From the theorem that imaginary roots of polynomials with real coefficients must appear in conjugate pairs, we know that $f(x)$ must have a real root x_1, and the two other roots x_2 and x_3 are either both real or conjugate complex numbers.

Necessity: Suppose that the real parts of the three roots are negative numbers, that is, $x_1 < 0$ and $x_2 + x_3 < 0$. Then $x_2 x_3 > 0$.

By Vieta's theorem, we see that

$$a_1 = -(x_1 + x_2 + x_3) > 0,$$
$$a_2 = x_1(x_2 + x_3) + x_2 x_3 > 0,$$
$$a_3 = -x_1 x_2 x_3 > 0.$$

Thus,

$$a_1 a_2 = -(x_1 + x_2 + x_3)(x_1 x_2 + x_1 x_3 + x_2 x_3)$$
$$= -x_1^2 x_2 - x_1^2 x_3 - x_1 x_2 x_3 - (x_2 + x_3) \cdot a_2$$
$$= -x_1 x_2 x_3 - x_1^2(x_2 + x_3) - (x_2 + x_3) \cdot a_2$$
$$> -x_1 x_2 x_3 = a_3.$$

This proves the necessity.

Sufficiency: Assume that $a_i > 0$ for $i = 1, 2$, and 3, and also $a_3 < a_1 a_2$.

Since each coefficient of $f(x)$ is a positive number, so $f(x)$ has no zero root and positive roots. If all the three roots ar negative numbers, then the proposition is proved.

If $x_1 < 0$ while $x_2 = a + bi$ and $x_3 = a - bi$ ($a, b \in \mathbf{R}$ and $b \neq 0$), then Vieta's theorem implies that

$$a_1 = -(x_1 + x_2 + x_3) = -(x_1 + 2a),$$
$$a_2 = 2ax_1 + a^2 + b^2,$$
$$a_3 = -x_1(a^2 + b^2).$$

Hence, $a_1 a_2 = -2ax_1^2 - x_1(a^2 + b^2) - 2aa_2$. If $a \geq 0$, then $a_1 a_2 \leq -x_1(a^2 + b^2) = a_3$, which contradicts the given condition. So, $a < 0$.

The sufficiency is thus proved. $\square$

Remark. For any nth order equation of one variable

$$a_n x^n + a_{n-1} x^{n-1} + \cdots + a_0 = 0 \ (a_0 \neq 0),$$

(1) the equation has exactly n roots in the set of all complex numbers;
(2) let $x_1, x_2, \ldots, x_n$ be the n roots of the equation, then they and the coefficients of the equation have the following relation (namely Vieta's theorem):

$$x_1 + x_2 + \cdots + x_n = -\frac{a_{n-1}}{a_n},$$

$$x_1 x_2 + x_1 x_3 + \cdots + x_{n-1} x_n = \frac{a_{n-2}}{a_n},$$

$$\cdots$$

$$x_1 x_2 \cdots x_n = (-1)^n \frac{a_0}{a_n}.$$

B. *Inequalities related to complex numbers*

Example 4. For n complex numbers $z_k = x_k + y_k i \ (k = 1, 2, \ldots, n)$, let r be the absolute value of the real part for a square root of $z_1^2 + z_2^2 + \cdots + z_n^2$. Prove: $r \leq |x_1| + |x_2| + \cdots + |x_n|$.

Analysis. Starting from the two equalities for the equal complex numbers, and using the Cauchy inequality and proof by contradiction, we can show the inequality in the problem.

Proof. Let $a + bi \ (a, b \in \mathbf{R})$ be a square root of $z_1^2 + z_2^2 + \cdots + z_n^2$. Then

$$(a + bi)^2 = (x_1 + y_1 i)^2 + (x_2 + y_2 i)^2 + \cdots + (x_n + y_n i)^2.$$

Comparing the real and imaginary parts of the both sides gives

$$x_1 y_1 + x_2 y_2 + \cdots + x_n y_n = ab, \tag{28.1}$$

$$(x_1^2 + x_2^2 + \cdots + x_n^2) - (y_1^2 + y_2^2 + \cdots + y_n^2) = a^2 - b^2. \tag{28.2}$$

In the following, we use proof by contradiction to prove the proposition.

If $r = |a| > |x_1| + |x_2| + \cdots + |x_n|$, then

$$a^2 > (|x_1| + |x_2| + \cdots + |x_n|)^2 \geq x_1^2 + x_2^2 + \cdots + x_n^2,$$

the combination of which and (28.2) imply that

$$b^2 > y_1^2 + y_2^2 + \cdots + y_n^2.$$

Consequently,

$$|ab| > \sqrt{x_1^2 + x_2^2 + \cdots + x_n^2} \sqrt{y_1^2 + y_2^2 + \cdots + y_n^2}. \tag{28.3}$$

On the other hand, (28.1) and the Cauchy inequality ensure that

$$|ab| = |x_1 y_1 + x_2 y_2 + \cdots + x_n y_n|$$

$$\leq \sqrt{x_1^2 + x_2^2 + \cdots + x_n^2} \sqrt{y_1^2 + y_2^2 + \cdots + y_n^2}.$$

This contradicts the inequality (28.3), so the initial hypothesis is not true, and therefore

$$r \leq |x_1| + |x_2| + \cdots + |x_n|. \qquad \square$$

Remark. It is difficult to get a good idea if trying to prove $r \leq |x_1| + |x_2| + \cdots + |x_n|$ directly, so we used a circuitous strategy via proof by contradiction. That is a frequently used method in the case that a direct proof cannot go smoothly.

Example 5. Suppose that a quadratic function $f(x) = ax^2 + bx + c$ of real coefficients satisfies the conditions that $0 < |c| \leq k$ and $|f(x)| \leq k$ for $x = \cos\theta + i\sin\theta$ (θ is any real number). Prove: For $r > 1$ there holds

$$|f(rx)| \leq (2r^2 - 1)k.$$

Proof. Let $r > 1$ be given. Then

$$f(rx) = \frac{r^2 + r}{2} f(x) + \frac{r^2 - r}{2} f(-x) - (r^2 - 1)c,$$

where $-x = \cos(\pi + \theta) + i\sin(\pi + \theta)$.

Since $|f(\pm x)| \leq k$, it follows that

$$
\begin{aligned}
f(rx) &\leq \frac{r^2+r}{2}|f(x)| + \frac{r^2-r}{2}|f(-x)| + (r^2-1)|c| \\
&\leq \left(\frac{r^2+r}{2} + \frac{r^2-r}{2} + r^2 - 1\right)k \\
&= (2r^2-1)k.
\end{aligned}
$$

$\square$

Example 6. Given a real number $r \in (0,1)$, prove: If n complex numbers $z_1, z_2, \ldots, z_n$ satisfy

$$|z_k - 1| \leq r \quad (k = 1, 2, \ldots, n),$$

then

$$|z_1 + z_2 + \cdots + z_n|\left|\frac{1}{z_1} + \frac{1}{z_2} + \cdots + \frac{1}{z_n}\right| \geq n^2(1 - r^2).$$

Proof. Write $z_k = x_k + y_k i$ $(x_k, y_k \in \mathbf{R})$ for $k = 1, 2, \ldots, n$.
We first show that

$$\frac{x_k^2}{x_k^2 + y_k^2} \geq 1 - r^2, \quad k = 1, 2, \ldots, n. \tag{28.4}$$

Denote $u = \frac{x_k^2}{x_k^2 + y_k^2}$. Then $u > 0$ and $y_k^2 = \left(\frac{1}{u} - 1\right)x_k^2$. Thus,

$$
\begin{aligned}
r^2 &\geq |z_k - 1|^2 = (x_k - 1)^2 + \left(\frac{1}{u} - 1\right)x_k^2 \\
&= \frac{1}{u}(x_k - u)^2 + 1 - u \geq 1 - u.
\end{aligned}
$$

Hence, $u \geq 1 - r^2$, that is, (28.4) is satisfied.
From $|x_k - 1| \leq r < 1$ we know that $x_k > 0$. Since

$$|z_1 + z_2 + \cdots + z_n| \geq |\Re(z_1 + z_2 + \cdots + z_n)| = \sum_{k=1}^{n} x_k.$$

On the other hand, $\frac{1}{z_k} = \frac{x_k - y_k i}{x_k^2 + y_k^2}$. Consequently,

$$\left|\frac{1}{z_1} + \frac{1}{z_2} + \cdots + \frac{1}{z_n}\right| \geq \left|\Re\left(\frac{1}{z_1} + \frac{1}{z_2} + \cdots + \frac{1}{z_n}\right)\right| = \sum_{k=1}^{n} \frac{x_k}{x_k^2 + y_k^2}.$$

Note that $x_k > 0$ for $k = 1, 2, \ldots, n$, so the Cauchy inequality implies that

$$|z_1 + z_2 + \cdots + z_n| \left| \frac{1}{z_1} + \frac{1}{z_2} + \cdots + \frac{1}{z_n} \right| \geq \sum_{k=1}^{n} x_k \sum_{k=1}^{n} \frac{x_k}{x_k^2 + y_k^2}$$

$$\geq \left(\sum_{k=1}^{n} \sqrt{\frac{x_k^2}{x_k^2 + y_k^2}} \right)^2$$

$$\geq \left(n\sqrt{1 - r^2} \right)^2 = n^2(1 - r^2). \quad \square$$

C. *Problems related to roots of unity*

Example 7. Let $\epsilon = \cos \frac{2}{n}\pi + i \sin \frac{2}{n}\pi$. Prove:

(1) $(1 - \epsilon)(1 - \epsilon^2) \cdots (1 - \epsilon^{n-1}) = n$.
(2) $\sin \frac{\pi}{n} \sin \frac{2}{n}\pi \cdots \sin \frac{(n-1)\pi}{n} = \frac{n}{2^{n-1}}$.

Analysis. The expression $\epsilon = \cos \frac{2}{n}\pi + i \sin \frac{2}{n}\pi$ means that ϵ is the nth root of unity ϵ_1, so $\epsilon_k = \epsilon^k$. Since $|1 - \epsilon_k| = 2 \sin \frac{k\pi}{n}$, the equalities can be shown accordingly.

Proof. The n roots of the equation $x^n - 1 = 0$ are

$$\epsilon_k = \cos \frac{2k}{n}\pi + i \sin \frac{2k}{n}\pi \quad (k = 0, 1, 2, \ldots, n - 1).$$

By the assumption, $\epsilon = \cos \frac{2}{n}\pi + i \sin \frac{2}{n}\pi$, so $\epsilon_k = \epsilon^k$.

Thus, from

$$x^n - 1 = (x - 1)(x - \epsilon)(x - \epsilon^2) \cdots (x - \epsilon^{n-1}),$$

we obtain that

$$(x - \epsilon)(x - \epsilon^2) \cdots (x - \epsilon^{n-1}) = \frac{x^n - 1}{x - 1} = x^{n-1} + x^{n-2} + \cdots + x + 1.$$

That is, $(x - \epsilon)(x - \epsilon^2) \cdots (x - \epsilon^{n-1}) = x^{n-1} + x^{n-2} + \cdots + x + 1$.

(1) Letting $x = 1$ gives

$$(1 - \epsilon)(1 - \epsilon^2) \cdots (1 - \epsilon^{n-1}) = n.$$

(2) Noting that
$$\left|1 - \epsilon^k\right| = 2\sin\frac{k\pi}{n},$$
we immediately get
$$2^{n-1}\sin\frac{\pi}{n}\sin\frac{2}{n}\pi\cdots\sin\frac{(n-1)\pi}{n} = n,$$
which proves (2). $\qquad\square$

Remark. The expression $\cos\frac{2k}{n}\pi + i\sin\frac{2k}{n}\pi$ reminded us of the definition and properties of the nth roots of unity. This was very helpful for solving the problem.

Example 8. Let $A_0 A_1 \cdots A_{n-1}$ be a regular n-polygon. Suppose that the radius of its circumcircle O is 1. Prove: The sum of the squares of the distances from any point P on the circle to every vertex is a fixed value.

Analysis. The point P and the points A_i ($i = 0, 1, \ldots, n-1$) are all on the unit circle, so the roots of unity can be used to solve the problem.

Proof. Use O as the origin to set up a rectangular coordinates system, so that the coordinates of A_0 are $(1, 0)$. Then the corresponding complex numbers of A_i are $\epsilon_k = \cos\frac{2k}{n}\pi + i\sin\frac{2k}{n}\pi$ ($k = 0, 1, 2, \ldots, n-1$). Let the corresponding complex number of the point P be z. Then $|z| = 1$. Consequently,

$$\sum_{k=0}^{n-1}|PA_k|^2 = \sum_{k=0}^{n-1}|z - \epsilon_k|^2 = \sum_{k=0}^{n-1}(z - \epsilon_k)(\overline{z} - \overline{\epsilon_k})$$

$$= \sum_{k=0}^{n-1}(1 - z\overline{\epsilon_k} - \overline{z}\epsilon_k + 1)$$

$$= 2n - z\sum_{k=0}^{n-1}\overline{\epsilon_k} - \overline{z}\sum_{k=0}^{n-1}\epsilon_k$$

$$= 2n. \qquad\square$$

Remark. All nth roots of unity just divide the unit circle of the complex plane into n equal arcs, in other words, $\epsilon_0 = 1, \epsilon_1, \epsilon_2, \ldots, \epsilon_{n-1}$ are the vertices of the inscribed regular n-polygon in the circle centered at the origin of radius 1. Therefore, using the roots of unity can solve many geometric problems related to regular polygons.

3. Exercises

Group A

I. Filling Problems

1. Let $a \in \mathbf{R}$ and complex numbers $z_1 = a+i$, $z_2 = 2a+2i$, and $z_3 = 3a+4i$. If $|z_1|, |z_2|$, and $|z_3|$ constitute a geometric sequence, then the value of the real number a is _____.

2. For complex numbers $z_1 = (2-a)+(1-b)i$, $z_2 = (3+2a)+(2+3b)i$, and $z_3 = (3-a)+(3-2b)i$, where $a, b \in \mathbf{R}$, when $|z_1| + |z_2| + |z_3|$ achieves the minimum value, $3a + 4b =$ _____.

3. If the principal value for the argument of a root z_0 of the equation $z^6 + z^3 + 1 = 0$ is in the interval $\left(\frac{\pi}{2}, \pi\right)$, then the value of $\arg z_0$ is _____.

4. Suppose that a complex number z satisfies $z^3 = 27$. Then the value of $z^5 + 3z^4 + 2242$ is _____.

5. Let A, B, C, and D represent complex numbers a, b, c, and d, respectively. If the four points A, B, C, and D are on the same circle, then the value of $\arg \frac{d-a}{b-a} + \arg \frac{b-c}{d-c}$ is _____.

6. Let x_1 and x_2 be the two roots of a quadratic equation $ax^2 + bx + c = 0$. If x_1 is an imaginary number and $\frac{x_1^2}{x_2}$ is a real number, then the value of
$$S = 1 + \frac{x_1}{x_2} + \left(\frac{x_1}{x_2}\right)^2 + \left(\frac{x_1}{x_2}\right)^4 + \left(\frac{x_1}{x_2}\right)^8 + \cdots + \left(\frac{x_1}{x_2}\right)^{2^{1999}} \text{ is } \underline{\hspace{2cm}}.$$

7. Let ϵ be a 7th imaginary root of unity. Then $\frac{\epsilon}{1+\epsilon^2} + \frac{\epsilon^2}{1+\epsilon^4} + \frac{\epsilon^3}{1+\epsilon^6} =$ _____.

II. Calculation Problems

8. Give two complex numbers $z_1 = 2 + i$ and $2z_2 = \frac{z_1-1}{2i+1-z_1}$. If the three interior angles A, B, and C of $\triangle ABC$ form an arithmetic sequence in succession, and a complex number $p = \cos A + 2i \cos^2 \frac{C}{2}$, find the range of the values for $|p + z_2|$.

9. Suppose that a sequence $\{a_n\}$ $(n = 0, 1, 2, \ldots)$ of rational numbers satisfies $a_n = \alpha x_1^n + \beta x_2^n \neq 0$ $(n = 0, 1, 2, \ldots)$ and $x_1 x_2 = 1$, where α and β are real numbers, and x_1 and x_2 are complex numbers. Prove:

(1) $x_1 + x_2$ is a rational number.
(2) If x_1 and x_2 are not real numbers, then $\alpha = \beta$.

Group B

10. Let $a_1, a_2, \ldots, a_n$ be mutually different real numbers. Prove: The equation $\frac{1}{x+a_1} + \frac{1}{x+a_2} + \cdots + \frac{1}{x+a_n} = 0$ has no imaginary roots.

11. Assume that $\theta_0, \theta_1, \ldots, \theta_n$ $(n \in \mathbf{N}_+)$ are all real numbers, and a is a complex root of the equation $z^n \cos\theta_n + z^{n-1} \cos\theta_{n-1} + \cdots + \cos\theta_0 = 2$. Prove: $|a| > \frac{1}{2}$.

12. Determine all the complex numbers α such that for any complex numbers z_1 and z_2 $(|z_1|, |z_2| < 1$ and $z_1 \neq z_2)$, it is always valid that

$$(z_1 + \alpha)^2 + \alpha\overline{z_1} \neq (z_2 + \alpha)^2 + \alpha\overline{z_2}.$$

Chapter 29

Mathematical Induction (II)

1. Key Points of Knowledge and Basic Methods

In this chapter, in addition to discussing further the two forms and basic proof techniques of mathematical induction introduced in Chapter 17, we shall introduce several additional special forms of mathematical induction.

A. *Method of reverse induction*

The method of reverse induction is also called the backward induction. Its principle is as follows:

Let $P(n)$ be a proposition related to natural numbers n. Suppose that

(1) $P(n)$ is true for infinitely many natural numbers n;
(2) under the assumption that $P(k+1)$ is true, $P(k)$ is deduced to be true, too.

Then the proposition $P(n)$ is true for all natural numbers n.

Using proof by contradiction, one can easily show that the principle of reverse induction is correct.

B. *Method of spiral induction*

The method of spiral induction is basically an induction method of a kind alternative transition. Its principle is the following:

Let $P(n)$ and $Q(n)$ be two propositions related to natural numbers n. Suppose that

(1) $P(0)$ is true;

(2) for any natural number k, if the proposition $P(k)$ is true, then the proposition $Q(k)$ is deduced to be true; and if the proposition $Q(k)$ is true, then the proposition $P(k+1)$ is deduced to be true.

Then the propositions $P(n)$ and $Q(n)$ are true for all natural numbers n.

The above method of spiral induction is with respect to two propositions. For three, four, or more propositions, there are corresponding methods of spiral induction.

Problems in mathematical competitions are often given in the form of one proposition (such as $P(n)$). To use the method of spiral induction, one needs to supplement the given proposition with another one (such as $Q(n)$). The second one can be called a "companion proposition" of the original proposition.

C. *Double mathematical induction*

Let $P(n, m)$ be a proposition related to mutually independent natural numbers n and m. Suppose that

(1) $P(0, m)$ is true for all natural numbers m and $P(n, 0)$ is true for all natural numbers n;
(2) if $P(n+1, m)$ and $P(n, m+1)$ are true, then it is deduced that $P(n+1, m+1)$ is also true.

Then the proposition $P(n, m)$ is true for all natural numbers n and m.

D. *Points of attention*

(1) The most basic and most often used forms of mathematical induction are still the first and second forms of mathematical induction. The other forms of mathematical induction are usually adapted ones while it is difficult to use the first or second forms of basic mathematical induction.
(2) Not every proposition related to natural numbers n can be proved with mathematical induction. Sometimes, blindly applying the inductive model often leads to a dead end, so flexibility is important during the process of solving problems.

2. Illustrative Examples

We start with applications of the most familiar basic forms of mathematical induction.

Example 1. A sequence $a_1, a_2, \ldots$ satisfies $a_1 = 0$ and $a_n = \max_{1 \le i \le n-1}\{i + a_i + a_{n-i}\}$ for $n \ge 2$. For example, $a_2 = 1$ and $a_3 = 3$. Find a_{200}.

Analysis. If the general term can be found, naturally one can calculate a_{200}. It is easy to see that $a_1 = 0, a_2 = 1, a_3 = 3, a_4 = 6$, and $a_5 = 10$. From these we guess that $a_n = \frac{n(n-1)}{2}$.

Solution. We use the second form of mathematical induction to show that
$$a_n = \frac{n(n-1)}{2}.$$
When $n = 1$, we have $a_1 = 0$, so the proposition is true.
Suppose that $a_n = \frac{n(n-1)}{2}$ for all $n \le k$.
When $n = k + 1$,

$$
\begin{aligned}
a_{k+1} &= \max_{1 \le i \le k}\left\{i + a_i + a_{k+1-i}\right\} \\
&= \max_{1 \le i \le k}\left\{i + \frac{i(i-1)}{2} + \frac{(k+1-i)(k-i)}{2}\right\} \\
&= \max_{1 \le i \le k}\left\{i^2 - ki + \frac{k^2+k}{2}\right\} \\
&= \frac{k^2+k}{2} + \max_{1 \le i \le k}\left\{i(i-k)\right\}.
\end{aligned}
$$

Obviously, $i(i-k) < 0$ when $1 \le i \le k-1$ and $i(i-k) = 0$ when $i = k$. Consequently, $\max_{1 \le i \le k}\{i(i-k)\} = 0$, and so
$$a_{k+1} = \frac{k^2+k}{2} = \frac{(k+1)[(k+1)-1]}{2}.$$
Thus, when $n = k+1$, the proposition is true.
In particular, $a_{200} = \frac{200 \cdot 199}{2} = 19900$. $\qquad \square$

Example 2. Suppose that p and $p + 2$ are both prime numbers with $p > 3$. A sequence $\{a_n\}$ is defined as $a_1 = 2$ and $a_n = a_{n-1} + \left\lceil \frac{p a_{n-1}}{n} \right\rceil$ for $n = 2, 3, \ldots$, where $\lceil x \rceil$ is the minimal integer that is not less than the real number x.

Prove: For $n = 3, 4, \ldots, p-1$, it is always true that $n \mid p a_{n-1} + 1$.

Proof. First, note that $\{a_n\}$ is a sequence of integers.
We use mathematical induction for n. When $n = 3$, by the condition $a_2 = 2 + p$, so $p a_2 + 1 = (p+1)^2$. Since p and $p + 2$ are both prime

numbers, and $p > 3$, it must be that $3 \mid p + 1$. Hence, $3 \mid pa_2 + 1$, namely the conclusion is satisfied when $n = 3$.

For $3 < n \leq p - 1$, suppose that $k \mid pa_{k-1} + 1$ for $k = 3, \ldots, n - 1$. Now,

$$\left\lceil \frac{pa_{k-1}}{k} \right\rceil = \frac{pa_{k-1} + 1}{k},$$

so

$$pa_{k-1} + 1 = p \left(a_{k-2} + \left\lceil \frac{pa_{k-2}}{k-1} \right\rceil \right) + 1$$

$$= p \left(a_{k-2} + \frac{pa_{k-2} + 1}{k-1} \right) + 1$$

$$= \frac{(pa_{k-2} + 1)(p + k - 1)}{k-1}.$$

Hence, for $3 < n \leq p - 1$,

$$pa_{n-1} + 1 = \frac{p + n - 1}{n - 1}(pa_{n-2} + 1)$$

$$= \frac{p + n - 1}{n - 1} \frac{p + n - 2}{n - 2}(pa_{n-3} + 1)$$

$$= \cdots = \frac{p + n - 1}{n - 1} \frac{p + n - 2}{n - 2} \cdots \frac{p + 3}{3}(pa_2 + 1).$$

Consequently,

$$pa_{n-1} + 1 = \frac{2n(p + 1)}{(p + n)(p + 2)} C_{p+n}^n.$$

It follows that (noting that C_{p+n}^n is an integer)

$$n \mid (p + n)(p + 2)(pa_{n-1} + 1). \tag{29.1}$$

Since $n < p$ and p is a prime number, we have

$$(n, n + p) = (n, p) = 1.$$

Also $p + 2$ is a prime number greater than n, so we see that $(n, p + 2) = 1$, thus n and $(p + n)(p + 2)$ are relatively prime. Therefore, $n \mid pa_{n-1} + 1$ from (29.1). By mathematical induction, this proposition has been proved.

$\square$

Remark. (1) Twin prime numbers exist.

(2) Here the natural numbers were restricted to between 3 and $p - 1$.

Example 3. Let a sequence $\{F_n\}$ be defined by: $F_{n+2} = F_{n+1} + F_n, F_1 = 1$, and $F_2 = 1$. Prove:

$$F_{n+1}^2 + F_n^2 = F_{2n+1}.$$

Analysis. We want to prove the conclusion by mathematical induction. The first step $(n = 1)$ is easy. Assume that the proposition is satisfied for $n = k$ $(k \geq 2)$. Then it suffices to show:

$$F_{k+2}^2 + F_{k+1}^2 = F_{2k+3}. \tag{29.2}$$

Substitute the recursive relation of the condition into (29.2) to see the effect:

$$(F_{k+1} + F_k)^2 + F_{k+1}^2 = F_{2k+3}, \tag{29.3}$$

that is,

$$\left(F_{k+1}^2 + F_k^2\right) + \left(2F_k F_{k+1} + F_{k+1}^2\right) = F_{2k+3}. \tag{29.4}$$

Note that the expression inside the first pair of parentheses of (29.4) is just F_{2k+1} (by inductive hypothesis). Thus, to prove (29.4), it is enough to show that

$$2F_k F_{k+1} + F_{k+1}^2 = F_{2k+2}. \tag{29.5}$$

The formula (29.5) is the same as

$$2F_k(F_{k-1} + F_k) + F_{k+1}^2 = F_{2k+2},$$

in other words,

$$\left(2F_k F_{k-1} + F_k^2\right) + \left(F_k^2 + F_{k+1}^2\right) = F_{2k+2}, \tag{29.6}$$

while the expression inside the second pair of parentheses of (29.6) is just F_{2k+1}. So, it is sufficient to show that

$$2F_k F_{k-1} + F_k^2 = F_{2k}. \tag{29.7}$$

The formulas (29.7) and (29.5) have the completely same structure. The original proposition is denoted as $P(n)$, and the claim $2F_{n+1}F_n + F_{n+1}^2 = F_{2n+2}$ is denoted as $Q(n)$. We can use the alternating relation between the formulas of (29.5) and (29.7) to get the alternating relation between $Q(n)$ and $P(n)$, so we may use the method of spiral induction to complete the proof of the problem.

Proof. Consider the propositions

$$P(n) : F_{n+1}^2 + F_n^2 = F_{2n+1}; \quad Q(n) : 2F_{n+1}F_n + F_{n+1}^2 = F_{2n+2}.$$

When $n = 1$, it is easy to see that $P(1)$ and $Q(1)$ are true.

Suppose that $P(k)$ and $Q(k)$ are satisfied when $n = k$, namely $F_{k+1}^2 + F_k^2 = F_{2k+1}$ and $2F_{k+1}F_k + F_{k+1}^2 = F_{2k+2}$. Then

$$\begin{aligned}
F_{k+2}^2 + F_{k+1}^2 &= (F_{k+1} + F_k)^2 + F_{k+1}^2 \\
&= \left(F_{k+1}^2 + 2F_{k+1}F_k\right) + \left(F_k^2 + F_{k+1}^2\right) \\
&= F_{2k+2} + F_{2k+1} = F_{2k+3},
\end{aligned}$$

in other words, $P(k+1)$ is satisfied. Since $P(k+1)$ and $Q(k)$ are both true, we obtain that

$$\begin{aligned}
2F_{k+2}F_{k+1} + F_{k+2}^2 &= 2(F_{k+1} + F_k)F_{k+1} + F_{2k+2}^2 \\
&= \left(F_{k+1}^2 + F_{k+2}^2\right) + \left(F_{k+1}^2 + 2F_{k+1}F_k\right) \\
&= F_{2k+3} + F_{2k+2} = F_{2k+4},
\end{aligned}$$

that is, $Q(k+1)$ is also satisfied. Hence, $P(n)$ and $Q(n)$ are true for all positive integers n. $\qquad\square$

Remark. Using the "method of spiral induction" often needs to combine the conditions of the problem to determine a "supplementary conclusion" (sometimes we need to construct such a conclusion). Constructing a good "supplementary conclusion" not only requires a good understanding of the condition, but also needs accumulations of usual knowledge.

When using mathematical induction to solve problems, often we are not sure how to go forward the transition from n to $n+1$ in many cases, or not able to reach an ideal result. At this time, we ought to pay an attention to the "method of reverse induction," which plays a good role in many occasions.

In the following, we look at two examples to illustrate the method of reverse induction.

Example 4. Suppose that a function $f : \mathbf{N}_+ \to [1, +\infty)$ satisfies:

(1) $f(2) = 2$;

(2) for any $m, n \in \mathbf{N}_+$, we have $f(mn) = f(m)f(n)$;

(3) when $m < n$, the inequality $f(m) < f(n)$ is valid.

Prove: For any positive integer n, it holds that $f(n) = n$.

Proof. The conditions (1) and (2) imply that $f(1) = 1$. Suppose now that $f(2^k) = 2^k$ for $k \in \mathbf{N}$. Then $f(2^{k+1}) = f(2^k)f(2) = 2^k \cdot 2 = 2^{k+1}$. Hence, $f(2^k) = 2^k$ for any $k \in \mathbf{N}$.

Now, we discuss the values of $f(n)$. Let $f(n) = l$. Then from (2) and mathematical induction, we see that $f(n^m) = l^m$ for any $m \in \mathbf{N}_+$.

Assume that $2^k \leq n^m < 2^{k+1}$. Then $f(2^k) \leq f(n^m) < f(2^{k+1})$ from (3). So, $2^k \leq l^m < 2^{k+1}$ by the previous conclusion. Comparing it with $2^k \leq n^m < 2^{k+1}$, we know that

$$\frac{1}{2} < \left(\frac{n}{l}\right)^m < 2, \tag{29.8}$$

which is true for any $m \in \mathbf{N}_+$.

If $n > l$, then we choose $m > \frac{l}{n-l}$, from which

$$\left(\frac{n}{l}\right)^m = \left(1 + \frac{n-l}{l}\right)^m \geq 1 + m\frac{n-l}{l} > 2,$$

which contradicts the formula (29.8). By the same way, if $n < l$, then with a choice of $m > \frac{n}{l-n}$, we get $\left(\frac{l}{n}\right)^m > 2$, namely $\left(\frac{n}{l}\right)^m < \frac{1}{2}$, which also contradicts (29.8). Hence, it can only be that $n = l$.

In summary, $f(n) = n$ for any $n \in \mathbf{N}_+$. $\qquad\square$

Example 5. Prove by the method of reverse induction: If a function $f(x)$ with its domain $[a, b]$ satisfies that for any $x, y \in [a, b]$, the inequality $f\left(\frac{x+y}{2}\right) \leq \frac{1}{2}[f(x) + f(y)]$ is valid, then for $x_1, x_2, \ldots, x_n \in [a, b]$, we have

$$f\left(\frac{x_1 + x_2 + \cdots + x_n}{n}\right) \leq \frac{1}{2}[f(x_1) + f(x_2) + \cdots + f(x_n)].$$

Analysis. One may use the same proof method for Example 4 to prove this proposition.

Proof. When $n = 1$ and 2, the inequality is obviously correct. Now, assume that the inequality is satisfied for $n = 2^k$. Then by the definition of $f(x)$, we know that

$$f\left(\frac{x_1 + \cdots + x_{2^{k+1}}}{2^{k+1}}\right) \le \frac{1}{2}\left[f\left(\frac{x_1 + \cdots + x_{2^k}}{2^k}\right) + f\left(\frac{x_{2^k+1} + \cdots + x_{2^{k+1}}}{2^k}\right)\right]$$

$$\le \frac{1}{2}\left[\frac{1}{2^k}\sum_{j=1}^{2^k} f(x_j) + \frac{1}{2^k}\sum_{j=1}^{2^k} f(x_{2^k+j})\right]$$

$$= \frac{1}{2^{k+1}}\sum_{j=1}^{2^{k+1}} f(x_j).$$

Thus, the inequality is satisfied for any $n = 2^k$ ($k \in \mathbf{N}_+$).

For a general $n \in \mathbf{N}_+$ ($n \ge 3$), let $2^k \le n < 2^{k+1}$ for some $k \in \mathbf{N}_+$. Denote

$$A = \frac{1}{n}(x_1 + x_2 + \cdots + x_n).$$

Then, since the inequality is true for 2^{k+1}, we see that

$$f\left(\frac{x_1 + \cdots + x_n + (2^{k+1} - n)A}{2^{k+1}}\right) \le \frac{1}{2^{k+1}}\left[\sum_{j=1}^{n} f(x_j) + (2^{k+1} - n)f(A)\right].$$

On the other hand,

$$\frac{1}{2^{k+1}}\left[x_1 + \cdots + x_n + \left(2^{k+1} - n\right)A\right] = \frac{1}{2^{k+1}}\left[nA + \left(2^{k+1} - n\right)A\right] = A.$$

Consequently,

$$2^{k+1}f(A) \le \sum_{j=1}^{n} f(x_j) + \left(2^{k+1} - n\right)f(A),$$

and so $f(A) \le \frac{1}{n}\sum_{j=1}^{n} f(x_j)$, namely the inequality is true for n.   □

Remark. From the above two examples, we may find that the key to an effective proof via reverse induction is the possibility of controlling the natural numbers n in a range. Then we have an inequality relation to use in the proof (for example the formula (29.8) in Example 4). At the same time the inequality is true for some particular n. Combining the two points will make the transition convenient.

In the following, we look at some examples with several natural number parameters. When solving such problems, one should pay an attention to choosing right objects for induction. When making induction to several parameters, first choose the easiest to deal with.

Example 6. Let $m, n \in \mathbf{N}_+$. Prove: For positive real numbers x_1, x_2, $\ldots, x_n$ and $y_1, y_2, \ldots, y_n$, if $x_i + y_i = 1$ with $i = 1, 2, \ldots, n$, then

$$(1 - x_1 x_2 \cdots x_n)^m + (1 - y_1^m)(1 - y_2^m) \cdots (1 - y_n^m) \geq 1. \qquad (29.9)$$

Analysis. There are two positive integer parameters m and n. First determine the object of induction. Note that choosing n as the induction object makes the transition easier, because the second additive term on the left hand side of (29.9) is easier to deal with during the induction transition.

Proof. Do the induction on n. When $n = 1$, the condition implies that

$$(1 - x_1)^m + (1 - y_1^m) = y_1^m + (1 - y_1^m) = 1,$$

so (29.9) is valid for $n = 1$.

Suppose that (29.9) is true for $n - 1$ ($n \geq 2$). Consider the case of n. First

$$(1 - x_1 x_2 \cdots x_n)^m + (1 - y_1^m)(1 - y_2^m) \cdots (1 - y_n^m)$$

$$= [1 - x_1 x_2 \cdots x_{n-1}(1 - y_n)]^m + (1 - y_1^m) \cdots (1 - y_n^m)$$

$$\geq (1 - x_1 x_2 \cdots x_{n-1} + x_1 x_2 \cdots x_{n-1} y_n)^m$$

$$\quad + [1 - (1 - x_1 x_2 \cdots x_{n-1})^m](1 - y_n^m).$$

If we let $a = 1 - x_1 x_2 \cdots x_{n-1}$ and $b = y_n$, then from the above inequality we know that, in order to prove (29.9) for n, it is enough to show that $(a + b - ab)^m + (1 - a^m)(1 - b^m) \geq 1$ is satisfied for any $a, b \in (0, 1)$. In other words,

$$(a + b - ab)^m \geq a^m + b^m - a^m b^m. \qquad (29.10)$$

Now, we prove (29.10) via mathematical induction on m.

Clearly, the inequality (29.10) is valid for $m = 1$. Now, assume that (29.10) is true for $m - 1$ ($m \geq 2$). Then

$$(a + b - ab)^m - a^m - b^m + a^m b^m$$

$$\geq (a^{m-1} + b^{m-1} - a^{m-1} b^{m-1})(a + b - ab) - a^m - b^m + a^m b^m$$

$$= 2a^m b^m + ab^{m-1} + ba^{m-1} - a^m b^{m-1} - a^{m-1} b^m - a^m b - ab^m$$

$$= (b^{m-1} - b^m)(a - a^m) + (a^{m-1} - a^m)(b - b^m),$$

and with the fact that $a, b \in (0, 1)$, we know the truth of (29.10) for m.

In summary, the formula (29.9) is satisfied. $\qquad \square$

Remark. This problem can also be dealt with via the method of "double induction," which may make the whole process more concise.

Example 7. Let nonnegative numbers $a_1, a_2, \ldots, a_{2006}$ satisfy:

$$a_i + a_j \leq a_{i+j} \leq a_i + a_j + 1$$

for all $1 \leq i, j \leq 2006$ such that $i + j \leq 2006$. Prove: There exists $x \in \mathbf{R}$ such that $a_n = [nx]$ for any $n \in \{1, 2, \ldots, 2006\}$.

Proof. First, we analyze the properties of $[nx]$. Note that $nx - 1 < a_n = [nx] \leq nx$, so $x \in \left[\frac{a_n}{n}, \frac{a_n+1}{n}\right)$. Let $I_n = \left[\frac{a_n}{n}, \frac{a_n+1}{n}\right)$.

Then we only need to show that $I_1 \cap I_2 \cap \cdots \cap I_{2006}$ is not an empty set. But this can be realized by proving that $\max_{1 \leq n \leq 2006} \frac{a_n}{n} < \min_{1 \leq n \leq 2006} \frac{a_n+1}{n}$.

Consequently, it is enough to show that for any $m, n \in \{1, 2, \ldots, 2006\}$, it is true that $\frac{a_n}{n} < \frac{a_m+1}{m}$, namely

$$m a_n < n(a_m + 1). \tag{29.11}$$

We use induction on $m + n$ to prove the above inequality as follows.

When $m + n = 2$, we have $m = n = 1$, and (29.11) is true in this case.

Suppose that (29.11) is satisfied for all positive integers m and n such that $m + n \leq k$. Then for $m + n = k + 1$, if $m = n$, then (29.11) is clearly true; if $m > n$, then by the inductive hypothesis, we know that

$$(m - n)a_n < n(a_{m-n} + 1).$$

The condition $a_i + a_j \leq a_{i+j}$ implies that $n(a_{m-n} + a_n) \leq na_m$, from which $ma_n < n(a_m + 1)$, namely (29.11) is true. On the other hand, if $m < n$, then $ma_{n-m} < (n - m)(a_m + 1)$ by the inductive hypothesis.

The condition $a_{i+j} \leq a_i + a_j + 1$ ensures that $ma_n \leq m(a_m + a_{n-m} + 1)$. Therefore,

$$m a_n < n(a_m + 1),$$

which means that (29.11) is satisfied.

To summarize, the inequality (29.11) is true for any $m, n \in \{1, 2, \ldots, 2006\}$. $\qquad\square$

Remark. The given condition should be considered to determine the direction of induction. Otherwise it is hard to obtain useful results.

Example 8. Suppose that there are $2n$ ($n \geq 2$) points in the space, in which any four of them are not in the same plane. If there are $n^2 + 1$ line segments connecting the points, prove: These line segments must constitute two triangles with a common side.

Proof. We denote the proposition to be proved as $P(n)$ and intend to construct another proposition $Q(n)$: Let there be $2n + 1$ ($n \geq 2$) points in the space, in which any four of them are not in the same plane. If there are $n^2 + n + 1$ line segments connecting the points, then these line segments must constitute two triangles with a common side.

We first show that $P(2)$ is true. For four points in the space not in the same plane with five line segments connecting them, the resulting figure must be a tetrahedron with one edge deleted. So, there must be two triangles with a common side, namely $P(2)$ is correct.

Suppose that $P(k)$ is true. Consider $2k + 1$ points that are connected by $k^2 + k + 1$ line segments. Let A be a point with the least number of the line segments connected to it (if there are more than one such points, just pick any one among them). Then the number of the line segments from the point A is less than or equal to k; otherwise, the number of the line segments would be greater than or equal to

$$\frac{1}{2}(2k + 1)(k + 1) = k^2 + k + \frac{1}{2}(k + 1) > k^2 + k + 1,$$

a contradiction. We remove the point A together with all the line segments connected to it. Then there are $2k$ points with at least $k^2 + 1$ line segments left. By the hypothesis that $P(k)$ is valid, we know that $Q(k)$ is also satisfied.

Suppose that $Q(k)$ is true. Consider $2k + 2$ points that are connected by $(k + 1)^2 + 1$ line segments. Let A be a point with the least number of the line segments connected to it. Then the number of the line segments from the point A is less than or equal to $k + 1$; otherwise, the number of the line segments would be greater than or equal to

$$\frac{1}{2}(2k + 2)(k + 2) = (k + 1)^2 + k + 1 > (k + 1)^2 + 1,$$

a contradiction. Remove the point A together with all the line segments connected to it. Then there are $2k + 1$ points with at least

$$(k + 1)^2 + 1 - (k + 1) = k^2 + k + 1$$

line segments left. By the hypothesis that $Q(k)$ is valid, we know that $P(k + 1)$ is also satisfied.

In summary, by the method of spiral induction, $P(n)$ and $Q(n)$ are both satisfied for all positive integers n $(n \geq 2)$. $\qquad\qquad\square$

3. Exercises

Group A

1. Assume that a sequence $\{a_n\}$ satisfies $a_0 = a, a_1 = b, a_2 = c$, and $a_n = a_{n-1} - 3a_{n-2} + 27a_{n-3}$ $(n \geq 3)$. If there exists a positive integer N' such that a_n are integers for $n \geq N'$, prove: $a, b,$ and c are rational numbers.
2. Prove: There exist infinitely many $n \in \mathbf{N}_+$ such that $n \mid 2^n + 2$.
3. Prove: There exist infinitely many $n \in \mathbf{N}_+$ such that $n \mid 2^n + 1$.
4. Suppose that a sequence $\{a_n\}$ satisfies $a_{2l} = 3l^2$ and $a_{2l-1} = 3l(l-1)+1$ for $l \in \mathbf{N}_+$. Denote by S_n the sum of the first n terms of the sequence. Prove:

$$S_{2l-1} = \frac{1}{2}l(4l^2 - 3l + 1), \quad S_{2l} = \frac{1}{2}l(4l^2 + 3l + 1).$$

5. Let f be a strictly increasing function defined on the set of all positive integers and satisfy: (1) $f(n)$ are all positive integers; (2) $f(2) = 2$; (3) $f(mn) = f(m)f(n)$ when m and n are relatively prime. Prove: $f(n) = n$ for all positive integers n.
6. For positive integers a and b such that $a > b$, let $\sin\theta = \frac{2ab}{a^2+b^2}$, where $\theta \in \left(0, \frac{\pi}{2}\right)$, and $A_n = (a^2 + b^2)^n \sin n\theta$. Prove: A_n are integers for all positive integers n.
7. Let m and n be arbitrary nonnegative integers. Prove: $\frac{(2m)!(2n)!}{m!n!(m+n)!}$ is an integer.

Group B

8. Let a sequence $\{a_n\}$ satisfy: $a_1 = 1$ and

$$a_{n+1} = \begin{cases} a_n - 2, & \text{if } a_n - 2 \notin \{a_1, a_2, \ldots, a_n\} \text{ and } a_n - 2 > 0, \\ a_n + 3, & \text{if otherwise.} \end{cases}$$

Prove: For any positive integer $k > 1$, there exists a positive integer n such that $a_n = a_{n-1} + 3 = k^2$.

9. Dye one of red, green, and blue colors to each vertex of a regular n-polygon, such that any two nearby vertices have different colors and every color appears at least once. Prove: It is possible to partition the

regular n-polygon into $n - 2$ triangles with some diagonals, so that the three vertices of each triangle have different colors.

10. For any $n \in \mathbf{N}_+$, denote $\rho(n) = k$, where k is a nonnegative integer satisfying $2^k \mid n$ and $2^{k+1} \nmid n$. Define a sequence $\{x_n\}$ as follows: $x_0 = 0$ and $\frac{1}{x_n} = 1 + 2\rho(n) - x_{n-1}$ for $n = 1, 2, \ldots$. Prove: Every nonnegative rational number appears exactly once in the sequence.

11. Find all positive integers m and n such that the $m \times n$-check chess plate can be partitioned into "L"-shaped (union of three checks) dominoes completely without repeat.

12. Suppose that a function $f(m, n)$ satisfies

$$f(m, n) \leq f(m, n - 1) + f(m - 1, n),$$

where m and n are positive integers and $m, n \geq 2$. If $f(1, n) = f(m, 1) = 1$, prove: $f(m, n) \leq C_{m+n-2}^{m-1}$.

Chapter 30

Proof by Contradiction

1. Key Points of Knowledge and Basic Methods

For some problems, a direct approach is relatively difficult, but using proof by contradiction is much easier to do. This is because when using proof by contradiction to show the proposition "if A then B," one increases one condition $\overline{B}$ artificially; here $\overline{B}$ represents the negation of condition B.

For a proposition with a conclusion of negative form, it is often proved via using proof by contradiction. Those propositions with conclusions that contain "at least" or "at most," "unique," "finite" or "infinite," and so on are also frequently proved with the approach of proof by contradiction. Even some propositions, which state conclusions with positive sentences at the first glance but essentially imply negative factors, may also be considered to use proof by contradiction for solutions; one example is that "mutually prime" is the same as "no common factors greater than 1.

2. Illustrative Examples

Example 1. Does there exist a real valued function $f(x)$ on $\mathbf{R}$ such that for any positive integer n, there holds that

$$3f(-n^2 + 3n + 1) - 2f(n)^2 = 5?$$

If it exists, find one $f(x)$; if it does not exist, give a proof.

Solution. It does not exist. Here is a proof.

Suppose that it exists. Let $n = 1$. Then

$$3f(3) = 2f(1)^2 + 5. \tag{30.1}$$

Let $n = 3$. Then

$$3f(1) = 2f(3)^2 + 5. \tag{30.2}$$

Thus, $f(1) > \frac{5}{3}$ and $f(3) > \frac{5}{3}$.

Subtracting (30.1) from (30.2) gives that

$$3[f(1) - f(3)] = 2[f(3)^2 - f(1)^2].$$

If $f(1) \neq f(3)$, then $f(1) + f(3) = -\frac{3}{2}$, a contradiction.

If $f(1) = f(3)$, then $3f(1) = 2f(1)^2 + 5$, and now $f(1)$ is not a real number, also a contradiction. $\qquad\qquad\square$

Remark. The process of proving a proposition with the help of using proof by contradiction is just "making a contradiction," realizing the purpose of the proof via "negation of negation."

Example 2. Prove: It does not exist a positive integer k such that $k + 4$ and $k^2 + 5k + 2$ are both cubic numbers.

Proof. Suppose that it does exist such a k, namely there are positive integers m and n such that

$$k + 4 = m^3 \quad \text{and} \quad k^2 + 5k + 2 = n^3.$$

Then

$$(mn)^3 = (k+4)(k^2 + 5k + 2) = k^3 + 9k^2 + 22k + 8$$

is a cubic number. From

$$(k+2)^3 = k^3 + 6k^2 + 12k + 8$$
$$< k^3 + 9k^2 + 22k + 8$$
$$< k^3 + 9k^2 + 27k + 27 = (k+3)^3,$$

we see that $(mn)^3$ is strictly between the cubic numbers of two consecutive positive integers, leading to a contradiction.

Therefore, such k does not exist. $\qquad\qquad\square$

Example 3. If a diagonal of a hexagon divides the hexagon into two quadrilaterals with inscribed circles, then the diagonal is called "good." How many good diagonals does a hexagon have at most?

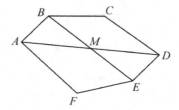

Figure 30.1 Figure for Example 3.

Solution. We first show that among two intersecting diagonals of a hexagon, there is at most one is good.

Suppose that two diagonals are both good. Without loss of generality, let AD and BE be the good diagonals of the hexagon $ABCDEF$ (see Figure 30.1). Then the quadrilaterals $ABCD$ and $BCDE$ both have inscribed circles.

Thus, the sums of the opposite sides of the two quadrilaterals are equal, respectively. In other words,

$$|AB| + |CD| = |BC| + |AD|,$$
$$|BC| + |DE| = |BE| + |CD|.$$

Adding up the two equalities gives

$$|AB| + |DE| = |AD| + |BE|. \tag{30.3}$$

Let M be the intersection point of AD and BE. Then from the triangle inequality,

$$|AD| + |BE| = (|AM| + |MD|) + (|BM| + |ME|)$$
$$= (|AM| + |BM|) + (|DM| + |EM|)$$
$$> |AB| + |DE|,$$

which contradicts the formula (30.3).

If a diagonal of a hexagon divides it into two quadrilaterals, then the end points of the diagonal are two opposite vertices of the hexagon. This diagonal is called a principal diagonal. Since any two principal diagonals intersect, so there are at most one good diagonal for a hexagon.

Next, there are many ways to construct a hexagon with a good diagonal. For example, as in Figure 30.2, first draw a circle, and then draw a quadrilateral $ABCD$ such that it makes the circle the inscribed one with

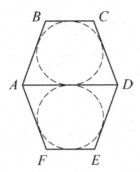

Figure 30.2   Figure for Example 3.

$AD \parallel BC$ and $|AD| > |BC|$. The symmetric points of B and C with respect to AD are F and E, respectively. Then AD is a good diagonal of the hexagon $ABCDEF$. $\Box$

Remark. A hexagon has only three principal diagonals. The situation that there is one good diagonal is obvious. The important thing that we wanted to discuss is of course "whether there exist two good diagonals." It was not difficult to obtain the equality (30.3). With Figure 30.1 at hand, the intuition told us that this equality seems not true. That is the origin of the contradiction!

Example 4. A sequence $\{p_n\}$ is defined as follows: $p_1 = 2$ and p_n is the maximal prime factor of $1 + p_1 p_2 \cdots p_{n-1}$ $(n > 1)$. Question: Does 11 appear in this sequence?

Solution. If there exists $n \in \mathbf{N}_+$ such that $p_n = 11$, then from $p_1 = 2, p_2 = 3$, and $p_3 = 7$, we know that $n \geq 4$. And furthermore, $1 + p_1 p_2 \cdots p_{n-1}, 2, 3$, and 7 are mutually prime, and the maximal prime factor of $1 + p_1 p_2 \cdots p_{n-1}$ is 11. It follows that $1 + p_1 p_2 \cdots p_{n-1}$ must be of the form

$$1 + p_1 p_2 \cdots p_{n-1} = 5^\alpha 11^\beta; \quad \alpha \geq 0, \ \beta \geq 1, \ \alpha, \beta \in \mathbf{N}. \tag{30.4}$$

Taking the module 4 operation to the both sides of (30.4) gives that $-1 \equiv (-1)^\beta \pmod 4$, so β is an odd number; taking the module 3 operation to the both sides of (30.4) ensures that $1 \equiv (-1)^{\alpha+\beta} \pmod 3$, and thus α is odd since β is odd.

Using the above conclusion and taking the module 7 operation to the both sides of (30.4), we have that

$$1 \equiv (-2)^\alpha 2^{2\beta} = -2^{\alpha+2\beta} \pmod{7}.$$

This requires that $2^{\alpha+2\beta} \equiv 6 \pmod 7$, but for any $n \in \mathbf{N}_+$, clearly $2^n \equiv 1, 2,$ or $4 \pmod 7$, which is a contradiction.

Hence, the equality (30.4) cannot be satisfied , that is, 11 does not appear in the sequence $\{p_n\}$. $\qquad\square$

Remark. When treating number theory problems using modular operations, sometimes we need to choose different modules to deduce the desired result by combining them together for the consideration. However, there is no unified method for choosing modules, so it is often done by experience and inspiration.

Example 5. Prove: The number 7 cannot be written as the sum of the squares of three rational numbers.

Proof. Suppose that one can find three rational numbers $x, y,$ and z such that

$$7 = x^2 + y^2 + z^2.$$

Viewing $x, y,$ and z as fractions, and getting rid of their denominators, we obtain a similar equality

$$7n^2 = a^2 + b^2 + c^2, \tag{30.5}$$

where $n, a, b, c,$ and d are all nonnegative integers, not all of which are 0.

If n is an even number, then there are no odd numbers among $a, b,$ and c, or there are two odd numbers among them. If there are exactly two odd numbers, then the sum of their squares has remainder 2 when divided by 4, while $7n^2$ has remainder 0 when divided by 4, a contradiction. Therefore, $a, b,$ and c are all even numbers. Then dividing the both sides of the equality (30.5) by 4 gives

$$7n_1^2 = a_1^2 + b_1^2 + c_1^2,$$

where $0 < a_1 + b_1 + c_1 < a + b + c$.

If n_1 is an even number, then one can repeat the previous step. But this situation can happen only finitely many times.

If n is an odd number, say $n = 2k + 1$, then

$$7n^2 = 7 \cdot 4k(k+1) = 8M + 7$$

for some positive integer M. Note that the remainder of a perfect square divided by 8 is $0, 1$, or 4, a contradiction.

Thus, the equality (30.5) is impossible. This shows that 7 cannot be written as the sum of the squares of three rational numbers.   □

Remark. Here, we used the infinite descending method. When dealing with problems of no solutions to indefinite equations, it is often to use the combination of proof by contradiction and the infinite descending method.

Example 6. Suppose that $a_1, a_2, \ldots, a_m$ $(m \geq 2)$ are all nonzero numbers, and for any integer k with $k = 0, 1, 2, \ldots, n$ $(n < m - 1)$,

$$a_1 + 2^k a_2 + \cdots + m^k a_m = 0.$$

Show that in the sequence $a_1, a_2, \ldots, a_m$, there exist at least $n + 1$ pairs of adjacent numbers with opposite signs.

Proof. Without loss of generality, assume that $a_m > 0$; otherwise we may multiply -1 to every number of $a_1, a_2, \ldots, a_m$. Let

$$b_i = \sum_{j=0}^{n} c_j i^j, \quad i = 1, \ldots, m,$$

where $c_0, c_1, \ldots, c_n$ are arbitrary real numbers. By the known condition,

$$\sum_{i=1}^{m} a_i b_i = \sum_{i=1}^{m} a_i \sum_{j=0}^{n} c_j i^j = \sum_{j=0}^{n} c_j \sum_{i=1}^{m} a_i i^j = 0. \tag{30.6}$$

Assume that there are k pairs of adjacent numbers with opposite signs in the sequence $a_1, a_2, \ldots, a_m$. Let $i_1, i_2, \ldots, i_k$ $(1 \leq i_1 < i_2 < \cdots < i_k \leq m)$ denote the subscripts of the first element in these number pairs.

If $k < n + 1$, then we choose

$$b_i = f(i) = (i - x_1)(i - x_2) \cdots (i - x_k),$$

where $x_l = i_l + \frac{1}{2}$ for $l = 1, 2, \ldots, k$, and the function $f(x) = (x - x_1)(x - x_2) \cdots (x - x_k)$.

The function $f(x)$ changes sign at x if and only if x is one of $x_1, x_2, \ldots, x_k$. By noting that

$$x_l = i_l + \frac{1}{2},$$

we have

$$b_i b_{i-1} < 0 \quad \Leftrightarrow \quad i = i_l, \quad l = 1, 2, \ldots, k.$$

Thus, the subscripts of the number pairs with changing sign in the sequences $a_1, a_2, \ldots, a_m$ and $b_1, b_2, \ldots, b_m$ are the same.

Also, since $a_m > 0$ and $b_m > 0$, so a_i and b_i have the same sign for $i = 1, 2, \ldots, m$, from which $a_i b_i > 0$. Therefore,

$$\sum_{i=1}^{m} a_i b_i > 0. \tag{30.7}$$

The two formulas (30.6) and (30.7) are contradictory, hence $k \geq n + 1$.

$\square$

Example 7. There are $2n$ points on a circle, which divide the circle into $2n$ equal parts. There are $n+1$ intervals of respective lengths $1, 2, \ldots, n+1$ that are placed on the circle. Suppose that the end points of such intervals are just the $2n$ points mentioned above. Prove: There must exist an interval that is in the interior of another interval completely.

Proof. Consider the interval of length 1. If it is already inside another interval, the proof is completed.

Next, we assume that there does not exist an interval containing this interval of length 1. Then we remove this interval and straighten the original circle into a line segment, thus the original proposition becomes: The n intervals

$$I_2 = [a_2, b_2], \quad \ldots, \quad I_{n+1} = [a_{n+1}, b_{n+1}]$$

with end points integers are subintervals of $[0, 2n - 1]$. Suppose that the length of the interval I_j is j. Show that there exists $i \neq j$ such that $I_i \subset I_j$.

We use proof by contradiction to do it.

If there does not exist a pair of intervals with the inclusion relation, then we define a function

$$f : \{2, 3, \ldots, n\} \rightarrow \{0, 1, \ldots, 2n - 1\} - \{a_{n+1}, a_{n+1} + 1, \ldots, b_{n+1}\},$$

$$f(j) = \begin{cases} a_j, & a_j < a_{n+1}, \\ b_j, & a_j \geq a_{n+1}. \end{cases}$$

For any j ($2 \leq j \leq n$), since there are no intervals inside I_n, if $a_j \geq a_{n+1}$, then $b_{n+1} < b_j$, hence the function f is well defined.

If $f(i) = f(j)$ for $i < j$, then I_i and I_j have the same end points. So, $I_i \subset I_j$, a contradiction.

Therefore, f is injective.

Note that the number of elements in the set

$$\{0, 1, \ldots, 2n - 1\} - \{a_{n+1}, a_{n+1} + 1, \ldots, b_{n+1}\}$$

is $2n - (n + 2) = n - 2$.

It follows that there does not exist an injective function satisfying the requirement, which contradicts the hypothesis. $\qquad\square$

Example 8. We are given an integral point triangle in a rectangular coordinates plane. Suppose that the length of one side of the integral point triangle is $\sqrt{n}$, where n has no square factors. Prove: The ratio of the radii of the circumcircle and the incircle of this triangle is an irrational number.

Proof. We use proof by contradiction to prove it. Suppose that $\frac{R}{r} = q$ is a rational number, where R and r are the radii of the circumcircle and the incircle of the integral point $\triangle ABC$, respectively.

Without loss of generality, we may assume that $|BC| = n$, and the coordinates of B, C, and A are $(0, 0), (x, y)$, and (z, t), respectively. Here, $x, y, z, t \in \mathbf{Z}$. Let the lengths of the three sides of $\triangle ABC$ be $\sqrt{u}, \sqrt{v}$, and $\sqrt{w}$. Here, $u, v, w \in \mathbf{N}_+$, and $n = u = x^2 + y^2, v = (x - z)^2 + (y - t)^2$, and $w = z^2 + t^2$. Since

$$q = \frac{R}{r} = \frac{abc}{4S} \frac{a + b + c}{2S} = \frac{abc(a + b + c)}{8S},$$

where S is the area of $\triangle ABC$, we see that S is a rational number. Thus,

$$\sqrt{uvw}(\sqrt{u} + \sqrt{v} + \sqrt{w}) = 8S^2 q \qquad (30.8)$$

is a rational number. By (30.8) we know that

$$u\sqrt{vw} + v\sqrt{uw} = 8S^2q - w\sqrt{uv}.$$

Squaring the both sides of the above equality ensures that $\sqrt{uv}$ is a rational number, so uv is a perfect square. By the same token, vw and wu are perfect squares. We write

$$u = a_1a_2^2, \quad v = b_1b_2^2, \quad w = c_1c_2^2,$$

where a_1, b_1, and c_1 are numbers without square factors, and $a_2, b_2, c_2 \in \mathbf{N}_+$. Since uv is a perfect square, we know that a_1b_1 is a perfect square. But since a_1 and b_1 are numbers without square factors, so the only possibility is that $a_1 = b_1$. Similarly, we can show that $b_1 = c_1$. Hence, $a_1 = b_1 = c_1$. Now, we may write

$$u = ma_2^2, \quad v = mb_2^2, \quad w = mc_2^2.$$

Note that $u = n$ and n is a number without square factors, from which $a_2 = 1$ and $m = n$. Consequently,

$$\begin{cases} x^2 + y^2 = m, \\ z^2 + t^2 = nb_2^2, \\ (x - z)^2 + (y - t)^2 = nc_2^2. \end{cases}$$

Using $a + b > c$ and $a + c > b$ for $\triangle ABC$, we obtain $1 + b_2 > c_2$ and $1 + c_2 > b_2$. This requires $b_2 = c_2$. It follows that

$$n = x^2 + y^2 = 2(xz + yt).$$

Denote $k = 2(xz - yt)$. Then

$$n^2 + k^2 = 4(x^2 + y^2)(z^2 + t^2) = 4n^2b_2^2.$$

So, $k^2 = n^2(4b_2^2 - 1)$, which means that $4b_2^2 - 1$ is a perfect square. This contradicts the fact that a square number $\not\equiv -1 \pmod 4$. Hence, the original proposition is true. $\qquad\square$

Remark. The definition of the so-called "a number without square factors" is: If for any $x \in \mathbf{N}_+$, the property $x^2 \mid n$ is not satisfied, then n is called a number without square factors.

3.　Exercises

1. Let $m \in \mathbf{N}_+$ be such that $m > 2$. Prove: There does not exist $n \in \mathbf{N}_+$ such that $2^m - 1 \mid 2^n + 1$.

2. Given an integer $n > 1$, if any two numbers of $1!, 2!, \ldots, n!$ have different remainders when divided by n, prove: n is a prime number.

3. Suppose that there are 12 numbers $a_1, a_2, \ldots, a_{12}$ satisfying:

$$\begin{cases} a_2(a_1 - a_2 + a_3) < 0, \\ a_3(a_2 - a_3 + a_4) < 0, \\ \cdots \\ a_{11}(a_{10} - a_{11} + a_{12}) < 0. \end{cases}$$

   Prove: One can find at least three positive numbers and three negative numbers among them.

4. Given $2n$ real numbers $a_1, a_2, \ldots, a_n$ and $b_1, b_2, \ldots, b_n$, prove: There exists an integer k $(1 \le k \le n)$ such that

$$\sum_{i=1}^{n} |a_i - a_k| \le \sum_{i=1}^{n} |b_i - a_k|.$$

5. Find the minimal prime number p (>3) such that there do not exist nonnegative integers a and b satisfying

$$\left| 3^a - 2^b \right| = p.$$

6. Let four mutually different real numbers x_1, x_2, x_3, and x_4 satisfy the inequality

$$(x_1 + x_2 + x_3 + x_4)\left(\frac{1}{x_1} + \frac{1}{x_2} + \frac{1}{x_3} + \frac{1}{x_4} \right) < 17.$$

   Prove: Choosing any three numbers from x_1, x_2, x_3, and x_4 as the side lengths can construct four different triangles.

7. A scoring rule of a round competition of 10 participants is: No ties, 1 point for the winner, and 0 point for the loser. Prove: The square sum of the scores of all participants does not exceed 285.

8. There are n points in a plane such that any straight line passing through any two of these points must pass through a third one. Prove: Such n points must be on the same straight line.

9. There are 20 members in a tennis club of some region who will have 14 competitions of singles with a rule that everyone must compete

at least once. Prove: There must be 6 singles with the 12 players all different people.

10. Suppose that a quadratic equation $ax^2 + bx + c = 0$ of positive coefficients has real roots. Prove:

$$\min\{a, b, c\} \leq \frac{1}{4}(a + b + c).$$

11. Define a function $f : (0, 1) \to (0, 1)$ by

$$f(x) = \begin{cases} x + \dfrac{1}{2}, & x < \dfrac{1}{2}, \\ x^2, & x \geq \dfrac{1}{2}. \end{cases}$$

Let real numbers a and b satisfy $0 < a < b < 1$, and define sequences $\{a_n\}$ and $\{b_n\}$ by $a_0 = a$ and $b_0 = b$, and $a_n = f(a_{n-1})$ and $b_n = f(b_{n-1})$ for any $n \geq 1$. Prove: There exists a positive integer n such that

$$(a_n - a_{n-1})(b_n - b_{n-1}) < 0.$$

12. Let $a_1, a_2, \ldots, a_n$ be real numbers satisfying $a_1 + a_2 + \cdots + a_n = \frac{1}{a_1} + \frac{1}{a_2} + \cdots + \frac{1}{a_n}$. Prove:

$$\frac{1}{n - 1 + a_1} + \frac{1}{n - 1 + a_2} + \cdots + \frac{1}{n - 1 + a_n} \geq 1.$$

Chapter 31

Construction Method

1. Key Points of Knowledge and Basic Methods

Geometric intuition helps us think about problems. If we can dig out the geometric meaning of the elements and relations in some algebraic problems and draw appropriate figures, we are able to achieve the goal of simplifying complexity and making difficulty easy via combinations of numbers and figures.

Construction is not restricted to geometric figures. Many mathematical problems can be solved by constructing and introducing proper mathematical models according to the characteristic of the problem, for example, via constructing functions, sequences, sets, drawers, special examples, counter examples, and etc.

2. Illustrative Examples

Example 1. Prove: For any positive integer n, there exist n consecutive positive integers that are not powers of prime numbers.

Analysis. This is a problem about existence. We can consider to solve it by the form of constructing a true example.

Proof. Let $a = (n+1)!$. We show that all the following n consecutive positive integers

$$a^2 + 2, \ a^2 + 3, \ \ldots, \ a^2 + (n+1)$$

are not powers of prime numbers.

First, $a^2 + k$ $(k = 2, 3, \ldots, n + 1)$ are not prime numbers because $k \mid a^2 + k$.

Second, if $a^2 + k = p^r$, where p is a prime number and $r \in \mathbf{N}_+$, then $p \mid k$ (since if k is not divisible by p, then $a^2 + k$ has a prime factor that is different from p, a contradiction). In addition, k cannot have another prime factor besides p (if $q \neq p$ is a prime factor of k, then $q \mid a^2 + k$, and so $q \mid p^r$, a contradiction). Hence, there exists $j \in \mathbf{N}_+$ such that $k = p^j$, from which $p^j \mid a^2 + k$.

But on the other hand, $p^{j+1} \nmid k$ and $p^{j+1} \nmid a^2$. This indicates that $a^2 + k = p^j$ (since $p^{j+1} \nmid a^2 + k$). As a result, it must be that $a^2 = 0$, which leads to a contradiction. This means that $a^2 + k = p^r$ cannot be satisfied.

Thus, $a^2 + k$ with $k = 2, 3, \ldots, n + 1$ are not powers of prime numbers.

$\square$

Example 2. For each given positive odd number n, prove: There exists a sequence $a_1, a_2, \ldots, a_{2017}$ of 2017 terms of positive integers satisfying $a_1 = 1$ and $a_{2017} = n$, and from the second term on, every term a_k is either equal to the sum of some previous term and a nonnegative integer power of 2, or is the minimal positive remainder of $b \pmod{c}$, where b and c are two previous terms.

Proof. We use the construction method to prove it.

If the given positive odd number $n = 1$, then we may choose the sequence $1, 2, 1, \ldots, 1, 1$, where

$$a_2 = a_1 + 2^0 = 2,$$
$$a_3 = 1 = a_1 \pmod{a_2},$$
$$a_k = 1 = a_1 \pmod{a_2} \ (k > 2).$$

Next, we consider the case $n > 1$.

Let $x = 2^t > n$ $(t \in \mathbf{Z}_+)$. Choose

$$a_1 = 1, a_2 = 1 + x = a_1 + 2^t,$$
$$a_3 = 1 + 2x = a_1 + 2^{t+1},$$
$$a_4 = x \equiv a_3 \pmod{a_2},$$
$$a_5 = (x + 1)^2 = a_3 + x^2,$$
$$a_6 = 1 + x^n = a_1 + x^n,$$
$$a_7 = n(x + 1) \equiv a_6 \pmod{a_5}.$$

(Here, we should note that if we let $x + 1 = m$, then

$$a_5 = m^2,$$
$$a_6 = x^n + 1 = (m - 1)^n + 1$$
$$\equiv (-1)^{n-1}nm + (-1)^n + 1$$
$$\equiv nm \pmod{m^2},$$

that is $a_6 \equiv \pmod{a_5}$.)

When $k \geq 7$, choose $a_k = n \equiv a_7 \pmod{a_4}$. $\qquad\square$

Remark. Here, the proof of the existence was done via the construction of an example.

Example 3. Let a, b, and c be positive numbers and $abc = 1$. Prove:

$$\left(a - 1 + \frac{1}{b}\right)\left(b - 1 + \frac{1}{c}\right)\left(c - 1 + \frac{1}{a}\right) \leq 1.$$

Analysis. The given condition can transform the inequality to be proved to the following one:

$$(a - 1 + ac)(a + 1 - ac)(ac + 1 - a) \leq a \cdot ac \cdot 1. \tag{31.1}$$

The inequality (31.1) is symmetric with respect to rotations of a, c, and 1, so without loss of generality, assume that $a \geq ac$ and $a \geq 1$.

If $a \geq ac + 1$, then it is easy to see that the left hand side of (31.1) is less than or equal to 0, so (31.1) is obviously satisfied.

If $a < ac+1$, then the expressions inside the three pairs of parentheses on the left hand side of (31.1) are all positive, so we may consider constructing a suitable figure to assist in the proof with the help of geometric intuition.

Proof. As in Figure 31.1, let $\angle C = 90°$ and CD be the median line of the side AB in $\triangle ABC$.

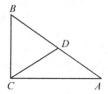

Figure 31.1 Figure for Example 3.

Assign $|BC| = \sqrt{a - 1 + ac}$ and $|AC| = \sqrt{a + 1 - ac}$. Then $|AB| = \sqrt{2a}$ and $|CD| = \frac{1}{2}\sqrt{2a}$.

From the area formula, we have $\frac{1}{2}|AC||BC| \leq \frac{1}{2}|AB||CD|$, and the equality is valid if and only if $|AC| = |BC|$. It follows that

$$\sqrt{(a + 1 - ac)(a - 1 + ac)} \leq \frac{1}{2}\sqrt{2a}\sqrt{2a} = a.$$

Similarly,

$$\sqrt{(a + 1 - ac)(ac + 1 - a)} \leq \frac{1}{2}\sqrt{2}\sqrt{2} = 1,$$

$$\sqrt{(a + 1 - a)(a - 1 + ac)} \leq \frac{1}{2}\sqrt{2ac}\sqrt{2ac} = ac.$$

Multiplying the above three inequalities gives that

$$\sqrt{(a - 1 + ac)^2(a + 1 - ac)^2(ac + 1 - a)^2} \leq a^2c,$$

in other words,

$$(a - 1 + ac)(a + 1 - ac)(ac + 1 - a) \leq a^2c,$$

which can be rewritten as

$$\left(a - 1 + \frac{1}{b}\right)\left(b - 1 + \frac{1}{c}\right)\left(c - 1 + \frac{1}{a}\right) \leq 1. \qquad \square$$

Example 4. Do there exist five consecutive positive integers such that their product is a perfect square?

Solution. There do not exist five consecutive positive integers satisfying the given property.

In fact, assume that $x_1x_2x_3x_4x_5 = y^2$, where $x_1, x_2, x_3, x_4,$ and x_5 are consecutive positive integers and $y \in \mathbf{N}_+$. Let $p \geq 5$ be a prime number and $p \mid y$. Then there is only one number among $x_1, x_2, x_3, x_4,$ and x_5 that is a multiple of p. Thus, the power index of p in the prime factor decomposition of this positive integer is an even number.

In the following, we consider the power indices of 2 and 3 in $x_1, x_2, x_3, x_4,$ and x_5. There are four cases according to the odd-even combinations.

(1) Both power indices of 2 and 3 are even (then from the previous discussion, this number is a perfect square);

(2) the power index of 2 is even while the power index of 3 is odd (then 3 times this number is a perfect square);

(3) the power index of 2 is odd while the power index of 3 is even (then 2 times this number is a perfect square);

(4) both power indices of 2 and 3 are odd (then six times this number is a perfect square).

By the drawer principle, there must be two numbers among x_1, x_2, x_3, x_4, and x_5 that belong to the same case, say that x_m and x_n belong to the same case. Then they must both be the numbers of the first case (since the difference of x_m and x_n does not exceed 4), and can only be $(x_m, x_n) = (1, 4)$, so such numbers x_1, x_2, x_3, x_4, and x_5 can only be $1, 2, 3, 4$, and 5. But then their product is 120, which is not a perfect square. This gives a contradiction.

As a consequence, there are no five consecutive positive integers satisfying the property. □

Remark. This is an example of constructing a drawer. How to construct a drawer is the key for applying the drawer principle. It is also easy to show that when $k \in \{2, 3, 4\}$, there do not exist k consecutive positive integers such that their product is a perfect square.

Example 5. Suppose that $x_1 \geq x_2 \geq x_3 \geq x_4 \geq 2$ and $x_2 + x_3 + x_4 \geq x_1$. Prove:

$$(x_1 + x_2 + x_3 + x_4)^2 \leq 4x_1 x_2 x_3 x_4.$$

Proof. Denote $a = x_2 + x_3 + x_4$ and $b = x_2 x_3 x_4$. Then the original inequality is $(x_1 + a)^2 \leq 4x_1 b$, that is

$$x_1^2 + 2(a - 2b)x_1 + a^2 \leq 0.$$

Let $f(x) = x^2 + 2(a - 2b)x + a^2$, so we need to show that $f(x_1) \leq 0$.

Since the discriminant $\Delta = 4(a - 2b)^2 - 4a^2 = 16b(b - a)$, and

$$\frac{a}{b} = \frac{x_2 + x_3 + x_4}{x_2 x_3 x_4} = \frac{1}{x_3 x_4} + \frac{1}{x_2 x_4} + \frac{1}{x_2 x_3}$$

$$\leq \frac{1}{4} + \frac{1}{4} + \frac{1}{4} = \frac{3}{4} < 1,$$

so $b > a$, from which $\Delta > 0$. Thus, the graph of $f(x)$ has two distinct intersection points with the x-axis. The two intersection points are easily found to be

$$u = 2b - a - 2\sqrt{b(b - a)}, \quad v = 2b - a + 2\sqrt{b(b - a)}.$$

We prove below that $x_1 \in [u, v]$. The inequalities $a \le 3x_1 \le 3a$ imply that $x_1 \in \left[\frac{a}{3}, a\right]$, so it is sufficient to show that $\left[\frac{a}{3}, a\right] \subset [u, v]$, namely $u \le \frac{a}{3}$ and $a \le v$. Since

$$v = 2b - a + 2\sqrt{b(b - a)} > 2b - a > a.$$
$$u = 2b - a - 2\sqrt{b(b - a)} = (\sqrt{b} - \sqrt{b - a})^2$$
$$= \left(\frac{a}{\sqrt{b} + \sqrt{b - a}}\right)^2 = a \Big/ \left(\sqrt{\frac{b}{a}} + \sqrt{\frac{b}{a} - 1}\right)^2$$
$$\le a \Big/ \left(\sqrt{\frac{4}{3}} + \sqrt{\frac{1}{3}}\right)^2 = \frac{a}{3},$$

it follows that $x_1 \in [u, v]$, which implies that $f(x_1) \le 0$. $\square$

Remark. By constructing a quadratic function and then using properties of quadratic functions to prove some inequalities, one can simplify the given problem.

Example 6. Suppose that every vertex of a regular pentagon corresponds to an integer in such a way that the sum of the five integers is positive. If the integers corresponding to three adjacent vertices among them are $x, y,$ and z with the middle $y < 0$, then do the following adjustment: the integers $x, y,$ and z will be changed to $x + y, -y,$ and $y + z$, respectively. If there is still one negative number among the resulting five integers, then continue this transformation. Question: Can such transformations be terminated necessarily after finitely many times?

Solution. The answer to the question is positive, in other words, such transformations must be terminated after finitely many steps.

For the sake of convenience, we write the five integers, which are arranged with a circular shape like the vertices of the regular pentagon, as

$$v, \ w, \ x, \ y, \ z,$$

where the numbers v and z are adjacent in the circular arrangement. According to the condition of the problem,

$$v + w + x + y + z > 0.$$

Without loss of generality, assume that $y < 0$. The circular sequence after the transformation becomes

$$v, \ w, \ x + y, \ -y, \ y + z.$$

Construct the function

$$f(x_1, x_2, x_3, x_4, x_5) = x_1^2 + x_2^2 + x_3^2 + x_4^2 + x_5^2 + (x_1 + x_2)^2$$
$$+ (x_2 + x_3)^2 + (x_3 + x_4)^2$$
$$+ (x_4 + x_5)^2 + (x_5 + x_1)^2.$$

Then the difference of the function values after the transformation is

$$f(v, w, x + y, -y, y + z) - f(v, w, x, y, z)$$
$$= \left[v^2 + w^2 + (x + y)^2 + (-y)^2 + (y + z)^2 + (v + w)^2 \right.$$
$$+ (w + x + y)^2 + x^2 + z^2 + (y + z + v)^2 \left. \right] - \left[v^2 + w^2 \right.$$
$$+ x^2 + y^2 + z^2 + (v + w)^2 + (w + x)^2 + (x + y)^2$$
$$+ (y + z)^2 (z + v)^2 \left. \right]$$
$$= 2y(v + w + x + y + z) < 0.$$

Since the function value of f is a nonnegative integer when the independent variables take integer values, so

$$f(v, w, x + y, -y, y + z) \le f(v, w, x, y, z) - 1.$$

In other words, every time after a transformation, the value of f is decreased at least by 1, Therefore, after finitely many times of transformations, no more transformations will be needed. $\qquad \square$

Remark. The current problem is a test problem of the 27th International Mathematics Olympiad, and was the one with the lowest score obtained by the participants. But the American participant Joseph Kerry received a Special Prize. The preliminary function that he constructed is

$$f(x_1, x_2, x_3, x_4, x_5) = \sum_{i=1}^{5} |x_i| + \sum_{i=1}^{5} |x_i + x_{i+1}|$$

$$+ \sum_{i=1}^{5} |x_i + x_{i+1} + x_{i+2}|$$

$$+ \sum_{i=1}^{5} |x_i + x_{i+1} + x_{i+2} + x_{i+3}|,$$

where $x_6 = x_1, x_7 = x_2$, and $x_8 = x_3$.

The interested reader can verify the effectiveness of the function for solving this problem.

Example 7. Let $S = \{1, 2, \ldots, 2000\}$. Prove: It must be possible to dye all the numbers of S with four different colors, such that any seven numbers that constitute an arithmetic sequence have different colors.

Proof. Since $2000 < 6 \cdot 7^3 = 2058$, so each number of $1, 2, \ldots, 2000$ can always be represented by a four-digit number $(dcba)_7$ (allowing generalized four-digit numbers with first digit 0) in the hepta-number system (namely with only $0, 1, 2, 3, 4, 5$, and 6 as digits of numbers). Construct the set

$$A_i = \{x \in S \mid x = (dcba)_7, a \neq i, b \neq i, c \neq i\},$$

where $i = 1, 2, 3$, and 4. We show below that there do not exist seven numbers in A_i ($1 \leq i \leq 4$) that form an arithmetic sequence. For this purpose, we first show that any seven numbers in the set

$$A = \{x \in S \mid x = (dcba)_7, a \neq 0, b \neq 0, c \neq 0\}$$

cannot constitute an arithmetic sequence.

If otherwise, let $x_1, x_2, \ldots, x_7$ be seven numbers in A that can form an arithmetic sequence. Assume that its common difference is $t > 0$. Since the last digits of these seven numbers are not 0, so none of them can be divided by 7 (since each number is represented in the hepta-number system). Thus, there exist $i, j \in \{1, 2, \ldots, 7\}$ with $i < j$ such that $7 \mid x_j - x_i$, namely

$7 \mid (j - i)t$. This means that $7 \mid t$. Let $t = 7k$, and let

$$x_1 = d_1 7^3 + c_1 7^2 + b_1 7 + a_1.$$

Then $x_i = d_1 7^3 + c_1 7^2 + b_1 7 + a_1 + 7(i-1)k$ for $i = 2, 3, \ldots, 7$.

If $7 \le t < 49$, then $(7, k) = 1$, so the remainders of $k, 2k, \ldots, 6k$ divided by 7 are one of $1, 2, \ldots, 6$, respectively. Thus, there exists $n \in \{1, 2, \ldots, 6\}$ such that

$$7 \mid b_1 + nk.$$

This indicates that the second digit from the right to the left of the representation of x_{n+1} in the system is 0, which contradicts the fact $x_{n+1} \notin A$.

If $49 \le t < 343$, then the same argument can show that there exists x_{n+1} with 0 as its third digit from the right to the left, a contradiction.

If $t \ge 343$, then $x_7 \ge 6 \cdot 343 > 2000$, also a contradiction.

This has shown that any seven numbers in A cannot constitute an arithmetic sequence. In order to show that any seven numbers in A_i ($1 \le i \le 4$) cannot constitute an arithmetic sequence, it is enough to note that when $\{x_1, x_2, \ldots, x_7\} \subset A_i$, if we let

$$y_j = x_j + 7^3 - (7^2 i + 7i + i), \quad j = 1, 2, \ldots, 7,$$

then $y_j \in A$ for $j = 1, 2, \ldots, 7$, so it is reduced to the case with A.

Lastly, we show that $A_1 \cup A_2 \cup A_3 \cup A_4 = S$.

Clearly, $A_1 \cup A_2 \cup A_3 \cup A_4 \subset S$. On the other hand, if $x \in S$ but $x \notin A_1 \cup A_2 \cup A_3 \cup A_4$, then the last three digits of the representation of x in the hepta-number system must be $1, 2, 3$, and 4 at the same time, which is impossible.

Hence, we may dye the numbers in A_1, A_2, A_3, A_4 with one of the four different colors respectively, satisfying the prescribed property of the problem. $\qquad\square$

Example 8. Let p be an odd prime number, and $a_1, a_2, \ldots a_{p-2}$ be positive integers (may be equal) such that $p \nmid a_k$ and $p \nmid a_k^k - 1$ for any $k \in \{1, 2, \ldots, p-2\}$. Prove: It is possible to choose several numbers from $a_1, a_2, \ldots a_{p-2}$ such that their product has a remainder 2 when divided by p.

Proof. We first construct $p - 2$ sequences:

$$\{b_{k,i}\}_{i=1}^{k} \quad (k = 2, 3, \ldots, p-1).$$

These $p - 2$ sequences satisfy:

(1) for all $1 \leq i \leq k$, the term $b_{k,i}$ equals either 1 or the product of some numbers in $a_1, a_2, \ldots a_{k-1}$;
(2) for all $1 \leq i < j \leq k$, it is always true that $b_{k,i} \not\equiv b_{k,j} \pmod{p}$.

Clearly, when $k = 2$, the group of numbers $b_{2,1} = 1$ and $b_{2,2} = a_1$ satisfies the above conditions.

Suppose that for some $k \geq 2$, the sequence $\{b_{k,i}\}_{i=1}^{k}$ satisfies the above two conditions.

Since $p \nmid a_k$, so for any $1 \leq i < j \leq k$, there holds that

$$a_k b_{k,i} \not\equiv a_k b_{k,j} \pmod{p}.$$

Also from $p \nmid a_k^k - 1$, we have that

$$\prod_{i=1}^{k} a_k b_{k,i} = a_k^k \prod_{i=1}^{k} b_{k,i} \not\equiv \prod_{i=1}^{k} b_{k,i} \pmod{p}.$$

Hence, the sets $\{b_{k,i} \pmod{p}\}_{i=1}^{k}$ and $\{a_k b_{k,i} \pmod{p}\}_{i=1}^{k}$ are different, so there must exist some i_0 ($1 \leq i_0 \leq k$) such that $a_k b_{k,i_0} \notin \{b_{k,i} \pmod{p}\}_{i=1}^{k}$.

Let $b_{k+1,i} = b_{k,i}$ ($i = 1, 2, \ldots, k$) and $b_{k+1,k+1} = a_k b_{k,i_0}$.

Then the sequence $\{b_{k+1,i}\}_{i=1}^{k+1}$ satisfies the above two conditions.

It is easy to see from the above construction that

$$\{b_{p-1,i} \pmod{p}\}_{i=1}^{p-1} = \{1, 2, \ldots, p-1\}.$$

It follows that there exists some $b_{p-1,i}$ such that its remainder is 2 when divided by p, and it is also the product of some numbers in $a_1, a_2, \ldots a_{p-2}$. $\square$

3. Exercises

Group A

I. Filling Problems

1. Assume that a, b, and c constitute a geometric sequence, and $a, \frac{b(b-1)}{2}$, and c constitute an arithmetic sequence. When $1 < a < 3 < c < 7$, the range of the values for c is _____.

2. Suppose that positive real numbers $x, y,$ and z satisfy

$$\begin{cases} x^2 + \dfrac{y^2}{3} = 4, \\[2mm] \dfrac{y^2}{3} + yz + z^2 = 4, \\[2mm] x^2 + z^2 + xz = 4. \end{cases}$$

   Then $2xy + yz + 3xz = $ _____ .

3. If a function is $f(x, y) = \sqrt{x^2 + y^2 - 6y + 9} + \sqrt{x^2 + y^2 + 2\sqrt{3}x + 3} + \sqrt{x^2 + y^2 - 2\sqrt{3}x + 3}$, then the minimum value of $f(x, y)$ is _____ .

II. Calculation Problems

4. Prove: There are infinitely many triples (a, b, c) of positive integers such that $a^2 + b^2, b^2 + c^2,$ and $c^2 + a^2$ are all perfect squares.
5. Does there exist a quadratic polynomial $P(x, y)$ of two variables such that for each nonnegative integer n, there is one and only one pair (k, m) of nonnegative integers satisfying $P(k, m) = n$?
6. Let $S = \{1, 2, \ldots, 280\}$. Find the minimal positive integer n such that every subset of S with n elements contains five mutually prime numbers.

Group B

7. Let $n \in \mathbf{N}_+$ and $x \in \mathbf{R}$. Prove:

$$[x] + \left[x + \frac{1}{n}\right] + \left[x + \frac{2}{n}\right] + \cdots + \left[x + \frac{n-1}{n}\right] = [nx].$$

8. A sequence $\{a_n\}$ is defined as follows: $a_n = [\sqrt{2}n]$ for $n = 0, 1, 2, \ldots$. Prove: The sequence contains infinitely many perfect squares.
9. Let $x_1, x_2, x_3, y_1, y_2, y_3 \in \mathbf{R}$ satisfy $x_1^2 + x_2^2 + x_3^2 \leq 1$. Prove:

$$(x_1 y_1 + x_2 y_2 + x_3 y_3 - 1)^2 \geq (x_1^2 + x_2^2 + x_3^2 - 1)(y_1^2 + y_2^2 + y_3^2 - 1).$$

10. Let a and b be given positive integers such that $(a, b) = 1$. Prove: In the arithmetic sequence $\{an + b\}$, there is an infinite subsequence such that its any two numbers are mutually prime.
11. Let $a_1, a_2, \ldots, a_{100}, b_1, b_2, \ldots, b_{100}$ be different real numbers. Fill them into a $100 \cdot 100$-check table according to the following rule: The number $a_i + b_j$ is placed in the check of row i and column j. If the product of

the numbers in every column is always equal to 1, prove: The product of the numbers in every row is always equal to -1,

12. Let $A \subset \mathbf{N}_+$ be an infinite set, in which each number a is the product of at most 1990 prime numbers. Prove: There must be an infinite subset B of A such that the greatest common factors of any two different numbers in B are the same.

Solutions

Solutions for Chapter 1

Maximum and Minimum Values

1. $\sqrt{2}$. By $\left(\frac{\sqrt{x}}{\sqrt{x+y}}\right)^2 + \left(\frac{\sqrt{y}}{\sqrt{x+y}}\right)^2 = 1$, we get $\frac{\sqrt{x}}{\sqrt{x+y}} + \frac{\sqrt{y}}{\sqrt{x+y}} \le \sqrt{2}$, and the equality holds when $x = y$.

2. 2. It is from
$$(x + y)(x + z) = yz + x(x + y + z)$$
$$\ge 2\sqrt{yzx(x + y + z)} = 2.$$

3. $\frac{25}{4}$. We have that $\left(a + \frac{1}{a}\right)\left(b + \frac{1}{b}\right) = ab + \frac{1}{ab} + \frac{a}{b} + \frac{b}{a} \ge ab + \frac{1}{ab} + 2 = ab + \frac{1}{16ab} + \frac{15}{16ab} + 2 \ge \frac{5}{2} + \frac{15}{4(a+b)^2} = \frac{25}{4}$.

4. $\frac{3}{2}$. It is easy to see that $t(1 - t) \le \frac{1}{4}$ for any real number t, and the equality is true if and only if $t = \frac{1}{2}$. Now, $M = x(1 - x) + 2y(1 - y) + 3z(1 - z) \le \frac{1}{4} + 2 \cdot \frac{1}{4} + 3 \cdot \frac{1}{4} = \frac{3}{2}$, and the equality holds if and only if $x = y = z = \frac{1}{2}$ and $w = -\frac{1}{2}$.

5. $\frac{13}{3}$. From $x + y + z = 5$ we get $(x + y)^2 = (5 - z)^2$, and from $xy + yz + zx = 3$, we have $xy = 3 - z(5 - z)$. Consequently,
$$(x - y)^2 = (x + y)^2 - 4xy = (5 - z)^2 - 4[3 - z(5 - z)]$$
$$= (13 - 3z)(1 + z) \ge 0.$$

Solving the above gives $-1 \le z \le \frac{13}{3}$, and $z = \frac{13}{3}$ when $x = y = \frac{1}{3}$.

6. $9 + 3\sqrt{15}$. Since
$$(x + y)^2 = 9\left(\sqrt{y + 2} + \sqrt{x + 1}\right)^2 \le 18[(y + 2) + (x + 1)],$$

we obtain $(x+y)^2 - 18(x+y) - 54 \leq 0$, from which $x + y \leq 9 + 3\sqrt{15}$, and the equality holds if and only if $x = 5 + \frac{3\sqrt{15}}{2}$ and $y = 4 + \frac{3\sqrt{15}}{2}$.

7. 0.   It is easy to see that $0 < a, b, c < 1$ and $a + b + c = 1$. Now, $\left(\frac{1}{a} + \frac{1}{b} + \frac{1}{c}\right) = (a + b + c)\left(\frac{1}{a} + \frac{1}{b} + \frac{1}{c}\right) \geq 9$ (the equality holds when $a = b = c = \frac{1}{3}$), namely $9abc \leq ab + bc + ca = \frac{1}{2}\left[(a+b+c)^2 - (a^2+b^2+c^2)\right] = \frac{1}{2}\left[1 - (a^2+b^2+c^2)\right]$. In other words,

$$a^2 + b^2 + c^2 + 18abc \leq 1.$$

8. Consider the three inequalities: (i) $x \geq S$; (ii) $y + \frac{1}{x} \geq S$; (iii) $\frac{1}{y} \geq S$. They are all satisfied and at least one equality is valid, too. Our purpose is to find the maximum value of S satisfying all the above conditions. Since $x, y \in \mathbf{R}_+$, so $S > 0$. By (iii), $y \leq \frac{1}{S}$. Also (i) implies that $\frac{1}{x} \leq \frac{1}{S}$, and (ii) ensures that

$$S \leq \frac{1}{x} + y \leq \frac{2}{S}.$$

In other words, $S^2 \leq 2$, so $S \leq \sqrt{2}$.

On the other hand, when $x = \sqrt{2}$ and $y = \frac{\sqrt{2}}{2}$, we have $S = \sqrt{2}$. Therefore, $S_{\max} = \sqrt{2}$.

9. Since $a = x_1 + x_2 + x_3 \geq 3\sqrt[3]{x_1 x_2 x_3} = 3\sqrt[3]{a}$, so $a \geq 3\sqrt{3}$. From

$$3(x_1 x_2 + x_2 x_3 + x_3 x_1) \leq (x_1 + x_2 + x_3)^2,$$

we have $3b \leq a$. Hence,

$$p = \frac{a^2 + 6b + 1}{a^2 + a} \leq \frac{a^2 + 2a + 1}{a^2 + a} = 1 + \frac{1}{a}$$

$$\leq 1 + \frac{1}{3\sqrt{3}} = \frac{9 + \sqrt{3}}{9},$$

and the equality is satisfied when $x_1 = x_2 = x_3 = \sqrt{3}$, namely $a = 3\sqrt{3}$ and $b = \sqrt{3}$. Hence,

$$p_{\max} = \frac{9 + \sqrt{3}}{9}.$$

10. We calculate

$$Q - R = \frac{x(y^2 - x^2) + y(z^2 - y^2) + z(x^2 - z^2)}{(x + y)(y + z)(z + x)}$$

$$= \frac{(x^2 - y^2)(z - x) - (y + z)(y - z)^2}{(x + y)(y + z)(z + x)} \tag{1.1}$$

$$= \frac{(y^2 - z^2)(x - y) - (x + z)(x - z)^2}{(x + y)(y + z)(z + x)} \tag{1.2}$$

$$= \frac{(z^2 - x^2)(y - z) - (x + y)(x - y)^2}{(x + y)(y + z)(z + x)}. \tag{1.3}$$

In the three numbers x, y, and z, if x is the maximum, then $Q \le R$ by (1.1); if y is the maximum, then $Q \le R$ by (1.2); if z is the maximum, then $Q \le R$ by (1.3). Hence, $Q \le R$ for any $x, y, z \in \mathbf{R}_+$. If we interchange x and z, then Q becomes P while R does not change, so $P \le R$. Thus, $f(x, y, z) = \max\{P, Q, R\} = R$. Now,

$$2R = \frac{2z}{x + y} + \frac{2x}{y + z} + \frac{2y}{z + x}$$

$$= \left(\frac{z + x}{x + y} + \frac{x + y}{y + z} + \frac{y + z}{z + x}\right) + \left(\frac{z + y}{x + y} + \frac{x + z}{y + z} + \frac{y + x}{z + x}\right) - 3$$

$$\ge 3\sqrt[3]{\frac{z + x}{x + y} \cdot \frac{x + y}{y + z} \cdot \frac{y + z}{z + x}} + 3\sqrt[3]{\frac{z + y}{x + y} \cdot \frac{x + z}{y + z} \cdot \frac{y + x}{z + x}} - 3 = 3,$$

where the equality holds if and only if $x = y = z$. So, $f(x, y, z)_{\min} = \frac{3}{2}$.

11. For $k = 1, 2, \ldots, 1990$, we have

$$|y_k - y_{k+1}| = \left|\frac{1}{k}(x_1 + x_2 + \cdots + x_k) - \frac{1}{k+1}(x_1 + x_2 + \cdots + x_{k+1})\right|$$

$$= \left|\frac{1}{k(k + 1)}(x_1 + x_2 + \cdots + x_k - kx_{k+1})\right|$$

$$\le \frac{1}{k(k + 1)}(|x_1 - x_2| + 2|x_2 - x_3| + \cdots + k|x_k - x_{k+1}|).$$

Thus,

$$|y_1 - y_2| + |y_2 - y_3| + \cdots + |y_{1990} - y_{1991}|$$

$$\leq |x_1 - x_2| \left(\frac{1}{1 \cdot 2} + \frac{1}{2 \cdot 3} + \cdots + \frac{1}{1990 \cdot 1991} \right)$$

$$+ 2|x_2 - x_3| \left(\frac{1}{2 \cdot 3} + \cdots + \frac{1}{1990 \cdot 1991} \right)$$

$$+ \cdots + 1990|x_{1990} - x_{1991}| \frac{1}{1990 \cdot 1991}$$

$$= |x_1 - x_2| \left(1 - \frac{1}{1991} \right) + |x_2 - x_3| \left(1 - \frac{2}{1991} \right)$$

$$+ \cdots + |x_{1990} - x_{1991}| \left(1 - \frac{1990}{1991} \right)$$

$$\leq 1991 \left(1 - \frac{1}{1991} \right) = 1990,$$

and the equality can be obtained when $x_1 = 1991$ and $x_2 = \cdots = x_{1991} = 0$. Hence, the maximum value is 1990.

12. Note that

$$\frac{a - bc}{a + bc} = \frac{a + bc - 2bc}{a + bc} = 1 - 2 \cdot \frac{bc}{a + bc}.$$

From $a + b + c \geq 1$, we obtain

$$\frac{bc}{a + bc} \geq \frac{bc}{a(a + b + c) + bc} = \frac{bc}{(a + b)(c + a)}.$$

It follows that

$$\frac{a - bc}{a + bc} + \frac{b - ca}{b + ca} + \frac{c - ab}{c + ab}$$

$$\leq 3 - \frac{2bc}{(a + b)(c + a)} - \frac{2ca}{(b + c)(a + b)} - \frac{2ab}{(c + a)(b + c)}.$$

One can show that

$$\frac{2bc}{(a + b)(c + a)} + \frac{2ca}{(b + c)(a + b)} + \frac{2ab}{(c + a)(b + c)} \geq \frac{3}{2}.$$

In fact, multiplying the above by $2(a + b)(b + c)(c + a)$ gives

$$4[bc(b + c) + ca(c + a) + ab(a + b)] \geq 3(a + b)(b + c)(c + a),$$

namely

$$(b^2c + bc^2) + (c^2a + ca^2) + (a^2b + ab^2) \geq 6abc.$$

But the last inequality is obviously true from the arithmetic mean-geometric mean inequality.

It is easy to verify that when $a = b = c = \frac{1}{3}$, the above inequality becomes an equality, so the maximum value is $\frac{3}{2}$.

Solutions for Chapter 2

Usual Methods for Proving Inequalities

1. $\frac{25}{2}$.   The original expression

$$= (a^2 + b^2)\left(1 + \frac{1}{a^2 b^2}\right) + 4$$

$$\geq \frac{1}{2}(a+b)^2\left[1 + \left(\frac{2}{a+b}\right)^4\right] + 4 = \frac{25}{2}.$$

2. $(x, y) = (0, 0), (0, 1),$ and $(1, 0)$.   Note that

$$5x(1 - x) + 5y(1 - y) - 8xy = 5(x + y) - 5(x^2 + y^2) - 8xy$$

$$= 5(x + y) - 5(x + y)^2 + 2xy$$

$$= 5(x + y)[1 - (x + y)] + 2xy$$

$$\geq 0.$$

So, $5x(1 - x) + 5y(1 - y) \geq 8xy$, and the equality is valid only if $(x, y) = (0, 0), (0, 1),$ and $(1, 0)$.

3. $\frac{n}{\sqrt[3]{2}}$.   For $i = 1, 2, \ldots, n$,

$$\sqrt[6]{a_i(1 - a_{i+1})} = 2^{\frac{4}{6}} \sqrt[6]{a_i(1 - a_{i+1})\frac{1}{2}\frac{1}{2}\frac{1}{2}\frac{1}{2}}$$

$$\leq 2^{\frac{2}{3}} \cdot \frac{1}{6}(a_i + 1 - a_{i+1} + 2)$$

$$= 2^{\frac{2}{3}} \cdot \frac{1}{6}(a_i - a_{i+1} + 3).$$

413

Thus,

$$\sum_{i=1}^{n} \sqrt[6]{a_i(1 - a_{i+1})} \leq 2^{\frac{2}{3}} \cdot \frac{1}{6} \sum_{i=1}^{n} (a_i - a_{i+1} + 3)$$

$$= 2^{\frac{2}{3}} \cdot \frac{n}{2} = \frac{n}{\sqrt[3]{2}},$$

where the equality holds when $a_1 = a_2 = \cdots = a_n = \frac{1}{2}$.

4. It follows from $a^3 + b^3 + c^3 + \frac{1}{a} + \frac{1}{b} + \frac{1}{c} \geq 2a + 2b + 2c \geq (a + b + c) + \left(\frac{1}{a} + \frac{1}{b} + \frac{1}{c}\right)$ that $a^3 + b^3 + c^3 \geq a + b + c$.

5. Since $x > \sqrt{2}$ and $y > \sqrt{2}$, we have $\frac{(x^2+y^2)^2}{4} \geq x^2 y^2$, $\frac{(x^2+y^2)^2}{2} \geq (x^2 + y^2)xy$, and $\frac{(x^2+y^2)^2}{4} > x^2 + y^2$. Summing up the three inequalities gives $(x^2 + y^2)^2 > (x^2 + y^2)(1 + xy) + x^2 y^2$, expansion of which proves the inequality after a simplification.

6. It is from

$$\frac{1}{1 + 2ab} + \frac{1}{1 + 2bc} + \frac{1}{1 + 2ca}$$

$$\geq \frac{1}{1 + a^2 + b^2} + \frac{1}{1 + b^2 + c^2} + \frac{1}{1 + c^2 + a^2}$$

$$\geq 3 \cdot \frac{3}{(1 + a^2 + b^2) + (1 + b^2 + c^2) + (1 + c^2 + a^2)} = 1.$$

7. Since $x > 0, y > 0$, and $n \in \mathbf{N}_+$, so $(x^{n-1}y + xy^{n-1}) - (x^n + y^n) = -(x - y)(x^{n-1} - y^{n-1}) \leq 0$, that is, $x^{n-1}y + xy^{n-1} \leq x^n + y^n$. Also, from $(1+x^2)(1+y^2) = 1+x^2+y^2+x^2y^2 \geq 1+2xy+x^2y^2 = (1+xy)^2$, we obtain

$$\frac{x^n}{1 + x^2} + \frac{y^n}{1 + y^2}$$

$$= \frac{x^n(1 + y^2) + y^n(1 + x^2)}{(1 + x^2)(1 + y^2)} = \frac{x^n + y^n + xy(x^{n-1}y + xy^{n-1})}{(1 + x^2)(1 + y^2)}$$

$$\leq \frac{x^n + y^n + xy(x^n + y^n)}{(1 + xy)^2} = \frac{x^n + y^n}{1 + xy}.$$

8. Let z be the number in x, y, and z with the maximum absolute value. Then $-1 \leq \frac{x}{z} \leq 1$ and $-1 \leq \frac{y}{z} \leq 1$ $(z \neq 0)$. The inequality is obviously satisfied for $z = 0$. When $z \neq 0$, the original inequality $\Leftrightarrow \left|\frac{x}{z}\right| + \left|\frac{y}{z}\right| + 1 - \left|\frac{x}{z} + \frac{y}{z}\right| - \left|\frac{y}{z} + 1\right| - \left|1 + \frac{y}{z}\right| + \left|\frac{x}{z} + \frac{y}{z} + 1\right| \geq 0 \Leftrightarrow \left|\frac{x}{z}\right| + \left|\frac{y}{z}\right| - \left|\frac{x}{z} + \frac{y}{z}\right| - \left(\frac{x}{z} + \frac{y}{z} + 1\right) + \left|\frac{x}{z} + \frac{y}{z} + 1\right| \geq 0$. Since $\left|\frac{x}{z} + \frac{y}{z}\right| \leq \left|\frac{x}{z}\right| + \left|\frac{y}{z}\right|$, the original inequality is satisfied.

9. Let $2a - \frac{1}{b}, 2b - \frac{1}{b}$, and $2c - \frac{1}{a}$ be all greater than 1. Since at least one of a, b, and c is positive, say $a > 0$, so $2c > 2c - \frac{1}{a} > 1$, from which $c > 0$. By the same token, $b > 0$. From $2b - \frac{1}{b} > 1$ we get $b > \frac{1}{2}(1 + \frac{1}{c})$. Also $\frac{2}{bc} - \frac{1}{b} > 1$ from $2a - \frac{1}{b} > 1$, namely $b < \frac{2}{c} - 1$. Hence, $\frac{1}{2}(1 + \frac{1}{c}) < b < \frac{2}{c} - 1$. Consequently, $c < 1$. Similarly, $a < 1$ and $b < 1$. That contradicts $abc = 1$.

10. The inequality is from $\sum_{k=1}^{n} \frac{a_k^2}{a_k + b_k} = \sum_{k=1}^{n} a_k - \sum_{k=1}^{n} \frac{a_k b_k}{a_k + b_k}$ and $\sum_{k=1}^{n} \frac{a_k b_k}{a_k + b_k} \leq \frac{1}{4} \sum_{k=1}^{n} \frac{(a_k + b_k)^2}{a_k + b_k} = \frac{1}{4} \sum_{k=1}^{n} (a_k + b_k) = \frac{1}{2} \sum_{k=1}^{n} a_k$.

11. Without loss of generality, assume that $a \geq b \geq c$. Then

$$\sqrt{a + b - c} - \sqrt{a} = \frac{(a + b - c) - a}{\sqrt{a + b - c} + \sqrt{a}} \leq \frac{b - c}{\sqrt{b} + \sqrt{c}} = \sqrt{b} - \sqrt{c}.$$

It follows that

$$\frac{\sqrt{a + b - c}}{\sqrt{a} + \sqrt{b} - \sqrt{c}} \leq 1. \tag{2.1}$$

Denote $p = \sqrt{a} + \sqrt{b}$ and $q = \sqrt{a} - \sqrt{b}$. Then $pq = a - b$ and $p \geq 2\sqrt{c}$. The Cauchy inequality implies that

$$\left(\frac{\sqrt{b + c - a}}{\sqrt{b} + \sqrt{c} - \sqrt{a}} + \frac{\sqrt{c + a - b}}{\sqrt{c} + \sqrt{a} - \sqrt{b}} \right)^2$$

$$= \left(\frac{\sqrt{c - pq}}{\sqrt{c} - q} + \frac{\sqrt{c + pq}}{\sqrt{c} + q} \right)^2$$

$$\leq \left(\frac{c - pq}{\sqrt{c} - q} + \frac{c + pq}{\sqrt{c} + q} \right) \left(\frac{1}{\sqrt{c} - q} + \frac{1}{\sqrt{c} + q} \right)$$

$$= \frac{2(c\sqrt{c} - pq^2)}{c - q^2} \cdot \frac{2\sqrt{c}}{c - q^2}$$

$$= 4 \cdot \frac{c^2 - \sqrt{c}pq^2}{(c - q^2)^2} \leq 4 \cdot \frac{c^2 - 2cq^2}{(c - q^2)^2} \leq 4.$$

Consequently,

$$\frac{\sqrt{b + c - a}}{\sqrt{b} + \sqrt{c} - \sqrt{a}} + \frac{\sqrt{c + a - b}}{\sqrt{c} + \sqrt{a} - \sqrt{b}} \leq 2. \tag{2.2}$$

Combining (2.1) and (2.2) proves the inequality.

12. If at least one of x, y, and z is not positive, say $z \leq 0$, then, since $x^2 + y^2 + z^2 = 2$, we have $x + y \leq \sqrt{2(x^2 + y^2)} \leq 2$ and $xy \leq \frac{1}{2}(x^2 + y^2) \leq 1$, from which

$$x + y + z - 2 - xyz = (x + y - z) + z(1 - xy) \leq 0.$$

Now, suppose that x, y, and z are all positive. Without loss of generality, assume that $x \leq y \leq z$. If $0 < z \leq 1$, then $2 + xyz - x - y - z = (1 - x)(1 - y) + (1 - z)(1 - xy) \geq 0$. If $z > 1$, then

$$x + y + z \leq \sqrt{2[z^2 + (x + y)^2]} = 2\sqrt{1 + xy} \leq 2 + xy < 2 + xyz.$$

Solutions for Chapter 3

Common Techniques for Proving Inequalities

1. $[3 - \sqrt{5}, 3 + \sqrt{5}]$.   From the condition, $(x - y)^2 + (2y)^2 = 4$, so by letting $x - y = 2\cos\theta$ and $y = \sin\theta$, we get $x = \sin\theta + 2\cos\theta$, $y = \sin\theta$. Thus,

$$x^2 + y^2 = (\sin\theta + 2\cos\theta)^2 + \sin^2\theta = 3 + \sqrt{5}\sin(\theta + \phi).$$

   Therefore, the range of the values for $x^2 + y^2$ is $[3 - \sqrt{5}, 3 + \sqrt{5}]$.

2. $-\frac{16}{9}$.   Let $r = x^2 + y^2$. Then $xy = 1 - x$. From $x^2 + y^2 \geq 2|xy|$ we obtain $\frac{2}{3} \leq r \leq 2$. Then we can find the maximum and minimum values of $F(x, y) = r(1 - r)$ easily.

3. $\frac{9}{16}$.   Let $x = b^t - 2$ and $y = a^t - 1$. Then $x > 0, y > 0$, and $1 \leq \frac{x}{y} \leq 2$. Thus,

$$\frac{b^t - 2}{a^{2t}} + \frac{a^t - 1}{b^{2t}} = \frac{x}{(y+1)^2} + \frac{y}{(x+2)^2} \leq \frac{x}{4y} + \frac{y}{8x}.$$

   Denote $p = \frac{x}{y}$. Then $f(p) = \frac{p}{4} + \frac{1}{8p}$ is a monotonically increasing function on $[1, 2]$. Hence,

$$f(p) \leq f(2) = \frac{9}{16}.$$

4. $-17 + 12\sqrt{2}$.   Let $a + 2b + c = x$, $a + b + 2c = y$, and $a + b + 3c = z$. Then $a + 3c = 2y - x$, $b = z + x - 2y$, and $c = z - y$. So,

$$\frac{a + 3c}{a + 2b + c} + \frac{4b}{a + b + 2c} - \frac{8c}{a + b + 3c}$$

$$= \frac{2y - x}{x} + \frac{4(z + x - 2y)}{y} + \frac{8(z - y)}{z}$$

$$= -17 + 2 \cdot \frac{y}{x} + 4 \cdot \frac{x}{y} + 4 \cdot \frac{z}{y} + 8 \cdot \frac{y}{z}$$

$$\geq -17 + 2\sqrt{8} + 2\sqrt{32} = -17 + 12\sqrt{2},$$

and the equality is satisfied when $a = 3 - 2\sqrt{2}, b = \sqrt{2} - 1$, and $c = \sqrt{2}$. Hence, the minimum value is $-17 + 12\sqrt{2}$.

5. $\frac{\sqrt{2}}{4}$.   Let $\frac{x^2}{1+x^2} = a, \frac{y^2}{1+y^2} = b$, and $\frac{z^2}{1+z^2} = c$. Then $a + b + c = 1$. Furthermore, $x^2 = \frac{a}{1-a} = \frac{a}{b+c}, y^2 = \frac{b}{c+a}$, and $z^2 = \frac{c}{a+b}$. Thus,

$$x^2 y^2 z^2 = \frac{a}{b+c} \frac{b}{c+a} \frac{c}{a+b}$$

$$\leq \frac{a}{2\sqrt{bc}} \frac{b}{2\sqrt{ca}} \frac{c}{2\sqrt{ab}} = \frac{1}{8}.$$

That is, $xyz \leq \frac{\sqrt{2}}{4}$, and the equality is valid when $a = b = c = \frac{1}{3}$, namely $x = y = z = \frac{\sqrt{2}}{2}$. Therefore, $(xyz)_{\max} = \frac{\sqrt{2}}{4}$.

6. Construct the function $f(x) = \frac{x}{1+x}$. We show below that $f(x)$ is an increasing function on $[0, +\infty)$.

For $0 \leq x_1 < x_2$, we have

$$f(x_2) - f(x_1) = \frac{x_2}{1+x_2} - \frac{x_1}{1+x_1} = \frac{x_2 - x_1}{(1+x_1)(1+x_2)} > 0.$$

So, $f(x)$ is increasing on $[0, +\infty)$. Since $|a+b+c| \leq |a| + |b| + |c|$,

$$f(|a+b+c|) \leq f(|a| + |b| + |c|).$$

It follows that

$$\frac{|a+b+c|}{1 + |a+b+c|}$$

$$\leq \frac{|a| + |b| + |c|}{1 + |a| + |b| + |c|}$$

$$= \frac{|a|}{1 + |a| + |b| + |c|} + \frac{|b|}{1 + |a| + |b| + |c|} + \frac{|c|}{1 + |a| + |b| + |c|}$$

$$\leq \frac{|a|}{1 + |a|} + \frac{|b|}{1 + |b|} + \frac{|c|}{1 + |c|}.$$

7. Thanks to the invariance under rotations, we may assume that

$$x_1 = \min\{x_1, x_2, x_3, x_4, x_5\}.$$

Denote $x_{5+k} = x_k$ for $k = 1, 2, 3, 4, 5$. Since

$$\left(\sum_{k=1}^{5} x_k\right)^2 - 4\sum_{k=1}^{5} x_k x_{k+1} = \sum_{k=1}^{5} x_k^2 - 2\sum_{k=1}^{5} x_k x_{k+1} + 2\sum_{k=1}^{5} x_k x_{k+2}$$

$$= (x_1 - x_2 + x_3 - x_4 + x_5)^2 - 4x_1 x_5 + 4x_1 x_4 + 4x_2 x_5$$

$$= (x_1 - x_2 + x_3 - x_4 + x_5)^2 + 4(x_2 - x_1)x_5 + 4x_1 x_2 > 0.$$

Hence, the original inequality is satisfied.

8. Without loss of generality, assume that $a \leq b \leq c$. It is enough to show that $\frac{a+b+c}{a+b+1} + (1-a)(1-b)(1-c) \leq 1$, that is,

$$(1-c)\left[(1-a)(1-b) - \frac{1}{a+b+1}\right] \leq 0.$$

From $0 \leq a \leq 1$ and $0 \leq b \leq 1$ we know that $(1-a)(1+a)(1-b)(1+b) \leq 1$, so

$$(1-a)(1-b) \leq \frac{1}{(1+a)(1+b)} \leq \frac{1}{a+b+1},$$

namely $(1-a)(1-b) - \frac{1}{a+b+1} \leq 0$. This completes the proof.

9. Denote $\frac{xy}{z} = a^2, \frac{yz}{x} = b^2$, and $\frac{zx}{y} = c^2$. Then $y = ab, z = bc$, and $x = ca$. The original inequality is equivalent to

$$\left(a^2 + b^2 + c^2\right)^3 > 8\left(a^3 b^3 + b^3 c^3 + c^3 a^3\right). \tag{3.1}$$

Now, we see that the left hand side of (3.1) equals

$$a^6 + b^6 + c^6 + 3\left[\left(a^4 b^2 + a^2 b^4\right) + \left(b^4 c^2 + b^2 c^4\right) + \left(c^4 a^2 + c^2 a^4\right)\right] + 6a^2 b^2 c^2,$$

which is greater than or equal to

$$4\left[\left(a^4 b^2 + a^2 b^4\right) + \left(b^4 c^2 + b^2 c^4\right) + \left(c^4 a^2 + c^2 a^4\right)\right] + 3a^2 b^2 c^2$$

by Shur's inequality. But $4[(a^4 b^2 + a^2 b^4) + (b^4 c^2 + b^2 c^4) + (c^4 a^2 + c^2 a^4)]$ is greater than or equal to the right hand side of (3.1), hence the original inequality is satisfied.

10. By symmetry, we can assume that $x \geq y \geq z \geq 0$. Then $x + y \geq \frac{2}{3}$ and $z \leq \frac{1}{3}$. Let

$$x + y = \frac{2}{3} + \delta, \quad z = \frac{1}{3} - \delta, \quad 0 \leq \delta \leq \frac{1}{3}.$$

We have

$$x^2 + y^2 + z^2 + 6xyz = (x+y)^2 + z^2 - 2xy(1-3z)$$

$$\leq \left(\frac{2}{3} - \delta\right)^2 + \left(\frac{1}{3} - \delta\right)^2 = 2\delta^2 + \frac{2}{3}\delta + \frac{5}{9}$$

$$\leq 2\left(\frac{1}{3}\right)^2 + \frac{2}{3} \cdot \frac{1}{3} + \frac{5}{9} = 1.$$

Then

$$x^2 + y^2 + z^2 + 6xyz = (x+y)^2 + z^2 - 2xy(1-3z)$$

$$\geq (x+y)^2 + z^2 - \frac{1}{2}(x+y)^2(1-3z)$$

$$= \left(\frac{2}{3} + \delta\right)^2 + \left(\frac{1}{3} - \delta\right)^2 - \frac{1}{2}\left(\frac{2}{3} + \delta\right)^2 \cdot 3\delta$$

$$= \frac{5}{9} - \frac{3}{2}\delta^2 \geq \frac{5}{9} - \frac{3}{2}\left(\frac{1}{3}\right)^2 = \frac{5}{9} - \frac{1}{18} = \frac{1}{2}.$$

11. As in Figure 3.1, let the terminal side OM of $\angle A$ intersect the unit circle at M. Draw $MN \perp Ox$ at M, and draw the tangent line QP at Q that intersects OM at P. Then from trigonometry, $\sin A = |NM|$, $\tan A = |QP|$, and $\angle A$ equals the length of the arc MQ. In the following, we use the computation of areas to prove that

$$\sin A + \tan A > 2\angle A.$$

The tangent line at M to the circle intersects PQ at T. Draw the chord MQ. Clearly, since $\angle PMT = 90°$, we see that $|PT| > |MT| = |QT|$,

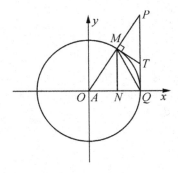

Figure 3.1   Figure for Exercise 11.

so

$$S_{\triangle PMT} > S_{\triangle MTQ} > S_{\text{bow-shaped region } MQ}.$$

Thus, $S_{\triangle OMQ} + S_{\triangle OPQ} > 2S_{\text{sector } OMQ}$, from which $|NM| + |QP| >$ 2 times the length of the arc MQ, in other words, $\sin A + \tan A > 2\angle A$. The conclusion is thus proved.

12. Without loss of generality, let $a \geq \max\{b, c\}$. Since

$$ba^3 + b^2c(b - c) = b(a - b)(a - c)(a + b - c) + 2bca^2 + (b^3 - b^2c - bc^2)a,$$

we have

$$\begin{aligned}
s &= a^2b(a - b) + b^2c(b - c) + c^2a(c - a) \\
&= ba^3 - (b^2 + c^2)a^2 + c^3a + b^2c(b - c) \\
&= b(a - b)(a - c)(a + b - c) - (b - c)^2a^2 + (b + c)(b - c)^2a \\
&= b(a - b)(a - c)(a + b - c) + a(b - c)^2(b + c - a).
\end{aligned}$$

From $a \geq b, a \geq c$, and $b + c \geq a$, we see that $s \geq 0$, namely the original inequality is satisfied. Clearly, the necessary and sufficient condition for the equality to be true is that $a = b = c$.

Solutions for Chapter 4

Arithmetic and Geometric Sequences

1. 2^{14}.   Let the common ratio be q. Then from

$$a_3 + a_4 + a_5 + a_6 + a_7 = q^2(a_1 + a_2 + a_3 + a_4 + a_5),$$

we find $q = \sqrt{2}$. By $q^2(a_1 + a_2 + a_3 + a_4 + a_5) = 12 + 14\sqrt{2}$, we get $a_1 = \sqrt{2}$. Hence, all the terms are $(\sqrt{2})^k$ for $k = 1, 2, \ldots, 7$. Therefore,

$$a_1 a_2 a_3 a_4 a_5 a_6 a_7 = (\sqrt{2})^1 (\sqrt{2})^2 \cdots (\sqrt{2})^7 = 2^{14}.$$

2. $\frac{7}{5}$.   From $a_{26} = a_1 + 25d$, we get the common difference $d = -\frac{4}{5}$. Now,

$$A = 7a_n + 21d = 7a_1 + 7(n-1)d + 21d = \frac{7}{5}(87 - 4n).$$

Since n is a positive integer, $|87 - 4n| \neq 0$, so $|87 - 4n| \geq 1$. Thus, $|A| \geq \frac{7}{5}$. When $n = 22$, we have $|A| = \frac{7}{5}$, Hence, the minimum value of $|A|$ is $\frac{7}{5}$.

3. $45 : 1$.   Since $S_{25} = 25a_{13}$, so from the given condition, $a_{13} : a_{23} = 1 : 5$. Hence, $a_{23} : a_{33} = 5 : 9$ and $a_{33} : a_{43} = 9 : 13$. Also $S_{65} = 65a_{33}$, therefore $S_{65} : a_{43} = 45 : 1$.

4. 2.   Since $3a_{n+1} - a_n = 0$, we have $\frac{a_{n+1}}{a_n} = \frac{1}{3}$. In other words, $\{a_n\}$ is a geometric sequence with 2 as the initial term and $\frac{1}{3}$ as the common ratio. So, $a_n = 2\left(\frac{1}{3}\right)^{n-1}$. Also from

$$b_n = \frac{1}{2}(a_n + a_{n+1}) = \frac{1}{2}\left[2\left(\frac{1}{3}\right)^{n-1} + 2\left(\frac{1}{3}\right)^n\right] = \frac{4}{3}\left(\frac{1}{3}\right)^{n-1},$$

we see that $\{b_n\}$ is a geometric sequence with common ratio $\frac{1}{3}$, and thus the sum of all of its terms is $S = \frac{\frac{4}{3}}{1 - \frac{1}{3}} = 2$.

5. $a_n = 1$ or $a_n = \frac{32}{5} - \frac{12}{5}n$.   Since $\{a_n\}$ is an arithmetic sequence,

$$S_n = na_1 + \frac{n(n-1)}{2}d.$$

Thus, $\frac{S_n}{n} = a_1 + \frac{n-1}{2}d$, in particular $\frac{S_3}{3} = a_1 + d$, $\frac{S_4}{4} = a_1 + \frac{3}{2}d$, and $\frac{S_5}{5} = a_1 + 2d$. Hence,

$$(a_1 + 2d)^2 = (a_1 + d)\left(a_1 + \frac{3}{2}d\right),$$

$$1 \cdot 2 = (a_1 + d) + \left(a_1 + \frac{3}{2}d\right).$$

Solving the above gives $a_1 = 1$ and $d = 0$, or $a_1 = 4$ and $d = -\frac{12}{5}$.

6. Since $a_n = a_{n-1}q$, we have $\lg a_n = \lg a_{n-1} + \lg q$. So, $\lg a_n - \lg a_{n-1} = \lg q$ (a constant). The sequence $\{\lg a_n\}$ is an arithmetic one and $\lg a_n > 0$. Hence, $\lg a_{n+1} = \frac{1}{2}(\lg a_n + \lg a_{n+2}) \geq \sqrt{\lg a_n \lg a_{n+2}}$, namely $(\lg a_{n+1})^2 \geq \lg a_n \lg a_{n+2}$. Consequently, $\frac{\lg a_{n+1}}{\lg a_n} \geq \frac{\lg a_{n+2}}{\lg a_{n+1}}$. In other words, $\log_{a_n} a_{n+1} \geq \log_{a_{n+1}} a_{n+2}$. Therefore, $\{\log_{a_n} a_{n+1}\}$ is a decreasing sequence.

7. Sufficiency: If there exists a nonnegative integer m such that $a_1 = q^m$, then $a_n = a_1 q^{n-1} = q^m q^{n-1} = q^{n+(m-1)}$ with $n \in \mathbf{N}_+$. Pick any two terms a_k and a_l in $\{a_n\}$ with $k, l \in \mathbf{N}_+$. Then

$$a_k a_l = q^{k+(m-1)} q^{l+(m-1)} = q^{(k+l+m-1)+(m-1)}.$$

Since $k, l \in \mathbf{N}_+$ and m is a nonnegative integer, $k + l + m - 1 \in \mathbf{N}_+$, so $a_{k+l+m-1} = q^{(k+l+m-1)+(m-1)}$, and thus $a_k a_l = a_{k+l+m-1}$. In other words, the product of any two terms of the geometric sequence $\{a_n\}$ is also a term of the sequence.

Necessity: Suppose that the product of any two terms of the geometric sequence $\{a_n\}$ is still a term of the sequence. Since $a_n = a_1 q^{n-1}$ for $n \in \mathbf{N}_+$, we have $a_k a_l = a_1^2 q^{k+l-2}$. Let $e \in \mathbf{N}_+$ be such that $a_k a_l = a_e$. Then $a_1^2 q^{k+l-2} = a_1 q^{e-1}$. The fact of $a_1 \neq 0$ and $q \neq 0$ implies that

$$a_1 = q^{e-1-(k+l-2)} = q^{(e+1)-(k+l)}.$$

Denote $m = (e+1) - (k+l)$. Then $a_1 = q^m$ with m an integer clearly. If $q = 1$, then $a_1 = 1^m = 1$, and the sequence $\{a_n\}$ is $1, 1, \ldots, 1, \ldots$. The integer m can be chosen to be any nonnegative integer, namely, there exists a nonnegative integer m such that $a_1 = q^m$. If $q \neq 1$, we use proof by contradiction to show that m must be a nonnegative integer. Suppose that m is a negative integer. Then $-m \in \mathbf{N}_+$. Since $a_1 = q^m$, so

$$a_n = a_1 q^{n-1} = q^m q^{n-1} = q^{n+(m-1)}, \ n \in \mathbf{N}_+.$$

Consequently, $a_{-m} = q^{-m+(m-1)}$, from which $a_1 a_{-m} = q^{m-1}$. Let $a_1 a_{-m} = a_p$, where $p \in \mathbf{N}_+$. Then $q^{m-1} = q^{p+(m-1)}$. Since $q \neq 0$ and $q \neq 1$, we must have $p = 0$, which contradicts $p \in \mathbf{N}_+$. Therefore, m is a nonnegative integer.

8. Let the common difference of $\{a_n\}$ be d and the common ratio of $\{b_n\}$ be q. Since $a_2 = b_2$, so $a_1 + d = a_1 q$, from which $d = a_1(q - 1)$. Consequently,

$$\begin{aligned}
a_n - b_n &= a_1 + (n-1)d - a_1 q^{n-1} \\
&= a_1 + (n-1)a_1(q-1) - a_1 q^{n-1} \\
&= a_1(1-q)\left[\frac{1-q^{n-1}}{1-q} - (n-1)\right] \\
&= a_1(1-q)[1 + q + q^2 + \cdots + q^{n-2} - (n-1)].
\end{aligned}$$

If $q > 1$, then $1+q+q^2+\cdots+q^{n-2} > n-1$, and so $a_n < b_n$ since $a_1 > 0$. If $q = 1$, then $a_n = b_n$. If $0 < q < 1$, then $1+q+q^2+\cdots+q^{n-2} < n-1$, so $a_n < b_n$. In summary, we have $a_n \leq b_n$.

9. Let the common difference be d and $a_{n+1} = a$. Then

$$S = a_{n+1} + a_{n+2} + \cdots + a_{2n+1} = (n+1)a + \frac{1}{2}n(n+1)d,$$

so $a + \frac{1}{2}nd = \frac{S}{n+1}$. Also

$$\begin{aligned}
M \geq a_1^2 + a_{n+1}^2 &= (a - nd)^2 + a^2 \\
&= \frac{4}{10}\left(a + \frac{1}{2}nd\right)^2 + \frac{1}{10}(4a - 3nd)^2 \\
&\geq \frac{4}{10}\left(\frac{S}{n+1}\right)^2,
\end{aligned}$$

thus

$$|S| \leq \frac{1}{2}\sqrt{10}(n+1)\sqrt{M},$$

and when $a = \frac{3}{\sqrt{10}}\sqrt{M}$ and $d = \frac{4}{\sqrt{10}}\frac{1}{n}\sqrt{M}$,

$$S = (n+1)\left(\frac{3}{\sqrt{10}}\sqrt{M} + \frac{2}{n}\frac{4}{\sqrt{10}}\frac{1}{n}\sqrt{M}\right)$$

$$= (n+1)\frac{5}{\sqrt{10}}\sqrt{M} = \frac{\sqrt{10}}{2}(n+1)\sqrt{M}.$$

Since at this time $4a = 3nd$, hence

$$a_1^2 + a_{n+1}^2 = \frac{4}{10}\left(\frac{S}{n+1}\right)^2 = \frac{4}{10}\frac{10}{4}M = M,$$

which means that the maximum value of S is $\frac{\sqrt{10}}{2}(n+1)\sqrt{M}$.

10. Let $A = \frac{a_{mk+2}}{a_{mk+1}}\frac{a_{2k+2}}{a_{2k+1}}\cdots\frac{a_{nk+2}}{a_{nk+1}}$. Since $a_{mk+2} > a_{mk+1} > 0$ and the common difference $d > 0$, we see that

$$\frac{a_{mk+2}}{a_{mk+1}} = \frac{a_{mk+1}+d}{a_{mk}+d} < \frac{a_{mk+1}}{a_{mk}} < \frac{a_{mk+1}-d}{a_{mk}-d} = \frac{a_{mk}}{a_{mk-1}}.$$

It follows that

$$A^k > \frac{a_{k+2}}{a_{k+1}}\frac{a_{k+3}}{a_{k+2}}\cdots\frac{a_{2k+1}}{a_{2k}}\frac{a_{2k+2}}{a_{2k+1}}\cdots\frac{a_{nk+2}}{a_{nk+1}}\frac{a_{nk+3}}{a_{nk+2}}\cdots\frac{a_{(n+1)k+1}}{a_{(n+1)k}}$$

$$= \prod_{i=k+1}^{(n+1)k}\frac{a_{i+1}}{a_i} = \frac{a_{(n+1)k+1}}{a_{k+1}}.$$

Also

$$A^k = \left(\frac{a_{k+2}}{a_{k+1}}\right)^k\left(\frac{a_{2k+2}}{a_{2k+1}}\right)^k\cdots\left(\frac{a_{nk+2}}{a_{nk+1}}\right)^k$$

$$< \frac{a_3}{a_2}\frac{a_4}{a_3}\cdots\frac{a_{k+1}}{a_k}\frac{a_{k+2}}{a_{k+1}}\cdots\frac{a_{(n-1)k+2}}{a_{(n-1)k+1}}\frac{a_{(n-1)k+3}}{a_{(n-1)k+2}}\cdots\frac{a_{nk+2}}{a_{nk+1}}$$

$$= \prod_{i=2}^{nk+1}\frac{a_{i+1}}{a_i} = \frac{a_{nk+2}}{a_2}.$$

Consequently, $\sqrt[k]{\frac{a_{(n+1)k+1}}{a_{k+1}}} < A < \sqrt[k]{\frac{a_{nk+2}}{a_2}}$.

11. Let the common difference of the sequence A be d. Then $1 \leq d \leq n-1$. When n is an even number, if $1 \leq d \leq \frac{n}{2}$, then the number of the sequences A that satisfy the condition with common difference d is d. If $\frac{n}{2} < d \leq n$, then the number of the sequences A that satisfy the condition with common difference d is $n - d$.

So, the total number of such sequences A is $\left(1 + 2 + \cdots + \frac{n}{2}\right) + \left\{1 + 2 + \cdots + \left[n - \left(\frac{n}{2} + 1\right)\right]\right\} = \frac{n^2}{4}$. When n is an odd number, a similar discussion shows that the total number of such sequences A is $\left(1 + 2 + \cdots + \frac{n-1}{2}\right) + \left[1 + 2 + \cdots + \left(n - \frac{n+1}{2}\right)\right] = \frac{n^2-1}{4}$. The two cases are unified to the conclusion: There are a total of $\left[\frac{n^2}{4}\right]$ such sequences A.

12. We may assume that the first terms of the two sequences are both a (otherwise, drop a finite number of initial terms from each of the two sequences, and then we can discuss the remaining sequences). From the condition, it is enough to show that there exist infinitely many pairs of positive integers (m, n) such that $a + mr = aq^n$. For this to be true, we just need to show that there exist infinitely many $n \in \mathbf{N}_+$ such that $aq^n \equiv a \pmod{r}$. In the following, we show that there exist infinitely many $n \in \mathbf{N}_+$ such that $q^n \equiv 1 \pmod{r}$ (then $aq^n \equiv a \pmod{r}$, thus proving the proposition). In fact, since there must exist two numbers among the following $r + 1$ numbers $1, q, q^2, \ldots, q^r$ that are congruent module r, namely there are $0 \le i < j \le r$ such that $q^j \equiv q^i \pmod{r}$, so $q^{j-i} \equiv 1 \pmod{r}$ by noting that $(q, r) = 1$. Thus, denoting $n_0 = j - i$, we see that for any $k \in \mathbf{N}_+$, with $n = kn_0$ it holds that $q^n \equiv (q^{n_0})^k \equiv 1^k \equiv 1 \pmod{r}$. Hence, the proposition is proved.

Solutions for Chapter 5

Arithmetic Sequences of Higher Order

1. 819.   Let the first term of $\Delta\{a_n\}$ be d. Then the condition implies that $\Delta\{a_n\}$ is $d, d+1, d+2, \ldots$, where the nth term is $d+(n-1)$. Thus, the sequence $\{a_n\}$ can be written as

$$a_1, \; a_1 + d, \; a_1 + d + (d+1), \; a_1 + d + (d+1) + (d+2), \; \ldots,$$

where the nth term is $a_n = a_1 + (n-1)d + \frac{1}{2}(n-1)(n-2)$. Since $a_{19} = a_{92} = 0$, so we have

$$\begin{cases} a_1 + 18d + 9 \cdot 17 = 0, \\ a_1 + 91d + 45 \cdot 91 = 0. \end{cases}$$

Solving the above system gives $a_1 = 819$.

2. $\frac{1}{4}n^2(n+1)^2$.   From

$$n^4 = \sum_{k=1}^{n} \left[k^4 - (k-1)^4\right] = 4\sum_{k=1}^{n} k^3 - 6\sum_{k=1}^{n} k^2 + 4\sum_{k=1}^{n} k - n,$$

we can get $\sum_{k=1}^{n} k^3 = \left[\frac{n(n+1)}{2}\right]^2$.

3. $a_n = \frac{1}{2}n^3 - \frac{1}{2}n^2 - n + 2$.

4. $a_n = \frac{1}{6}(2n^3 + 3n^2 - 5n + 12)$.

5. $S_n = \frac{1}{5}n^5$.

6. $a_n = n^3$.

7. Let the right hand side of the original equality $= a_n$. Make the difference equation $a_n - a_{n-1} = \frac{1}{(2n-1)2n}$, in other words,

$$
\begin{cases}
a_1 = \dfrac{1}{2}, \\[2mm]
a_k - a_{k-1} = \dfrac{1}{(2k-1)2k}, & k = 2, 3, \ldots, n.
\end{cases}
$$

Summing up the n equalities, we obtain $a_n = \frac{1}{1\cdot 2} + \frac{1}{3\cdot 4} + \cdots + \frac{1}{(2n-1)2n}$. Hence, the original equality is satisfied.

8. Let the first order difference sequence of $\{a_n\}$ be $\{b_n\}$. Then $b_1 = a_2 - a_1$ and $b_n = b_1 q^{n-1}$. Thus, $\sum_{i=1}^{n-1} b_i = b_1 \sum_{i=1}^{n-1} q^{i-1} = b_1 \frac{1-q^{n-1}}{1-q}$. Hence, $a_n = a_1 + \sum_{i=1}^{n-1}(a_{i+1} - a_i) = a_1 + \sum_{i=1}^{n-1} b_i = a_1 + b_1 \frac{1-q^{n-1}}{1-q} = a_1 + (a_2 - a_1)\frac{1-q^{n-1}}{1-q}$.

9. Denote $T_n = a + 3a^2 + 5a^3 + \cdots + (2n-1)a^n$. Then $aT_n = a^2 + 3a^3 + \cdots + (2n-3)a^n + (2n-1)a^{n+1}$. Thus, $(1-a)T_n = 2\frac{a-a^{n+1}}{1-a} - a - (2n-1)a^{n+1}$. Consequently, $T_n = \frac{(2n-1)a^{n+1}+a}{a-1} + 2\frac{a-a^{n+1}}{(1-a)^2}$. Similarly, $(1-a)S_n = T_n - n^2 a^{n+1}$, and it follows that

$$
S_n = 2\frac{a - a^{n+1}}{(1-a)^3} - \frac{(2n-1)a^{n+1} + a}{(1-a)^2} - \frac{n^2 a^{n+1}}{1-a}.
$$

10. First show that the two sequences are both second order arithmetic sequences, and then we can get

$$
a_{1m} = m^2 + m - 1, \quad a_{1n} = n^2 - n + 1.
$$

11. $\frac{1}{6}(2n^3 - 3n^2 + n + 24)$.

12. We first find the general term formula for $\{a_n\}$. To avoid solving a system of three general linear equations for the three coefficients of a quadratic polynomial of n, we write $a_n = a(n-1)(n-2) + b(n-1) + c$. Then $a_1 = c, a_2 = b + c$, and $a_3 = 2a + 2b + c$. So, $S_3 = a_1 + a_2 + a_3 = 2a + 3b + 3c$. Thus, the given conditions imply that $c = 1, b + c = 3$, and $2a + 3b + 3c = 13$, the solutions of which are easily obtained to be $a = 2, b = 2$, and $c = 1$. Hence, $a_n = 2(n-1)(n-2) + 2(n-1) + 1 = 2(n-1)^2 + 1$.

Therefore,

$$S_n = \sum_{k=1}^{n} a_k = 2\sum_{k=1}^{n}(k-1)^2 + \sum_{k=1}^{n} 1$$
$$= 2\frac{n(n-1)(2n-1)}{6} + n = \frac{n(2n^2 - 3n + 4)}{3}.$$

Solutions for Chapter 6

Sequence Summation

1. n. Since

$$a_n = \sqrt{1 + \frac{1}{n^2} + \frac{1}{(n+1)^2}} = \sqrt{\frac{[n(n+1)]^2 + (n+1)^2 + n^2}{n^2(n+1)^2}}$$

$$= \frac{n(n+1)+1}{n(n+1)} = 1 + \frac{1}{n} - \frac{1}{n+1},$$

we have

$$S_n = \left(1 + \frac{1}{1} - \frac{1}{2}\right) + \left(1 + \frac{1}{2} - \frac{1}{3}\right) + \cdots + \left(1 + \frac{1}{n} - \frac{1}{n+1}\right)$$

$$= n + 1 - \frac{1}{n+1}.$$

Thus, $\lfloor S_n \rfloor = n$.

2. $\frac{2002}{2003}$. The two roots of $(n^2 + n)x^2 - (2n+1)x + 1 = 0$ are $x_1 = \frac{1}{n}$ and $x_2 = \frac{1}{n+1}$, so $|A_n B_n| = \left|\frac{1}{n} - \frac{1}{n+1}\right| = \frac{1}{n} - \frac{1}{n+1}$. Consequently,

$$|A_1 B_1| + |A_2 B_2| + \cdots + |A_{2002} B_{2002}| = 1 - \frac{1}{2003} = \frac{2002}{2003}.$$

3. $\frac{(n+1)!-1}{(n+1)!}$. It follows from

$$\frac{n}{(n+1)!} = \frac{(n+1)-1}{(n+1)!} = \frac{1}{n!} - \frac{1}{(n+1)!}.$$

4. $(n+1)! - 1$. Since

$$n \cdot n! = [(n+1) - 1] \cdot n! = (n+1)! - n!,$$

so $\sum_{k=1}^{n} k \cdot k! = (n+1)! - 1$.

5. $\dfrac{(-1)^{n-1}na^{n+1}}{1+a} + \dfrac{a[1-(-a)^n]}{(1+a)^2}$.

6. $\dfrac{5}{4} - \dfrac{2n+5}{2(n+1)(n+2)}$. Using the summation method by combining terms and putting together the fractions of the same denominators can find the total sum.

7. $\dfrac{9}{2}$.   By Vieta's theorem, $c_n = a_n + a_{n+1}$ and $a_n a_{n+1} = \left(\frac{1}{3}\right)^n$, so $a_{n+1}a_{n+2} = \left(\frac{1}{3}\right)^{n+1}$. Hence, $\dfrac{a_{n+2}}{a_n} = \frac{1}{3}$, and thus

$$\sum_{n=1}^{+\infty} c_n = \sum_{n=1}^{+\infty} a_n + \sum_{n=2}^{+\infty} a_n = \sum_{k=0}^{+\infty} a_{2k+1} + 2\sum_{k=1}^{+\infty} a_{2k} + \sum_{k=1}^{+\infty} a_{2k+1}$$

$$= 2\sum_{k=0}^{+\infty} a_{2k+1} + 2\sum_{k=1}^{+\infty} a_{2k} - a_1 = 2 \cdot \frac{2}{1-\frac{1}{3}} + 2 \cdot \frac{\frac{1}{6}}{1-\frac{1}{3}} - 2 = \frac{9}{2}.$$

8. From $\tan\left(\arctan\frac{1}{n} - \arctan\frac{1}{n+1}\right) = \dfrac{\frac{1}{n}-\frac{1}{n+1}}{1+\frac{1}{n}\frac{1}{n+1}} = \dfrac{1}{1+n+n^2}$, we obtain that

$$\arctan\frac{1}{1+n+n^2} = \arctan\frac{1}{n} - \arctan\frac{1}{n+1}.$$

Using the summation method by splitting terms gives the answer $\frac{\pi}{4} - \arctan\frac{1}{n+1}$ to the problem.

9. Since

$$a_n = \frac{4n-1}{n(n+2)}3^{n-1} = \frac{1}{2}\left(\frac{9}{n+2} - \frac{1}{n}\right)3^{n-1}$$

$$= \frac{1}{2}\left(\frac{3^{n+1}}{n+2} - \frac{3^{n-1}}{n}\right),$$

we see that $S_n = \frac{1}{2}\left(\frac{3^{n+1}}{n+2} + \frac{3^n}{n+1} - \frac{5}{2}\right)$.

10. Note that

$$\frac{2}{\sqrt{k}+\sqrt{k+1}} < \frac{1}{\sqrt{k}} < \frac{2}{\sqrt{k-1}+\sqrt{k}}$$

for $k = 1, 2, \ldots$. Then we have

$$\sum_{k=1}^{10^6} \frac{2}{\sqrt{k}+\sqrt{k+1}} < x < 1 + \sum_{k=2}^{10^6} \frac{2}{\sqrt{k-1}+\sqrt{k}},$$

in other words,

$$2\sum_{k=1}^{10^6}(\sqrt{k+1}-\sqrt{k}) < x < 1+2\sum_{k=1}^{10^6}(\sqrt{k}-\sqrt{k-1}).$$

It follows that

$$2(\sqrt{10^6+1}-1) < x < 1+2(\sqrt{10^6}-\sqrt{1}).$$

Consequently $1998 < 2(\sqrt{10^6+1}-1) < x < 1999$, therefore $[x] = 1998$.

11. We first find the number of the positive integers k that satisfy $f(k) = m$ (for a fixed positive integer m). According to the condition of the problem, $m - \frac{1}{2} < \sqrt[4]{k} < m + \frac{1}{2}$, that is,

$$m^4 - 2m^3 + \frac{3}{2}m^2 - \frac{1}{2}m + \frac{1}{16} < k < m^4 + 2m^3 + \frac{3}{2}m^2 + \frac{1}{2}m + \frac{1}{16}.$$

This indicates that the number of the positive integers k that satisfy $f(k) = m$ is exactly $4m^3 + m$. When $m = 7$ in particular, $m^4 - 2m^3 + \frac{3m^2-m}{2} < 2002 < m^4 + 2m^3 + \frac{3m^2+m}{2}$, and

$$2002 - \left(7^4 - 2\cdot 7^3 + \frac{3\cdot 7^2 - 7}{2}\right) = 217.$$

Therefore,

$$\sum_{k=1}^{2002}\frac{1}{f(k)} = \sum_{m=1}^{6}\frac{1}{m}(4m^3+m) + \frac{1}{7}\cdot 217$$

$$= 4\sum_{m=1}^{6}m^2 + 6 + 31 = 4\cdot 7\cdot 13 + 37 = 401.$$

12. (1) Subtracting $4S_{n-1} = 3a_{n-1} + 2^n$ ($n \in \mathbf{N}_+$) from $4S_n = 3a_n + 2^{n+1}$ ($n \geq 0$ with $n \in \mathbf{Z}$) gives that $4a_n = 3a_n - 3a_{n-1} + 2^n$, from which

$$a_n = 2^n - 3a_{n-1} \ (n \in \mathbf{N}_+).$$

(2) Dividing 2^n from the both sides of $a_n = 2^n - 3a_{n-1}$, we obtain that

$$\frac{a_n}{2^n} = -\frac{3}{2} \cdot \frac{a_{n-1}}{2^{n-1}} + 1$$

$$\Rightarrow \frac{a_n}{2^n} - \frac{2}{5} = -\frac{3}{2}\left(\frac{a_{n-1}}{2^{n-1}} - \frac{2}{5}\right)$$

$$\Rightarrow \frac{a_n}{2^n} - \frac{2}{5} = \left(-\frac{3}{2}\right)^n \left(\frac{a_0}{2^0} - \frac{2}{5}\right)$$

$$\Rightarrow a_n = 2^n \left[\frac{2}{5} + \left(-\frac{3}{2}\right)^n \left(a_0 - \frac{2}{5}\right)\right]$$

$$= 2^n \left(-\frac{3}{2}\right)^n \left(a_0 - \frac{2}{5}\right) + \frac{2^{n+1}}{5}$$

$$= (-3)^n \left(a_0 - \frac{2}{5}\right) + \frac{2^{n+1}}{5}$$

and consequently,

$$a_{n+1} - a_n = \left[(-3)^{n+1}\left(a_0 - \frac{2}{5}\right) + \frac{2^{n+2}}{5}\right]$$

$$- \left[(-3)^n \left(a_0 - \frac{2}{5}\right) + \frac{2^{n+1}}{5}\right]$$

$$= \left(a_0 - \frac{2}{5}\right)\left[(-3)^{n+1} - (-3)^n\right] + \frac{2^{n+1}}{5}$$

$$= \frac{2^{n+1}}{5}\left[-10\left(-\frac{3}{2}\right)^n \left(a_0 - \frac{2}{5}\right) + 1\right].$$

If $a_0 > \frac{5}{2}$, then for even numbers n large enough,

$$-10\left(-\frac{3}{2}\right)^n \left(a_0 - \frac{2}{5}\right) + 1 < 0 \Rightarrow a_{n+1} < a_n.$$

If $a_0 < \frac{5}{2}$, then for odd numbers n large enough,

$$-10\left(-\frac{3}{2}\right)^n \left(a_0 - \frac{2}{5}\right) + 1 < 0 \Rightarrow a_{n+1} < a_n.$$

To summarize, $a_0 = \frac{2}{5}$.

Solutions for Chapter 7

Synthetic Problems for Sequences

1. $-\frac{3}{2}$.   Let $a = \frac{b}{q}, c = bq, \log_a b = x - d, \log_b c = x$, and $\log_c a = x + d$. Then $a^{x-d} = b, b^x = c$, and $c^{x+d} = a$, which implies that

$$\left(\frac{b}{q}\right)^{x-d} = b, \tag{7.1}$$

$$b^x = bq, \tag{7.2}$$

$$(bq)^{x+d} = \frac{b}{q}. \tag{7.3}$$

(7.1) times (7.3) gives $b^{2x}q^{2d} = b^2 q^{-1}$. Combining it with (7.2), we get $b^2 q^{2d+2} = b^2 q^{-1}$, thus $q^{2d+3} = 1$. Since $q \neq 1$. so $d = -\frac{3}{2}$.

2. 18.   Denote

$$S_n = 1 \cdot 2 + 2 \cdot 3 + \cdots + n(n+1)$$

$$= (1^2 + 2^2 + \cdots + n^2) + (1 + 2 + \cdots + n)$$

$$= \frac{n(n+1)(n+2)}{3}.$$

If 2014 is in the $(n+1)$st group, then $S_n < 2014 < S_{n+1}$ since there are S_n numbers before this group. From

$$S_{17} = 1938, \quad S_{18} = 2280,$$

we see that $S_{17} < 2014 < S_{18}$, hence 2014 is inside the 18th group.

3. 10200.   From the given condition,

$$T_{100} = \sum_{n=1}^{100} f(n) = -2^2 + 4^2 - 6^2 + 8^2 - \cdots - 98^2 + 100^2$$

$$= 4(3 + 7 + \cdots + 99) = 5100.$$

Thus, $S_{100} = 2T_{100} - f(1) + f(101) = 10200$.

4. (1) Let the common difference of the arithmetic sequence $\{a_n\}$ be d. From $\frac{S_{2n}}{S_n} = \frac{4n+2}{n+1}$, we see that $\frac{a_1+a_2}{a_1} = 3$, so $a_2 = 2$, from which $d = a_2 - a_1 = 1$. It follows from

$$\frac{4n+2}{n+1} = \frac{S_{2n}}{S_n} = \frac{\frac{a_n+nd+a_1}{2}2n}{\frac{a_n+a_1}{2}n}$$

$$= \frac{2(a_n + nd + a_1)}{a_n + a_1} = \frac{2(a_n + n + 1)}{a_n + 1}$$

that $a_n = n$.

(2) From $b_n = a_n p^{a_n}$, we have $b_n = np^n$. So $T_n = p + 2p^2 + 3p^3 + \cdots + (n-1)p^{n-1} + np^n$. When $p = 1$, we obtain $T_n = \frac{n(n+1)}{2}$. When $p \neq 1$,

$$pT_n = p^2 + 2p^3 + 3p^4 + \cdots + (n-1)p^n + np^{n+1},$$

$$(1-p)T_n = p + p^2 + p^3 + \cdots + p^{n-1} + p^n - np^{n+1}$$

$$= \frac{p(1-p^n)}{1-p} - np^{n+1}.$$

That is,

$$T_n = \begin{cases} \dfrac{n(n+1)}{2}, & p = 1, \\[2ex] \dfrac{p(1-p^n)}{(1-p)^2} - \dfrac{np^{n+1}}{1-p}, & p \neq 1. \end{cases}$$

5. (1) From $S_{14} = 98$, we have $2a_1 + 13d = 14$. Also $a_{11} = a_1 + 10d = 0$. Solving the above two equations gives $d = -2$ and $a_1 = 20$. Hence, the general term formula of $\{a_n\}$ is $a_n = 22 - 2n$ for $n = 1, 2, 3, \ldots$.

(2) From $S_{14} \leq 77, a_{11} > 0$, and $a_1 \geq 6$, we see that $2a_1 + 13d \leq 11, a_1 + 10d > 0$, and $a_1 \geq 6$. In other words,

$$2a_1 + 13d \leq 11, \tag{7.4}$$

$$-2a_1 - 20d > 0, \tag{7.5}$$

$$-2a_1 \leq -12. \tag{7.6}$$

(7.4) plus (7.5) gives $-7d < 11$, namely $d > -\frac{11}{7}$. (7.4) plus (7.6) gives $13d \leq -1$, namely $d \leq -\frac{1}{13}$. So $-\frac{11}{7} < d \leq -\frac{1}{13}$. Since $d \in \mathbf{Z}$, thus $d = -1$. Substituting it into (7.4) and (7.5), we get $10 < a_1 \leq 12$. Since $a_1 \in \mathbf{Z}$, we have $a_1 = 11$ or $a_1 = 12$. Therefore, the general terms of all the possible sequences $\{a_n\}$ are $a_n = 12 - n$ and $a_n = 13 - n$ with $n = 1, 2, 3, \ldots$.

6. (1) The given condition implies that $\frac{S_n}{n} = 3n - 2$, namely $S_n = 3n^2 - 2n$. When $n \geq 2$, we find that $a_n = S_n - S_{n-1} = 3n^2 - 2n - [3(n-1)^2 - 2(n-1)] = 6n - 5$. When $n = 1$, we get $a_1 = S_1 = 3 \cdot 1^2 - 2 \cdot 1 = 6 \cdot 1 - 5$. Thus, $a_n = 6n - 5$ ($n \in \mathbf{N}_+$).

   (2) From (1), $b_n = \frac{3}{a_n a_{n+1}} = \frac{3}{(6n-5)[6(n+1)-5]} = \frac{1}{2}(\frac{1}{6n-5} - \frac{1}{6n+1})$ and $T_n = \sum_{i=1}^{n} b_i = \frac{1}{2}[(1 - \frac{1}{7}) + (\frac{1}{7} - \frac{1}{13}) + \cdots + (\frac{1}{6n-5} - \frac{1}{6n+1})] = \frac{1}{2}(1 - \frac{1}{6n+1})$. So, the m that makes $\frac{1}{2}(1 - \frac{1}{6n+1}) < \frac{m}{20}$ ($n \in \mathbf{N}_+$) valid must satisfy $\frac{1}{2} \leq \frac{m}{20}$, namely $m \geq 10$. Hence, the minimal positive integer m satisfying the requirement is 10.

7. Since $a_1 = \frac{1}{2}$ and $a_{n+1} = a_n^2 + a_n$ for $n \in \mathbf{N}_+$, so $a_{n+1} = a_n(a_n + 1)$. Consequently,

$$b_n = \frac{1}{1 + a_n} = \frac{a_n^2}{a_n a_{n+1}} = \frac{a_{n+1} - a_n}{a_n a_{n+1}} = \frac{1}{a_n} - \frac{1}{a_{n+1}},$$

$$P_n = b_1 b_2 \cdots b_n = \frac{a_1}{a_2} \frac{a_2}{a_3} \cdots \frac{a_n}{a_{n+1}} = \frac{1}{2 a_{n+1}},$$

$$S_n = b_1 + b_2 + \cdots + b_n$$

$$= \left(\frac{1}{a_1} - \frac{1}{a_2}\right) + \left(\frac{1}{a_2} - \frac{1}{a_3}\right) + \cdots + \left(\frac{1}{a_n} - \frac{1}{a_{n+1}}\right)$$

$$= 2 - \frac{1}{a_{n+1}}.$$

Thus, $2P_n + S_n = \frac{1}{a_{n+1}} + \left(2 - \frac{1}{a_{n+1}}\right) = 2$.

8. (1) From the given condition, $a_{n+1} = a_n^2 + 2a_n$, thus $a_{n+1} + 1 = (a_n + 1)^2 > 1$ since $a_1 = 2$. Taking the logarithm both sides gives $\lg(1 + a_{n+1}) = 2\lg(1 + a_n)$, namely $\frac{\lg(1+a_{n+1})}{\lg(1+a_n)} = 2$. Hence, $\{\lg(1 + a_n)\}$ is a geometric sequence of common ratio 2.

   (2) From (1), $\lg(1 + a_n) = 2^{n-1}\lg(1 + a_1) = 2^{n-1}\lg 3 = \lg 3^{2^{n-1}}$, so $1 + a_n = 3^{2^{n-1}}$, from which $a_n = 3^{2^{n-1}} - 1$. Consequently, $T_n = (1 + a_1)(1 + a_2) \cdots (1 + a_n) = 3^{2^0} 3^{2^1} 3^{2^2} \cdots 3^{2^{n-1}} = 3^{1+2+2^2+\cdots+2^{n-1}} = 3^{2^n - 1}$.

(3) Since $a_{n+1} = a_n^2 + 2a_n = a_n(a_n + 2)$, we have $\frac{1}{a_{n+1}} = \frac{1}{2}(\frac{1}{a_n} - \frac{1}{a_n+2})$, from which $\frac{1}{a_n+2} = \frac{1}{a_n} - \frac{2}{a_{n+1}}$. Since $b_n = \frac{1}{a_n} + \frac{1}{a_n+2}$, so $b_n = 2(\frac{1}{a_n} - \frac{1}{a_{n+1}})$. Thus, $S_n = b_1 + b_2 + \cdots + b_n = 2(\frac{1}{a_1} - \frac{1}{a_2} + \frac{1}{a_2} - \frac{1}{a_3} + \cdots + \frac{1}{a_n} - \frac{1}{a_{n+1}}) = 2(\frac{1}{a_1} - \frac{1}{a_{n+1}})$. Since $a_n = 3^{2^{n-1}} - 1, a_1 = 2$, and $a_{n+1} = 3^{2^n} - 1$, we have $S_n = 1 - \frac{2}{3^{2^n}-1}$. Since $T_n = 3^{2^n-1}$, it follows that

$$S_n + \frac{2}{3T_n - 1} = 1.$$

9. **Necessity:** Let $\{a_n\}$ be an arithmetic sequence of common difference d_1. Then

$$b_{n+1} - b_n = (a_{n+1} - a_{n+3}) - (a_n - a_{n+2})$$
$$= (a_{n+1} - a_n) - (a_{n+3} - a_{n+2}) = d_1 - d_1 = 0,$$

so $b_n \le b_{n+1}$ $(n = 1, 2, 3, \ldots)$. Also $c_{n+1} - c_n = (a_{n+1} - a_n) + 2(a_{n+2} - a_{n+1}) + 3(a_{n+3} - a_{n+2}) = 6d_1$ (a constant) $(n = 1, 2, 3, \ldots)$. Thus, $\{c_n\}$ is an arithmetic sequence.

Sufficiency: Suppose that $\{c_n\}$ is an arithmetic sequence of common difference d_2, and $b_n \le b_{n+1}$ $(n = 1, 2, 3, \ldots)$. Since

$$c_n = a_n + 2a_{n+1} + 3a_{n+2}, \tag{7.7}$$

$$c_{n+2} = a_{n+2} + 2a_{n+3} + 3a_{n+4}, \tag{7.8}$$

$(7.7) - (7.8)$ gives that $c_n - c_{n+2} = (a_n - a_{n+2}) + 2(a_{n+1} - a_{n+3}) + 3(a_{n+2} - a_{n+4}) = b_n + 2b_{n+1} + 3b_{n+2}$. Since $c_n - c_{n+2} = (c_n - c_{n+1}) + (c_{n+1} - c_{n+2}) = -2d_2$, we have

$$b_n + 2b_{n+1} + 3b_{n+2} = -2d_2, \tag{7.9}$$

and thus

$$b_{n+1} + 2b_{n+2} + 3b_{n+3} = -2d_2. \tag{7.10}$$

Subtracting (7.9) from (7.10) implies that

$$(b_{n+1} - b_n) + 2(b_{n+2} - b_{n+1}) + 3(b_{n+3} - b_{n+2}) = 0. \tag{7.11}$$

Since $b_{n+1} - b_n \ge 0, b_{n+2} - b_{n+1} \ge 0$, and $b_{n+3} - b_{n+2} \ge 0$, (7.11) ensures that $b_{n+1} - b_n = 0$ $(n = 1, 2, 3, \ldots)$. Let $b_n = d_3$ $(n = 1, 2, 3, \ldots)$. Then

$a_n - a_{n+2} = d_3$ (a constant). Hence,

$$c_n = a_n + 2a_{n+1} + 3a_{n+2} = 4a_n + 2a_{n+1} - 3d_3, \qquad (7.12)$$

from which

$$c_{n+1} = 4a_{n+1} + 2a_{n+2} - 3d_3 = 4a_{n+1} + 2a_n - 5d_3. \qquad (7.13)$$

By $(7.13) - (7.12)$, we obtain that $c_{n+1} - c_n = 2(a_{n+1} - a_n) - 2d_3$. Therefore,

$$a_{n+1} - a_n = \frac{1}{2}(c_{n+1} - c_n) + d_3$$

$$= \frac{1}{2}d_2 + d_3 \text{ (a constant) } (n = 1, 2, 3, \ldots).$$

In other words, the sequence $\{a_n\}$ is an arithmetic one. In summary, a necessary and sufficient condition for $\{a_n\}$ to be an arithmetic sequence is that $\{c_n\}$ is an arithmetic sequence and $b_n \leq b_{n+1}$.

10. It is obvious that the sequences are monotonically increasing from the second term on: $x_{n+2} > x_{n+1}^2 > x_{n+1}$ and $y_{n+2} > y_{n+1}$. Since $x_3 > 1 + 1^2 = 2$ and $y_3 > 1^2 + 1 = 2$, every term of each sequence will be greater than 2 from the third term on. Similarly, $x_n > 3$ and $y_n > 3$ when $n > 3$. Note that $x_{n+2} > x_{n+1}^2 > x_n^4$ when $n > 1$. On the other hand, $y_{n+2} = y_n^2 + y_{n+1} = y_n^2 + y_n + y_{n-1}^2 < 3y_n^2 < y_n^3$ when $n > 3$. Thus, $\frac{\lg x_{n+2}}{\lg y_{n+2}} > \frac{4\lg x_n}{3\lg y_n}$ when $n > 3$. Consequently, $\frac{\lg x_{2k}}{\lg y_{2k}} > (\frac{4}{3})^{k-1}\frac{\lg x_2}{\lg y_2}$. When k is large enough, the right hand side of the last inequality is greater than 1, hence $x_{2k} > y_{2k}$.

11. First we show that, each integer appears at most one time in this sequence. If some integer k appears at least twice in the sequence, then by letting $a_i = a_j = k$ $(i < j)$, there are two numbers a_i and a_j in $a_1, a_2, \ldots, a_j$ whose remainders will be the same when divided by j, a contradiction. Next, let the maximal number and minimal number of $a_1, a_2, \ldots a_k$ be x_k and y_k respectively $(k = 1, 2, \ldots)$, and we claim that $x_k - y_k \leq k - 1$. If $x_k - y_k \geq k$ for some k, and let $a_i = x_k, a_j = y_k$, and $a_i - a_j = l \geq k$. Then $i, j \leq k \leq l$, and so there are two numbers a_i and a_j in $a_1, a_2, \ldots, a_l$ whose remainders are the same when divided by l, a contradiction. We show below that for each integer t satisfying $y_k \leq t \leq x_k$, there exists a positive integer s with $1 \leq s \leq k$ such that $a_s = t$. If it is not so, then $a_1, a_2, \ldots, a_k$ can only take values inside $\{u \in \mathbf{Z}_+ \mid y_k \leq u \leq x_k, u \neq t\}$. But this latter set has only $x_k - y_k$ numbers while $x_k - y_k \leq k - 1 < k$, thus there must be

two same numbers among $a_1, a_2, \ldots, a_k$. This contradicts the conclusion proved earlier. For any integer m, since there are infinitely many positive integers in the sequence and only finitely many positive integers not exceeding $|m|$, there must exist a positive integer p such that $a_p > |m|$. By the same token, there must exist a positive integer q such that $a_q < -|m|$. Let $r = \max\{p, q\}$. Then $x_r > |m|$ and $y_r < -|m|$. Therefore, $y_r < m < x_r$. By the previous conclusion, there must exist a positive integer s such that $a_s = m$. hence, each integer must appear in the sequence. To summarize, every integer appears exactly one time in the sequence.

12. (1) Since

$$\frac{x_{n+1} - 1}{x_{n+1} + 1} = \frac{f(x_n) - 1}{f(x_n) + 1} = \frac{\frac{x_n^3 + 3x_n}{3x_n^2 + 1} - 1}{\frac{x_n^3 + 3x_n}{3x_n^2 + 1} + 1}$$

$$= \frac{x_n^3 - 3x_n^2 + 3x_n - 1}{x_n^3 + 3x_n^2 + 3x_n + 1} = \left(\frac{x_n - 1}{x_n + 1}\right)^3,$$

we have $\log_3 \frac{x_{n+1} - 1}{x_{n+1} + 1} = 3\log_3 \frac{x_n - 1}{x_n + 1}$, namely $b_n = 3b_{n-1}$, so $\{b_n\}$ is a geometric sequence.

Also $x_2 = \frac{14}{13}$ from the condition, so

$$b_1 = \log_3 \frac{\frac{14}{13} - 1}{\frac{14}{13} + 1} = \log_3 \frac{1}{27} = -3.$$

Thus, $b_n = -3^n$. Hence, the general term formula for the sequence $\{b_n\}$ is $b_n = -3^n$.

(2) By (1), $b_n = -3^n$, so $c_n = n3^n$,

$$T_n = 1 \cdot 3^1 + 2 \cdot 3^2 + 3 \cdot 3^3 + \cdots + n3^n,$$

$$3T_n = 1 \cdot 3^2 + 2 \cdot 3^3 + 3 \cdot 3^4 + \cdots + n3^{n+1}.$$

After subtraction, we obtain

$$-2T_n = 3 + 3^2 + 3^3 + \cdots + 3^n - n3^{n+1} = \frac{3^{n+1} - 3}{2} - n3^{n+1}.$$

That is, $T_n = \frac{(2n-1)3^{n+1} + 3}{4}$. Therefore, the formula for the sum of the first n terms of the sequence $\{c_n\}$ is

$$T_n = \frac{(2n - 1)3^{n+1} + 3}{4} \quad (n \in \mathbf{N}_+).$$

Solutions for Chapter 8

Coordinates Systems

1. $\frac{1}{7}$.   Take the vertex A of $\triangle ABC$ as the origin and the straight line containing AB as the x-axis to set a rectangular coordinates system.
2. $(4, -4)$.   Using the middle point formula gives the result.
3. $\frac{2}{3}\sqrt{3(a^2 + b^2 + c^2)}$.   Set up a rectangular coordinates system with the vertex on l_2 as its origin and l_2 as its x-axis. Let the coordinates of the other two vertices be (m, a) and $(n, -b)$ respectively. Using the fact that all sides of an equilateral triangle are equal can solve the problem.
4. It is easy to see that the maximal common factor of n and $n + 3$ is

$$(n, n + 3) = \begin{cases} 3, & \text{if } 3 \mid n, \\ 1, & \text{if } 3 \nmid n. \end{cases}$$

When $(n, n+3) = 1$, there are no integral points in OA_n. Otherwise, let (m, l) be an integral point in OA_n with $1 \le m < n$ and $1 \le l < n+3$. Then from $\frac{m}{l} = \frac{n}{n+3}$, namely $m(n + 3) = ln$, we deduce that $n \mid m$, which contradicts the assumption that $m < n$.

When $(n, n + 3) = 3$, let $n = 3k$. Then there are two integral points $(k, k + 1)$ and $(2k, 2k + 2)$ in OA_n. Thus,

$$\sum_{i=1}^{1990} f(i) = 2\left[\frac{1990}{3}\right] = 1326,$$

where $[x]$ represents the largest integer not exceeding x.

5. Set up a rectangular coordinates system with A as the origin and the straight line that contains AT as the x-axis. Then it is enough to show that the slope of the straight line MN equals 0.

6. Let the coordinates of every check be (i, j) with $1 \leq i, j \leq 8$. The sum of the coordinates is an odd number for black checks, and the sum of the coordinates is an even number for checks with marks. So the number of black checks with marks is an even number.

7. With B as the origin and BB' as the x-axis, set up a rectangular coordinates system. Then using Theorem 4 can prove the conclusion.

8. Using the inner center Q of $\triangle ABC$ as the origin, we set up a rectangular coordinates system. Let $(x_P, y_P), (x_A, y_A), (x_B, y_B)$, and (x_C, y_C) be the coordinates of P, A, B, and C. By the coordinates formula for the inner center,

$$ax_A + bx_B + cx_C = 0, \quad ay_A + by_B + cy_C = 0.$$

Then

$$a|PA|^2 = a(x_P - x_A)^2 + a(y_P - y_A)^2$$
$$= a(x_P^2 + y_P^2) + a(x_A^2 + y_A^2) - 2ax_P x_A - 2ay_P y_A.$$

By the same token,

$$b|PB|^2 = b(x_P^2 + y_P^2) + b(x_B^2 + y_B^2) - 2bx_P x_B - 2by_P y_B,$$
$$c|PC|^2 = c(x_P^2 + y_P^2) + c(x_C^2 + y_C^2) - 2cx_P x_C - 2cy_P y_C.$$

Adding the both sides of the above three equalities gives

$$a|PA|^2 + b|PB|^2 + c|PC|^2$$
$$= |QP|^2(a + b + c) + a|QA|^2 + b|QB|^2 + c|QC|^2.$$

9. With l as the x-axis and its perpendicular line passing through the point P as the y-axis, set up a rectangular coordinates system. Let the coordinates of P, Q, and R be $(0, a), (c, d)$, and (x, y) with $a, d, y \in \mathbf{R}_+$, respectively. Then

$$\delta(R) = |PR| + |RQ| + |RS|$$
$$= \sqrt{x^2 + (y - a)^2} + \sqrt{(x - c)^2 + (y - d)^2} + y.$$

We view $\sqrt{x^2 + (y - a)^2} + \sqrt{(x - c)^2 + (y - d)^2}$ as the sum of the distances of the moving point $M(x, 0)$ to two fixed points $A(0, a - y)$ and $B(c, y - d)$ (y is considered a constant temporarily). By the fact that the sum of any two sides of a triangle is not less than the third side

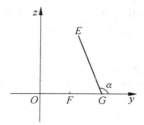

Figure 8.1 Figure for Exercise 9.

(the situation of co-linear three points is considered a special case), we have

$$\delta(R) \geq \sqrt{c^2 + (2y - a - d)^2} + y,$$

where the equality holds when M is an interior point of AB, that is, when

$$\frac{2y - a - d}{c} = \frac{y - a}{x} \tag{8.1}$$

is satisfied. View

$$z = \sqrt{(2y - a - d)^2 + c^2} + \sqrt{(2y - y)^2 + 0^2}$$

as the sum of the distances of $G(2y, 0)$ to points $E(a+d, c)$ and $F(y, 0)$. As in Figure 8.1, let $\angle yGE = \alpha$ and $|EG| = t$. Then

$$t = \frac{c}{\sin \alpha},$$

$$y = \frac{a + d - c \cot \alpha}{2}. \tag{8.2}$$

Therefore,

$$z = |EG| + |FG| = t + y$$

$$= \frac{a + d}{2} + c \left(\frac{c}{\sin \alpha} - \frac{\cot \alpha}{2} \right)$$

$$= \frac{a + d}{2} + \frac{1}{2} c \left(\frac{3}{2} \tan \frac{\alpha}{2} + \frac{1}{2} \cot \frac{\alpha}{2} \right)$$

$$\geq \frac{a + d}{2} + c \sqrt{\frac{3}{2} \tan \frac{\alpha}{2} \cdot \frac{1}{2} \cot \frac{\alpha}{2}}$$

$$= \frac{a + d}{2} + \frac{\sqrt{3}}{2} c,$$

where the equality is satisfied if and only if $\frac{3}{2}\tan\frac{\alpha}{2} = \frac{1}{2}\cot\frac{\alpha}{2}$, namely $\cot\alpha = \frac{\sqrt{3}}{3}$. Hence, from (8.1) and (8.2) we find that the coordinates of R are

$$\left(\frac{\sqrt{3}}{2}(a-d) + \frac{c}{2}, \frac{a+d}{2} - \frac{\sqrt{3}}{6}c \right).$$

10. Let $x_1 = (n+1)!, x_2 = (n+2)!, \ldots, x_n = (2n)!$, and let the points (x_i, y_i) $(1 \le i \le n)$ be on the parabola $y = x^2$. Then (x_i, y_i) are all integral points. The set of such n points satisfies the requirement of the problem.

11. Clearly P is not on the axes or the straight lines $y = \pm x$. Then among the four possible moves, only one move is towards the origin, while the other three moves are farther away from the origin. If the move passes through the successive points $P_0, P_1, \ldots, P_n = P_0$, where P_i is the farthest to O, then $|OP_{i-1}| < |OP_i|$ and $|OP_{i+1}| < |OP_i|$. Since there is only one move making it nearer to O, we have $P_{i-1} = P_{i+1}$. But this contradicts the rule that no return is allowed.

12. (1) Let the vertices of the polygon be $P_i(x_i, y_i)$ $(i = 1, 2, \ldots, 100)$. Then its area is

$$S = \frac{1}{2}\sum_{i=1}^{100} \begin{vmatrix} x_i & y_i \\ x_{i+1} & y_{i+1} \end{vmatrix} \quad (x_{101} = x_1 \quad \text{and} \quad y_{101} = y_1).$$

Without loss of generality,, assume that the first side $P_1 P_2$ is parallel to the y-axis. Then $x_{2i-1} = x_{2i}$, $y_{2i} = y_{2i+1}$, $x_{2i+1} - x_{2i} \equiv 1$, and $y_{2i+2} - y_{2i+1} \equiv 1 \pmod 2$ for $i = 1, 2, \ldots, 50$. Thus,

$$S = \frac{1}{2}(x_1 y_2 + x_2 y_3 + \cdots + x_{99} y_{100} - x_2 y_1 - x_3 y_2 - \cdots - x_1 y_{100})$$

$$= (x_1 - x_3)y_2 + (x_3 - x_5)y_4 + \cdots + (x_{99} - x_1)y_{100}$$

$$\equiv y_2 + y_4 + \cdots + y_{100}$$

$$\equiv 50y_2 + 25 \equiv 1 \pmod 2.$$

In other words, the area of P is an odd number.

Solutions for Chapter 9

Straight Lines

1. P and Q. Since N is on l, we have $b + \frac{1}{c} = 1$, thus $P\left(\frac{1}{c}, b\right)$ is on l. Also M is on l, so $a + \frac{1}{b} = 1$. Eliminating b from the above, we know that $c + \frac{1}{a} = 1$, thus $Q\left(\frac{1}{a}, c\right)$ is also on l.

2. 1. Let the equation of the straight line be $y = k(x - \sqrt{2})$. The case that $k = 0$ satisfies the condition. When $k \neq 0$, if the straight line passes through two rational points (x_1, y_1) and (x_2, y_2), then $\frac{y_1}{y_2} = \frac{x_1 - \sqrt{2}}{x_2 - \sqrt{2}}$, which implies that $\sqrt{2} = \frac{x_1 y_2 - x_2 y_1}{y_2 - y_1}$ is a rational number, a contradiction.

3. $\frac{3}{5}$. From geometry, the minimal distance from a point to a straight line is the same as the length of the perpendicular line segment from the point.

4. $7x - y - 3 = 0$.

5. $4\sqrt{5}$. Let the symmetric point of the point A with respect to the straight line l be A_1 and the symmetric point of A with respect to the y-axis be A_2. Then the minimum value of the perimeter of $\triangle ABC$ is $|A_1 A_2|$.

6. 7 (square units). Using the intuition from the geometric figure.

7. -2. If $\overrightarrow{a} \parallel \overrightarrow{b}$, then $\frac{2x-y}{-1} = \frac{m}{1}$, namely $m = -2x + y$. So $y = 2x + m$. Draw the feasible region (the shaded part of Figure 9.1) corresponding to the inequality, and translate the straight line $y = 2x + m$. From the translation we know that when the straight line passes through the point B, its intercept is the minimal, and the current value of m is the minimum value.

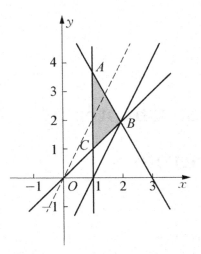

Figure 9.1   Figure for Exercise 7.

Solving $x = y$ and $y = 6 - 2x$ gives $x = 2$ and $y = 2$. That is, the coordinates of B are $(2, 2)$. Substituting into $m = -2x + y$, we obtain $m = -4 + 2 = -2$. In other words, the minimum value of the objective function $m = -2x + y$ is -2.

8. The coordinates of the point Q are $(2, 8)$, and the minimum value of the area of the triangle is 40.

9. Let $K_1 = \{(x, y) \mid |x| + |3y| - 6 \le 0\}$. First consider the part of K_1 in the first quadrant, the points of which satisfy $x + 3y \le 6$. Such points correspond to $\triangle OCD$ and its interior in the figure. By symmetry, the region corresponding to K_1 is the rhombus $ABCD$ and its interior.

Similarly, let $K_2 = \{(x, y) \mid |3x| + |y| - 6 \le 0\}$. Then the region corresponding to K_2 is the rhombus $EFGH$ and its interior.

From the definition of the point set K, we know that the region corresponding to K is the part that is covered by exactly one of K_1 and K_2. Therefore, what the problem wants to find is the area S of the shaded region in Figure 9.2.

Since the equation of the straight line CD is $x + 3y = 6$ and the equation of the straight line GH is $3x + y = 6$, the coordinates of their intersection point P are $\left(\frac{3}{2}, \frac{3}{2}\right)$. By symmetry,

$$S = 8S_{\triangle CPG} = 8 \cdot \frac{1}{2} \cdot 4 \cdot \frac{3}{2} = 24.$$

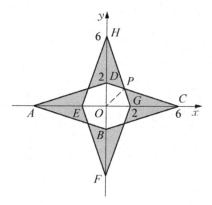

Figure 9.2 Figure for Exercise 9.

10. The maximum value of $f(x, y)$ is the same as the maximal vertical intercept of the straight line $y = -ax + b$ passing through the region $|x| + |y| \leq 1$. When $0 < a \leq 1$, we have $f(x, y)_{\max} = f(0, 1) = 1$, and $f(x, y)_{\max} = f(1, 0) = a$ when $a > 1$.

11. With A as the origin and the bisector of $\angle BAC$ as the x-axis, set up a rectangular coordinates system. Let $\angle BAE = \angle CAE = \theta$ and $\tan \theta = k$. It is easy to know that

$$|AF| = m = \frac{13\sqrt{12^4 + 60^2 k^2}}{144 - 25k^2}.$$

When $k = 1$, namely $m = \frac{2028}{119}$, we see that $\angle A$ is a right angle; when $k < 1$, namely $13 < m < \frac{2028}{119}$, the angle $\angle A$ is an acute one; when $k > 1$, namely $m > \frac{2028}{119}$, we get an obtuse angle $\angle A$.

12. Draw $OB \perp a$ at B. Set up a rectangular coordinates system with B as its origin and the straight line a as its x-axis. Let the radius of the circle O' be r, $|BO'| = x$, and $|BO| = m$. Then $m > R$ and $x^2 + m^2 = (R + r)^2$. Let the coordinates of A be $(0, n)$. It is easy to obtain

$$\tan \angle MAN = \frac{2nr}{R^2 + 2Rr - m^2 + n^2},$$

and the above expression is a constant if and only if $R^2 + n^2 = m^2$. Hence, points A that satisfy the condition exist, In fact there are two such points, and their coordinates are $(0, \pm\sqrt{m^2 - R^2})$.

Solutions for Chapter 10

Circles

1. $\frac{\pi}{2}$.   The inequality $(x-y)\left(y-\frac{1}{x}\right) \geq 0$ can be reduced to

$$\begin{cases} y - x \geq 0, \\ y - \dfrac{1}{x} \geq 0 \end{cases} \quad \text{or} \quad \begin{cases} y - x \leq 0, \\ y - \dfrac{1}{x} \leq 0. \end{cases}$$

Let B denote the set of all points on the circle $(x-1)^2 + (y-1)^2 = 1$ or inside the circle. The plane region represented by $A \cap B$ is the shaded region in Figure 10.1. Since the curve $y = \frac{1}{x}$ and the circle $(x-1)^2 + (y-1)^2 = 1$ are both symmetric with respect to the straight line $y = x$, so the area of the shaded region is half of that of circle.

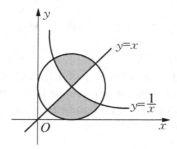

Figure 10.1 Figure for Exercise 1.

2. $\left[0, \frac{12}{5}\right]$.   The trajectory equation of the point M is $x^2 + (y+1)^2 = 4$. The center of the circle is $B(0, -1)$ and the center of the moving circle

is $C(a, 2a-4)$. The distance of the two centers is less than the sum of the two radii.

3. Tangent.

4. $(4, 6)$.

5. $\sqrt{3}\left(\pi - \arctan \frac{2\sqrt{30}}{7}\right)$.

6. $x^2 + y^2 - cx + \frac{c^2}{2} - \frac{r^2}{2} = 0$.

7. The equation of the circle is $(x-2)^2+(y-2)^2 = 1$, and the equation of its symmetric circle with respect to the x-axis is $(x-2)^2+(y+2)^2 = 1$. Let the equation of the desired straight line be $y - 3 = k(x+3)$. Then the distance of the point $(2, -2)$ to this straight line is 1, so $d = \frac{|5k+5|}{\sqrt{k^2+1}} = 1$, solving which gives $k = -\frac{3}{4}$ or $k = -\frac{4}{3}$. Hence, the equations of the desired straight lines are $4x + 3y + 3 = 0$ or $3x + 4y - 3 = 0$.

8. Let the center of the circle be $P(a, b)$ and the radius be r. By the condition, $r^2 = 1 + a^2$ and $r^2 = 2b^2$, so $a^2 + 1 = 2b^2$. Also the distance from P to l is $d = \frac{|a - 2b|}{\sqrt{5}}$, thus $d^2 = \frac{a^2 - 4ab + 4b^2}{5} \geq \frac{a^2 + 4b^2 - 2(a^2 + b^2)}{5} = \frac{2b^2 - a^2}{5} = \frac{1}{5}$, and the equality holds if and only if $a = b$. Therefore, $a = 1$ and $b = 1$, or $a = -1$ and $b = -1$.

   Hence, the equations of the circles are $(x - 1)^2 + (y - 1)^2 = 2$ or $(x + 1)^2 + (y + 1)^2 = 2$.

9. Denote $P(x_1, y_1)$ and $Q(x_2, y_2)$. Since $OP \perp OQ$, we have $x_1 x_2 + y_1 y_2 = 0$. Substituting $y = \frac{-x+3}{2}$ into $x^2 + y^2 + x - 6y + F = 0$ leads to $5x^2 + 10x + 4F - 27 = 0$. Then $x_1 + x_2 = -2$ and $x_1 x_2 = \frac{4F - 27}{5}$. Also, since $y_1 y_2 = \frac{1}{4}(3 - x_1)(3 - x_2) = \frac{1}{4}[9 - 3(x_1 + x_2) + x_1 x_2]$, so

$$9 - 3(x_1 + x_2) + x_1 x_2 + 4x_1 x_2 = 0.$$

Thus, we find that $F = 3$.

10. As in Figure 10.2, let the equation of PQ be $y = kx$ (it is easy to see that the case that the slope does not exist does not satisfy the condition of the problem), and denote the points $P(x_1, y_1), Q(x_2, y_2)$ $(x_1 x_2 \neq 0)$, and $S(m, n)$. Then

$$y_1 = kx_1, \tag{10.1}$$

$$y_2 = kx_2, \tag{10.2}$$

$$n = km. \tag{10.3}$$

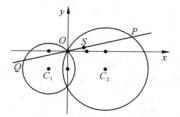

Figure 10.2   Figure for Exercise 10.

Substituting (10.1) and (10.2) respectively into the equations of the corresponding circles, we obtain

$$\begin{cases} x_1^2 + k^2 x_1^2 + 2x_1 + 2kx_1 = 0, \\ x_2^2 + k^2 x_2^2 - 4x_2 + 2kx_2 = 0. \end{cases}$$

Since $x_1 x_2 \neq 0$, the solutions of the above are

$$x_1 = \frac{-2 - 2k}{1 + k^2}, \quad x_2 = \frac{-2k + 4}{1 + k^2}.$$

The fact that S is the middle point of PQ implies

$$m = \frac{x_1 + x_2}{2} = \frac{1 - 2k}{1 + k^2}.$$

From (10.3), $k = \frac{n}{m}$, thus

$$m = \frac{1 - \frac{2n}{m}}{1 + \left(\frac{n}{m}\right)^2},$$

in other words,

$$m^2 + n^2 - m + 2n = 0.$$

Hence, the equation of the trajectory is $x^2 + y^2 - x + 2y = 0$ (with the point $(0, -2)$ removed).

11. (1) Since the circle C passes through the origin O, so $|OC|^2 = t^2 + \frac{4}{t^2}$, and the equation of the circle C is $(x - t)^2 + \left(y - \frac{2}{t}\right)^2 = t^2 + \frac{4}{t^2}$, namely $x^2 + y^2 - 2tx - \frac{4}{t}y = 0$. Thus, we get the points $A(2t, 0)$ and $B\left(0, \frac{4}{t}\right)$.

It follows that $S = \frac{1}{2}|OA||OB| = \frac{1}{2}|2t| \left|\frac{4}{t}\right| = 4$, which is a fixed number.

(2) Since $|OM| = |ON|$ and $|CM| = |CN|$, the straight line OC is the perpendicular bisector of the line segment MN, so $k_{OC} = -\frac{1}{k_{MN}} = \frac{1}{2}$, and the equation of the straight line OC is $y = \frac{1}{2}x$. Consequently, $\frac{2}{t} = \frac{1}{2}t$, the solutions of which are $t = \pm 2$.

When $t = 2$, the center of the circle is $C(2,1)$ with radius $|OC| = \sqrt{5}$, and the distance from the center C of the circle to the straight line l is $d = \frac{1}{\sqrt{5}} < \sqrt{5}$, satisfying the requirement.

When $t = -2$, the center of the circle is $C(-2,-1)$ with radius $|OC| = \sqrt{5}$, and the distance from the center C of the circle to the straight line l is $d = \frac{9}{\sqrt{5}} > \sqrt{5}$, not satisfying the requirement.

In summary, the standard equation of the circle C is $(x-2)^2 + (y-1)^2 = 5$.

12. As in Figure 10.3, let O be the origin and the east direction be the positive direction of the x-axis of a rectangular coordinates system.

At the moment t (in hour), the coordinates of the center $P'(x', y')$ of the typhoon are

$$\begin{cases} x' = 300 \cdot \dfrac{\sqrt{2}}{10} - 20 \cdot \dfrac{\sqrt{2}}{2}t, \\[3mm] y' = -300 \cdot \dfrac{7\sqrt{2}}{10} + 20 \cdot \dfrac{\sqrt{2}}{2}t. \end{cases}$$

Now, the radius of the circle is $10t + 60$, and the region influenced by the typhoon is represented by

$$(x - x')^2 + (y - y')^2 \leq (10t + 60)^2.$$

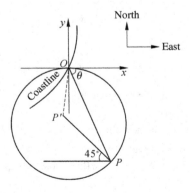

Figure 10.3   Figure for Exercise 12.

If the city O is invaded by the typhoon at the moment t, then

$$(0 - x')^2 + (0 - y')^2 \le (10t + 60)^2,$$

that is, $\left(-300 \cdot \frac{\sqrt{2}}{10} + 20 \cdot \frac{\sqrt{2}}{2}t\right)^2 + \left(300 \cdot \frac{7\sqrt{2}}{10} - 20 \cdot \frac{\sqrt{2}}{2}t\right)^2 \le (10t+60)^2$. In other words, $t^2 - 36t + 288 \le 0$, solving which gives $12 \le t \le 24$. Hence, 12 hours later the city will be attacked by the typhoon.

Solutions for Chapter 11

Ellipses

1. 1.

2. $\left(\frac{\sqrt{5}}{5}, \frac{3}{5}\right)$. From the given condition, we see immediately that $b < \frac{b}{2} + c < a$. Hence, $\frac{b}{2} < c$ and $\frac{b}{2} < a - c$, squaring of which gives $\frac{b^2}{4} = \frac{a^2 - c^2}{4} < c^2$ and $\frac{a^2 - c^2}{4} < (a-c)^2$, namely $a^2 - 5c^2 < 0$ and $3a^2 - 8ac + 5c^2 > 0$. Thus, $e^2 > \frac{1}{5}$ and $5e^2 - 8e + 3 > 0$, and consequently $\frac{\sqrt{5}}{5} < e < \frac{3}{5}$.

3. $90°$. From $\frac{c}{a} = \frac{\sqrt{5}-1}{2}$, we get $c^2 + ac - a^2 = 0$. Also $|AB|^2 = a^2 + b^2$ and $|BF|^2 = a^2$, so $|AB|^2 + |BF|^2 = 2a^2 + b^2 = 3a^2 - c^2$. On the other hand,

$$|AF|^2 = (a+c)^2 = a^2 + 2ac + c^2 = 3a^2 - c^2 = |AB|^2 + |BF|^2.$$

Therefore, $\angle ABF = 90°$.

4. bc. $S_{\triangle ABF} = \frac{1}{2}|y_A - y_B|c \leq \frac{1}{2} \cdot 2bc = bc$.

5. It is easy to see that the coordinates of B are $(0, k)$. Denote $A(x_0, y_0)$ and $D(m, 0)$. Then $\overrightarrow{BA} = (x_0, y_0 - k)$ and $\overrightarrow{BD} = (m, -k)$. Since the three points B, A, and D are co-linear, so

$$|BA||BD| = \overrightarrow{BA} \cdot \overrightarrow{BD} = mx_0 - ky_0 + k^2. \tag{11.1}$$

The equation of the straight line BD is $\frac{x}{m} + \frac{y}{k} = 1$. Since $A(x_0, y_0)$ is on the straight line, we have $\frac{x_0}{m} + \frac{y_0}{k} = 1$, namely $m = \frac{kx_0}{k - y_0}$. Substituting it into (11.1) gives that

$$|BA||BD| = \frac{kx_0^2}{k - y_0} - ky_0 + k^2 = \frac{k(x_0^2 + y_0^2) - k^2 y_0}{k - y_0} + k^2.$$

Since A is also on the circle O, we have $x_0^2 + y_0^2 = k^2$, and it follows that

$$|BA||BD| = \frac{k^2(k - y_0)}{k - y_0} + k^2 = 2k^2 \quad \text{(a fixed number).}$$

6. As in Figure 11.1, denote $A(-a, 0), B(a, 0)$, and $P(x_0, y_0)$. We may assume that $y_0 > 0$. Since $k_{AP} = \frac{y_0}{x_0 + a}$ and $k_{BP} = \frac{y_0}{x_0 - a}$, by the difference angle formula, $\frac{k_{BP} - k_{AP}}{1 + k_{BP}k_{AP}} = \tan \frac{2\pi}{3}$, that is,

$$\frac{\frac{y_0}{x_0 - a} - \frac{y_0}{x_0 + a}}{1 + \frac{y_0}{x_0 - a} \cdot \frac{y_0}{x_0 + a}} = -\sqrt{3}.$$

Simplifying the above equality gives

$$\sqrt{3}(x_0^2 + y_0^2 - a^2) + 2ay_0 = 0. \tag{11.2}$$

Since $P(x_0, y_0)$ is on the ellipse, $\frac{x_0^2}{a^2} + \frac{y_0^2}{b^2} = 1$, that is, $x_0^2 = a^2\left(1 - \frac{y_0^2}{b^2}\right)$. Substituting the above into (11.2) gives $-\sqrt{3} \cdot \frac{a^2 - b^2}{b^2} \cdot y_0^2 + 2ay_0 = 0$.

Since $0 < y_0 \le b$, so $y_0 = \frac{2ab^2}{\sqrt{3}c^2} \le b$, namely $\frac{ab}{c^2} \le \frac{\sqrt{3}}{2}$. Thus, $\frac{a^2}{c^2}\frac{\sqrt{a^2 - c^2}}{a} \le \frac{\sqrt{3}}{2}$. That is, $\frac{1}{e^2}\sqrt{1 - e^2} \le \frac{\sqrt{3}}{2}$. Solving it gives that $\frac{\sqrt{6}}{3} \le e < 1$.

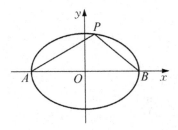

Figure 11.1   Figure for Exercise 6.

7. Denote $M(x_0, y_0)$. Then the equation of the tangent line to the ellipse at the point M is $\frac{x_0 x}{a^2} + \frac{y_0 y}{b^2} = 1$. Since $OP \parallel l$, the equation of the straight line OP is $y = -\frac{b^2 x_0}{a^2 y_0}$. The equation of the straight line MF is $y = \frac{y_0}{x_0 - c}(x - c)$. Solving the above two equations gives $N\left(\frac{a^2 y_0^2 c}{b^2(a^2 - cx_0)}, -\frac{x_0 y_0 c}{a^2 - cx_0}\right)$. By the distance formula between two points and $y_0^2 = b^2\left(1 - \frac{x_0^2}{a^2}\right)$, we obtain that $|MN| = a$.

8. As Figure 11.2 shows, draw PP' to be perpendicular to the directrix l at P', and draw QQ' to be perpendicular to the directrix l at Q', choose the middle point M of the chord PQ, and draw the perpendicular bisector of PQ that intersects l at R. Then the statement that $\triangle PQR$ is an equilateral triangle is equivalent to that the equality $|MR| = \frac{\sqrt{3}}{2}|PQ|$ is valid.

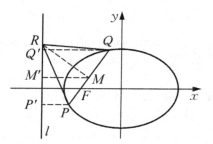

Figure 11.2 Figure for Exercise 8.

Suppose that there exists such an equilateral triangle. Clearly $|MR| > |MM'|$ (where $MM' \perp l$ with foot M' of the perpendicular). By the second definition of an ellipse,

$$|MM'| = \frac{1}{2}(|PP'| + |QQ'|)$$

$$= \frac{1}{2}\left(\frac{|PF|}{e} + \frac{|QF|}{e}\right) = \frac{|PQ|}{2e}.$$

Therefore, $\frac{\sqrt{3}}{2}|PQ| > \frac{|PQ|}{2e}$, namely $e > \frac{\sqrt{3}}{3}$. Thus, $e \in \left(\frac{\sqrt{3}}{3}, 1\right)$.

Also, since $\cos \angle RMM' = \frac{|MM'|}{|RM|} = \frac{|PQ|}{2e}\frac{2}{\sqrt{3}|PQ|} = \frac{1}{\sqrt{3}e}$, we have $\cot \angle RMM' = \frac{1}{\sqrt{3e^2 - 1}}$. Consequently,

$$k_{PQ} = \tan \angle QFX = \tan \angle FMM' = \pm \cot \angle RMM' = \pm \frac{1}{\sqrt{3e^2 - 1}}.$$

9. Let the equation of the inner ellipse be $\frac{x^2}{a^2} + \frac{y^2}{b^2} = 1$ $(a > b > 0)$ and the equation of the outer ellipse be $\frac{x^2}{(ma)^2} + \frac{y^2}{(mb)^2} = 1$. Then we have $A(ma, 0)$ and $B(0, mb)$. Let the equation of the tangent line AC be $y = k_1(x - ma)$ and the equation of the tangent line BD be $y - mb = k_2 x$. Eliminating y from the combined equations $(bx)^2 + (ay)^2 = (ab)^2$

and $y = k_1(x - ma)$ gives

$$(b^2 + a^2 k_1^2)x^2 - 2ma^3 k_1^2 x + m^2 a^4 k_1^2 - a^2 b^2 = 0.$$

Simplify the equation $\Delta = 0$, where Δ is the discriminant of the above quadratic equation. Then we obtain $k_1^2 = \frac{b^2}{a^2} \frac{1}{m^2-1}$. By the same token, eliminating y from $(bx)^2 + (ay)^2 = (ab)^2$ and $y = k_2 x + mb$ implies

$$(b^2 + a^2 k_2^2)x^2 + 2mba^2 k_2 x + m^2 a^2 b^2 - a^2 b^2 = 0.$$

A Simplification of $\Delta = 0$ gives $k_2^2 = \frac{b^2}{a^2}(m^2 - 1)$. The condition $k_1^2 k_2^2 = \left(-\frac{9}{16}\right)^2$ implies that $\frac{b^2}{a^2} = \frac{9}{16}$, from which $e^2 = \frac{c^2}{a^2} = 1 - \frac{b^2}{a^2} = \frac{7}{16}$. Hence, $e = \frac{\sqrt{7}}{4}$.

10. First consider a special case. When $MN \perp$ the x-axis, we may write $M\left(x_0, \frac{b}{a}\sqrt{a^2 - x_0^2}\right)$ and $N\left(x_0, -\frac{b}{a}\sqrt{a^2 - x_0^2}\right)$. Since $MA \perp NA$, we have $\frac{b}{a}\sqrt{a^2 - x_0^2} = x_0 - (-a)$, and simplifying it gives that $(a^2 + b^2)x_0^2 + 2a^3 x_0 + a^2(a^2 - b^2) = 0$. The solutions are $x_0 = -\frac{a(a^2-b^2)}{a^2+b^2}$ or $x_0 = -a$ (omitted). So, the equation of the straight line MN is $x = -\frac{a(a^2-b^2)}{a^2+b^2}$.

When MN is not perpendicular to the x-axis, let the equation of MN be $y = kx + m$. Combining it with $b^2 x^2 + a^2 y^2 = a^2 b^2$ and eliminating y give that $(b^2 + a^2 k^2)x^2 + 2a^2 kmx + a^2(m^2 - b^2) = 0$. Denote $M(x_1, y_1)$ and $N(x_2, y_2)$. From $\overrightarrow{AM} \cdot \overrightarrow{AN} = 0$, we see that $(1 + k^2)x_1 x_2 + (a + km)(x_1 + x_2) + a^2 + m^2 = 0$. Substituting $x_1 + x_2 = -\frac{2a^2 km}{b^2 + a^2 k^2}$ and $x_1 x_2 = \frac{a^2(m^2 - b^2)}{b^2 + a^2 k^2}$ into the above equality, we get after a simplification that

$$(a^2 + b^2)m^2 - 2a^3 km + a^2 k^2(a^2 - b^2) = 0,$$

the solutions of which are $m = ak$ or $m = \frac{ak(a^2-b^2)}{a^2+b^2}$. When $m = ak$, the equation of MN is $y = kx + ak = k(x + a)$ that passes through $(-a, 0)$, not satisfying the condition of the problem. When $m = \frac{ak(a^2-b^2)}{a^2+b^2}$, the equation of MN is

$$y = kx + \frac{ak(a^2 - b^2)}{a^2 + b^2} = k\left[x + \frac{a(a^2 - b^2)}{a^2 + b^2}\right],$$

which passes through the point $\left(-\frac{a(a^2-b^2)}{a^2+b^2}, 0\right)$.

To summarize, the straight line MN always passes through the fixed point $\left(-\frac{a(a^2-b^2)}{a^2+b^2}, 0\right)$.

11. Suppose that there exists such a circumscribed quadrilateral of the circle (and also an inscribed quadrilateral of the ellipse). From the

knowledge of plane geometry, this parallelogram must be a rhombus. Then the center of the circle is the center of the rhombus.

First consider a special case. When the coordinates of the point P are $(a, 0)$, the coordinates of the other three vertices of the rhombus are $(-a, 0), (0, b)$, and $(0, -b)$. The equation of one side of the rhombus is $\frac{x}{a} + \frac{y}{b} = 1$, that is $bx + ay = ab$. Since the rhombus is tangent to the circle C_1, we have $\frac{ab}{\sqrt{b^2 + a^2}} = 1$, namely $\frac{1}{a^2} + \frac{1}{b^2} = 1$, which is the condition for a and b to satisfy. This proves the necessity.

We show the sufficiency below. Let $\frac{1}{a^2} + \frac{1}{b^2} = 1$, and P be any point on the ellipse C_2. Draw a straight line PO that intersects the ellipse C_2 at points P and R. Draw at O a straight line that is perpendicular to PR, which intersects the ellipse C_2 at points Q and S. Then $PQRS$ is an inscribed rhombus to the ellipse C_2. Let $|OP| = r_1$ and $|OQ| = r_2$. Then substituting the coordinates $P(r_1 \cos \theta, r_1 \sin \theta)$ and $Q\left(r_2 \cos\left(\theta + \frac{\pi}{2}\right), r_2 \sin\left(\theta + \frac{\pi}{2}\right)\right)$ into the equation of the ellipse, we obtain that

$$\frac{(r_1 \cos \theta)^2}{a^2} + \frac{(r_1 \sin \theta)^2}{b^2} = 1,$$

$$\frac{\left[r_2 \cos\left(\theta + \frac{\pi}{2}\right)\right]^2}{a^2} + \frac{\left[r_2 \sin\left(\theta + \frac{\pi}{2}\right)\right]^2}{b^2} = 1.$$

Then

$$\frac{1}{|OP|^2} + \frac{1}{|OQ|^2} = \frac{1}{r_1^2} + \frac{1}{r_2^2}$$

$$= \left(\frac{\cos^2 \theta}{a^2} + \frac{\sin^2 \theta}{b^2}\right) + \left[\frac{\cos^2\left(\theta + \frac{\pi}{2}\right)}{a^2} + \frac{\sin^2\left(\theta + \frac{\pi}{2}\right)}{b^2}\right]$$

$$= \frac{1}{a^2} + \frac{1}{b^2}.$$

Also in the right $\triangle POQ$, let the distance from O to PQ be h. Then $|PQ|h = |PO||QO|$, and

$$\frac{1}{h^2} = \frac{|PQ|^2}{|PO|^2|QO|^2} = \frac{|PO|^2 + |QO|^2}{|PO|^2|QO|^2}$$

$$= \frac{1}{|PO|^2} + \frac{1}{|QO|^2} = \frac{1}{a^2} + \frac{1}{b^2} = 1.$$

Thus, $h = 1$, in other words, PQ and the circle C_1 are tangent.

The same argument can show that the other sides are also tangent to the circle C_1.

Therefore, a necessary and sufficient condition is that $\frac{1}{a^2} + \frac{1}{b^2} = 1$.

12. Suppose that such a constant d exists. We start from a special case. Let $l \in A_d$ be a special straight line $x = d$ that intersects the ellipse at two points P and Q.

Draw a circle O_d with the origin O as its center and d as its radius, which intersects the positive half axis of the x-axis at a point G. Obviously, the circle O_d and the straight line PQ are tangent at the point G, and $|GP| = |GQ|$.

By the condition, there exist straight lines l_1 and l_2 that pass through the points P and Q respectively, and are tangent to the circle O_d with tangent points P_0 and Q_0 respectively. Then OP_0 and OQ_0 are perpendicular to the parallel lines l_1 and l_2, so P_0Q_0 is a diameter of the circle O_d. Thus, OG is the median line of the trapezoid P_0Q_0QP. From $OG \perp PQ$, we know that $l_1 \perp PQ$ and $l_2 \perp PQ$. Consequently, the coordinates of the point P are (d, d), and l_1 and l_2 are $y = d$ and $y = -d$ respectively.

Since $P(d, d)$ is on the ellipse, if such a constant d exists, then the equation of the ellipse ensures that $\frac{d^2}{a^2} + \frac{d^2}{b^2} = 1$, solving which gives $d = \frac{ab}{\sqrt{a^2+b^2}}$.

In the following, we prove that $d = \frac{ab}{\sqrt{a^2+b^2}}$ is what we wanted to find.

First we show that if $l \in A_d$ and intersects the ellipse at the two points P and Q, then it must be that $OP \perp OQ$.

Let the equation of the straight line l be $y = kx + m$. Then the distance from the origin O to l is $\frac{|m|}{\sqrt{k^2+1}}$, so we have $\frac{ab}{\sqrt{a^2+b^2}} = \frac{|m|}{\sqrt{k^2+1}}$, that is,

$$(k^2 + 1)a^2b^2 = m^2(a^2 + b^2).$$

Substituting the equation $y = kx + m$ of the straight line l into the equation of the ellipse implies that

$$(a^2k^2 + b^2)x^2 + 2a^2kmx + a^2(m^2 - b^2) = 0.$$

Denote $P(x_1, y_1)$ and $Q(x_2, y_2)$. Then by Vieta's theorem,

$$x_1 + x_2 = -\frac{2a^2km}{a^2k^2 + b^2}, \quad x_1x_2 = \frac{a^2(m^2 - b^2)}{a^2k^2 + b^2}.$$

It follows that

$$x_1x_2 + y_1y_2$$

$$= x_1x_2 + (kx_1 + m)(kx_2 + m)$$

$$= (k^2 + 1)x_1x_2 + mk(x_1 + x_2) + m^2$$

$$= \frac{k^2a^2m^2 + a^2m^2 - (k^2 + 1)a^2b^2 - 2k^2a^2m^2 + k^2a^2m^2 + b^2m^2}{a^2k^2 + b^2}$$

$$= \frac{(a^2 + b^2)m^2 - (k^2 + 1)a^2b^2}{a^2k^2 + b^2} = 0.$$

In other words, $OP \perp OQ$.

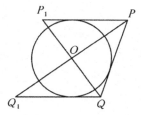

Figure 11.3   Figure for Exercise 12.

It is easy to show that if the slope of the straight line l does not exist, then $OP \perp OQ$.

Suppose that l_1 intersects the ellipse at P and P_1, and l_2 intersects the ellipse at Q and Q_1. Then as Figure 11.3 indicates, $OP \perp OQ, OP \perp OP_1$, and $OQ \perp OQ_1$, and $\angle OPP_1 = \angle OPQ$ and $\angle OQQ_1 = \angle OQP$. Hence, $\angle P_1PQ + \angle Q_1QP = \pi$, namely l_1 and l_2 are parallel.

In summary, there exists a unique number $d = \frac{ab}{\sqrt{a^2+b^2}}$ such that for any $l \in A_d$ there are $l_1, l_2 \in A_d$ that satisfy the conditions posed in the problem.

Solutions for Chapter 12

Hyperbolas

1. 2.   Let $|PF_1| = 4t$. Then $|PF_2| = 3t$, so $4t - 3t = |PF_1| - |PF_2| = 2$, from which $t = 2$, $|PF_1| = 8$, and $|PF_2| = 6$. Combining the above with $|F_1F_2| = 10$, we find that $\triangle PF_1F_2$ is a right triangle with $PF_1 \perp PF_2$. Therefore, the radius of the incircle of $\triangle PF_1F_2$ is $r = \frac{6+8-10}{2} = 2$.

2. 4.   By the symmetry of the figure, there are following two possibilities of making exactly three straight lines: (1) Two lines intersecting the right branch and one line intersecting both the left and right branches. The latter is the x-axis and the length of the line segment is 2. But the minimum value of the length of the line segment from the intersection of any straight line passing through the right focus with the hyperbola is $2\frac{b}{a} = 4$, which does not satisfy the requirement in the problem. (2) One line intersecting the right branch and two lines intersecting both the left and right branches. Then $\lambda = 4$ satisfies the requirement.

3. $\left[\frac{\sqrt{5}}{2}, \sqrt{5}\right]$.   Let the equation of the straight line be $\frac{x}{a} + \frac{y}{b} = 1$, that is $bx + ay - ab = 0$. From the distance formula from a point to a straight line and the fact $a > 1$, we find that the distance from the point $(1, 0)$ to the straight line l is $d_1 = \frac{ba - b}{\sqrt{a^2 + b^2}}$.

The distance from the point $(-1, 0)$ to the straight line l is $d_2 = \frac{ba + b}{\sqrt{a^2 + b^2}}$.

So, $s = d_1 + d_2 = \frac{2ab}{\sqrt{a^2 + b^2}} = \frac{2ab}{c}$.

From $s \geq \frac{4}{5}c$, we get $\frac{2ab}{c} \geq \frac{4}{5}c$, namely $5a\sqrt{c^2 - a^2} \geq 2c^2$.

Since $e = \frac{c}{a}$, we find that $5\sqrt{e^2 - 1} \geq 2e^2$. Thus, $\frac{5}{4} \leq e^2 \leq 5$ ($e > 1$).

Hence, $\frac{\sqrt{5}}{2} \leq e \leq \sqrt{5}$.

4. $\frac{a(l+2a)}{2\sqrt{a^2+b^2}}$.   First, when the length of the chord with the two end points on the right branch is greater than or equal to $\frac{2b^2}{a}$, this chord may pass through the right focus. This is because that among all the chords passing through the right focus, when the focal chord is perpendicular to the real axis, the length of the chord is minimal according to the second definition of hyperbolas, and the minimum value is $\frac{2b^2}{a}$ by an easy calculation. When a chord of length l passes through the right focus, if we let the distances from A and B to the right directrix be d_1 and d_2 respectively, then $l = |AF| + |BF| = ed_1 + ed_2$ (here F is the right focus of the hyperbola and e is the eccentricity of the hyperbola). Thus,

$$x_M = \frac{1}{2}(d_1 + d_2) + \frac{a^2}{c} = \frac{1}{2}\frac{l}{e} + \frac{a^2}{c} = \frac{a(l+2a)}{2\sqrt{a^2+b^2}}.$$

When a chord of length l does not pass through the right focus,

$$|AF| + |BF| > |AB| = l,$$

namely $ed_1 + ed_2 > l$. Consequently,

$$x_M = \frac{1}{2}(d_1 + d_2) + \frac{a^2}{c} > \frac{l}{2e} + \frac{a^2}{c} = \frac{a(l+2a)}{2\sqrt{a^2+b^2}}.$$

5. $(2, 2\sqrt{2}]$.   Without loss of generality, assume that P is on the right branch. Write $P(x_0, y_0)$. Then $x_0 \geq a$. By the computational formula for the focal radius of a hyperbola, $|PF_1| = ex_0 + a$ and $|PF_2| = ex_0 - a$, where $e = \sqrt{2}$. Hence, $\frac{|PF_1|+|PF_2|}{|PO|} = \frac{2ex_0}{\sqrt{x_0^2+y_0^2}} = \frac{2\sqrt{2}x_0}{\sqrt{2x_0^2-a^2}} = \frac{2\sqrt{2}}{\sqrt{2-\frac{a^2}{x_0^2}}} \in (2, 2\sqrt{2}]$.

6. Suppose that such hyperbolas exist. We may assume that their equations are $3(x-m)^2 - (y-m)^2 = k$. By substituting $y = x$ into the above, we get $2(x-m)^2 = k$. Let $k > 0$. Then $x = m \pm \sqrt{\frac{k}{2}}$, so the length of the line segment intersected by l is $\sqrt{2}\left|m + \sqrt{\frac{k}{2}} - \left(m - \sqrt{\frac{k}{2}}\right)\right| = 2\sqrt{k} = 2\sqrt{2}$. Hence, $k = 2$.

Let $x = 0$ in $3(x-m)^2 - (y-m)^2 = k$. Then $y = m \pm \sqrt{3m^2 - k}$, from which the length of the line segment obtained via the intersection of l and the hyperbola is $2\sqrt{3m^2 - k} = 2\sqrt{3m^2 - 2} = 2\sqrt{2}$. Thus, $m = \pm\frac{2\sqrt{3}}{3}$. Therefore, there exist such hyperbolas $3\left(x + \frac{2\sqrt{3}}{3}\right)^2 - \left(y + \frac{2\sqrt{3}}{3}\right)^2 = 2$, or $3\left(x - \frac{2\sqrt{3}}{3}\right)^2 - \left(y - \frac{2\sqrt{3}}{3}\right)^2 = 2$.

7. (1) Let the equation of l be $y = kx + 1$. Eliminating y from $x^2 - \frac{y^2}{3} = 1$ and $y = kx + 1$, we have $(3 - k^2)x^2 - 2kx - 4 = 0$.

Since the straight line l and the hyperbola C have two different intersection points,

$$\begin{cases} 3 - k^2 \neq 0, \\ \Delta = 4k^2 + 16(3 - k^2) > 0, \end{cases}$$

whose solution set is $-2 < k < 2$, and $k \neq \pm\sqrt{3}$. Hence, the range of the values for k is

$$(-2, -\sqrt{3}) \cup (-\sqrt{3}, \sqrt{3}) \cup (\sqrt{3}, 2).$$

(2) Denote $A(x_1, y_1)$ and $B(x_2, y_2)$. Then $x_1 + x_2 = \frac{2k}{3-k^2}$ and $x_1 x_2 = \frac{-4}{3-k^2}$. The coordinates of the right focus are $F_2(2, 0)$, so $|AF_2| = \sqrt{(x_1 - 2)^2 + y_1^2} = \sqrt{x_1^2 - 4x_1 + 4 + (3x_1^2 - 3)} = |2x_1 - 1|$ and $|BF_2| = |2x_2 - 1|$.

On the other hand, $(2x_1 - 1)(2x_2 - 1) = 4x_1 x_2 - 2(x_1 + x_2) + 1 = \frac{-16}{3-k^2} - \frac{4k}{3-k^2} + 1 = -\frac{k^2 + 4k + 13}{3 - k^2}$. Thus, when $k^2 < 3$, we see that $(2x_1 - 1)(2x_2 - 1) < 0$, and so

$$|AF_2| + |BF_2| = |2x_1 - 1| + |2x_2 - 1|$$

$$= 2|x_1 - x_2| = 2\sqrt{(x_1 + x_2)^2 - 4x_1 x_2}$$

$$= \frac{4\sqrt{3}\sqrt{4 - k^2}}{3 - k^2}.$$

The condition $|AF_2| + |BF_2| = 6$ implies that $\frac{4\sqrt{3}\sqrt{4-k^2}}{3-k^2} = 6$, from which $k^2 = 1$ or $k^2 = \frac{11}{3} > 3$ (omitted). Hence, $k = \pm 1$.

When $3 < k^2 < 4$, clearly $(2x_1 - 1)(2x_2 - 1) > 0$. Then

$$|AF_2| + |BF_2| = |2x_1 - 1| + |2x_2 - 1|$$

$$= 2|x_1 + x_2 - 1| = 2\left| \frac{2k}{3 - k^2} - 1 \right|.$$

Now, from $|AF_2| + |BF_2| = 6$ we have $2\left| \frac{2k}{3-k^2} - 1 \right| = 6$, whose solutions are $k = -2, k = \frac{3}{2}$, or $k = \frac{1 \pm \sqrt{13}}{2}$. None of them satisfies the condition, so omitted. Thus, for this case, no k exists to satisfy the given condition.

To summarize, the values of k are 1 or -1.

8. (1) The equation of the straight line containing the chord at contact AB is $x_0 x - y_0 y = 1$.

   From $x = m$ we have $l_{AB} : mx - y_0 y = 1$. Substituting $M\left(\frac{1}{m}, 0\right)$ into the straight line equation verifies the conclusion.

   (2) Denote $A(x_1, y_1)$. Then the equation of the perpendicular line AN is $y - y_1 = -x + x_1$.

   From $y - y_1 = -x + x_1$ and $x - y = 0$ we get $N\left(\frac{x_1 + y_1}{2}, \frac{x_1 + y_1}{2}\right)$. Denote the center of gravity by $G(x, y)$. Then

   $$\begin{cases} x = \dfrac{1}{3}\left(x_1 + \dfrac{1}{m} + \dfrac{x_1 + y_1}{2}\right), \\ y = \dfrac{1}{3}\left(y_1 + 0 + \dfrac{x_1 + y_1}{2}\right) \end{cases} \Rightarrow \begin{cases} x_1 = \dfrac{9x - 3y - \frac{3}{m}}{4}, \\ y_1 = \dfrac{9y - 3x + \frac{1}{m}}{4}. \end{cases}$$

   Since $x_1^2 - y_1^2 = 1$, the equation of the curve drawn from the center of gravity G is $\left(x - \frac{1}{3m}\right)^2 - y^2 = \frac{2}{9}$.

9. (1) Since the eccentricity of the hyperbola is 2, we have $c = 2a$ and $b = \sqrt{3}a$. So, the equation of the hyperbola can be reduced to $\frac{x^2}{a^2} - \frac{y^2}{3a^2} = 1$. Since the equation of the straight line is $y = x + m$, eliminating y from the two equations together gives

   $$2x^2 - 2mx - m^2 - 3a^2 = 0. \tag{12.1}$$

   Denote $A(x_1, y_1)$ and $B(x_2, y_2)$. Then $x_1 + x_2 = m$ and $x_1 x_2 = \frac{-m^2 - 3a^2}{2}$.

   Since $\overrightarrow{AP} = 3\overrightarrow{PB}$, so $(-x_1, m - y_1) = 3(x_2, y_2 - m)$. Thus, $x_1 = -3x_2$, from which and $x_1 + x_2 = m$ we get $x_1 = \frac{3}{2}m$ and $x_2 = -\frac{1}{2}m$. Substituting them into $x_1 x_2 = \frac{-m^2 - 3a^2}{2}$ results in the equation $-\frac{3}{4}m^2 = \frac{-m^2 - 3a^2}{2}$. After a simplification it is $m^2 = 6a^2$. Now,

   $$\overrightarrow{OA} \cdot \overrightarrow{OB} = x_1 x_2 + y_1 y_2 = x_1 x_2 + (x_1 + m)(x_2 + m)$$

   $$= 2x_1 x_2 + m(x_1 + x_2) + m^2 = m^2 - 3a^2 = 3a^2.$$

   Since $\overrightarrow{OA} \cdot \overrightarrow{OB} = 3$, we find $a^2 = 1$ and then $m = \sqrt{6}$. Substituting into (12.1) and manipulating the resulting expression, we obtain that $2x^2 - 2\sqrt{6}x - 9 = 0$. Clearly this equation has two different real roots, meaning that $a^2 = 1$ satisfies the requirement. Hence, the equation of the hyperbola is $x^2 - \frac{y^2}{3} = 1$.

(2) Suppose that the point M exists, written as $M(t, 0)$. From (1), the right focus of the hyperbola is $F(2, 0)$. Let $Q(x_0, y_0)$ $(x_0 \geq 1)$ be a point on the right branch of the hyperbola.

When $x_0 \neq 2$, we have $\tan \angle QFM = -k_{QF} = -\frac{y_0}{x_0 - 2}$ and $\tan \angle QMF = k_{QM} = \frac{y_0}{x_0 - t}$.

Since $\angle QFM = 2\angle QMF$, so $-\frac{y_0}{x_0 - 2} = \frac{2\frac{y_0}{x_0 - t}}{1 - \left(\frac{y_0}{x_0 - t}\right)^2}$. Substituting $y_0^2 = 3x_0^2 - 3$ into it, we have after a simplification that

$$-2x_0^2 + (4 + 2t)x_0 - 4t = -2x_0^2 - 2tx_0 + t^2 + 3.$$

Thus, solving $4 + 2t = -2t$ and $-4t = t^2 + 3$ together gives $t = -1$.

When $x_0 = 2$, we see that $\angle QFM = 90°$, and $\angle QMF = 45°$ for $t = -1$, satisfying $\angle QFM = 2\angle QMF$.

Therefore, $t = -1$ fulfills the requirement, and the point M that satisfies the condition exists, whose coordinates are $(-1, 0)$.

10. Let the coordinates of P, Q, R, and S be (x_i, y_i) for $i = 1, 2, 3$, and 4, respectively. Combining $y = mx + b$ and $x^2 + y^2 = 1$ implies

$$(1 + m^2)x^2 + 2mbx + b^2 - 1 = 0, \tag{12.2}$$

from which $x_1 + x_2 = -\frac{2mb}{1 + m^2}$. Combining $y = mx + b$ and $x^2 - y^2 = 1$ implies

$$(1 - m^2)x^2 - 2mbx - (b^2 + 1) = 0, \tag{12.3}$$

from which $x_3 + x_4 = \frac{2mb}{1 - m^2}$. Since P and Q divide the line segment RS into three equal parts, so the middle point of PQ is the middle point of RS, thus $x_1 + x_2 = x_3 + x_4$. Consequently, $-\frac{2mb}{1 + m^2} = \frac{2mb}{1 - m^2}$, whose solutions are $m = 0$ or $b = 0$. (1) When $m = 0$, the equations (12.2) and (12.3) are just $x^2 + b^2 - 1 = 0$ and $x^2 - b^2 - 1 = 0$, respectively. From $|PQ| = \frac{1}{3}|RS|$, we see that $|x_1 - x_2| = \frac{1}{3}|x_3 - x_4|$, namely $2\sqrt{1 - b^2} = \frac{1}{3} \cdot 2\sqrt{1 + b^2}$, the solutions of which are $b = \pm\frac{2}{5}\sqrt{5}$. (2) When $b = 0$, similarly we get $m = \pm\frac{2}{5}\sqrt{5}$.

11. As Figure 12.1 shows, without loss of generality, assume that $P(x_0, y_0)$ is on the right branch of the hyperbola $\frac{x^2}{a^2} - \frac{y^2}{b^2} = 1$ ($a > 0$ and $b > 0$). Then the equation of the tangent line PQ at the point P is $\frac{x_0 x}{a^2} - \frac{y_0 y}{b^2} = 1$. Letting $y = 0$ gives the coordinates $\left(\frac{a^2}{x_0}, 0\right)$ of the intersection point Q of the tangent line and the x-axis.

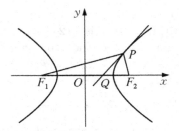

Figure 12.1 Figure for Exercise 11.

By the formulas of focal radii, $|PF_1| = ex_0 + a$ and $|PF_2| = ex_0 - a$, where $e = \frac{c}{a}$ is the eccentricity of the hyperbola.

Now, $|F_1Q| = \frac{a^2}{x_0} + c$ and $|F_2Q| = c - \frac{a^2}{x_0}$, so $\frac{|F_1Q|}{|F_2Q|} = \frac{a^2 + cx_0}{cx_0 - a^2}$ while $\frac{|PF_1|}{|PF_2|} = \frac{ex_0 + a}{ex_0 - a} = \frac{cx_0 + a^2}{cx_0 - a^2}$. It follows that $\frac{|F_1Q|}{|F_2Q|} = \frac{|PF_1|}{|PF_2|}$, in other words, PQ bisects $\angle F_1PF_2$, thus proving the proposition.

12. Let $P(x_0, y_0)$ be a point on the hyperbola $\frac{x^2}{a^2} - \frac{y^2}{b^2} = 1$. Then the equation of its polar line (that is, the straight line containing the two tangent points) with respect to the hyperbola $\frac{x^2}{a^2} - \frac{y^2}{b^2} = k$ is $\frac{x_0 x}{a^2} - \frac{y_0 y}{b^2} = k$, and the horizontal coordinates of its intersection points A and B with the two asymptotes $\frac{x}{a} - \frac{y}{b} = 0$ and $\frac{x}{a} + \frac{y}{b} = 0$ are

$$x_1 = \frac{ak}{\frac{x_0}{a} - \frac{y_0}{b}} \quad \text{and} \quad x_2 = \frac{ak}{\frac{x_0}{a} + \frac{y_0}{b}},$$

respectively.

The area of $\triangle OAB$ is

$$\left| \frac{1}{2} \begin{vmatrix} 0 & 0 & 1 \\ x_1 & y_1 & 1 \\ x_2 & y_2 & 1 \end{vmatrix} \right| = \left| \frac{1}{2} \begin{vmatrix} x_1 & y_1 \\ x_2 & y_2 \end{vmatrix} \right|$$

$$= \frac{1}{2}|x_1 y_2 - x_2 y_1| = \frac{1}{2}\left| x_1 \cdot \left(-\frac{b}{a} x_2\right) - x_2 \cdot \frac{b}{a} x_1 \right|$$

$$= \frac{b}{a}|x_1 x_2| = \frac{b}{a}\left| \frac{ak}{\frac{x_0}{a} - \frac{y_0}{b}} \cdot \frac{ak}{\frac{x_0}{a} + \frac{y_0}{b}} \right|$$

$$= \left| \frac{abk^2}{\frac{x_0^2}{a^2} - \frac{y_0^2}{b^2}} \right| = abk^2 \text{ (a fixed number)}.$$

This completes the proof.

Solutions for Chapter 13

Parabolas

1. $st = 1$.   The coordinates of the vertex of the parabola $y = ax^2 + bx + c$ are $(-\frac{b}{2a}, \frac{4a-b^2}{4a})$. Let $s = -\frac{b}{2a}$ and $t = \frac{4a-b^2}{4a}$. Then we have $\frac{b}{a} = -2s$ and $t = 1 - \frac{b^2}{4a} = 1 + \frac{bs}{2}$.

   Since $a \neq 0$, so the condition that a and b satisfy is equivalent to $8 + 4 \cdot \frac{b}{a} = b(\frac{b}{a})^2$. Thus,

$$8 + 4(-2s) = b(-2s)^2 = 4s(2t - 2).$$

   Hence, $st = 1$.

2. $(-\infty, -3] \cup [1, +\infty)$.

3. $y = x + 2$.   Denote $P_1(x_1, y_1)$ and $P_2(x_2, y_2)$. Put them into the equation $x^2 = y$ of the parabola respectively and make a subtraction. Then we have $(x_1 + x_2)(x_1 - x_2) = y_1 - y_2$. Let the middle point of $P_1 P_2$ be $M(x_0, y_0)$. Since the slope of $P_1 P_2$ is 1, so $2x_0 = 1$, that is $x_0 = \frac{1}{2}$. Since $x_0 + y_0 = 3$, we obtain $y_0 = \frac{5}{2}$. Hence, the equation of the straight line $P_1 P_2$ is $y - \frac{5}{2} = x - \frac{1}{2}$, namely $y = x + 2$.

4. $\frac{\sqrt{2}}{2}$.   The coordinates of the point F are $(\frac{1}{2}, 0)$. Write $A(x_1, y_1)$ and $B(x_2, y_2)$. Then $x_1 = \frac{y_1^2}{2}$ and $x_2 = \frac{y_2^2}{2}$. Consequently, $-1 = \overrightarrow{OA} \cdot \overrightarrow{OB} = x_1 x_2 + y_1 y_2 = \frac{1}{4}(y_1 y_2)^2 + y_1 y_2$, so $y_1 y_2 = -2$. Hence,

$$S_{\triangle OFA} + S_{\triangle OFB} = \frac{1}{2}|OF||y_1| + \frac{1}{2}|OF||y_2|$$

$$= \frac{1}{4}(|y_1| + |y_2|)$$

$$\geq \frac{1}{2}\sqrt{|y_1 y_2|} = \frac{\sqrt{2}}{2},$$

where the equality is satisfied if and only if $|y_1| = |y_2| = \sqrt{2}$, that is, if and only if the coordinates of A and B are $A(1, \sqrt{2})$ and $B(1, -\sqrt{2})$, or $A(1, -\sqrt{2})$ and $B(1, \sqrt{2})$. Thus, the minimum value of $S_{\triangle OFA} + S_{\triangle OFB}$ is $\frac{\sqrt{2}}{2}$.

5. Since the equation of the circle P is $(x-1)^2 + y^2 = 1$, the length of its diameter is $|BC| = 2$ and the center of the circle is $P(1,0)$.

   Let the equation of the straight line l be $ky = x - 1$, namely $x = ky + 1$. Substituting it into the equation of the parabola gives $y^2 = 4ky + 4$. Denote $A(x_1, y_1)$ and $B(x_2, y_2)$. Then $y_1 + y_2 = 4k$ and $y_1 y_2 = -4$, from which

$$(y_1 - y_2)^2 = (y_1 + y_2)^2 - 4y_1 y_2 = 16(k^2 + 1).$$

Thus,

$$|AD|^2 = (y_1 - y_2)^2 + (x_1 - x_2)^2$$

$$= (y_1 - y_2)^2 + \left(\frac{y_1^2 - y_2^2}{4}\right)^2 = 16(k^2 + 1)^2,$$

and so

$$|AD| = 4(k^2 + 1).$$

Since $|AB|, |BC|,$ and $|CD|$ constitute an arithmetic sequence,

$$2|BC| = |AB| + |CD| = |AD| - |BC|,$$

and so $|AD| = 3|BC| = 6$, that is $4(k^2 + 1) = 6$. Hence, $k = \pm\frac{\sqrt{2}}{2}$, which gives the equations of the straight lines l as $x = \frac{\sqrt{2}}{2}y + 1$ or $x = -\frac{\sqrt{2}}{2}y + 1$.

6. It is easy to have the coordinates $\left(\frac{p}{2}, 0\right)$ of B. Denote $A(2pt^2, 2pt)$. Then $\overrightarrow{AB} = \left(\frac{p}{2} - 2pt^2, -2pt\right)$ and $\overrightarrow{AC} = (c - 2pt^2, -2pt)$. Since $\angle BAC$ is an acute angle, $\overrightarrow{AB} \cdot \overrightarrow{AC} > 0$, namely

$$\left(\frac{p}{2} - 2pt^2\right)(c - 2pt^2) + 4p^2 t^2 > 0,$$

where $c > 0$ and $c \neq \frac{p}{2}$. Simplifying the above gives $\left(\frac{1}{2} - 2t^2\right)c + 4pt^4 + 3pt^2 > 0$, which is always valid for all $t \in \mathbf{R}$ if and only if $\angle BAC$ is an acute angle for all points A on the parabola.

   Let $m = t^2 \geq 0$. Then $f(m) = 4pm^2 + (3p - 2c)m + \frac{c}{2} > 0$ is satisfied for any $m \in [0, +\infty)$.

Since the symmetric axis of the graph of $f(m)$ is $m = -\frac{3p-2c}{8p}$, we discuss two different cases as follows:

(1) $-\frac{3p-2c}{8p} \leq 0$, namely when $0 < c \leq \frac{3}{2}p$, the desired property is satisfied.

(2) $-\frac{3p-2c}{8p} > 0$, namely when $c > \frac{3}{2}p$, it holds that the discriminant $\Delta = (3p - 2c)^2 - 8pc < 0$, from which $\frac{3}{2}p < c < \frac{9}{2}p$.

To summarize, $c \in \left(0, \frac{9}{2}p\right)$.

7. Let the coordinates of M be (x_0, y_0), and let the coordinates of M_1 and M_2 be (x_1, y_1) and (x_2, y_2), respectively. Since A, M, and M_1 are co-linear, $\frac{y_1-y_0}{x_1-x_0} = \frac{y_0-b}{x_0-b}$. Putting $x_0 = \frac{y_0^2}{2p}$ and $x_1 = \frac{y_1^2}{2p}$ into the above equality, we get

$$y_1 = \frac{by_0 - 2pa}{y_0 - b}. \tag{13.1}$$

By the same token, since B, M, and M_1 are co-linear,

$$y_2 = \frac{2pa}{y_0}. \tag{13.2}$$

Thus, the equation of the straight line M_1M_2 is $y - y_2 = \frac{y_1-y_2}{x_1-x_2}(x - x_2)$. Putting $x_1 = \frac{y_1^2}{2p}$ and $x_2 = \frac{y_2^2}{2p}$ into it gives $2px - (y_1 + y_2)y + y_1y_2 = 0$. Substituting (13.1) and (13.2) into the above and after a simplification, we have $y_0^2(2px - by) + 2pby_0(a - x) + 2pa(by - 2pa) = 0$ ($\star$). Let the coefficients of y_0 (including the constant term) be zero, that is, let $2px - by = 0, a - x = 0$, and $by - 2pa = 0$. Then their solutions are $x = a$ and $y = \frac{2pa}{b}$. This means that when $x = a$ and $y = \frac{2pa}{b}$, the equation ($\star$) is always satisfied for any y_0. In other words, the straight line M_1M_2 always passes through the fixed point $\left(a, \frac{2pa}{b}\right)$.

8. By the symmetry of a parabola, we can write $P(2t^2, 2t)$ $(t > 0)$.

Since the equation of the circle is $x^2 + y^2 - 2x = 0$, so the equation of the tangent chord MN is $2t^2x + 2ty - (x + 2t^2) = 0$, that is,

$$\left(2t^2 - 1\right)x + 2ty - 2t^2 = 0.$$

Thus, the double tangent lines PB and PC are just the conic sections that pass through all the common points of the double straight line

$$\left[2t^2x + 2ty - \left(x + 2t^2\right)\right]^2 = 0$$

and the circle, whose equation is

$$(x - 1)^2 + y^2 - 1 + \lambda\left[2t^2x + 2ty - \left(x + 2t^2\right)\right]^2 = 0.$$

Substituting the coordinates of the point P into the above expression gives that $\lambda = -\frac{1}{4t^4}$, and consequently the equation of the double tangent lines PB and PC is

$$(x-1)^2 + y^2 - 1 - \frac{1}{4t^4}\left[2t^2x + 2ty - (x+2t^2)\right]^2 = 0.$$

Let $x = 0$ in the above, giving $y = \frac{1}{t}y - 1$ or $y = 1 - \frac{1}{t}y$.

Since there is only one tangent line PB that intersects the axis when $t = 1$ and the circle $(x-1)^2 + y^2 = 1$ is an external tangent circle of $\triangle PBC$ when $0 < t < 1$, we must have $t > 1$, Consequently,

$$y_B = \frac{t}{1+t}, \quad y_C = \frac{t}{1-t}, \quad |BC| = \frac{t}{1+t} - \frac{t}{1-t} = \frac{2t^2}{t^2-1}.$$

Therefore,

$$S_{\triangle PBC} = \frac{1}{2}|BC||x_P| = \frac{1}{2}\frac{2t^2}{t^2-1}2t^2 = \frac{2t^2}{t^2-1}$$

$$= 2\left[2 + (t^2-1) + \frac{1}{t^2-1}\right] \geq 8,$$

and if and only if $t = \sqrt{2}$, the equality holds in the above inequality. Hence, the desired minimum value is 8.

9. Assume that $A(x_1, y_1)$ and $B(x_2, y_2)$ are symmetric with respect to l and let the middle point of AB be $M(x_0, y_0)$. The equation of AB is $y = -\frac{1}{m}x + b$. Substituting it into $y = x^2$ gives $x^2 + \frac{1}{m}x - b = 0$. So $x_0 = \frac{x_1+x_2}{2} = -\frac{1}{2m}$ and $y_0 = -\frac{1}{m}x_0 + b = \frac{1}{2m^2} + b$.

Also the point M is on the straight line $y = m(x-3)$, so we have $\frac{1}{2m^2} + b = m\left(-\frac{1}{m} - 3\right)$. Thus,

$$4b = -\frac{2}{m^2} - 12m - 2.$$

From $\Delta = \frac{1}{m^2} + 4b = \frac{1}{m^2} - \frac{2}{m^2} - 12m - 2 > 0$, we obtain that $m < -\frac{1}{2}$ (the requirement is obviously satisfied when $m = 0$).

10. Denote $A(x_1, y_1), B(x_2, y_2), C(x_3, y_3)$, and $D(x_4, y_4)$. Let the equation of the circle containing A, B, C and D be $x^2 + y^2 + dx + ey + f = 0$. Combining it with $y^2 = 2px$ and eliminating x, we get

$$\frac{y^4}{4p^2} + y^2 + \frac{dy^2}{2p} + ey + f = 0.$$

By Vieta's theorem, the four roots y_1, y_2, y_3, and y_4 of the equation satisfy

$$y_1 + y_2 + y_3 + y_4 = 0.$$

Now,

$$k_{AB} = \frac{y_1 - y_2}{x_1 - x_2} = \frac{y_1 - y_2}{\frac{y_1^2}{2p} - \frac{y_2^2}{2p}} = \frac{2p}{y_1 + y_2} = 1.$$

So,

$$k_{CD} = \frac{2p}{y_3 + y_4} = \frac{2p}{-(y_1 + y_2)} = -1.$$

Hence, $AB \perp CD$.

11. (1) Substituting the equation $y = x - 1$ of l into $y^2 = 4x$ gives $x^2 - 6x + 1 = 0$. Denote $A(x_1, y_1)$ and $B(x_2, y_2)$. Then $x_1 + x_2 = 6$ and $x_1 x_2 = 1$. So,

$$\overrightarrow{OA} \cdot \overrightarrow{OB} = x_1 x_2 + y_1 y_2 = 2x_1 x_2 - (x_1 + x_2) + 1 = -3,$$

$$|\overrightarrow{OA}||\overrightarrow{OB}| = \sqrt{x_1^2 + y_1^2}\sqrt{x_2^2 + y_2^2}$$

$$= \sqrt{x_1 x_2 [x_1 x_2 + 4(x_1 + x_2) + 16]} = \sqrt{41},$$

$$\cos\langle \overrightarrow{OA}, \overrightarrow{OB} \rangle = -\frac{3}{41}\sqrt{41}.$$

Hence, the angle of $\overrightarrow{OA}$ and $\overrightarrow{OB}$ is $\pi - \arccos \frac{3}{41}\sqrt{41}$.

(2) The given condition $\overrightarrow{FB} = \lambda \overrightarrow{AF}$ means that $(x_2 - 1, y_2) = \lambda(1 - x_1, -y_1)$, in other words,

$$x_2 - 1 = \lambda(1 - x_1), \tag{13.3}$$

$$y_2 = -\lambda y_1, \tag{13.4}$$

From (13.4), $y_2^2 = \lambda^2 y_1^2$. Also $y_1^2 = 4x_1$ and $y_2^2 = 4x_2$. It follows that

$$x_2 = \lambda^2 x_1. \tag{13.5}$$

Solving (13.3) and (13.5) together gives $x_2 = \lambda$. From the condition, $\lambda > 0$, so we have the points $B(\lambda, 2\sqrt{\lambda})$ or $B(\lambda, -2\sqrt{\lambda})$. Since the focus is $F(1, 0)$, the straight lines l are $(\lambda - 1)y = 2\sqrt{\lambda}(x - 1)$ or $(\lambda - 1)y = -2\sqrt{\lambda}(x - 1)$.

Therefore, the intercepts of l in the y-axis are $\frac{2\sqrt{\lambda}}{\lambda - 1}$ or $\frac{-2\sqrt{\lambda}}{\lambda - 1}$.

From $\frac{2\sqrt{\lambda}}{\lambda-1} = \frac{2}{\sqrt{\lambda}+1} + \frac{2}{\lambda-1}$ (or using a derivative argument), we know that $\frac{2\sqrt{\lambda}}{\lambda-1}$ is decreasing on $[4, 9]$.

Hence, the range of the intercepts of l in the y-axis is $\left[-\frac{4}{3}, -\frac{3}{4}\right] \cup \left[\frac{3}{4}, \frac{4}{3}\right]$.

12. The intersection point (x_0, y_0) of the parabola $y^2 = nx - 1$ and the straight line $y = x$ satisfies $x_0 = y_0$ and $x_0^2 = nx_0 - 1$. So, $x_0 + \frac{1}{x_0} = n$.

If (x_0^m, y_0^m) is the intersection point of the parabola $y^2 = kx - 1$ with the straight line $y = x$, then $x_0^{2m} = kx_0^m - 1$, namely $k = x_0^m + \frac{1}{x_0^m}$.

Denote $k_m = x_0^m + \frac{1}{x_0^m}$. Then $k_{m+1} = k_m(x_0 + \frac{1}{x_0}) - k_{m-1} = nk_m - k_{m-1}$ $(m \geq 2)$. Since $k_1 = x_0^1 + \frac{1}{x_0^1} = n$ is a positive integer, $k_2 = x_0^2 + \frac{1}{x_0^2} = \left(x_0 + \frac{1}{x_0}\right)^2 - 2 = n^2 - 2$ is also a positive integer, and the above recursive relation implies that k_m must be a positive integer. Since $x_0 + \frac{1}{x_0} = n$, we see that $x_0 > 0$, from which $k_m = x_m + \frac{1}{x_m} \geq 2$. Hence, $k_m = x_m + \frac{1}{x_m}$ are positive integers not less than 2 for all positive integers m.

Therefore, for any positive integer m, there exists $k = x_m + \frac{1}{x_m}$ such that $y^2 = kx - 1$ and $y = x$ intersect at (x_0^m, y_0^m).

Solutions for Chapter 14

Parametric Equations

1. $\left(x + \frac{2}{3}\right)^2 + \left(y - \frac{4}{3}\right)^2 = \frac{64}{9}$.

2. $\frac{x^2}{a^2} - \frac{y^2}{b^2} = 1$. Write $M(a\cos\theta, 0)$ with $\theta \in [0, 2\pi)$. Then we have $P(a\cos\theta, b\sin\theta)$ and $P'(a\cos\theta, -b\sin\theta)$. So, the equations of PA and $P'A'$ are $\frac{y}{x+a} = \frac{b\sin\theta}{a(\cos\theta+1)}$ and $\frac{y}{x-a} = \frac{-b\sin\theta}{a(\cos\theta-1)}$, respectively. Multiplying the two equations gives $\frac{y^2}{x^2-a^2} = \frac{b^2}{a^2}$, namely $\frac{x^2}{a^2} - \frac{y^2}{b^2} = 1$.

3. 12. Let the parametric equations of the straight line be

$$\begin{cases} x = -2 - \dfrac{\sqrt{2}}{2}t, \\[2mm] y = 3 + \dfrac{\sqrt{2}}{2}t, \end{cases} \quad (t \text{ is a parameter}).$$

Substituting them into the equation of the circle gives $t^2 + 5\sqrt{2}t - 12 = 0$. Then $|PA||PB| = |t_1 t_2| = 12$.

4. 20. Denote $P(4\cos\alpha, 2\sin\alpha)$ and $Q(4\cos\beta, 2\sin\beta)$. Since the product of the slopes of OP and OQ is $-\frac{1}{4}$, so $\frac{2\sin\alpha}{4\cos\alpha} \cdot \frac{2\sin\beta}{4\cos\beta} = -\frac{1}{4}$, which is simplified to $\cos(\alpha - \beta) = 0$. Consequently,

$$\alpha = 2k\pi \pm \frac{\pi}{2} + \beta.$$

Thus, $|OP|^2 + |OQ|^2 = 16\cos^2\alpha + 4\sin^2\alpha + 16\cos^2\beta + 4\sin^2\beta = 20$.

5. Denote $P(x_0, y_0)$, $A(2\cos\theta, \sqrt{2}\sin\theta)$, and $B(2\cos\theta, -\sqrt{2}\sin\theta)$. Then $x_0 = 2\cos\theta$. From $|PA||PB| = 1$, we have $|\sqrt{2}\sin\theta - y_0||y_0 + \sqrt{2}\sin\theta| = 1$, that is $y_0^2 = 2\sin^2\theta \pm 1 = 2 - 2\cos^2\theta \pm 1$. Substituting $\cos\theta = \frac{x_0}{2}$ into it gives $y_0^2 = 2 - \frac{1}{2}x_0^2 \pm 1$, namely $\frac{x_0^2}{6} + \frac{y_0^2}{3} = 1$ or $\frac{x_0^2}{2} + y_0^2 = 1$. Since $x_0 = 2\cos\theta$, and l intersects the ellipse at two points,

so $-2 < x_0 < 2$. Hence, the trajectory of the point P has the equation $\frac{x^2}{6} + \frac{y^2}{3} = 1 \, (-2 < x < 2)$ or $\frac{x^2}{2} + y^2 = 1$.

6. The parametric equations of the ellipse are $x = a\cos\theta$ and $y = b\sin\theta$. The coordinates of the points A and N are

$$A(-a, 0), \quad N(a\cos\theta, b\sin\theta),$$

respectively. Then the parametric equations of the straight line AN are $x = -a + t\cos\theta$ and $y = t\sin\theta$. Substituting them into the equation of the ellipse, we obtain

$$(b^2\cos^2\theta + a^2\sin^2\theta)t^2 - 2ab^2 t\cos\theta = 0,$$

$$|AN| = t = \frac{2ab^2\cos\theta}{b^2\cos^2\theta + a^2\sin^2\theta}, \quad |AM| = \frac{a}{\cos\theta} \ \left(\theta \neq \frac{\pi}{2}\right).$$

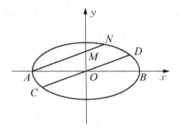

Figure 14.1   Figure for Exercise 6.

Therefore,

$$|AM||AN| = \frac{2a^2b^2}{b^2\cos^2\theta + a^2\sin^2\theta}.$$

From $AN \parallel CD$, the coordinates of C and D are $C(-|OD|\cos\theta, -|OD|\sin\theta)$ and $D(|OD|\cos\theta, |OD|\sin\theta)$.

As in Figure 14.1, since C and D are on the ellipse,

$$|CO|^2 = \frac{a^2b^2}{b^2\cos^2\theta + a^2\sin^2\theta},$$

while

$$|CO||CD| = 2|CO|^2 = \frac{2a^2b^2}{b^2\cos^2\theta + a^2\sin^2\theta}.$$

Therefore,

$$|AM||AN| = |CO||CD|.$$

7. From the condition, the right focus of the hyperbola is $F(c, 0)$.

Let the inclination angle of the straight line l be α and the parametric equations of l be $x = c + t\cos\alpha$ and $y = t\sin\alpha$, where t is a parameter. Substituting them into the equation of the hyperbola, we have

$$(b^2\cos^2\alpha - a^2\sin^2\alpha)t^2 + 2b^2ct\cos\alpha + b^4 = 0,$$

namely

$$(c^2\cos^2\alpha - a^2)t^2 + (2b^2c\cos\alpha)t + b^4 = 0.$$

Then

$$t_1 t_2 = \frac{b^4}{c^2\cos^2\alpha - a^2},$$

$$|t_1 - t_2| = \frac{2b^2\sqrt{c^2\cos^2\alpha - (c^2\cos^2\alpha - a^2)}}{|c^2\cos^2\alpha - a^2|}$$

$$= \frac{2ab^2}{|c^2\cos^2\alpha - a^2|}.$$

Since $|OF||AB| = |FA||FB|$, that is,

$$\frac{2ab^2 c}{|c^2\cos^2\alpha - a^2|} = \frac{b^4}{|c^2\cos^2\alpha - a^2|},$$

it follows that $2ac = b^2$, namely $e^2 - 2e - 1 = 0$. Since $e > 1$, so $e = 1 + \sqrt{2}$.

8. (1) According to the definition of hyperbolas, $|MA| - |MB| = 2\cos\theta$, so the trajectory of the point M is the branch of a hyperbola with the foci A and B and the real axis length $2\cos\theta$, which is near B.

   (2) Set up an appropriate coordinates system. Then the equation of the circle is $(x + \cos\theta)^2 + y^2 = 4\cos^2\theta$. When $|MN|$ is of the minimum value, we have $M(\cos\theta, 0), N(\sin\theta, 0)$, and $f(\theta) = |\cos\theta - \sin\theta| = |\sqrt{2}\sin(\theta - \frac{\pi}{4})|$. From $\theta \in (0, \frac{\pi}{2})$, we know that $0 < f(\theta) < 1$.

9. (1) Write $P(a\sec\theta, b\tan\theta)$ and let the equations of the two asymptotes of the hyperbola be $l_1 : bx - ay = 0$ and $l_2 : bx + ay = 0$, respectively.

The condition $PQ \parallel l_2$ implies that $k_{PQ} = -\frac{b}{a}$. We may assume that the parametric equations of the straight line PQ are

$$\begin{cases} x = a \sec\theta - \dfrac{a}{\sqrt{a^2 + b^2}}t, \\[2mm] y = b\tan\theta + \dfrac{b}{\sqrt{a^2 + b^2}}t, \end{cases} \qquad (t \text{ is a parameter}).$$

Substituting them into the equation of l_1 gives

$$b\left(a\sec\theta - \frac{a}{\sqrt{a^2 + b^2}}t\right) - a\left(b\tan\theta + \frac{b}{\sqrt{a^2 + b^2}}\right) = 0,$$

whose solution is $t = \frac{1}{2}\sqrt{a^2 + b^2}(\sec\theta - \tan\theta)$. Thus,

$$|PQ| = |t| = \frac{1}{2}\sqrt{a^2 + b^2}|\sec\theta - \tan\theta|.$$

By the same token, $|PR| = \frac{1}{2}\sqrt{a^2 + b^2}|\sec\theta + \tan\theta|$. Hence,

$$|PQ||PR| = \frac{1}{4}(a^2 + b^2)|\sec^2\theta - \tan^2\theta| = \frac{1}{4}(a^2 + b^2).$$

(2) The area formula is from the following:

$$S_{\square PQOR} = |OR||OQ|\sin\angle ROQ = |PQ||PR|\frac{2\tan\angle xOQ}{1 + \tan^2\angle xOQ}$$

$$= \frac{1}{4}(a^2 + b^2)\frac{2\cdot\frac{b}{a}}{1 + \left(\frac{b}{a}\right)^2} = \frac{1}{2}ab.$$

10. (1) Denote $P(a\cos\theta, b\sin\theta)$. Then the equation of the polar line (that is, the tangent chord) of the point P with respect to the circle $x^2 + y^2 = b^2$ is $a\cos\theta \cdot x + b\sin\theta \cdot y = b^2$. Letting $y = 0$ gives $x = \frac{b^2}{a\cos\theta}$, and letting $x = 0$ gives $y = \frac{b}{\sin\theta}$. In other words, $|OM| = \frac{b^2}{a|\cos\theta|}$ and $|ON| = \frac{b}{|\sin\theta|}$. Consequently,

$$S_{\triangle OMN} = \frac{1}{2}|OM||ON| = \frac{b^3}{a|\sin 2\theta|} \geq \frac{b^3}{a}.$$

(2) Suppose that there exist such two tangent lines. Since $OA \perp PA$, $OB \perp PB$, and $|OA| = |OB|$, we know that $OAPB$ is a square, so $|OP| = \sqrt{2}b$. Conversely, if $|OP| = \sqrt{2}b$, then $OAPB$ is a square.

Hence, there must be $|OP| = \sqrt{2}b$, namely $a^2 \cos^2 \theta + b^2 \sin^2 \theta = 2b^2$, from which $\cos^2 \theta = \frac{b^2}{a^2-b^2}$, and so $\frac{b^2}{a^2-b^2} \leq 1$ since $\cos^2 \theta \leq 1$. That is, there exists such a point P only when $a \geq \sqrt{2}b$.

When $a = \sqrt{2}b$, we see that $\cos \theta = \pm 1$. Then the coordinates of the point P are $(a, 0)$ or $(-a, 0)$.

When $a > \sqrt{2}b$, we have $\cos \theta = \pm \frac{b^2}{\sqrt{a^2-b^2}}$, and then $\sin \theta = \pm \frac{\sqrt{a^2-2b^2}}{\sqrt{a^2-b^2}}$. Namely the coordinates of the point P are

$$\left(\frac{ab}{\sqrt{a^2-b^2}}, \pm \frac{b\sqrt{a^2-2b^2}}{\sqrt{a^2-b^2}} \right) \text{ or } \left(-\frac{ab}{\sqrt{a^2-b^2}}, \pm \frac{b\sqrt{a^2-2b^2}}{\sqrt{a^2-b^2}} \right).$$

When $a < \sqrt{2}b$, there does not exist such a point P on the ellipse.

11. Denote $P(x_0, y_0)$. Let the parametric equations of l_1 be $x = x_0 + t \cos \alpha$ and $y = y_0 + t \sin \alpha$ with t a parameter, and the parametric equations of l_2 be $x = x_0 + m \cos \beta$ and $y = y_0 + m \sin \beta$ with m a parameter. Substituting the equations for l_1 into the equation of the ellipse, we get

$$(b^2 \cos^2 \alpha + a^2 \sin^2 \alpha)t^2 + 2(b^2 x_0 \cos \alpha + a^2 y_0 \sin \alpha)t$$
$$+ b^2 x_0^2 + a^2 y_0^2 - a^2 b^2 = 0.$$

Then

$$t_1 t_2 = \frac{b^2 x_0^2 + a^2 y_0^2 - a^2 b^2}{b^2 \cos^2 \alpha + a^2 \sin^2 \alpha}.$$

By the same token,

$$m_1 m_2 = \frac{b^2 x_0^2 + a^2 y_0^2 - a^2 b^2}{b^2 \cos^2 \beta + a^2 \sin^2 \beta}.$$

Since $\alpha + \beta = \pi$, we have $|\sin \alpha| = |\sin \beta|$ and $|\cos \alpha| = |\cos \beta|$. Thus, $|t_1 t_2| = |m_1 m_2|$, that is, $|PA||PB| = |PC||PD|$. By the knowledge of plane geometry, the four points A, B, C, and D are on the same circle.

12. Write $M(a \sec \theta, b \tan \theta)$ and $N(a \sec \alpha, b \tan \alpha)$. Then the equation of the straight line MN is

$$y = \frac{b \tan \theta - b \tan \alpha}{a \sec \theta - a \sec \alpha}(x - a \sec \theta) + b \tan \theta.$$

Substituting the point $F(c, 0)$ into the above, we obtain in succession that

$$\frac{b\tan\theta - b\tan\alpha}{a\sec\theta - a\sec\alpha}(c - a\sec\theta) + b\tan\theta = 0,$$

$$a\sec\theta - c = \frac{a\tan\theta(\cos\alpha - \cos\theta)}{\sin\theta\cos\alpha - \cos\theta\sin\alpha},$$

$$\frac{c}{a} = \frac{\tan\theta\cos\alpha - \sin\alpha - \tan\theta\cos\alpha + \sin\theta}{\sin(\theta - \alpha)}$$

$$= \frac{\sin\theta - \sin\alpha}{\sin(\theta - \alpha)}.$$

With the point $A(-a, 0)$ in mind, we know that the equation of the straight line AM is $y = \frac{b\tan\theta}{a\sec\theta + a}(x + a)$. By substituting $x = \frac{a^2}{c}$ into it, we have

$$y = \frac{(a + c)b\tan\theta}{c(\sec\theta + 1)}.$$

Therefore, the coordinates of the point P are $\left(\frac{a^2}{c}, \frac{(a+c)b\tan\theta}{c(\sec\theta+1)}\right)$.

Similarly, the coordinates of the point Q are $\left(\frac{a^2}{c}, \frac{(a+c)b\tan\alpha}{c(\sec\alpha+1)}\right)$.

Let the middle point of the side PQ be R. To make the conclusion be true, we need the equality $|PQ| = 2|RF|$. In other words,

$$\left|\frac{a+c}{2c}\left(\frac{b\tan\theta}{\sec\theta + 1} - \frac{b\tan\alpha}{\sec\alpha + 1}\right)\right|$$

$$= \sqrt{\left(\frac{a^2}{c} - c\right)^2 + \left(\frac{a+c}{2c}\right)^2\left(\frac{b\tan\theta}{\sec\theta + 1} + \frac{b\tan\alpha}{\sec\alpha + 1}\right)^2}$$

$$\Leftarrow \frac{4b^2\tan\theta\tan\alpha}{(\sec\theta + 1)(\sec\alpha + 1)}\left(\frac{a+c}{2c}\right)^2 = -\left(\frac{a^2 - c^2}{c}\right)^2$$

$$\Leftarrow -\frac{b^2}{(a+c)^2} = \frac{\tan\theta\tan\alpha}{(\sec\theta + 1)(\sec\alpha + 1)}$$

$$= \frac{\sin\theta\sin\alpha}{(\cos\theta + 1)(\cos\alpha + 1)} = \tan\frac{\theta}{2}\tan\frac{\alpha}{2}.$$

On the other hand,

$$\frac{c}{a} = \frac{\sin\theta - \sin\alpha}{\sin(\theta - \alpha)}$$

$$= \frac{2\sin\frac{\theta-\alpha}{2}\cos\frac{\theta+\alpha}{2}}{2\sin\frac{\theta-\alpha}{2}\cos\frac{\theta-\alpha}{2}} = \frac{\cos\frac{\theta+\alpha}{2}}{\cos\frac{\theta-\alpha}{2}}$$

$$= \frac{1 - \tan\frac{\theta}{2}\tan\frac{\alpha}{2}}{1 + \tan\frac{\theta}{2}\tan\frac{\alpha}{2}}.$$

Therefore, $\tan\frac{\theta}{2}\tan\frac{\alpha}{2} = -\frac{c^2-a^2}{(a+c)^2} = -\frac{b^2}{(a+c)^2}$, and so the conclusion is proved.

Solutions for Chapter 15

Families of Curves

1. $39x + 13y - 25 = 0$.   Let the equation of the straight line be

$$(2x - 3y - 1) + \lambda(3x + 2y - 2) = 0,$$

which is rewritten as

$$(2 + 3\lambda)x + (-3 + 3\lambda)y + (-1 - 2\lambda) = 0. \tag{15.1}$$

Since the slope of the known straight line $y + 3x = 0$ is -3, we have $-\frac{2+3\lambda}{-3+2\lambda} = -3$, from which $\lambda = \frac{11}{3}$. Substituting $\lambda = \frac{11}{3}$ into (15.1) gives $39x + 13y - 25 = 0$.

2. $x^2 + y^2 - 11x + 3y - 30 = 0$.   Let the equation of the circle be $(x - 8)^2 + (y - 6)^2 + \lambda(x + 3y - 26) = 0$. Substituting $(-2, -4)$ into it and solving for λ give $\lambda = 5$.

3. $\frac{x^2}{10} + \frac{y^2}{6} = 1$ or $\frac{2x^2}{5} - \frac{2y^2}{3} = 1$.

4. $\frac{(x-2)^2}{6} + \frac{(y-1)^2}{12} = 1$.   Let the desired equation of the curve be $(x - 4)^2 + n(y - 3)^2 + \mu(2x + y - 1) = 0$. Substituting the coordinates of Q and R into the equation gives $n = \frac{1}{2}$ and $\mu = 2$. Putting them back into the equation, we obtain $\frac{(x-2)^2}{6} + \frac{(y-1)^2}{12} = 1$.

5. From $\frac{x^2}{a^2} + \frac{y^2}{b^2} = 1 \geq 2\sqrt{\frac{x^2y^2}{a^2b^2}} = \frac{2xy}{ab}$, we see that the area of the quadrilateral $S = 4|xy| \leq 2ab$, and if and only if $x = \frac{\sqrt{2}}{2}b$ and $y = \frac{\sqrt{2}}{2}a$, the area achieves its maximum value. Let the equation of the hyperbola be $\frac{y^2}{m^2} - \frac{x^2}{a^2-b^2-m^2} = 1$. Substituting $\left(\frac{\sqrt{2}}{2}b, \frac{\sqrt{2}}{2}a\right)$ into it, we have $\frac{2y^2}{a^2-b^2} - \frac{2x^2}{a^2-b^2} = 1$. Thus, the coordinates of the four vertices are $\left(\frac{\sqrt{2}}{2}b, \frac{\sqrt{2}}{2}a\right), \left(\frac{\sqrt{2}}{2}b, -\frac{\sqrt{2}}{2}a\right), \left(-\frac{\sqrt{2}}{2}b, \frac{\sqrt{2}}{2}a\right)$, and $\left(-\frac{\sqrt{2}}{2}b, -\frac{\sqrt{2}}{2}a\right)$.

6. Set up a coordinates system, and denote $A(0, a), B(b, 0), C(c, 0)$, and $H(0, h)$. Then

$$BH : \frac{x}{b} + \frac{y}{h} = 1,$$

$$AC : \frac{x}{c} + \frac{y}{a} = 1.$$

The family of the straight lines passing through the intersection point of BH and AC is given by

$$\lambda \left(\frac{x}{b} + \frac{y}{h} - 1 \right) + \mu \left(\frac{x}{c} + \frac{y}{a} - 1 \right) = 0.$$

Let $\lambda = 1$ and $\mu = -1$. Then

$$x \left(\frac{1}{b} - \frac{1}{c} \right) + y \left(\frac{1}{h} - \frac{1}{a} \right) = 0.$$

Since this straight line passes through the origin, it is just the straight line DE. By the same token, the equation of DF is

$$x \left(\frac{1}{c} - \frac{1}{b} \right) + y \left(\frac{1}{h} - \frac{1}{a} \right) = 0.$$

Clearly, the slopes of DE and DF are opposite numbers, so AD bisects the angle formed by ED and DF.

7. With A as the origin and AB as the x-axis, set up a coordinates system. Denote $B(a, 0)$ and $M(t, 0)$ $(0 < t < a)$. Then the equations of the circumcircles of the two squares are $x^2 + y^2 - tx - ty = 0$ and

$$x^2 + y^2 - (a + t)x - (a - t)y + at = 0.$$

Subtracting them gives the equation of MN as $ax + (a - 2t)y - at = 0$, namely

$$ax + ay - t(2y + a) = 0.$$

Hence, such MN always pass through the fixed point $\left(\frac{a}{2}, -\frac{a}{2} \right)$.

8. Write $A(x_1, y_1)$ and $B(x_2, y_2)$. Then

$$3x_1^2 + y_1^2 = \lambda, \tag{15.2}$$

$$3x_2^2 + y_2^2 = \lambda. \tag{15.3}$$

$(15.2) - (15.3)$ gives

$$3(x_1 - x_2)(x_1 + x_2) + (y_1 - y_2)(y_1 + y_2) = 0,$$

namely

$$\frac{y_1 - y_2}{x_1 - x_2} = -\frac{3(x_1 + x_2)}{y_1 + y_2} = -1,$$

which implies the equation $x + y - 4 = 0$ of AB and the equation $x - y + 2 = 0$ of CD. Since $N(1,3)$ is on the ellipse, so $\lambda > 12$.

The equation of the conic section passing through A, B, C, and D is

$$(3x^2 + y^2 - \lambda) + \mu(x + y - 4)(x - y + 2) = 0.$$

Its simplified version is

$$(3 + \mu)x^2 + (1 - \mu)y - 2\mu x + 6\mu y - 8\mu - \lambda = 0.$$

Letting $3 + \mu = 1 - \mu$ gives $\mu = -1$, and substituting it into the above expression, we obtain

$$x^2 + y^2 - x + 3y - 4 + \frac{\lambda}{2} = 0.$$

Hence, for any $\lambda > 12$, the four points A, B, C, and D are on the same circle.

9. The family of the curves pass through the origin. While $y = 2x$ also passes through the origin, so the lengths of the chords obtained by the intersections of the family of the curves to $y = 2x$ depend on the coordinates of the other intersection points. Substituting $y = 2x$ into the equation of the curves gives $(2\sin\theta - \cos\theta + 3)x^2 - (8\sin\theta + \cos\theta + 1)x = 0$. Also, $2\sin\theta - \cos\theta + 3 = \sqrt{5}\sin\left(\theta - \arctan\frac{1}{2}\right) + 3 \neq 0$. When $x \neq 0$, we have $x = \frac{8\sin\theta + \cos\theta + 1}{2\sin\theta - \cos\theta + 3}$. Let $\sin\theta = \frac{2u}{1+u^2}$ and $\cos\theta = \frac{1-u^2}{1+u^2}$. Then $x = \frac{8u+1}{2u^2+2u+1}$. Hence, $2xu^2 + 2(x - 4)u + (x - 1) = 0$. From $u \in \mathbf{R}$ we know that, when $x \neq 0$, the discriminant

$$\begin{aligned} \Delta &= [2(x - 4)]^2 - 8x(x - 1) \\ &= 4(-x^2 - 6x + 16) \\ &\geq 0, \end{aligned}$$

that is $x^2 + 6x - 16 \leq 0$ and $x \neq 0$, and namely $-8 \leq x \leq 2$ and $x \neq 0$. Therefore, $|x|_{\max} = 8$. From $y = 2x$ we know that $|y|_{\max} = 16$. Hence, the maximum value of the chord length is $\sqrt{8^2 + 16^2} = 8\sqrt{5}$.

10. Let the equations of the three circles be respectively

$$x^2 + y^2 + D_1 x + E_1 y + F_1 = 0, \tag{15.4}$$

$$x^2 + y^2 + D_2 x + E_2 y + F_2 = 0, \tag{15.5}$$

$$x^2 + y^2 + D_3 x + E_3 y + F_3 = 0, \tag{15.6}$$

(15.4) − (15.5) gives

$$(D_1 - D_2)x + (E_1 - E_2)y + F_1 - F_2 = 0, \tag{15.7}$$

(15.5) − (15.6) gives

$$(D_2 - D_3)x + (E_2 - E_3)y + F_2 - F_3 = 0, \tag{15.8}$$

and (15.6) − (15.4) gives

$$(D_3 - D_1)x + (E_3 - E_1)y + F_3 - F_1 = 0. \tag{15.9}$$

If (15.7) and (15.8) have an intersection point, then adding (15.7) and (15.8) results in (15.9), so the three lines have a common point.

If (15.7) and (15.8) have no intersection point, namely they are parallel, then there exists a real number k such that

$$D_1 - D_2 = k(D_2 - D_3),$$

$$E_1 - E_2 = k(E_2 - E_3),$$

$$F_1 - F_2 \neq k(F_2 - F_3).$$

So, $D_3 - D_1 = D_3 - D_2 + D_2 - D_1 = (1 + k)(D_3 - D_2).$

By the same token, $E_3 - E_1 = (1 + k)(E_3 - E_2)$ and $F_3 - F_1 \neq (1 + k)(F_3 - F_2).$

Consequently, (15.8) and (15.9) are parallel, so the three lines are mutually parallel.

11. With O as the origin and EF as the x-axis, set up a rectangular coordinates system. Denote $E(-x_0, 0)$ and $F(x, 0)$, and let the radius of the circle M be r. Then the equation of the circle is $(x-x_0)^2 + (y-r)^2 = r^2$, that is,

$$x^2 + y^2 - 2x_0 x - 2ry + x_0^2 = 0. \tag{15.10}$$

We may assume that the equations of the two straight lines AB and CD passing through E are $h_1 y = x + x_0$ and $h_2 y = x + x_0$ respectively, the combination of which is

$$(x - h_1 y + x_0)(x - h_2 y + x_0) = 0. \tag{15.11}$$

And the equations of the two straight lines BD and AC are $y = kx$ and $ax + by + c = 0$ respectively, the combination of which is

$$(y - kx)(ax + by + c) = 0. \tag{15.12}$$

The family of the curves generated by (15.10) and (15.11) pass through $A, B, C,$ and D, while the curve (15.12) also passes through $A, B, C,$ and D, so (15.12) is one of the family of the curves.

Comparing the constant terms of the family of the curves generated by (15.10) and (15.11) with (15.12), we find that (15.12) can be obtained by subtracting (15.10) and (15.11), after which there is no x^2 term. Consequently, $a = 0$ in (15.12). In other words, $AC \parallel EF$.

12. As in Figure 15.1, with A as the origin and the straight line AC as the x-axis, set up a rectangular coordinates system. Denote the coordinates of the points $C, F, D,$ and B as $(c, 0), (f, 0), (x_D, kx_D),$ and $(x_B, -kx_B)$. Then the equation of the straight line DF is

$$x - f + \frac{f - x_D}{kx_D} y = 0, \tag{15.13}$$

and the equation of the straight line BC is

$$x - c + \frac{c - x_B}{-kx_B} y = 0. \tag{15.14}$$

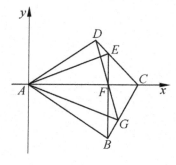

Figure 15.1 Figure for Exercise 12.

Subtracting (15.14) multiplied by f from (15.13) multiplied by c, we get

$$(c-f)x + \frac{1}{k}\left[cf\left(\frac{1}{x_D} + \frac{1}{x_B}\right) - (c+f)\right]y = 0. \qquad (15.15)$$

The straight line represented by (15.15) passes through the origin A, and also the intersection point G of DF and BC, thus (15.15) is the equation of the straight line AG. Similarly, we can obtain the equation

$$(c-f)x - \frac{1}{k}\left[cf\left(\frac{1}{x_D} + \frac{1}{x_B}\right) - (c+f)\right]y = 0. \qquad (15.16)$$

of the straight line AE. The slopes of (15.15) and (15.16) are opposite numbers, hence $\angle GAC = \angle EAC$.

Solutions for Chapter 16

Derivatives

1. $\frac{3\sqrt{3}}{3}$.   The range of the x values is $|x| \leq 1$. Since we want to find the maximum value, we can assume that $0 < x \leq 1$. Let $x = \sin\alpha$ with $\alpha \in \left(0, \frac{\pi}{2}\right]$. Then $y = \sin\alpha(1 + \cos\alpha)$ and $y' = \cos\alpha + \cos 2\alpha$. Letting $y' = 0$ gives $2\alpha = \pi - \alpha$, so $\alpha = \frac{\pi}{3}$ and the corresponding values of x and y are $x = \frac{\sqrt{3}}{2}$ and $y = \frac{3\sqrt{3}}{3}$.

2. $\frac{13}{16}\pi^2$.   Since $\{a_n\}$ is an arithmetic sequence with common difference $\frac{\pi}{8}$, and $f(a_1) + f(a_2) + \cdots + f(a_5) = (2a_1 - \cos a_1) + (2a_2 - \cos a_2) + \cdots + (2a_5 - \cos a_5) = 5\pi$, namely

$$2(a_1 + a_2 + \cdots + a_5) - (\cos a_1 + \cos a_2 + \cdots + \cos a_5) = 5\pi,$$

so we have

$$10a_3 - \left[\cos\left(a_3 - \frac{\pi}{4}\right) + \cos\left(a_3 - \frac{\pi}{8}\right) + \cos a_3\right.$$
$$\left. + \cos\left(a_3 + \frac{\pi}{8}\right) + \cos\left(a_3 + \frac{\pi}{4}\right)\right] = 5\pi.$$

Denote $g(x) = 10x - \left(2\cos\frac{\pi}{4} + 2\cos\frac{\pi}{8} + 1\right)\cos x - 5\pi$. Then $g'(x) = 10 + \left(2\cos\frac{\pi}{4} + 2\cos\frac{\pi}{8} + 1\right)\sin x > 0$, that is, $g(x)$ is a monotonically increasing function on $\mathbf{R}$, and it has a unique zero $x = \frac{\pi}{2}$. Thus, $a_3 = \frac{\pi}{2}$. Therefore,

$$f(a_3)^2 - a_1 a_5 = \left(\pi - \cos\frac{\pi}{2}\right)^2 - \left(\frac{\pi}{2} - \frac{\pi}{4}\right)\left(\frac{\pi}{2} + \frac{\pi}{4}\right) = \frac{13}{16}\pi^2.$$

3. $\sqrt{2}$.   Since the curves of $y = e^x$ and $y = \ln x$ are symmetric with respect to the straight line $y = x$. The desired minimum value of $|PQ|$ is 2 times the minimal distance of the point on the curve $y = e^x$ to the

straight line $y = x$. Let $P(x, e^x)$ be any point on $y = e^x$. Then the distance of P to the straight line $y = x$ is $d(x) = \frac{|e^x - x|}{\sqrt{2}} = \frac{e^x - x}{\sqrt{2}}$. We have $d'(x) = \frac{e^x - 1}{\sqrt{2}}$.

When $x > 0$, we see that $d'(x) > 0$, so $d(x)$ is monotonically increasing; when $x < 0$, we get $d'(x) < 0$, so $d(x)$ is monotonically decreasing. Thus, $d(x)_{\min} = d(0) = \frac{\sqrt{2}}{2}$, namely $|PQ|_{\min} = \sqrt{2}$.

4. $(0, 2)$.   Let $g(x) = f(x) - \frac{1}{2}x$. Then $g'(x) = f'(x) - \frac{1}{2} < 0$, so $g(x)$ is a decreasing function on **R**.

Since $f(\log_2 x) > \frac{\log_2 x + 1}{2}$, we obtain that

$$f(\log_2 x) - \frac{1}{2}\log_2 x > \frac{1}{2} = f(1) - \frac{1}{2},$$

that is, $g(\log_2 x) > g(1)$. Hence, $\log_2 x < 1$, from which $0 < x < 2$.

5. (1) The inequality to be proved is equivalent to $e^{-x} < \frac{1}{x+1}$, namely $-x < -\ln(x+1)$. Let $h(x) = x - \ln(x+1)$. Then $h'(x) = 1 - \frac{1}{x+1}$, so $h(x)$ is monotonically increasing on $[0, +\infty)$, and thus $h(x) > h(0) = 0$ when $x > 0$.

(2) Denote $g(x) = -\ln\frac{1-e^{-x}}{x}$. Then $a_{n+1} = g(a_n)$. To show that $\{a_n\}$ is decreasing, it is enough to show that $g(x) < x$ when $x > 0$. In fact, $g(x) < x$ is equivalent to $\ln\frac{1-e^{-x}}{x} > -x$, namely

$$1 - e^{-x} > xe^{-x}.$$

Note that $f(x) = 1 - e^{-x}$. So, the above inequality is also equivalent to $f(x) > x[1 - f(x)]$, that is, $f(x) > \frac{x}{x+1}$, which was already proved in (1).

To show that $a_n < \frac{1}{2^n}$, it is sufficient to show that $g(x) < \frac{x}{2}$ when $x > 0$. This is equivalent to

$$\frac{1 - e^{-x}}{x} > e^{-\frac{x}{2}},$$

namely $e^{\frac{x}{2}} - e^{-\frac{x}{2}} > x$. For this purpose, let $F(x) = e^{\frac{x}{2}} - e^{-\frac{x}{2}} - x$. Then

$$F'(x) = \frac{1}{2}(e^{\frac{x}{2}} + e^{-\frac{x}{2}}) - 1 \geq 0,$$

and the equality holds if and only if $e^{\frac{x}{2}} = e^{-\frac{x}{2}} = 1$, that is $x = 0$. Thus, $F(x)$ is monotonically increasing on $[0, +\infty)$. Hence, $F(x) > F(0) = 0$, and so $e^{\frac{x}{2}} - e^{-\frac{x}{2}} > x$ is proved.

6. First we show: If $m > n \geq 3$, then $n^m > m^n$, that is, $\frac{n^m - m^n}{n+m} > 0$.

   Let $f(x) = \frac{\ln x}{x}$. Note that $f'(x) = \frac{1 - \ln x}{x^2} < 0$ when $x > e$. Then $f(x)$ is monotonically decreasing on the interval $(e, +\infty)$. Thus, $\frac{\ln n}{n} > \frac{\ln m}{m}$, namely $m \ln n > n \ln m$. Hence, $n^m = e^{m \ln n} > e^{n \ln m} = m^n$.

   When $n = 2$, it is enough to choose $m = 10$.

   Assume $n > 2$. Then $n^m - m^n \equiv n^m - (-n)^n = n^n(n^{m-n} - (-1)^n) \pmod{m+n}$. Choose $m = kn^n - n$ ($k \in \mathbf{Z}_+$). So, $m+n = kn^n$, and consequently $(m + n) \mid (n^m - m^n) \Leftrightarrow k \mid [n^{m-n} - (-1)^n]$.

   (1) When n is an odd number, $n^{m-n} - (-1)^n$ is an even number. It is enough to choose $k = 2$. Now, $m = 2n^n - n$.
   (2) When n is an even number, $n^{m-n} - (-1)^n = n^{m-n} - 1$ can be divided by $n - 1$. It is enough to choose $k = n - 1$. Then $m = (n-1)n^n - n$.

7. (1) Since $f'(x) = \frac{1}{1+x} - 2 + ax$, so $f'(0) = -1$. Thus, the equation of the tangent line l to the curve C at the point P is $y = -x + 1$. Also the straight line l and the curve C have one and only one common point P, hence the equation $f(x) = -x + 1$ with respect to x, namely

$$\ln(1 + x) - x + \frac{1}{2}ax^2 = 0, \qquad (16.1)$$

has a unique real number solution $x = 0$ in the interval $(-1, +\infty)$.

   Let $g(x) = \ln(1 + x) - x + \frac{1}{2}ax^2$ $(x > -1)$. Then

$$g'(x) = \frac{1}{1 + x} - 1 + ax$$

$$= \frac{ax\left(x - \frac{1-a}{a}\right)}{1 + x} \quad (x > -1).$$

If $a = 1$, then $g'(x) = \frac{x^2}{1+x} \geq 0$, so the function $g(x)$ is monotonically increasing in the interval $(-1, +\infty)$. Thus, the equation (16.1) has a unique real number solution, satisfying the requirement.

   When $a > 1$, we have $-1 < \frac{1-a}{a} < 0$.

   If $-1 < x < \frac{1-a}{a}$, then $g'(x) > 0$; if $\frac{1-a}{a} < x < 0$, then $g'(x) < 0$. Hence, the function $g(x)$ is monotonically increasing in the interval $\left(-1, \frac{1-a}{a}\right)$ and monotonically decreasing in the interval $\left(\frac{1-a}{a}, 0\right)$.

   Hence, $g(x)_{\max} = g\left(\frac{1-a}{a}\right) > g(0) = 0$.

   When $x \to -1$, we see that $g(x) \to -\infty$, from which the function $g(x)$ has another zero in the interval $\left(-1, \frac{1-a}{a}\right)$, not satisfying the requirement.

   In summary, $a = 1$.

(2) From the given condition, $f'(x) = \frac{ax^2+(a-2)x-1}{1+x}$ $(x > -1)$.

A sufficient condition for the existence of an interval $[x_1, x_2]$ on which the function $f(x)$ is monotonically decreasing is that the strict inequality $f'(x) < 0$ is satisfied for all $x \in [x_1, x_2]$ $(x_1 > -1)$, that is,

$$h(x) = ax^2 + (a-2)x - 1 < 0.$$

Since the discriminant $\Delta = (a-2)^2 + 4a = a^2 + 4 > 0$, the x coordinate of the symmetric axis $x = -\frac{a-2}{2a} > -1$, and $h(-1) = 1 > 0$, so the equation $h(x) = 0$ has two different real roots x_1 and x_2 in the interval $(-1, +\infty)$. Then

$$x_1 + x_2 = -\frac{a-2}{a}, \ \ x_1 x_2 = -\frac{1}{a},$$

$$(x_2 - x_1)^2 = (x_1 + x_2)^2 - 4x_1 x_2 = \left(-\frac{a-2}{a}\right)^2 + \frac{4}{a} = 1 + \frac{4}{a^2}.$$

Also $a \geq 1$, hence $1 < (x_2 - x_1)^2 \leq 5 \Rightarrow 1 < x_2 - x_1 \leq \sqrt{5}$. Therefore, the range of the values for $x_2 - x_1$ is $(1, \sqrt{5}]$.

8. (1) From $y = \frac{x^2}{8} + b$ we get $y' = \frac{x}{4}$. From $y_G = y_F = b + 2$, we have $x_G = 4$. The slope of the tangent line to the parabola at the point G is 1, the equation of the straight line $F_1 G$ is $y - (b+2) = x - 4$, and the point $F_1(b, 0)$ is on the line. So $-(b+2) = b-4$, whose solution is $b = 1$. Therefore, the equation of the ellipse is $\frac{x^2}{2} + y^2 = 1$ and the equation of the parabola is $x^2 = 8(y - 1)$.

(2) Draw two perpendicular lines to the x-axis at A and B respectively, which intersect the parabola at points P_1 and P_2. These are the two points on the parabola that make $\triangle PAB$ a right triangle.

If $\angle APB = 90°$, then $\frac{y_P}{x_P - \sqrt{2}} \cdot \frac{y_P}{x_P + \sqrt{2}} = -1$, that is, $y_P^2 + x_P^2 - 2 = 0$. Also $y_P^2 + 8y_P - 10 = 0$ from $x_P^2 = 8(y_P - 1)$. Solving the above gives $y_P = -4 \pm \sqrt{26}$. Since $y_P \geq 1$, we have $y_P = \sqrt{26} - 4$. Hence, there are two points on the parabola that make $\angle APB = 90°$.

To summarize, there are totally four points on the parabola that make $\triangle APB$ a right triangle.

9. (1) First, $f'(x) = \frac{1}{x} - \frac{a}{x^2} = \frac{x-a}{x^2}$ for $x > 0$.

When $a \leq 0$, we see that $f'(x) > 0$, so $f(x)$ is monotonically increasing in $(0, +\infty)$, and consequently there is no minimum value, not satisfying the meaning of the problem.

When $a > 0$, if $0 < x < a$, then $f'(x) < 0$; if $x > a$, then $f'(x) > 0$. Thus, the function $f(x)$ is monotonically increasing in $(0, a)$ and monotonically decreasing in $(a, +\infty)$.

Hence, $f(x)_{\min} = f(a) = \ln a - a + 1$.

Let $g(a) = \ln a - a + 1$ $(a > 0)$. Then

$$g'(a) = \frac{1}{a} - 1 = \frac{1-a}{a}.$$

If $0 < a < 1$, then $g'(a) > 0$; if $a > 1$, then $g'(a) < 0$. So, the function $g(a)$ is monotonically increasing in $(0,1)$ and monotonically decreasing in $(1, +\infty)$.

Thus, $g(a) \le g(1) = 0$, and the equality holds if and only if $a = 1$.

Hence, $f(x)$ achieves its minimum value 0 when $a = 1$.

(2) From (1), $f(x) = \ln x + \frac{1}{x} - 1$, so

$$a_{n+1} = f(a_n) + 2 = \ln a_n + \frac{1}{a_n} + 1.$$

From $a_1 = 1$, we get $a_2 = 2$, and so $a_3 = \ln 2 + \frac{3}{2}$. Since $\frac{1}{2} < \ln 2 < 1$, from which $2 < a_3 < 3$.

In the following, we use mathematical induction to show that $2 < a_n < 3$ for $n \ge 3$.

(I) When $n = 3$, the conclusion was already prove above.

(II) Suppose that when $n = k$ $(k \ge 3)$, we have $2 < a_k < 3$. Then for $n = k + 1$,

$$a_{k+1} = \ln a_k + \frac{1}{a_k} + 1.$$

From (1) we know that $h(x) = f(x) + 2 = \ln x + \frac{1}{x} + 1$ is monotonically increasing in $(2, 3)$. Thus, $h(2) < h(a_k) < h(3)$, that is,

$$\ln 2 + \frac{3}{2} < h(a_k) < \ln 3 + \frac{1}{3} + 1.$$

Since $\ln 2 > \frac{1}{2}$ and $\ln 3 < \frac{5}{3}$, so $2 < h(a_k) < 3$, namely $2 < a_{k+1} < 3$. That is, the conclusion is true for $n = k + 1$.

By (I) and (II), $2 < a_n < 3$ for all integers $n \ge 3$.

It follows that $[a_1] = 1$ and $[a_n] = 2$ for $n \ge 2$.

Therefore, $S_n = [a_1] + [a_2] + \cdots + [a_n] = 1 + 2(n-1) = 2n - 1$.

10. (1) Taking derivative gives that

$$f'(x) = \frac{2\ln(1+x)}{1+x} - \frac{x^2 + x}{(1+x)^2}$$

$$= \frac{2(1+x)\ln(1+x) - x^2 - 2x}{(1+x)^2}.$$

Let $g(x) = 2(1+x)\ln(1+x) - x^2 - 2x$. Then $g'(x) = 2\ln(1+x) - 2x$.

Let $h(x) = 2\ln(1 + x) - 2x$. Then $h'(x) = \frac{2}{1+x} - 2 = \frac{-2x}{1+x}$.

Thus, $h'(x) > 0$ when $x \in (-1, 0)$, so $h(x)$ is an increasing function on $(-1, 0)$; also $h'(x) < 0$ when $x \in (0, +\infty)$, so $h(x)$ is a decreasing function on $(0, +\infty)$.

Hence, $h(x) \le h(0) = 0$ when $x \in (-1, +\infty)$, and $g(x)$ is a decreasing function on $(-1, +\infty)$.

Also, since $g(0) = 0$, so $g(x) > g(0) = 0$ when $x \in (-1, 0)$, that is, $f'(x) > 0$; also $g(x) < g(0) = 0$ when $x \in (0, +\infty)$, that is, $f'(x) < 0$.

It follows that for $f(x)$, the interval of increase is $(-1, 0)$ and the interval of decrease is $(0, +\infty)$.

(2) Let $G(x) = \frac{1}{\ln(1+x)} - \frac{1}{x}$ for $x \in (0, 1]$. Then

$$G'(x) = \frac{(1 + x)[\ln(1 + x)]^2 - x^2}{x^2(1 + x)[\ln(1 + x)]^2}.$$

By (1), $f(x) < f(0) = 0$ with $x \in (0, 1]$, namely $[\ln(1 + x)]^2 - \frac{x^2}{1+x} < 0$. Thus, $G'(x) < 0$, and so $G(x)$ is a decreasing function on $(0, 1]$.

Since $a \le \frac{1}{\ln\left(1+\frac{1}{n}\right)} - n$ for any $n \in \mathbf{N}_+$, by letting $n = 1$ we have

$$a \le \frac{1}{\ln(1 + 1)} - 1 = \frac{1}{\ln 2} - 1.$$

Hence, the maximum value for a that makes $\left(1 + \frac{1}{n}\right)^{n+a} \le e$ be satisfied for all $n \in \mathbf{N}_+$ is $\frac{1}{\ln 2} - 1$.

11. The domain of the function $f(x)$ is $(-1, +\infty)$.

(1) When $a = 1$, since $f(x) = \ln(x + 1) + \frac{2}{x+1} + x - 2$, we have $f'(x) = \frac{x(x+3)}{(x+1)^2}$. We find that $f'(x) < 0$ when $-1 < x < 0$ and $f'(x) > 0$ when $x > 0$. So, $f(x)$ is a decreasing function on the interval $(-1, 0]$ and increasing function on the interval $[0, +\infty)$. Thus, for $a = 1$, the minimum value of $f(x)$ is $f(0) = 0$.

(2) (i) From (1), $f(x) \ge 0$ is always satisfied when $a = 1$, namely

$$\ln(x + 1) + \frac{2}{x + 1} + x - 2 \ge 0$$

for all $x \in (-1, +\infty)$. So, when $a \ge 1$ and $x \in [0, 2]$,

$$f(x) = \ln(x + 1) + \frac{2}{x + 1} + x - 2 + (a - 1)x$$

$$\ge \ln(x + 1) + \frac{2}{x + 1} + x - 2 \ge 0.$$

Hence, $a \ge 1$ satisfies the requirement.

(ii) When $0 < a < 1$, we have $f'(x) = \frac{ax^2+(2a+1)x+a-1}{(x+1)^2}$ $(x > -1)$.

Since for the equation $ax^2 + (2a + 1)x + a - 1 = 0$, the discriminant $\Delta = 8a + 1 > 0$, so there are two different real roots, denoted as x_1 and x_2 with $x_1 < x_2$.

From $x_1 x_2 = \frac{a-1}{a} < 0$, we know that $x_1 < 0 < x_2$ (the two roots of $f'(x) = 0$ are one positive and one negative), so $f'(x) < 0$ when $0 < x < x_2$. Thus, $f(x)$ is a decreasing function on $[0, x_2]$.

If $0 < x_2 < 2$, then $f(x_2) < f(0) = 0$, which contradicts the fact that $f(x) \geq 0$ for $x \in [0,2]$. If $x_2 \geq 2$, then $f(2) < f(0) = 0$, also a contradiction to the same fact. Therefore, $0 < a < 1$ does not satisfy the requirement.

Summarizing (i) and (ii), we can conclude that the range of the values for a is $[1, +\infty)$.

12. By the condition of the problem, $a^x = \frac{y-1}{y+1} > 0$, so

$$g(x) = \log_a \frac{x - 1}{x + 1}, \quad x \in (-\infty, -1) \cup (1, +\infty).$$

From $\log_a \frac{t}{(x^2-1)(7-x)} = \log_a \frac{x-1}{x+1}$,

$$t = (x - 1)^2(7 - x) = -x^3 + 9x^2 - 15x + 7, \quad x \in [2,6].$$

Then $t' = -3x^2 + 18x - 15 = -3(x - 1)(x - 5)$.

When $x = 2$, we have $t = 5$; and $t = 32$ when $x = 5$; also $t = 25$ when $x = 6$. So, $t \in [5, 32]$.

(2) We see that

$$\sum_{k=2}^{n} g(k) = \ln \frac{1}{3} + \ln \frac{2}{4} + \ln \frac{3}{5} + \cdots + \ln \frac{n-1}{n+1}$$

$$= \ln \frac{2}{n(n+1)} = -\ln \frac{n(n+1)}{2}.$$

Let $u(z) = -\ln z^2 - \frac{1-z^2}{z} = -2\ln z + z - \frac{1}{z}$ for $z > 0$. Then

$$u'(z) = \left(1 - \frac{1}{z}\right)^2 \geq 0.$$

Thus, $u(z)$ is an increasing function on $(0, +\infty)$.

Since $\sqrt{\frac{n(n+1)}{2}} > 1$, so $u\left(\sqrt{\frac{n(n+1)}{2}}\right) > u(1) = 0$, that is, $\ln \frac{2}{n(n+1)} - \frac{1-\frac{n(n+1)}{2}}{\sqrt{\frac{n(n+1)}{2}}} > 0$. In other words, $\sum_{k=2}^{n} g(k) > \frac{2-n-n^2}{\sqrt{2n(n+1)}}$.

(3) Let $a = \frac{1}{1+p}$. Then $p \geq 1$ and $1 < f(1) = \frac{1+a}{1-a} = 1 + \frac{2}{p} \leq 3$. When $n = 1$, we see that $|f(1) - 1| = \frac{2}{p} \leq 2 < 4$. When $n \geq 2$, let $k \geq 2$ with $k \in \mathbf{N}_+$. Then

$$f(k) = \frac{(1+p)^k + 1}{(1+p)^k - 1} = 1 + \frac{2}{(1+p)^k - 1}$$

$$= 1 + \frac{2}{C_k^1 p + C_k^2 p^2 + \cdots + C_k^k p^k}.$$

Hence,

$$1 < f(k) \leq 1 + \frac{2}{C_k^1 + C_k^2} = 1 + \frac{4}{k(k+1)} = 1 + \frac{4}{k} - \frac{4}{k-1}.$$

Consequently,

$$n - 1 < \sum_{k=2}^{n} f(k) \leq n - 1 + \frac{4}{2} - \frac{4}{n-1} = n + 1 - \frac{4}{n-1} < n + 1.$$

Hence, $n < \sum_{k=1}^{n} f(k) < f(1) + n + 1 \leq n + 4$.

In summary, $\left| \sum_{k=1}^{n} f(k) - n \right| < 4$.

Solutions for Chapter 17

Mathematical Induction (I)

1. 7.   We can prove that $a_n = 4m - 1$ $(m \in \mathbf{N})$ for any natural number n. Then

$$a_{1993} = 3^{a_{1992}} = 3^{4m-1} = (3^4)^{m-1} \cdot 3^3 = 81^{m-1} \cdot 27,$$

so the last digit of a_{1993} is 7.

2. $g(n) = 2^{n+1} - 1$.   First calculate $g(1), g(2)$, and $g(3)$, from which we guess that $g(n) = 2^{n+1} - 1$. Then prove the conjecture by mathematical induction.

3. $a_n = 3^{2^{n-1}} + \frac{1}{3^{2^{n-1}}} - 3$.   Let $a_n = \frac{1}{t_n} + t_n - 3$, and substitute it into the known relation, resulting in $\frac{1}{t_{n+1}} + t_{n+1} = \frac{1}{t_n^2} + t_n^2 \Rightarrow t_{n+1} = t_n^2$. Also $a_1 = 3 + \frac{1}{3} - 3$ $(t_1 = 3)$. Mathematical induction gives

$$a_n = 3^{2^{n-1}} + \frac{1}{3^{2^{n-1}}} - 3.$$

4. From the given condition,

$$a_3 = \frac{x \cdot 1 + 1}{x + 1} + (2 + 2) = 5, \tag{17.1}$$

$$a_4 = \frac{x \cdot 5 + 1}{x + 5} + (3 + 2) = 10 - \frac{24}{x + 5}.$$

Since $x, a_4 \in \mathbf{Z}_+$, so $x = 1, 3, 7, 19$.

When $x = 1$, we have $a_4 = 6$ and

$$a_5 = \frac{6 \cdot 5 + 1}{6 + 5} + (4 + 2) = \frac{31}{11} + 6 \notin \mathbf{Z}_+.$$

When $x = 7$, we have $a_4 = 8$, and

$$a_5 = \frac{8 \cdot 5 + 1}{8 + 5} + 6 = \frac{41}{13} + 6 \notin \mathbf{Z}_+.$$

When $x = 19$, we have $a_4 = 9$ and

$$a_5 = \frac{9 \cdot 5 + 1}{9 + 5} + 6 = \frac{23}{7} + 6 \notin \mathbf{Z}_+.$$

When $x = 3$, namely $a_2 = 3$, we have $a_4 = 7$ and

$$a_5 = \frac{7 \cdot 5 + 1}{7 + 5} + 6 = 9 \in \mathbf{Z}_+.$$

Conjecture: When $x = 3$, namely $a_2 = 3$, we have that $\{a_n\}$ is a sequence of positive odd numbers.

We prove it with mathematical induction as follows.

When $n = 3$, by (17.1) $a_3 = 5$, so the conclusion is true.

Suppose that when $n \leq k, a_1, a_2, \ldots, a_k$ are the first k positive odd numbers in succession, that is, $a_n = 2n - 1$ for $n = 1, 2, \ldots, k$.

When $n = k + 1$, from

$$a_{k+1} = \frac{(2k - 1)(2k - 3) + 1}{(2k - 1) + (2k - 3)} + k + 2$$

$$= \frac{4k^2 - 8k + 4}{4k - 4} + k + 2 = k - 1 + k + 2$$

$$= 2k + 1 = 2(k + 1) - 1,$$

the conclusion is also true.

In summary, $\{a_n\}$ is a sequence of positive odd numbers.

5. Suppose that such semicircles mutually cut into at most $f(n)$ circular arcs. It is easy to see that $f(1) = 1, f(2) = 4$, and $f(3) = 9$. This leads to a conjecture: $f(n) = n^2$. In the following, we prove it by mathematical induction.

(1) When $n = 1$, the conclusion is obviously true.

(2) Assume that the conjecture is true for $n = k$, namely $f(k) = k^2$.

When $n = k + 1$, we make the $(k + 1)$st semicircle that intersects the previous k semicircles, which increases at most k intersection points. The $(k + 1)$st semicircle is itself divided into $k + 1$ circular arcs, and the previous k semicircles are cut into one more arc each, so

$$f(k + 1) = f(k) + k + k + 1 = k^2 + 2k + 1 = (k + 1)^2,$$

that is, the conclusion is true when $n = k + 1$. Hence, $f(n) = n^2$ for all positive integers n.

6. From $0 < a_2 \le a_1 - a_1^2$ we get $0 < a_1 < 1$.

When $n = 2$, from $a_2 \le a_1 - a_1^2 = -\left(a_1 - \frac{1}{2}\right)^2 + \frac{1}{4} \le \frac{1}{4} = \frac{1}{2+2}$, we know the validity of the inequality.

Assume that the proposition is true for $n = k$ $(k \ge 2)$, namely $0 < a_k \le \frac{1}{k+2} \le \frac{1}{4}$. From $a_{k+1} \le a_k - a_k^2 = -\left(a_k - \frac{1}{2}\right)^2 + \frac{1}{4} \le -\left(\frac{1}{k+2} - \frac{1}{2}\right)^2 + \frac{1}{4} = \frac{k+1}{(k+2)^2} < \frac{k+1}{k^2+4k+3} = \frac{1}{(k+1)+2}$, we see that the inequality is satisfied for $n = k + 1$.

7. The conclusion is obviously true when $n = 1$. Suppose that the proposition is satisfied for $n = k$. For 2^{k+1} balls, if the number of the balls in each pile is an even number, then every shift moves an even number of balls, so any such shift will move by "binding" two balls, thus not changing the nature of the shift, and is consequently equivalent to the shift of 2^k balls. By the inductive hypothesis, it is possible to realize the purpose. If there exists a pile of odd numbers of balls, then from the fact that the total number of the balls is even, the number of piles with odd numbers of balls is even, we first make them in pairs to move once each, so that now each pile has an even number of balls. This becomes the previous case, so the conclusion is true when $n = k + 1$.

8. By mathematical induction we can show that the numbers in the nth line constitute an arithmetic sequence of common difference 2^{n-1}.

When $n = 2$, the second line consists of consecutive odd numbers.

Assume that the conclusion is valid for $n - 1$.

Let x, y, and z be three consecutive numbers in the $(n - 1)$st line. Then $x + y$ and $y + z$ are two adjacent numbers in the nth line. By the inductive hypothesis, $y - x = z - y = 2^{n-2}$. So,

$$(y + z) - (x + y) = (z - y) + (y - x)$$
$$= 2^{n-2} + 2^{n-2} = 2^{n-1}.$$

Let a_n be the first number of the nth line.

For all $n \in \{1, 2, \ldots, 2016\}$, we have

$$a_n = a_{n-1} + a_{n-1} + 2^{n-2} = 2a_{n-1} + 2^{n-2}.$$

Using the above relation and noting that $a_1 = 1$, we see easily that

$$a_n = (n + 1)2^{n-2}.$$

Hence, the desired number is $a_{2016} = 2^{2014} \cdot 2017$.

9. We use the induction method for m $(< n)$.

The proposition is true for $m = 1$.

Suppose that the proposition is valid for all positive integers less than k $(< n)$. Since $k < n$, let $n = qk - r$, where q is a positive integer and $0 \le r < k$.

When $r = 0$, we have $n = qk$, and then $\frac{k}{n} = \frac{k}{qk} = \frac{1}{q}$, so the proposition is true.

When $r > 0$, we obtain $\frac{k}{n} = \frac{n+r}{nq} = \frac{1}{q} + \frac{r}{nq}$. By the inductive hypothesis,

$$\frac{r}{n} = \frac{1}{n_1} + \frac{1}{n_2} + \cdots + \frac{1}{n_t},$$

where $n_1 < n_2 < \cdots < n_t$ are mutually different positive integers. Thus,

$$\frac{k}{n} = \frac{1}{q} + \frac{1}{qn_1} + \frac{1}{qn_2} + \cdots + \frac{1}{qn_t},$$

where $q < qn_1 < qn_2 < \cdots < qn_t$. Therefore, the proposition is also satisfied.

10. If $x \ge -1$ satisfies the condition of the problem, we may assume that $a_1 = a_2 = \cdots = a_n = 1$. By

$$\prod_{i=1}^{n} \frac{a_i + x}{2} \le \frac{a_1 a_2 \cdots a_n + x}{2}, \tag{17.2}$$

we see that $\left(\frac{1+x}{2}\right)^n \le \frac{1+x}{2}$. Hence, $x \le 1$.

In the following, we prove that when $n \ge 2$, for $x \in [-1, 1]$, (17.2) is satisfied, where for all $n \in \mathbf{N}_+$,

$$a_i \ge 1 \ (i = 1, 2, \ldots, n).$$

When $n = 2$, the inequality (17.2) is

$$\frac{a_1 + x}{2} \cdot \frac{a_2 + x}{2} \le \frac{a_1 a_2 + x}{2}, \tag{17.3}$$

which is equivalent to

$$x^2 + (a_1 + a_2 - 2)x - a_1 a_2 \le 0.$$

Since

$$x^2 + (a_1 + a_2 - 2)x - a_1a_2$$

$$\leq 1 + (a_1 + a_2 - 2) - a_1a_2$$

$$= a_1 + a_2 - a_1a_2 - 1$$

$$= -(a_1 - 1)(a_2 - 1) \leq 0,$$

we see that (17.2) is satisfied when $n = 2$.

Assume that (17.2) is true for some $n \geq 2$.

By (17.2) and (17.3), we have

$$\prod_{i=1}^{n+1} \frac{a_i + x}{2} \leq \frac{a_1a_2 \cdots a_n + x}{2} \cdot \frac{a_{n+1} + x}{2} \leq \frac{a_1a_2 \cdots a_{n+1} + x}{2}.$$

Therefore, $x \in [-1, 1]$.

11. When $n = 3, 4$, and 5, the conclusion is true, and the respective divisions are as Figures 17.1, 17.2, and 17.3 show.

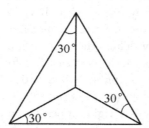

Figure 17.1 Figure 1 for Exercise 11: $n = 3$.

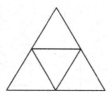

Figure 17.2 Figure 2 for Exercise 11: $n = 4$.

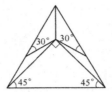

Figure 17.3   Figure 3 for Exercise 11: $n = 5$.

Obviously, the above three methods of division have no mutual relations. When $n = 5$, there appears an isosceles right triangle in the division. Since any isosceles right triangle can be evenly partitioned into two smaller isosceles right triangles, this even partition can continue indefinitely from $n = 5$ on. Thus, starting from $n = 5$, we can use mathematical induction to prove the proposition.

12. We generalize the proposition by considering a general $(6n + 1) \times (6n + 1)$ check table.

When $n = 1$, for the 7×7 check table, because of symmetry, it is enough to consider the table such that the removed check is one in the shaded region shown by Figure 17.4. Dividing the situation into three cases with their concrete coverage demonstrated by Figures 17.5, 17.6, and 17.7, respectively. Thus, the proposition is true when $n = 1$.

Assume that the proposition is satisfied for $n = k$. Since $(6k + 7) \times (6k + 7) = (6k + 1) \times (6k + 1) + 2 \times 6 \times (6k + 1) + 6 \times 6$, when $n = k+1$, among the four $(6k+1) \times (6k+1)$ check tables at the left upper corner, left down corner, right upper corner, and right down corner (may have overlapping), at least one of them contains the removed check. First, choose one of the $(6k + 1) \times (6k + 1)$ check tables that

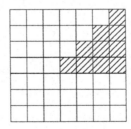

Figure 17.4   Figure 1 for Exercise 12.

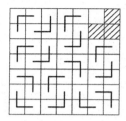

Figure 17.5   Figure 2 for Exercise 12.

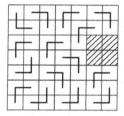

Figure 17.6   Figure 3 for Exercise 12.

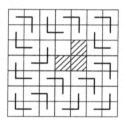

Figure 17.7   Figure 4 for Exercise 12.

contains the removed check. By the inductive hypothesis, it can be covered by several L-shaped figures. For the remaining part, it is easy to see that it can be divided into several L-shaped figures. That is, the conclusion is valid for $n = k + 1$.

Solutions for Chapter 18

Complex Numbers

1. $\sqrt{5}$.   Let $z = a + bi$ with $a, b \in \mathbf{R}$. From the condition,

$$(a + 9) + bi = 10a + (-10b + 22)i.$$

Comparing the real and imaginary parts of the two sides, we have $a + 9 = 10a$ and $b = -10b + 22$. Solving the system of two equations gives $a = 1$ and $b = 2$. Thus, $z = 1 + 2i$, and so $|z| = \sqrt{5}$.

2. $2015 + 1007i$.   The given condition implies that $z_{n+2} = \overline{z_{n+1}} + 1 + (n+1)i = \overline{\overline{z_n} + 1 + ni} + (n+1)i = z_n + 2 + i$ for any positive integer n. Hence,

$$z_{2015} = z_1 + 1007 \cdot (2 + i) = 2015 + 1007i.$$

3. 0.   Since $|z_1 - \overline{z_2}|^2 = (z_1 - \overline{z_2})(\overline{z_1} - z_2) = |z_1|^2 + |z_2|^2 - z_1 z_2 - \overline{z_1 z_2}$ and $|1 - \overline{z_1 z_2}|^2 = (1 - \overline{z_1 z_2})(1 - z_1 z_2) = 1 - z_1 z_2 - \overline{z_1 z_2} + |z_1|^2 |z_2|^2$, the condition gives that $|z_1|^2 + |z_2|^2 = 1 + |z_1|^2 |z_2|^2$, from which $(|z_1| - 1)(|z_2| - 1)(|z_1| + 1)(|z_2| + 1) = 0$. Hence, $(|z_1| - 1)(|z_2| - 1) = 0$.

4. $\frac{1}{2} + \frac{\sqrt{3}}{2}i$.   Write $z + 1 = \frac{\sqrt{3}}{2}r_1 + \frac{1}{2}r_1 i$ and $z - 1 = -\frac{1}{2}r_2 + \frac{\sqrt{3}}{2}r_2 i$. Solving the two equations $-1 + \frac{\sqrt{3}}{2}r_1 = 1 - \frac{1}{2}r_2$ and $\frac{1}{2}r_1 = \frac{\sqrt{3}}{2}r_2$ gives $r_1 = \sqrt{3}$ and $r_2 = 1$.

5. 12.   Since $(a + bi)^2 = b + ai$, we have $a^2 - b^2 = b$ and $2ab = a$, from which $a = \frac{\sqrt{3}}{2}$ and $b = \frac{1}{2}$. Now, the common ratio $q = a + bi = \cos\frac{\pi}{6} + i\sin\frac{\pi}{6}$. The condition $z_1 + z_2 + \cdots + z_n = 0$ implies that $q^n = 1$, namely $\cos\frac{n}{6}\pi = 1$ and $\sin\frac{n}{6}\pi = 0$. Thus, $\frac{n}{6}\pi = 2k\pi$ with $k \in \mathbf{Z}$, and so $n = 12k$ with $k \in \mathbf{Z}$.

6. 9^{2000}.   We find $9 = |z_1 + z_2|^2 = |z_1|^2 + |z_2|^2 + z_1\overline{z_2} + \overline{z_1}z_2$ and $27 = |z_1 - z_2|^2 = |z_1|^2 + |z_2|^2 - (z_1\overline{z_2} + \overline{z_1}z_2)$. Since $|z_1| = 3$, $|z_2| = 3$, so

$z_1\overline{z_2} + \overline{z_1}z_2 = -9$ and $|z_1\overline{z_2}| = |\overline{z_1}z_2| = 9$. Let $z_1\overline{z_2} = 9(\cos\theta + i\sin\theta)$. Then $\overline{z_1}z_2 = 9(\cos\theta - i\sin\theta)$. From $-9 = z_1\overline{z_2} + \overline{z_1}z_2 = 18\cos\theta$, we obtain that $\cos\theta = -\frac{1}{2}$. Then $z_1\overline{z_2} = 9\omega$ or $z_1\overline{z_2} = 9\omega^2$ (here $\omega = -\frac{1}{2} + \frac{\sqrt{3}}{2}i$). When $z_1\overline{z_2} = 9\omega$, we get $(z_1\overline{z_2})^{2000} + (\overline{z_1}z_2)^{2000} = -9^{2000}$; when $z_1\overline{z_2} = 9\omega^2$, we have the same result.

7. Since z_1 and z_2 are conjugate complex numbers, we assume that $z = z_1$ and $\overline{z} = z_2$. Also from the condition, $\frac{z_1^3}{z_2^3}$ is a real number, so $\frac{z^3}{\overline{z}^3} = \overline{\left(\frac{z^3}{\overline{z}^3}\right)} = \frac{\overline{z}^3}{z^3}$, namely $z^3 = \overline{z}^3$, which can be written as $(z - \overline{z})(z^2 + z\overline{z} + \overline{z}^2) = 0$.

From $|z - \overline{z}| = |z_1 - z_2| = \sqrt{6} \neq 0$, we see that $z - \overline{z} \neq 0$, that is

$$z^2 + z\overline{z} + \overline{z}^2 = 0. \tag{18.1}$$

Also $|z - \overline{z}| = \sqrt{6}$, namely $6 = |z - \overline{z}|^2 = (z - \overline{z})(\overline{z} - z)$. In other words,

$$6 = 2z\overline{z} - z^2 - \overline{z}^2. \tag{18.2}$$

Adding (18.1) and (18.2) gives $3z\overline{z} = 6$, namely $|z|^2 = 2$. Hence, $|z| = \sqrt{2}$.

8. Since $z_1 = 2\cos\frac{\alpha}{2}\left(\cos\frac{\alpha}{2} + i\sin\frac{\alpha}{2}\right)$ with $0 < \alpha < \pi$, so $|z_1| = 2\cos\frac{\alpha}{2}$ and $\arg z_1 = \frac{\alpha}{2}$. Also, since $z_2 = 2\sin\frac{\beta}{2}\left[\cos\left(\frac{\pi}{2} - \frac{\beta}{2}\right) + i\sin\left(\frac{\pi}{2} - \frac{\beta}{2}\right)\right]$, from $\pi < \beta < 2\pi$, we have $-\frac{\pi}{2} < \frac{\pi}{2} - \frac{\beta}{2} < 0$. Hence,

$$|z_2| = 2\sin\frac{\beta}{2}, \quad \arg z_2 = \frac{\pi}{2} - \frac{\beta}{2} + 2\pi = \frac{5\pi}{2} - \frac{\beta}{2}.$$

Consequently, $\frac{\alpha}{2} + \frac{5\pi}{2} - \frac{\beta}{2} = \frac{13\pi}{6}$, that is,

$$\frac{\alpha}{2} = \frac{\beta}{2} - \frac{\pi}{3}. \tag{18.3}$$

Now, the condition $|z_1||z_2| = \sqrt{3}$ implies that

$$2\cos\frac{\alpha}{2} \cdot 2\sin\frac{\beta}{2} = \sqrt{3}. \tag{18.4}$$

Substituting (18.3) into (18.4) gives that $4\cos\left(\frac{\beta}{2} - \frac{\pi}{3}\right)\sin\frac{\beta}{2} = \sqrt{3}$, in other words, $2\left[\sin\left(\beta - \frac{\pi}{3}\right) + \sin\frac{\pi}{3}\right] = \sqrt{3}$. Hence, $\sin\left(\beta - \frac{\pi}{3}\right) = 0$. Since $\pi < \beta < 2\pi$, we have $\beta = \frac{4\pi}{3}$ and $\alpha = \frac{2\pi}{3}$.

9. It is enough to show that $\overline{p} = p$, which can be shown by using $z\overline{z} = |z|^2$.

10. We see that $c = (a^3 - 3ab^2) + (3a^2b - b^3 - 107)i$. From $c \in \mathbf{N}$, we have $3a^2b - b^3 - 107 = 0$, that is, $b(3a^2 - b^2) = 107$. Since 107 is

a prime number, a calculation gives that $b = 1$ and $a = 6$. Therefore, $c = a^3 - 3ab^2 = 198$.

11. The answer is $\sqrt{(a+b)^2 + c^2}$ or $\sqrt{(b+c)^2 + a^2}$ or $\sqrt{(c+a)^2 + b^2}$. From the second equality of the condition we see that $\frac{z_1}{z_2} + \frac{z_2}{z_3} + \frac{z_3}{z_1} \in \mathbf{R}$, so

$$\frac{z_1}{z_2} + \frac{z_2}{z_3} + \frac{z_3}{z_1} = \overline{\frac{z_1}{z_2}} + \overline{\frac{z_2}{z_3}} + \overline{\frac{z_3}{z_1}} = \frac{\overline{z_1}}{\overline{z_2}} + \frac{\overline{z_2}}{\overline{z_3}} + \frac{\overline{z_3}}{\overline{z_1}}. \tag{18.5}$$

Also $|z_1| = |z_2| = |z_3| = 1$ implies that $\overline{z_k} = \frac{1}{z_k}$ ($k = 1, 2, 3$). Substituting into (18.5), we obtain that $\frac{z_1}{z_2} + \frac{z_2}{z_3} + \frac{z_3}{z_1} = \frac{z_2}{z_1} + \frac{z_3}{z_2} + \frac{z_1}{z_3}$, that is $z_1^2 z_3 + z_2^2 z_1 + z_3^2 z_2 = z_2^2 z_3 + z_3^2 z_1 + z_1^2 z_2$, from which

$$(z_1 - z_2)(z_2 - z_3)(z_3 - z_1) = 0.$$

It follows that $z_1 = z_2$, $z_2 = z_3$, or $z_3 = z_1$. If $z_1 = z_2$, then substituting into the second equality, we see that $\frac{z_3}{z_1} = \pm i$ and

$$|az_1 + bz_2 + cz_3| = |z_1||a + b \pm ci| = \sqrt{(a+b)^2 + c^2}.$$

Similarly, the cases $z_2 = z_3$ and $z_3 = z_1$ both give the same result.

12. Construct complex numbers $z_1 = x + yi$ and $z_2 = a + bi$. Then $|z_1| \le 1$ and $|z_2| \le \sqrt{2}$. Since

$$z_1^2 z_2 = \left[a\left(x^2 - y^2\right) - 2bxy \right] + \left[b\left(x^2 - y^2\right) + 2axy \right] i,$$

from $\left| \mathrm{Im} \left(z_1^2 z_2 \right) \right| \le |z_1^2 z_2|$, we can prove the inequality.

Solutions for Chapter 19

Geometric Meaning of Complex Operations

1. $\sqrt{2}$. The points corresponding to $z_1, z_2, z_1 + z_2$, and the origin constitute the four vertices of a square.
2. $\left[0, \frac{\pi}{4}\right] \cup \left[\frac{7\pi}{4}, 2\pi\right)$.
3. $2 + 2\sqrt{3}$. As in Figure 19.1, denote the points $Z_1(-3, -3\sqrt{3})$ and $Z_2(\sqrt{3}, 1)$. The moving point Z is on the circle $x^2 + (y - 2)^2 = 3$. It is easy to see that the points Z_1 and Z_2 are outside the circle. The equation of the straight line $Z_1 Z_2$ is $x - \sqrt{3}y = 0$, and the straight line $Z_1 Z_2$ and the circle are tangent at the point $T\left(\frac{\sqrt{3}}{2}, \frac{1}{2}\right)$. In other words, $z = \frac{\sqrt{3}}{2} + \frac{1}{2}i$ makes $|z - z_1| + |z - z_2|$ the minimum value $2 + 2\sqrt{3}$.

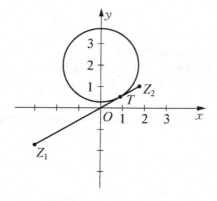

Figure 19.1 Figure for Exercise 3.

4. $(2, +\infty)$.   First, $a^2 - 3a + 2 \geq 0$ and $a > 0$ make $a \in (0, 1] \cup [2, +\infty)$. Second, since in the complex plane, the former condition represents a circle and the latter condition represents a disk that does not include the boundary, they must have intersection points. So, the requirements are $\sqrt{a^2 - 3a + 2} + a > 2$ and $\sqrt{a^2 - 3a + 2} < 2 + a$. Solving them gives $a > 2$.

5. 4.   Let $z_1 = \cos\theta + i\sin\theta$. Then

$$z_k = (\cos\theta + i\sin\theta)\left[\cos\frac{2(k-1)\pi}{20} + i\sin\frac{2(k-1)\pi}{20}\right], \quad 1 \leq k \leq 20.$$

From $1995 = 20 \cdot 99 + 15$, we have

$$z_k^{1995} = (\cos 1995\theta + i\sin 1995\theta)\left(\cos\frac{3}{2}\pi + i\sin\frac{3}{2}\pi\right)^{k-1}$$

$$= (\cos 1995\theta + i\sin 1995\theta)(-i)^{k-1}$$

for $k = 1, 2, \ldots, 20$. There are totally four different values.

6. Since $w = z_1 - z_2 = (z_1 - 8i) - z_2 + 8i$, so $|w - 8i| = |z_1 - 8i - z_2| \in [2, 6]$. Thus, the point set representation corresponding to w is the ring centered at $(0, 8)$ of inner circle radius 2 and outer circle radius 6, whose area is 32π.

7. Let the trajectory of the point B be $|z_B - a| = r$ with $a \in \mathbf{C}$ and $r > 0$. Let the complex number corresponding to the point A be z_1. Then $\overrightarrow{AC} \cdot \sqrt{2}e^{i\cdot\frac{\pi}{4}} = \overrightarrow{AB}$. Hence, $(z_C - z_1)(1 + i) = z_B - z_1$. Consequently,

$$|(z_C - z_1)(1 + i) + z_1 - a| = r, \quad \left|z_C - \left(z_1 + \frac{a - z_1}{1 + i}\right)\right| = \frac{\sqrt{2}}{2}r.$$

This means that C is on a circle of radius $\frac{\sqrt{2}}{2}r$, the center of which is the right angle vertex of an isosceles right triangle obtained clockwise by the side with A and the center of the circle containing B as its end points.

8. The sets A and B of the complex numbers are respectively the disks represented by

$$(x - 2)^2 + y^2 \leq 4 \quad \text{and} \quad (x - b)^2 + (y - 1)^2 \leq 1.$$

(1) If $A \cap B = \emptyset$, then $b > 2 + 2\sqrt{2}$ or $b < 2 - 2\sqrt{2}$.
(2) If $A \cap B = B$, then $b = 2$.

9. From the condition, the vector corresponding to the complex number $-z_1$ is the diagonal vector $\overrightarrow{OZ_1'}$ of the parallelogram $OZ_2Z_1'Z_3$. Now,

$$\cos(\beta - \gamma) = -\cos \angle OZ_2Z_1' = 1 + \frac{3}{2(k-1)^2 - 2}.$$

Also from $||z_1| - |z_2|| \le |z_1 + z_2| \le |z_1| + |z_2|$, we have $\frac{1}{2} \le k \le \frac{3}{2}$. So, $[\cos(\beta - \gamma)]_{max} = -\frac{1}{2}$ when $k = 1$, and $[\cos(\beta - \gamma)]_{min} = -1$ when $k = \frac{1}{2}$ or $\frac{3}{2}$.

10. It is easy to see that $z_1 = \overline{z_2}$. Let $z_1 = \alpha + \beta i$. Then $z_2 = \alpha - \beta i$ ($\alpha, \beta \in \mathbf{R}$). Since the discriminant $\Delta < 0$, we have $a^2 < b$, so $x = -a \pm \sqrt{b^2 - a^2}i$. Also from the given condition, $|\alpha| < |\beta|$. Thus, $2a^2 < b$. Hence, the existence range of the complex numbers $z = a + bi$ is the interior of the region above the graph of the parabola $b = 2a^2$ (figure omitted).

11. By the coordinates formula for the center of gravity, $3z = z_1 + z_2 = (r_1 + r_2)\cos\theta + i(r_1 - r_2)\sin\theta$. So, $|3z|^2 = (r_1 + r_2)^2\cos^2\theta + (r_1 - r_2)^2\sin^2\theta = (r_1 - r_2)^2 + 4r_1r_2\cos^2\theta$. Also, since the area of $\triangle OZ_1Z_2$ is S, we get $r_1r_2 = \frac{2S}{\sin 2\theta}$. Hence, $|3z|^2 = (r_1 - r_2)^2 + 4S\cot\theta$, and $|z|_{min} = \frac{2}{3}\sqrt{S \cdot \cot\theta}$ achieved when $r_1 = r_2 = \sqrt{\frac{2S}{\sin 2\theta}}$.

12. Let $\frac{a_2}{a_1} = \frac{a_3}{a_2} = \frac{a_4}{a_3} = \frac{a_5}{a_4} = t$. Then $a_1 + a_2 + a_3 + a_4 + a_5 = a_1(1 + t + t^2 + t^3 + t^4)$ and $4\left(\frac{1}{a_1} + \frac{1}{a_2} + \frac{1}{a_3} + \frac{1}{a_4} + \frac{1}{a_5}\right) = \frac{4}{a_1}\left(1 + \frac{1}{t} + \frac{1}{t^2} + \frac{1}{t^3} + \frac{1}{t^4}\right)$. If $1 + t + t^2 + t^3 + t^4 = 1$, then $t^5 = 1$ and $|a_1| = |a_1t| = |a_2| = |a_3| = |a_4| = |a_5|$, so a_1, a_2, a_3, a_4, and a_5 lie on the same circle; if $a_1 = \frac{4}{a_1t^4}$, then $a_3 = a_1t^2 = \pm 2$; if $a_3 = -2$, then let $a_i' = -a_i$ and $S' = -S$. Thus, $a_3' = -2$ and $|S'| \le 2$. So, we may let

$$a_3 = 2, \ a_1 = \frac{2}{t^2}, \ a_2 = \frac{2}{t}, \ a_4 = 2t, \ a_5 = 2t^2,$$

$$2\left(\frac{1}{t^2} + \frac{1}{t} + 1 + t + t^2\right) = S.$$

Let $x = t + \frac{1}{t}$. Then $x^2 + x = \frac{S}{2} + 1$. Denote $x = a + bi$ ($a, b \in \mathbf{R}$). Then $a^2 - b^2 + a = \frac{S}{2} + 1$ and $2ab + b = 0$. Thus, $a = -\frac{1}{2}$ or $b = 0$. If $a = -\frac{1}{2}$, then $a^2 + a < 0$ and $-b^2 \le 0$. But $|S| \le 2$ and $\frac{S}{2} + 1 \ge 0$, a contradiction; if $b = 0$, then $t + \frac{1}{t} = a \in \mathbf{R}$. Let $t = re^{i\theta}$ ($r > 0$).

Then $\frac{1}{t} = \frac{1}{r}e^{i(-\theta)}$. But the imaginary part of $t + \frac{1}{t}$ is 0, so $r = \frac{1}{r}$, from which $r = 1$. Consequently, $|a_2| = |a_1 t| = |a_1 r| = |a_1|$. By the same token,

$$|a_1| = |a_2| = |a_3| = |a_4| = |a_5|.$$

Therefore, a_1, a_2, a_3, a_4, and a_5 are on the same circle.

Solutions for Chapter 20

Mean Value Inequalities

1. 4.   Since $\frac{(x+2)^2}{y} + \frac{(y+2)^2}{x} = \left(\frac{x^2}{y} + \frac{y^2}{x} + \frac{4}{y} + \frac{4}{x}\right) + \left(\frac{4x}{y} + \frac{4y}{x}\right) \geq$
$4\left(\frac{x^2}{y} \cdot \frac{y^2}{x} \cdot \frac{4}{y} \cdot \frac{4}{x}\right)^{\frac{1}{4}} + 2 \cdot 4 = 16$, and the condition for the equality
to be satisfied is $x = y = 2$, we obtain that $x + y = 4$.

2. 4.   Since $a^2 = (3x+1) + (3y+1) + (3z+1) + 2\sqrt{(3x+1)(3y+1)} +$
$2\sqrt{(3y+1)(3z+1)} + 2\sqrt{(3x+1)(3z+1)} \leq 3[(3x+1) + (3y+1) +$
$(3z+1)] = 18$, we have $a \leq \sqrt{18} < 5$. Also, since $0 \leq x, y, z \leq 1$, so
$x \geq x^2, y \geq y^2$, and $z \geq z^2$. Consequently,

$$a \geq \sqrt{x^2 + 2x + 1} + \sqrt{y^2 + 2y + 1} + \sqrt{z^2 + 2z + 1}$$
$$= (x+1) + (y+1) + (z+1) = 4,$$

thus, from $4 \leq a < 5$, we get $[a] = 4$.

3. $\sqrt{2} - \frac{1}{2}$.   Without loss of generality, assume that $a \geq d$ and $b \geq c$.
Then

$$\frac{b}{c+d} + \frac{c}{a+b} = \frac{b+c}{c+d} + c\left(\frac{1}{a+b} - \frac{1}{c+d}\right)$$

$$\geq \frac{\frac{1}{2}(a+b+c+d)}{c+d} + (c+d)\left(\frac{1}{a+b} - \frac{1}{c+d}\right)$$

$$= \frac{a+b}{2(c+d)} + \frac{c+d}{a+b} - \frac{1}{2} \geq \sqrt{2} - \frac{1}{2}.$$

4. We have $\sqrt[3]{abc} = \sqrt[3]{\frac{a}{4} \cdot b \cdot 4c} \leq \frac{a}{12} + \frac{b}{3} + \frac{4c}{3}$ and $\sqrt{ab} = \sqrt{\frac{a}{2} \cdot 2b} < \frac{a}{4} + b$,
so $a + \sqrt{ab} + \sqrt[3]{abc} \leq \frac{4}{3}(a+b+c)$.

5. Let $p = xyzw$. From the first equation, we have $1 = \frac{x+y+z+w}{4} \geq \sqrt[4]{xyzw} = \sqrt[4]{p}$, that is $p \leq 1$. By the second equation, $1 = \frac{1}{5}\left(\frac{1}{x} + \frac{1}{y} + \frac{1}{z} + \frac{1}{w} + \frac{1}{xyzw}\right) \geq \sqrt[5]{\frac{1}{(xyzw)^2}} = \frac{1}{\sqrt[5]{p^2}}$, that is $p \geq 1$. Hence, $p = 1$.

     The condition for the equality implies that $x = y = z = w = xyzw = 1$. Therefore, the solution of the original system of equations is

$$(x, y, z, w) = (1, 1, 1, 1).$$

6. Since $\frac{1}{x^3} + \frac{1}{y^3} + \frac{1}{z^3} \geq 3\sqrt[3]{\frac{1}{x^3y^3z^3}} = 3$, so $\frac{x^6+2}{x^3} + \frac{y^6+2}{y^3} + \frac{z^6+2}{z^3} \geq x^3 + y^3 + z^3 + \frac{1}{x^3} + \frac{1}{y^3} + \frac{1}{z^3} + 3 = \left(x^3 + \frac{1}{y^3} + 1\right) + \left(y^3 + \frac{1}{z^3} + 1\right) + \left(z^3 + \frac{1}{x^3} + 1\right) \geq 3\sqrt[3]{\frac{x^3}{y^3}} + 3\sqrt[3]{\frac{y^3}{z^3}} + 3\sqrt[3]{\frac{z^3}{x^3}} = 3\left(\frac{x}{y} + \frac{y}{z} + \frac{z}{x}\right).$

7. We have

$$\frac{1}{n}\left(1 + \frac{1}{2} + \frac{1}{3} + \cdots + \frac{1}{n} + n\right)$$

$$= \frac{1}{n}\left(2 + \frac{3}{2} + \frac{4}{3} + \cdots + \frac{n+1}{n}\right)$$

$$> \sqrt[n]{n+1}.$$

8. From the given condition, $1 = 2a^2 + 2b^2 + 2c^2$, so the original inequality is equivalent to

$$\frac{a^2 + 2b^2 + 3c^2}{ac + 2bc} + \frac{b^2 + 2c^2 + 3a^2}{ab + 2ca} + \frac{c^2 + 2a^2 + 3b^2}{cb + 2ab} \geq 6. \quad (20.1)$$

Since $a^2 + 2b^2 + 3c^2 = a^2 + c^2 + 2(b^2 + c^2) \geq 2ac + 4bc = 2(ac + 2bc)$,

$$\frac{a^2 + 2b^2 + 3c^2}{ac + 2bc} \geq \frac{2(ac + 2bc)}{ac + 2bc} = 2.$$

By the same token, $\frac{b^2+2c^2+3a^2}{ab+2ca} \geq 2$ and $\frac{c^2+2a^2+3b^2}{cb+2ab} \geq 2$.
Adding the above three inequalities gives (20.1).

9. First $S^2 = \frac{x^2y^2}{z^2} + \frac{y^2z^2}{x^2} + \frac{z^2x^2}{y^2} + 2(x^2 + y^2 + z^2)$.

From the arithmetic mean-geometric mean inequality,

$$\frac{1}{2}\left(\frac{x^2y^2}{z^2} + \frac{y^2z^2}{x^2}\right) \geq y^2,$$

$$\frac{1}{2}\left(\frac{y^2z^2}{x^2} + \frac{z^2x^2}{y^2}\right) \geq z^2,$$

$$\frac{1}{2}\left(\frac{z^2x^2}{y^2} + \frac{x^2y^2}{z^2}\right) \geq z^2.$$

Hence, $S^2 \geq 3(x^2 + y^2 + z^2) = 3$, namely $S \geq \sqrt{3}$. When $x = y = z = \frac{\sqrt{3}}{3}$ that satisfy $x^2 + y^2 + z^2 = 1$, we find that $S = \sqrt{3}$. Thus, the minimum value of S is $\sqrt{3}$.

10. Note that the equality in the inequality is valid when $a = b = c = \frac{1}{3}$. Deform the original inequality to

$$\left(3a + \frac{3}{a}\right)\left(3b + \frac{3}{b}\right)\left(3c + \frac{3}{c}\right) \geq 1000.$$

After a careful observation, we see that for the equality in the inequality to be satisfied, it is enough to have $3a = \frac{3}{ma} = 1$, from which $m = 9$. So, using the arithmetic mean-geometric mean inequality, we obtain that

$$\left(3a + \frac{3}{a}\right)\left(3b + \frac{3}{b}\right)\left(3c + \frac{3}{c}\right)$$

$$= \left(3a + \frac{1}{3a} + \cdots + \frac{1}{3a}\right)\left(3b + \frac{1}{3b} + \cdots + \frac{1}{3b}\right)\left(3c + \frac{1}{3c} + \cdots + \frac{1}{3c}\right)$$

$$\geq 10\sqrt[10]{\left(\frac{1}{3a}\right)^8} \cdot 10\sqrt[10]{\left(\frac{1}{3b}\right)^8} \cdot 10\sqrt[10]{\left(\frac{1}{3c}\right)^8}.$$

$$= 1000\left(\frac{1}{3a \cdot 3b \cdot 3c}\right)^{\frac{4}{5}}.$$

Therefore, $\left(a + \frac{1}{a}\right)\left(b + \frac{1}{b}\right)\left(c + \frac{1}{c}\right) \geq \frac{1000}{27}$.

11. By the power mean inequality,

$$\left(\frac{x^{k+1} + y^{k+1} + z^{k+1}}{3}\right)^{\frac{1}{k+1}} \geq \left(\frac{x^k + y^k + z^k}{3}\right)^{\frac{1}{k}} = \left(\frac{1}{3}\right)^{\frac{1}{k}}.$$

Hence, $x^{k+1} + y^{k+1} + z^{k+1} \geq 3 \left(\frac{1}{3}\right)^{\frac{k+1}{k}} = 3^{-\frac{1}{k}}$. The equality is true when $x = y = z = 3^{-\frac{1}{k}}$. Thus, the minimum value is $3^{-\frac{1}{k}}$.

12. Denote the left hand side of the inequality as S. When $a_1 = a_2 = \cdots = a_n$, we have $S = 1$. In the following, we show that $S \geq 1$. Let

$$x_k = \frac{\sqrt{a_k^{n-1}}}{\sqrt{a_k^{n-1} + (n^2 - 1)\frac{a_1 a_2 \cdots a_n}{a_k}}} > 0, \ k = 1, 2, \ldots, n.$$

Then

$$\frac{1}{x_k^2} = 1 + (n^2 - 1)\frac{a_1 a_2 \cdots a_n}{a_k},$$

and so

$$\prod_{k=1}^n \left(\frac{1}{x_k^2} - 1\right) = (n^2 - 1)^n \tag{20.2}$$

with $S = x_1 + x_2 + \cdots + x_n$.

Suppose that $S < 1$ on the contrary, namely $1 > x_1 + x_2 + \cdots + x_n$. Then by (20.2),

$$(n^2 - 1)^n = \frac{\prod_{k=1}^n (1 - x_k)(1 + x_k)}{\prod_{k=1}^n x_k^2}$$

$$> \frac{\prod_{k=1}^n \left[\sum_{i=1, i \neq k}^n x_i \cdot \left(\sum_{i=1}^n x_i + x_k\right)\right]}{\prod_{k=1}^n x_k^2}$$

$$\geq \frac{1}{\prod_{k=1}^n x_k^2} \prod_{k=1}^n (n-1) \sqrt[n-1]{x_1 \cdots x_{k-1} x_{k+1} \cdots x_n} \cdot$$

$$(n+1) \sqrt[n+1]{x_1 x_2 \cdots x_n \cdot x_k}$$

$$= \frac{1}{\prod_{k=1}^n x_k^2} \cdot (n^2 - 1)^n \left(\prod_{k=1}^n x_k\right) \cdot \left(\prod_{k=1}^n x_k\right)$$

$$= (n^2 - 1)^n,$$

which is a contradiction. Hence, $S \geq 1$.

In summary, $\lambda_{\max} = 1$.

Solutions for Chapter 21

The Cauchy Inequality

1. 3, 1, and 4.   Adding the two equations gives $(2x)^2 + (3y+3)^2 + (z+2)^2 = 108$, and the first equation can be deformed to $2x + (3y+3) + (z+2) = 18$. So, using the condition for the equality to be satisfied in the Cauchy inequality, we find $x = 3, y = 1$, and $z = 4$.

2. $\sqrt{2}$.   By the Cauchy inequality,

$$\sqrt{x^2 - x^4} - \sqrt{2x^2 - x^4} \le \sqrt{(x^2 + 2 - x^2)(1 - x^2 + x^2)} = \sqrt{2},$$

and the equality holds if and only if $x^2 = \frac{2}{3}$.

3. 1.   From $a + 2b + 3c + 4d = \sqrt{10}$, we get

$$(1-t)a + (2-t)b + (3-t)c + (4-t)d + t(a+b+c+d) = \sqrt{10}.$$

Then by the Cauchy inequality,

$$\left[(1-t)^2 + (2-t)^2 + (3-t)^2 + (4-t)^2 + t^2\right]\left[a^2 + b^2 + c^2 + d^2\right.$$
$$\left. + (a+b+c+d)^2\right] \ge 10.$$

Hence,

$$a^2 + b^2 + c^2 + d^2 + (a+b+c+d)^2$$
$$\ge \frac{10}{5t^2 - 20t + 30} = \frac{10}{5(t-2)^2 + 10} \ge \frac{10}{10} = 1,$$

and the equality is valid if and only if $t = 2, a = -\frac{\sqrt{10}}{10}, b = 0, c = \frac{\sqrt{10}}{10}$, and $d = \frac{\sqrt{10}}{5}$. Hence, the minimum value of $a^2 + b^2 + c^2 + d^2 + (a+b+c+d)^2$ is 1.

4. 40.   By the Cauchy inequality,

$$(a_1^2 + a_2^2 + a_3^2)(a_2^2 + a_3^2 + a_4^2) \geq (a_1a_2 + a_2a_3 + a_3a_4)^2.$$

A necessary and sufficient condition for the equality to be true is $\frac{a_1}{a_2} = \frac{a_2}{a_3} = \frac{a_3}{a_4}$, that is, a_1, a_2, a_3, and a_4 form a geometric sequence. Hence, the problem is equivalent to calculating the number of the geometric sequences a_1, a_2, a_3, a_4 that satisfy

$$\{a_1, a_2, a_3, a_4\} \subseteq \{1, 2, \ldots, 100\}.$$

Assume that the common ratio $q \neq 1$ for the geometric sequence and q is a rational number, denoted as $q = \frac{n}{m}$, where m and n are relatively prime positive integers with $m \neq n$.

First consider the case $n > m$.

Now, $a_4 = a_1 \left(\frac{n}{m}\right)^3 = \frac{a_1 n^3}{m^3}$. Note that m^3 and n^3 are relatively prime. So, $l = \frac{a_1}{m^3}$ is a positive integer. Thus, a_1, a_2, a_3, and a_4 are respectively equal to $m^3 l, m^2 nl, mn^2 l$, and $n^3 l$, all of which are positive integers. This means that for any given $q = \frac{n}{m} > 1$, the number of the geometric sequences a_1, a_2, a_3, a_4 with common ratio q satisfying the condition is the same as the number of the positive integers l satisfying the inequality $n^3 l \leq 100$, that is $\left[\frac{100}{n^3}\right]$.

Since $5^3 > 100$, so it is enough to consider only the cases $q = 2, 3, \frac{3}{2}, 4$, and $\frac{4}{3}$. The number of the corresponding geometric sequences is $\left[\frac{100}{8}\right] + \left[\frac{100}{27}\right] + \left[\frac{100}{27}\right] + \left[\frac{100}{64}\right] + \left[\frac{100}{64}\right] = 12 + 3 + 3 + 1 + 1 = 20$.

When $n < m$, from the symmetry, there are also 20 geometric sequences a_1, a_2, a_3, a_4 satisfying the condition.

In summary, there are 40 ordered numbers (a_1, a_2, a_3, a_4) that satisfy the given condition.

5. By the Cauchy inequality,

$$\sqrt{a}\cos^2\theta + \sqrt{b}\sin^2\theta$$
$$\leq \left[(\sqrt{a}\cos\theta)^2 + (\sqrt{b}\sin\theta)^2\right]^{\frac{1}{2}} \cdot \left(\cos^2\theta + \sin^2\theta\right)^{\frac{1}{2}}$$
$$= \left(a\cos^2\theta + b\sin^2\theta\right)^{\frac{1}{2}} < \sqrt{c}.$$

6. The Cauchy inequality implies that

$$\sqrt{(a^2+1)(b^2+1)(c^2+1)}$$
$$= \sqrt{[(a+b)^2 + (ab-1)^2](c^2+1)}$$
$$\geq (a+b)c + (ab-1).$$

The same way gives

$$\sqrt{(b^2+1)(c^2+1)(d^2+1)} \geq (b+c)d+(bc-1),$$
$$\sqrt{(c^2+1)(d^2+1)(a^2+1)} \geq (c+d)a+(cd-1),$$
$$\sqrt{(d^2+1)(a^2+1)(b^2+1)} \geq (d+a)b+(da-1).$$

Adding the above four inequalities implies that

$$\sqrt{(a^2+1)(b^2+1)(c^2+1)} + \sqrt{(b^2+1)(c^2+1)(d^2+1)}$$
$$+ \sqrt{(c^2+1)(d^2+1)(a^2+1)} + \sqrt{(d^2+1)(a^2+1)(b^2+1)}$$
$$\geq 2(ab+bc+cd+da)-4,$$

and the equality is satisfied when $a = b = c = d = \sqrt{3}$. Hence, the minimum value of k is 4.

7. The answer that the minimum value equals $\frac{3}{4}$ comes from

$$\frac{a^3}{b+c+d} + \frac{b^3}{a+c+d} + \frac{c^3}{a+b+d} + \frac{d^3}{a+b+c}$$

$$= \frac{a^4}{a(b+c+d)} + \frac{b^4}{b(a+c+d)} + \frac{c^4}{c(a+b+d)} + \frac{d^4}{d(a+b+c)}$$

$$\geq \frac{(a^2+b^2+c^2+d^2)^2}{2(ab+ac+ad+bc+bd+cd)}$$

$$\geq \frac{2}{3} \frac{(a^2+b^2+c^2+d^2)(ab+ac+ad+bc+bd+cd)}{2(ab+ac+ad+bc+bd+cd)}$$

$$= \frac{1}{3}(a^2+b^2+c^2+d^2) \geq \frac{1}{12}(a+b+c+d)^2$$

$$= \frac{3^2}{12} = \frac{3}{4}.$$

8. Since $a_i > 0$ $(i = 1, 2, \ldots, n)$, by the Cauchy inequality,

$$(a_1 + a_2 + \cdots + a_k)\left(\frac{1^2}{a_1} + \frac{2^2}{a_2} + \cdots + \frac{k^2}{a_k}\right)$$

$$\geq (1 + 2 + \cdots + k)^2 = \frac{k^2(k+1)^2}{4},$$

so $\frac{k^2}{a_1+a_2+\cdots+a_k} \le \frac{4}{(k+1)^2}\left(\frac{1^2}{a_1} + \frac{2^2}{a_2} + \cdots + \frac{k^2}{a_k}\right)$. Consequently,

$$\sum_{k=1}^{n} \frac{k^2}{a_1 + a_2 + \cdots + a_k} \le 4\sum_{k=1}^{n} \frac{1}{(k+1)^2}\left(\frac{1^2}{a_1} + \frac{2^2}{a_2} + \cdots + \frac{k^2}{a_k}\right).$$

Since

$$\sum_{k=1}^{n} \frac{1}{(k+1)^2}\left(\frac{1^2}{a_1} + \frac{2^2}{a_2} + \cdots + \frac{k^2}{a_k}\right)$$

$$= \frac{1}{2^2}\cdot\frac{1^2}{a_1} + \frac{1}{3^2}\left(\frac{1^2}{a_1} + \frac{2^2}{a_2}\right) + \cdots + \frac{1}{(n+1)^2}\left(\frac{1^2}{a_1} + \frac{2^2}{a_2} + \cdots + \frac{n^2}{a_n}\right)$$

$$= \left[\frac{1}{2^2} + \frac{1}{3^2} + \cdots + \frac{1}{(n+1)^2}\right]\frac{1^2}{a_1} + \left[\frac{1}{3^2} + \frac{1}{4^2} + \cdots\right.$$

$$\left. + \frac{1}{(n+1)^2}\right]\frac{2^2}{a_2} + \cdots + \frac{1}{(n+1)^2}\frac{n^2}{a_n},$$

and

$$\frac{1}{(i+1)^2} + \frac{1}{(i+2)^2} + \cdots + \frac{1}{(n+1)^2}$$

$$< \frac{1}{i(i+1)} + \frac{1}{(i+1)(i+2)} + \cdots + \frac{1}{n(n+1)}$$

$$= \left(\frac{1}{i} - \frac{1}{i+1}\right) + \left(\frac{1}{i+1} - \frac{1}{i+2}\right) + \cdots + \left(\frac{1}{n} - \frac{1}{n+1}\right)$$

$$= \frac{1}{i} - \frac{1}{n+1} < \frac{1}{i}, \ i = 1, 2, \ldots, n,$$

it follows that

$$\sum_{k=1}^{n} \frac{1}{(k+1)^2}\left(\frac{1^2}{a_1} + \frac{2^2}{a_2} + \cdots + \frac{k^2}{a_k}\right)$$

$$< \frac{1}{1}\cdot\frac{1^2}{a_1} + \frac{1}{2}\cdot\frac{2^2}{a_2} + \cdots + \frac{1}{n}\frac{n^2}{a_n}$$

$$= \frac{1}{a_1} + \frac{2}{a_2} + \cdots + \frac{n}{a_n} = \sum_{k=1}^{n} \frac{k}{a_k}.$$

Therefore, $\sum_{k=1}^{n} \frac{k^2}{a_1+a_2+\cdots+a_k} < 4\sum_{k=1}^{n} \frac{k}{a_k}$.

9. First, we have

$$\sqrt{r_1} + \sqrt{r_2} + \sqrt{r_3}$$

$$= \sqrt{a_1 r_1} \cdot \sqrt{\frac{1}{a_1}} + \sqrt{a_2 r_2} \cdot \sqrt{\frac{1}{a_2}} + \sqrt{a_3 r_3} \cdot \sqrt{\frac{1}{a_3}}$$

$$\leq (a_1 r_1 + a_2 r_2 + a_3 r_3)^{\frac{1}{2}} \left(\frac{1}{a_1} + \frac{1}{a_2} + \frac{1}{a_3} \right)^{\frac{1}{2}}.$$

Denote S the area of $\triangle ABC$. Then

$$a_1 r_1 + a_2 r_2 + a_3 r_3 = 2S = 2 \cdot \frac{a_1 a_2 a_3}{4R}.$$

It follows that

$$\sqrt{r_1} + \sqrt{r_2} + \sqrt{r_3} \leq \frac{1}{\sqrt{2R}} (a_2 a_3 + a_3 a_1 + a_1 a_2)^{\frac{1}{2}}$$

$$\leq \frac{1}{\sqrt{2R}} \left[(a_2^2 + a_3^2 + a_1^2)^{\frac{1}{2}} (a_3^2 + a_1^2 + a_2^2)^{\frac{1}{2}} \right]^{\frac{1}{2}}$$

$$= \frac{1}{\sqrt{2R}} (a_1^2 + a_2^2 + a_3^2)^{\frac{1}{2}}.$$

The condition for the equality to be valid is: $\triangle ABC$ is an equilateral triangle and P is its inner center.

10. Let $\sum_{i=1}^{n} \frac{1}{a_i} = a$. Then $\sum_{i=1}^{n} \frac{1+a_i}{a_i} = n + a$. By the Cauchy inequality,

$$\left(\sum_{i=1}^{n} \frac{a_i}{1+a_i} \right) \cdot \left(\sum_{i=1}^{n} \frac{1+a_i}{a_i} \right) \geq n^2.$$

So, $\sum_{i=1}^{n} \frac{a_i}{1+a_i} \geq \frac{n^2}{n+a}$, and

$$\sum_{i=1}^{n} \frac{1}{a_i + 1} = \sum_{i=1}^{n} \left(1 - \frac{a_i}{a_i + 1} \right) = n - \sum_{i=1}^{n} \frac{a_i}{a_i + 1}$$

$$\leq n - \frac{n^2}{n+a} = \frac{na}{n+a}.$$

Hence, the left hand side of the original expression $\geq \frac{1}{\frac{na}{n+a}} - \frac{1}{a} = \frac{a}{na} = \frac{1}{n}$.

11. Since $1 - \sin^{2n} x = \cos^2 x \left[1 + \sin^2 x + \sin^4 x + \cdots + \sin^{2(n-1)} x \right]$ and $1 - \cos^{2n} x = \sin^2 x \left[1 + \cos^2 x + \cos^4 x + \cdots + \cos^{2(n-1)} x \right]$, so by the

Cauchy inequality,

$$\frac{1 - \sin^{2n} x}{\sin^{2n} x} \cdot \frac{1 - \cos^{2n} x}{\cos^{2n} x}$$

$$\geq \frac{1}{\sin^{2n-2} x} \left(1 + \sin^2 x + \sin^4 x + \cdots + \sin^{2n-2} x \right)$$

$$\cdot \frac{1}{\cos^{2n-2} x} \left(1 + \cos^2 x + \cos^4 x + \cdots + \cos^{2n-2} x \right)$$

$$= \left(1 + \frac{1}{\sin^2 x} + \frac{1}{\sin^4 x} + \cdots + \frac{1}{\sin^{2n-2} x} \right)$$

$$\cdot \left(1 + \frac{1}{\cos^2 x} + \frac{1}{\cos^4 x} + \cdots + \frac{1}{\cos^{2n-2} x} \right)$$

$$\geq \left[1 + \frac{1}{\sin x \cos x} + \frac{1}{(\sin x \cos x)^2} + \cdots + \frac{1}{(\sin x \cos x)^{n-1}} \right]^2$$

$$= \left(1 + \frac{2}{\sin 2x} + \frac{4}{\sin^2 2x} + \cdots + \frac{2^{n-1}}{\sin^{n-1} 2x} \right)^2$$

$$\geq \left(1 + 2 + 2^2 + \cdots + 2^{n-1} \right)^2 = (2^n - 1)^2.$$

12. For any $1 \leq m \leq n$, the Cauchy inequality ensures that

$$\left(\sum_{k=m}^{n} \frac{a_k^2}{k} \right) \cdot \left(\sum_{k=m}^{n} k \right) \geq \left(\sum_{k=m}^{n} a_k \right)^2 \geq \left(\sum_{k=m}^{n} k \right)^2.$$

So $\sum_{k=m}^{n} \frac{a_k^2}{k} \geq \sum_{k=m}^{n} k$, and their summation with $m = 1, 2, \ldots, n$ gives

$$\sum_{m=1}^{n} \sum_{k=m}^{n} \frac{a_k^2}{k} \geq \sum_{m=1}^{n} \sum_{k=m}^{n} k.$$

On the other hand,

$$\sum_{m=1}^{n} \sum_{k=m}^{n} \frac{a_k^2}{k} = \sum_{k=1}^{n} \sum_{m=1}^{k} \frac{a_k^2}{k} = \sum_{k=1}^{n} a_k^2,$$

$$\sum_{m=1}^{n} \sum_{k=m}^{n} k = \sum_{k=1}^{n} \sum_{m=1}^{k} k = \sum_{k=1}^{n} k^2 = \frac{1}{6} n(n + 1)(2n + 1).$$

Therefore, $\sum_{k=1}^{n} a_k^2 \geq \frac{1}{6} n(n + 1)(2n + 1)$.

Solutions for Chapter 22

Rearrangement Inequalities

1. n.   We may assume that $a_1 \geq a_2 \geq \cdots \geq a_n > 0$. Then $\frac{1}{a_1} \leq \frac{1}{a_2} \leq \cdots \leq \frac{1}{a_n}$. Thus, $n = a_1 \cdot \frac{1}{a_1} + a_2 \cdot \frac{1}{a_2} + \cdots + a_n \cdot \frac{1}{a_n} \leq \frac{a_1}{b_1} + \frac{a_2}{b_2} + \cdots + \frac{a_n}{b_n}$. When $a_i = b_i$ for all i, the equality is satisfied.

2. $\frac{\sqrt{3}}{R}$.   Denote $M = \frac{\sin A}{h_a} + \frac{\sin B}{h_b} + \frac{\sin C}{h_c}$. By Chebyshev's inequality, $M \geq \frac{1}{3}\left(\frac{1}{h_a} + \frac{1}{h_b} + \frac{1}{h_c}\right)(\sin A + \sin B + \sin C)$. Also by a geometric inequality, $\frac{1}{h_a} + \frac{1}{h_b} + \frac{1}{h_c} \geq \frac{2}{\sqrt{3}}\left(\frac{1}{a} + \frac{1}{b} + \frac{1}{c}\right)$, and by the law of sines, $M \geq \frac{1}{3} \cdot \frac{2}{\sqrt{3}}\left(\frac{1}{a} + \frac{1}{b} + \frac{1}{c}\right)\left(\frac{a}{2R} + \frac{b}{2R} + \frac{c}{2R}\right) = \frac{1}{3\sqrt{3}R}\left(\frac{1}{a} + \frac{1}{b} + \frac{1}{c}\right)(a + b + c) \geq \frac{1}{3\sqrt{3}R} \cdot 3^2 = \frac{\sqrt{3}}{R}$.

3. The original inequality is equivalent to

$$\frac{z^2}{x+y} + \frac{x^2}{y+z} + \frac{y^2}{z+x} \geq \frac{x^2}{x+y} + \frac{y^2}{y+z} + \frac{z^2}{z+x}.$$

We may assume that $x \leq y \leq z$. Then $x^2 \leq y^2 \leq z^2, x + y \leq x + z \leq y + z$, and $\frac{1}{x+y} \geq \frac{1}{x+z} \geq \frac{1}{y+z}$. The conclusion follows from the rearrangement inequalities.

4. The original inequality is equivalent to $\sum_{i=1}^{n} x_i y_i \geq \sum_{i=1}^{n} x_i z_i$. The left hand side of the inequality is a sum of same order numbers, while the right hand side is a sum of random order numbers, so the proposition is true.

5. Take a logarithm, and then use the rearrangement inequalities.

6. Order $a_1, a_2, \ldots, a_{n-1}$ from the smallest to the largest to form $1 \leq c_1 < c_2 < \cdots < c_{n-1} \leq n$. Also order $a_2, a_3, \ldots, a_n$ from the smallest to the largest to form $1 \leq b_1 < c_2 < \cdots < b_{n-1} \leq n$. So, $1 \geq \frac{1}{b_1} > \frac{1}{b_2} > \cdots > \frac{1}{b_{n-1}} \geq \frac{1}{n}$. By the rearrangement inequalities, the right hand side of the

original inequality $\geq \frac{c_1}{b_1} + \frac{c_2}{b_2} + \cdots + \frac{c_{n-1}}{b_{n-1}} \geq 1 \cdot \frac{1}{2} + 2 \cdot \frac{1}{3} + \cdots + (n-1)\frac{1}{n} =$ the left hand side.

7. Let $b_1, b_2, \ldots, b_n$ be a new permutation of $a_1, a_2, \ldots, a_n$, satisfying $b_1 < b_2 < \cdots < b_n$. Since $1 > \frac{1}{2^2} > \frac{1}{3^2} > \cdots > \frac{1}{n^2}$, so $a_1 + \frac{a_2}{2^2} + \cdots + \frac{a_n}{n^2} \geq b_1 + \frac{b_2}{2^2} + \cdots + \frac{b_n}{n^2}$. Since $b_1, b_2, \ldots, b_n$ are mutually different natural numbers, we see that $b_1 \geq 1, b_2 \geq 2, \ldots, b_n \geq n$. Hence, the original inequality is satisfied.

8. Let $A_i = a_1 a_2 \cdots a_{i-1} a_{i+1} \cdots a_n$. Since $0 \leq b_i \leq p, \sum_{i=1}^n b_i = 1$, and $\frac{1}{2} \leq p \leq 1$, by the rearrangement inequalities, $\sum_{i=1}^n b_i A_i \leq pA_i + (1-p)A_i \leq p(A_1 + A_2)$. By the arithmetic mean-geometric mean inequality,
$$A_1 + A_2 = a_3 a_4 \cdots a_n(a_2 + a_1) \leq \left(\frac{1}{n-1}\sum_{i=1}^n a_i\right)^{n-1} = \text{the right hand}$$
side.

9. By Chebyshev's inequality,
$$\sum_{i=1}^n \frac{a_i^r}{b_i^s} \geq \frac{1}{n}\sum_{i=1}^n a_i^r \sum_{i=1}^n \frac{1}{b_i^s} \geq \frac{1}{n} \cdot n^{1-r}\left(\sum_{i=1}^n a_i\right)^r \frac{n^{s+1}}{\left(\sum_{i=1}^n b_i\right)^s},$$
which is the right hand side.

10. Without loss of generality, assume that $x_1 \geq x_2 \geq \cdots \geq x_n > 0$. Then $0 < \frac{1}{1+x_1} \leq \frac{1}{1+x_2} \leq \cdots \leq \frac{1}{1+x_n}$. Chebyshev's inequality implies that
$$\frac{1}{n}\left(\frac{x_1}{1+x_1} + \frac{x_2}{1+x_2} + \cdots + \frac{x_n}{1+x_n}\right)$$
$$\leq \frac{1}{n}(x_1 + x_2 + \cdots + x_n) \cdot \frac{1}{n}\left(\frac{1}{1+x_1} + \frac{1}{1+x_2} + \cdots + \frac{1}{1+x_n}\right).$$

Also from the given condition, $\frac{1}{1+x_1} + \frac{1}{1+x_2} + \cdots + \frac{1}{1+x_n} = n - 1$, hence $x_1 + x_2 + \cdots + x_n \geq \frac{n}{n-1}$.

11. When $n = 1$, the inequality to be proved is symmetric with respect to $a, b,$ and c. We may assume that $a \geq b \geq c$. Then $a + b \geq a + c \geq b + c$, so $\frac{1}{b+c} \geq \frac{1}{c+a} \geq \frac{1}{a+b}$. From Chebyshev's inequality,
$$\frac{a}{b+c}(b+c) + \frac{b}{a+c}(a+c) + \frac{c}{a+b}(a+b)$$
$$\leq \frac{1}{3}\left(\frac{a}{b+c} + \frac{b}{a+c} + \frac{c}{a+b}\right)(b+c+a+c+a+b).$$

Hence, $\frac{a}{b+c} + \frac{b}{a+c} + \frac{c}{a+b} \geq \left(\frac{2}{3}\right)^{-1}$.

Suppose that the proposition is true for $n = k$, that is

$$\frac{a^k}{b+c} + \frac{b^k}{c+a} + \frac{c^k}{a+b} \geq \left(\frac{2}{3}\right)^{k-2} \cdot S^{k-1}.$$

When $n = k+1$, again assume that $a \geq b \geq c$. Then $\frac{a^k}{b+c} \geq \frac{b^k}{c+a} \geq \frac{c^k}{a+b}$.
Using Chebyshev's inequality and the inductive hypothesis, we get

$$\frac{a^{k+1}}{b+c} + \frac{b^{k+1}}{c+a} + \frac{c^{k+1}}{a+b} = a \cdot \frac{a^k}{b+c} + b \cdot \frac{b^k}{c+a} + c \cdot \frac{c^k}{a+b}$$

$$\geq \frac{1}{3}(a+b+c)\left(\frac{a^k}{b+c} + \frac{b^k}{c+a} + \frac{c^k}{a+b}\right)$$

$$\geq \frac{2}{3}S\left(\frac{2}{3}\right)^{k-2} \cdot S^{k-1} = \left(\frac{2}{3}\right)^{k-1} \cdot S^k.$$

12. Let

$$M = \sum_{i=1}^{n} \frac{a_i^{k+1}}{a_i^k + a_i^{k-1}a_{i+1} + \cdots + a_i a_{i+1}^{k-1} + a_{i+1}^k},$$

$$N = \sum_{i=1}^{n} \frac{a_{i+1}^{k+1}}{a_i^k + a_i^{k-1}a_{i+1} + \cdots + a_i a_{i+1}^{k-1} + a_{i+1}^k}.$$

Then

$$M - N = \sum_{i=1}^{n} \frac{a_i^{k+1} - a_{i+1}^{k+1}}{a_i^k + a_i^{k-1}a_{i+1} + \cdots + a_i a_{i+1}^{k-1} + a_{i+1}^k}$$

$$= \sum_{i=1}^{n} (a_i - a_{i+1}) = 0. \tag{22.1}$$

For any given positive integer k, by the rearrangement inequalities,

$$a^{k+1} + b^{k+1} \geq a^{k+1-r}b^r + a^r b^{k+1-r},$$

where $r = 0, 1, \ldots, k$.
Adding these $k + 1$ inequalities and factorizing the expression, we have

$$(a + b)\left(a^k + a^{k-1}b + \cdots + ab^{k-1} + b^k\right) \leq (k + 1)\left(a^{k+1} + b^{k+1}\right).$$

Hence,

$$\frac{a^{k+1} + b^{k+1}}{a^k + a^{k-1}b + \cdots + ab^{k-1} + b^k} \geq \frac{a+b}{k+1}.$$

Thus, for $i = 1, 2, \ldots, n$,

$$\frac{a_i^{k+1} + a_{i+1}^{k+1}}{a_i^k + a_i^{k-1}a_{i+1} + \cdots + a_i a_{i+1}^{k-1} + a_{i+1}^k} \geq \frac{a_i + a_{i+1}}{k+1}.$$

Summing up the above inequalities gives

$$M + N = \sum_{i=1}^{n} \frac{a_i^{k+1} + a_{i+1}^{k+1}}{a_i^k + a_i^{k-1}a_{i+1} + \cdots + a_i a_{i+1}^{k-1} + a_{i+1}^k}$$

$$\geq \sum_{i=1}^{n} \frac{a_i + a_{i+1}}{k+1} = 2 \sum_{i=1}^{n} \frac{a_i}{k+1}. \tag{22.2}$$

Now, (22.1) and (22.2) imply that

$$M = N \geq \frac{1}{k+1} \sum_{i=1}^{n} a_i.$$

Solutions for Chapter 23

Convex Functions and Jensen's Inequality

1. $\frac{3}{2}\sqrt{3}$.   By Jensen's inequality,

$$\sin A + \sin B + \sin C \leq 3\sin\frac{A+B+C}{3} = \frac{3}{2}\sqrt{3}.$$

2. $A < B$.   Since $y = \sqrt[3]{x}$ is a concave function, Jensen's inequality implies that

$$\frac{\sqrt[3]{3-\sqrt[3]{3}} + \sqrt[3]{3+\sqrt[3]{3}}}{2} \leq \sqrt[3]{\frac{3-\sqrt[3]{3}+3+\sqrt[3]{3}}{2}} = \sqrt[3]{3}.$$

Since $3 - \sqrt[3]{3} \neq 3 + \sqrt[3]{3}$, it follows that $\frac{\sqrt[3]{3-\sqrt[3]{3}} + \sqrt[3]{3+\sqrt[3]{3}}}{2} < \sqrt[3]{3}$.

3. Let $f(x) = \ln\left(x + \frac{1}{x}\right)$. We show that for any $a, b \in (0,1)$,

$$\frac{\ln\left(a + \frac{1}{a}\right) + \ln\left(b + \frac{1}{b}\right)}{2} \geq \ln\left(\frac{a+b}{2} + \frac{2}{a+b}\right),$$

namely

$$\left(a + \frac{1}{a}\right)\left(b + \frac{1}{b}\right) \geq \left(\frac{a+b}{2} + \frac{2}{a+b}\right)^2,$$

that is,

$$ab + \frac{1}{ab} + \frac{a}{b} + \frac{b}{a} \geq \left(\frac{a+b}{2}\right)^2 + \frac{1}{\left(\frac{a+b}{2}\right)^2} + 2. \tag{23.1}$$

Since $\frac{a}{b} + \frac{b}{a} \geq 2$ and $ab \leq \left(\frac{a+b}{2}\right)^2$, and since $y = x + \frac{1}{x}$ is a monotonically decreasing function on $(0,1]$, so

$$ab + \frac{1}{ab} \geq \left(\frac{a+b}{2}\right)^2 + \frac{1}{\left(\frac{a+b}{2}\right)^2},$$

which proves (23.1). Thus, $f(x) = \ln\left(x + \frac{1}{x}\right)$ is a convex function on $(0, 1)$. By Jensen's inequality,

$$\frac{1}{n}\sum_{i=1}^{n}\ln\left(a_i + \frac{1}{a_i}\right) \geq \ln\left(\frac{\sum_{i=1}^{n}a_i}{n} + \frac{n}{\sum_{i=1}^{n}a_i}\right) = \ln\left(n + \frac{1}{n}\right).$$

Therefore, $\prod_{i=1}^{n}\left(a_i + \frac{1}{a_i}\right) \geq \left(n + \frac{1}{n}\right)^n$.

4. Since $f(x) = \left(x + \frac{a}{x}\right)^p$ ($p \in \mathbf{N}$ and a is a positive constant) is a convex function on $(0, +\infty)$, by Jensen's inequality,

$$\frac{1}{n}\left[\left(x_1 + \frac{a}{x_1}\right)^p + \left(x_2 + \frac{a}{x_2}\right)^p + \cdots + \left(x_n + \frac{a}{x_n}\right)^p\right]$$

$$\geq \left[\frac{1}{n}(x_1 + x_2 + \cdots + x_n) + \frac{a}{\frac{1}{n}(x_1 + x_2 + \cdots + x_n)}\right]^p$$

$$= \left(\frac{n^2a + 1}{n}\right)^p.$$

5. Since $f(x) = \sqrt{x}$ ($x > 0$) is a strictly concave function, so

$$\frac{1}{3}(\sqrt{n} + \sqrt{n+1} + \sqrt{n+2}) < \sqrt{\frac{1}{3}(n + n+1 + n+2)} = \sqrt{n+1}.$$

Also, since $\sqrt{n} + \sqrt{n+1} + \sqrt{n+2} > 3\sqrt[3]{\sqrt{n}\cdot\sqrt{n+1}\cdot\sqrt{n+2}}$, it is enough to show that $n(n+1)(n+2) > \left(n + \frac{8}{9}\right)^3$.

6. First use Jensen's inequality, and then use the mean value inequality.

7. As Figure 23.1 shows, we have $|PA|\sin\alpha = |PB|\sin\beta'$, $|PB|\sin\beta = |PC|\sin\gamma'$, and $|PC|\sin\gamma = |PA|\sin\alpha'$. So

$$\sin\alpha\sin\beta\sin\gamma = \sin\alpha'\sin\beta'\sin\gamma'.$$

Since $y = \sin x$ is a concave function on $(0, \pi)$, by Jensen's inequality together with the arithmetic mean-geometric mean inequality,

$$(\sin\alpha\sin\beta\sin\gamma)^2 = \sin\alpha\sin\beta\sin\gamma\sin\alpha'\sin\beta'\sin\gamma'$$

$$\leq \sin^6\frac{\alpha + \beta + \gamma + \alpha' + \beta' + \gamma'}{6} = \sin^6 30° = \left(\frac{1}{2}\right)^6.$$

Namely, there must be one among $\sin\alpha, \sin\beta$, and $\sin\gamma$ that is $\leq \frac{1}{2}$, say $\sin\alpha \leq \frac{1}{2}$. Thus, when α is an acute angle, $\alpha \leq 30°$; when α is an obtuse angle, $\alpha \geq 150°$, and $\beta, \gamma \leq 30°$.

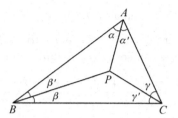

Figure 23.1 Figure for Exercise 7.

8. Let $f(x) = \tan\frac{x}{m}$ with $x \in (0, \pi)$. Then for any two points x_1 and x_2,

$$f(x_1) + f(x_2) = \tan\frac{x_1}{m} + \tan\frac{x_2}{m} = \frac{\sin\frac{x_1+x_2}{m}}{\cos\frac{x_1}{m}\cos\frac{x_2}{m}}$$

$$= \frac{2\sin\frac{x_1+x_2}{m}}{\cos\frac{x_1+x_2}{m} + \cos\frac{x_1-x_2}{m}} \geq \frac{2\sin\frac{x_1+x_2}{m}}{\cos\frac{x_1+x_2}{m} + 1}$$

$$= 2\tan\frac{x_1+x_2}{2m} = 2f\left(\frac{x_1+x_2}{2}\right),$$

so $f(x) = \tan\frac{x}{m}$ is a convex function on $(0, \pi)$. Choose $x_1 = \frac{A}{m}, x_2 = \frac{B}{m}$, and $x_3 = \frac{C}{m}$. According to Jensen's inequality,

$$\tan\frac{A}{m} + \tan\frac{B}{m} + \tan\frac{C}{m} \geq 3\tan\frac{\frac{A}{m} + \frac{B}{m} + \frac{C}{m}}{3} = 3\tan\frac{\pi}{3m}.$$

9. Let $g = \left(\frac{1}{x^2} + x\right)\left(\frac{1}{y^2} + y\right)\left(\frac{1}{z^2} + z\right)$. Then

$$\ln g = \ln\left(\frac{1}{x^2} + x\right) + \ln\left(\frac{1}{y^2} + y\right) + \ln\left(\frac{1}{z^2} + z\right).$$

Define the function $f(x) = \ln\left(\frac{1}{x^2} + x\right)$ $(0 < x < 1)$. Then $f'(x) = \frac{x^3-2}{x^4+x}$ and

$$f''(x) = \frac{-x^6 + 10x^3 + 2}{(x^4 + x)^2}$$

$$= \frac{-[x^3 - (5 + \sqrt{27})][x^3 - (5 - \sqrt{27})]}{(x^4 + x)^2} > 0.$$

Consequently, $f(x)$ is a convex function on $(0, 1)$, and so by the definition of convex functions, $\ln g \geq 3\ln\left(3^2 + \frac{1}{3}\right) = \ln\left(\frac{28}{3}\right)^3$, from which $g \geq \left(\frac{28}{3}\right)^3$.

10. If $x > 0$ and $nx = x^n$, then $x = n^{\frac{1}{n-1}}$.

(1) When $x_1 x_2 \cdots x_n \geq n^{\frac{n}{n-1}}$, since x^α $(\alpha \geq 1)$ is a strictly convex function on $(0, +\infty)$, so by the definition of convex functions,

$$
\frac{x_1^\alpha + x_2^\alpha + \cdots + x_n^\alpha}{x_1 x_2 \cdots x_n} \geq \frac{n \left(\frac{x_1 + x_2 + \cdots + x_n}{n} \right)^\alpha}{x_1 x_2 \cdots x_n}
$$

$$
\geq n^{1-\alpha} (x_1 x_2 \cdots x_n)^{\alpha - 1}
$$

$$
\geq n^{1-\alpha} \left(n^{\frac{n}{n-1}} \right)^{\alpha - 1} = n^{\frac{\alpha - 1}{n-1}}.
$$

The equality in the above is satisfied if and only if $x_1 = x_2 = \cdots = x_n = n^{\frac{1}{n-1}}$.

(2) When $x_1 x_2 \cdots x_n \leq n^{\frac{n}{n-1}}$, from the arithmetic mean-geometric mean inequality,

$$
\frac{x_1^\alpha + x_2^\alpha + \cdots + x_n^\alpha}{x_1 x_2 \cdots x_n} \geq \frac{n(x_1 x_2 \cdots x_n)^{\frac{\alpha}{n}}}{x_1 x_2 \cdots x_n}
$$

$$
= \frac{n}{(x_1 x_2 \cdots x_n)^{1 - \frac{\alpha}{n}}} \geq \frac{n}{\left(n^{\frac{n}{n-1}} \right)^{1 - \frac{\alpha}{n}}} = n^{\frac{\alpha - 1}{n-1}}.
$$

The equality in the above is satisfied if and only if $x_1 = x_2 = \cdots = x_n = n^{\frac{1}{n-1}}$.

11. By symmetry, we may assume that $x \leq y \leq z$. Then

$$
\frac{1}{(1+y)(1+z)} \leq \frac{1}{(1+z)(1+x)} \leq \frac{1}{(1+x)(1+y)}.
$$

Chebyshev's inequality implies that

$$
\frac{x^3}{(1+y)(1+z)} + \frac{y^3}{(1+z)(1+x)} + \frac{z^3}{(1+x)(1+y)}
$$

$$
\geq \frac{1}{3} (x^3 + y^3 + z^3) \left[\frac{1}{(1+y)(1+z)} \right.
$$

$$
+ \frac{1}{(1+z)(1+x)} + \left. \frac{1}{(1+x)(1+y)} \right]
$$

$$
= \frac{1}{3} (x^3 + y^3 + z^3) \frac{3 + (x + y + z)}{(1+x)(1+y)(1+z)}.
$$

Let $\frac{1}{3}(x + y + z) = a$. Then, since the function t^3 is strictly convex on $(0, +\infty)$, by Jensen's inequality and the arithmetic mean-geometric mean inequality,

$$\frac{1}{3}(x^3 + y^3 + z^3) \geq a^3,$$

$$x + y + z \geq 3\sqrt[3]{xyz} = 3,$$

$$(1+x)(1+y)(1+z) \leq \left[\frac{(1+x) + (1+y) + (1+z)}{3}\right]^3 = (1+a)^3.$$

Therefore,

$$\frac{x^3}{(1+y)(1+z)} + \frac{y^3}{(1+z)(1+x)} + \frac{z^3}{(1+x)(1+y)} \geq a^3 \cdot \frac{3+3}{(1+a)^3}.$$

Thus, it is enough to show that $\frac{6a^3}{(1+a)^3} \geq \frac{3}{4}$.

Since $a \geq 1$, the above inequality is obviously satisfied, and the equality holds if and only if $x = y = z = 1$.

12. Let $x_k = a_k^\beta$. Then $a_k^\alpha = x_k^{\frac{\alpha}{\beta}}$ and $\frac{\alpha}{\beta} > 1$. Consider the convexity of the function $f(x) = x^{\frac{\alpha}{\beta}}$ $(x > 0)$. By Jensen's inequality,

$$f\left(\frac{x_1 + x_2 + \cdots + x_n}{n}\right) \leq \frac{1}{n}[f(x_1) + f(x_2) + \cdots + f(x_n)],$$

that is,

$$\left(\frac{x_1 + x_2 + \cdots + x_n}{n}\right)^{\frac{\alpha}{\beta}} \leq \frac{x_1^{\frac{\alpha}{\beta}} + x_2^{\frac{\alpha}{\beta}} + \cdots + x_n^{\frac{\alpha}{\beta}}}{n}.$$

In other words,

$$\left(\frac{x_1^{\frac{\alpha}{\beta}} + x_2^{\frac{\alpha}{\beta}} + \cdots + x_n^{\frac{\alpha}{\beta}}}{n}\right)^{\frac{1}{\alpha}} \geq \left(\frac{x_1 + x_2 + \cdots + x_n}{n}\right)^{\frac{1}{\beta}}.$$

Hence,

$$\left(\frac{a_1^\alpha + a_2^\alpha + \cdots + a_n^\alpha}{n}\right)^{\frac{1}{\alpha}} \geq \left(\frac{a_1^\beta + a_2^\beta + \cdots + a_n^\beta}{n}\right)^{\frac{1}{\beta}}.$$

Solutions for Chapter 24

Recursive Sequences

1. 2.   From $a_{n+1} = a_n^2 + a_n$, we have $a_{n+1} = a_n(a_n + 1)$, so

$$\frac{1}{a_{n+1}} = \frac{1}{a_n(a_n+1)} = \frac{1}{a_n} - \frac{1}{a_n+1}.$$

Thus, $\frac{1}{a_n+1} = \frac{1}{a_n} - \frac{1}{a_{n+1}}$. Consequently,

$$\sum_{n=1}^{2016} \frac{1}{a_n+1} = \sum_{n=1}^{2016} \left(\frac{1}{a_n} - \frac{1}{a_{n+1}} \right) = \frac{1}{a_1} - \frac{1}{a_{2017}} = 3 - \frac{1}{a_{2017}}.$$

Note that $\{a_n\}$ is a monotonically increasing sequence, and it is easy to find that $a_2 > \frac{1}{3}, a_3 > \frac{1}{2}$, and $a_5 > 1$. Thus, $a_{2017} > 1$. Hence, the integral part of $\sum_{n=1}^{2016} \frac{1}{a_n+1}$ is 2.

2. $2n - 1$.   The recursive relation can be deformed to $\frac{a_{n+1} - a_n}{1 + a_n a_{n+1}} = \frac{1}{2n^2}$. We let $b_n = \arctan a_n$. Then $b_1 = \frac{\pi}{4}$ and $\tan(b_{n+1} - b_n) = \frac{1}{2n^2}$, namely $b_{n+1} - b_n = \arctan \frac{1}{2n^2}$. It follows that $b_n - b_1 = \sum_{k=1}^{n-1} \arctan \frac{1}{2k^2} = \sum_{k=1}^{n-1} \left(\arctan \frac{1}{2k-1} - \arctan \frac{1}{2k+1} \right) = \arctan 1 - \arctan \frac{1}{2n-1}$. Therefore, $b_n = \frac{\pi}{2} - \arctan \frac{1}{2n-1}$, and so

$$a_n = \tan b_n = \cot \left(\arctan \frac{1}{2n-1} \right) = 2n - 1.$$

3. Let $b_n = a_n - na_{n-1}$. Then the given condition implies $b_n = 4b_{n-1} - 4b_{n-2}$, where $b_1 = 1$ and $b_2 = 0$. By the method of characteristic roots we know that $b_n = \left(1 - \frac{n}{2} \right) \cdot 2^n$, so $a_n - na_{n-1} = \left(1 - \frac{n}{2} \right) \cdot 2^n = 2^n - n \cdot 2^{n-1}$. Let $c_n = a_n - 2^n$. Then $c_n = nc_{n-1}$, where $c_0 = 1$ and $c_1 = 1$. Thus, $c_n = (n!)c_0 = n!$. Hence, $a_n = 2^n + n!$.

4. We use the fixed point method to do the problem. Solving the equation $\lambda = \frac{\lambda}{2} + \frac{1}{\lambda}$ gives $\lambda = \pm\sqrt{2}$. So, we have $a_{n+1} - \sqrt{2} = \frac{a_n}{2} + \frac{1}{a_n} - \sqrt{2} = \frac{(a_n - \sqrt{2})^2}{2a_n}$ and $a_{n+1} + \sqrt{2} = \frac{a_n}{2} + \frac{1}{a_n} + \sqrt{2} = \frac{(a_n + \sqrt{2})^2}{2a_n}$ (noting $a_n > 0$ from the recursive expression and mathematical induction, so the denominator is nonzero). Dividing the two equalities gives that $\frac{a_{n+1} - \sqrt{2}}{a_{n+1} + \sqrt{2}} = \left(\frac{a_n - \sqrt{2}}{a_n + \sqrt{2}}\right)^2 = \cdots = \left(\frac{a_1 - \sqrt{2}}{a_1 + \sqrt{2}}\right)^{2^n}$. Consequently,

$\frac{a_{n+1} - \sqrt{2}}{a_{n+1} + \sqrt{2}} = \left(\frac{2 - \sqrt{2}}{2 + \sqrt{2}}\right)^{2^n} = (3 - 2\sqrt{2})^{2^n}$, whose solution is $a_{n+1} = \sqrt{2} \cdot \frac{(3 - 2\sqrt{2})^{2^n} + 1}{1 - (3 - 2\sqrt{2})^{2^n}}$. Therefore, the general term of the sequence $\{a_n\}$ is $a_n = \sqrt{2} \cdot \frac{1 + (3 - 2\sqrt{2})^{2^{n-1}}}{1 - (3 - 2\sqrt{2})^{2^{n-1}}}$.

5. By the condition we know that $a_9 = 21a_2 + 13a_1$. So, the proposition is reduced to finding the minimal positive integer k such that the indeterminate equation $13x + 21y = k$ has two pairs of positive integer solutions $x \le y$. For this purpose, let

$$k = 13a + 21b = 13x + 21y,$$

where $a \le b, x \le y, a \le x$, and $(a, b) \ne (x, y)$. Then $13(x - a) = 21(b - y)$. Note that $(13, 21) = 1$, and $x - a$ and $b - y$ are not all zero, so $x - a > 0$ and then $b - y > 0$. It follows that $21 \mid x - a$ and $13 \mid b - y$, from which $x \ge a + 21$, and thus $k = 13x + 21y \ge 34x \ge 34(a + 21) \ge 34 \cdot 22 = 748$. Also, when $k = 748$, the sequences $\{a_n\}$ with initial value conditions $(a_1, a_2) = (1, 35)$ and $(22, 22)$ respectively both satisfy $a_9 = 748$. Hence, the minimum value of k is 748.

6. The given condition ensures that the positive integers larger than a_1 all appear in $\{a_n\}_{n \ge 2}$. We choose the smallest positive integer n such that $a_n > a_1$. Then $2a_n - a_1 > a_1$, and so the number $2a_n - a_1$ appears in $\{a_n\}_{n \ge 2}$. Thus, there exists $m \in \mathbf{N}_+$ such that $a_m = 2a_n - a_1$. Clearly $m > n$ because of the minimality of n. Therefore, the proposition is proved.

7. From the condition, $x_n = \frac{2(2n-1)}{n} \cdot x_{n-1}$, so

$$x_n = \frac{2(2n-1)}{n} \cdot \frac{2(2n-3)}{n-1} x_{n-2} = \cdots$$

$$= \frac{2(2n-1)}{n} \cdot \frac{2(2n-3)}{n-1} \cdots \cdots \frac{2 \cdot 3}{2} x_1$$

$$= \frac{2^n \cdot [(2n-1)!!]}{n!},$$

where $n! = 1 \cdot 2 \cdots \cdots n$ and $(2n-1)!! = 1 \cdot 3 \cdots \cdots (2n-1)$. This indicates that $x_n = \frac{2n(2n-1)(2n-2)(2n-3)\cdots\cdots 2\cdot 1}{n!n!} = \frac{2n(2n-1)\cdots\cdots(n+1)}{n!}$. Since the product of any n consecutive positive integers is a multiple of $n!$, so x_n is a positive integer.

8. It is easy to see from the condition that for any $n \in \mathbf{N}_+$, it is always the case that $x_n > 0$. Hence,

$$x_{n+1} = \frac{x_n^4 + 1}{5x_n} = \frac{x_n^3}{5} + \frac{1}{15x_n} + \frac{1}{15x_n} + \frac{1}{15x_n} \geq 4\sqrt[4]{\frac{1}{5 \cdot 15^3}} = \frac{3\sqrt[4]{3}}{15} > \frac{1}{5}.$$

On the other hand, $x_2 = \frac{17}{10} < 2$ when $n = 2$. Assume that $x_k < 2$ when $n = k$ $(k > 1)$. Now, if $x_k \geq 1$, then $x_{k+1} = \frac{x_k^4+1}{5x_k} < \frac{2^3}{5} + \frac{1}{5\cdot 1} < 2$. If $x_k < 1$, then by combining $x_k \geq \frac{1}{5}$, we see that

$$x_{k+1} = \frac{x_k^4 + 1}{5x_k} < \frac{1}{5} + \frac{1}{5 \cdot \frac{1}{5}} < 2.$$

Hence, in any case, $x_{k+1} < 2$. To summarize, the proposition is proved.

9. Since $a_1 > 0$ and $b_1 > 0$, by the recursive formula, we know that $a_n > 0$ and $b_n > 0$. Now,

$$\begin{cases} a_{n+1} + 1 = \dfrac{(a_n + 1)(b_n + 1)}{b_n}, \\ b_{n+1} + 1 = \dfrac{(a_n + 1)(b_n + 1)}{a_n}, \end{cases}$$

from which

$$\begin{cases} \dfrac{1}{a_{n+1} + 1} = \dfrac{b_n}{(a_n + 1)(b_n + 1)}, \\ \dfrac{1}{b_{n+1} + 1} = \dfrac{a_n}{(a_n + 1)(b_n + 1)}. \end{cases}$$

Consequently,

$$\frac{1}{a_{n+1} + 1} - \frac{1}{b_{n+1} + 1} = \frac{b_n}{(a_n + 1)(b_n + 1)} - \frac{a_n}{(a_n + 1)(b_n + 1)}$$

$$= \frac{(b_n + 1) - (a_n + 1)}{(a_n + 1)(b_n + 1)} = \frac{1}{a_n + 1} - \frac{1}{b_n + 1}$$

$$= \cdots = \frac{1}{a_1 + 1} - \frac{1}{b_1 + 1} = \frac{1}{6}.$$

Therefore, $\frac{1}{a_{2015}+1} > \frac{1}{6}$, so $a_{2015} < 5$.

10. From the given condition,

$$a_{n+1} = 7a_n + 6b_n - 3,$$

$$b_n = 8a_{n-1} + 7b_{n-1} - 4,$$

$$a_n = 7a_{n-1} + 6b_{n-1} - 3.$$

Eliminating b_n and b_{n-1} from the above three equalities, we get

$$a_{n+1} = 14a_n - a_{n-1} - 6. \tag{24.1}$$

Thus, $a_n = 14a_{n-1} - a_{n-2} - 6$. Eliminating the constant term -6 from the previous two equalities gives

$$a_{n+1} = 15a_n - 15a_{n-1} + a_{n-2},$$

whose characteristic equation is $q^3 = 15q^2 - 15q + 1$. So, the characteristic roots are $q_1 = 1, q_2 = 7 + 4\sqrt{3}$, and $q_3 = 7 - 4\sqrt{3}$. Hence,

$$a_n = A_1 \cdot 1^n + A_2(7 + 4\sqrt{3})^n + A_3(7 - 4\sqrt{3})^n.$$

The initial values are $a_0 = 1, a_1 = 7 - 3 = 4$, and $a_2 = 14 \cdot 4 - 1 - 6 = 49$ by (24.1). Thus, we have the system of equations

$$\begin{cases} A_1 + A_2 + A_3 = 1, \\ A_1 + (7 + 4\sqrt{3})A_2 + (7 - 4\sqrt{3})A_3 = 4, \\ A_1 + (7 + 4\sqrt{3})^2 A_2 + (7 - 4\sqrt{3})^2 A_3 = 49, \end{cases}$$

the solutions of which are $A_1 = \frac{1}{2}$ and $A_2 = A_3 = \frac{1}{4}$. Therefore,

$$a_n = \frac{1}{4} \left[2 + (7 + 4\sqrt{3})^n + (7 - 4\sqrt{3})^n \right]$$

$$= \left[\frac{(2 + \sqrt{3})^n + (2 - \sqrt{3})^n}{2} \right]^2 = \left(\sum_{k=0}^{\left[\frac{n}{2}\right]} C_n^{2k} 2^{n-3k} 3^k \right)^2$$

is a perfect square.

11. (1) From the condition, $a_1 = 5$ and $\{a_n\}$ is strictly increasing. Rewrite the recursive formula as

$$2a_{n+1} - 7a_n = \sqrt{45a_n^2 - 36}.$$

Squaring the both sides and simplifying the resulting expression give

$$a_{n+1}^2 - 7a_n a_{n+1} + a_n^2 + 9 = 0, \qquad (24.2)$$

thus,

$$a_n^2 - 7a_{n-1}a_n + a_{n-1}^2 + 9 = 0. \qquad (24.3)$$

By $(24.2) - (24.3)$, we obtain

$$(a_{n+1} - a_{n-1})(a_{n+1} + a_{n-1} - 7a_n) = 0.$$

Since $a_{n+1} > a_n$, so $a_{n+1} + a_{n-1} - 7a_n = 0$, from which

$$a_{n+1} = 7a_n - a_{n-1}. \qquad (24.4)$$

From (24.4), $a_0 = 1, a_1 = 5$, and mathematical induction, we find that a_n is a positive integer for any $n \in \mathbf{N}$.

(2) Completing the square to the both sides of (24.2) gives $(a_{n+1} + a_n)^2 = 9(a_n a_{n+1} - 1)$. Hence,

$$a_n a_{n+1} - 1 = \left(\frac{a_{n+1} + a_n}{3}\right)^2. \qquad (24.5)$$

From (24.4), we have

$$a_{n+1} + a_n = 9a_n - (a_n + a_{n-1}).$$

It follows that

$$a_{n+1} + a_n \equiv -(a_n + a_{n-1}) \equiv \cdots \equiv (-1)^n(a_1 + a_0) \equiv 0 \pmod 3.$$

Therefore, $\frac{a_{n+1}+a_n}{3}$ is a positive integer, so $a_n a_{n+1} - 1$ is a perfect square.

12. This sequence is not bounded.

First, each term of the sequence $\{a_n\}$ is nonnegative and there are no two equal nearby terms.

In fact, if there exists n such that $a_n = a_{n+1} = c$, then $a_{n-1} = 0$ and $a_{n-2} = a_{n-3} = c$. Continuing in this way up to either $a_1 = a_0 = c$ or there is one 0 in a_0 and a_1, we are lead to a contradiction in both cases.

The above discussion concludes that $a_n > 0$ for all $n \geq 0$.

Second, by definition we see that $a_{n+2} = a_{n+1} + a_n$ or $a_{n+2} = a_{n+1} - a_n$. If it is the former, then $a_{n+2} > a_{n+1}$; if it is the latter, then $a_{n+2} < a_{n+1}$, from which $a_{n+1} > a_n$ since $a_{n+2} > 0$. On the other hand, if $a_{n+2} = a_{n+1} - a_n$, then $a_{n+3} = a_{n+2} + a_{n+1}$ (otherwise, if $a_{n+3} = a_{n+2} - a_{n+1}$, then $a_{n+3} = -a_n$, a contradiction). So, if $a_{n+2} < a_{n+1}$, then $a_{n+3} > a_{n+1}$. Now, remove all a_n that satisfy $a_n < a_{n-1}$ and $a_n < a_{n+1}$. Denote the remaining sequence as $\{b_m\}$. Then $\{b_m\}$ is a strictly increasing sequence. If we can show that $\{b_m\}$ is an unbounded sequence, then $\{a_n\}$ is also an unbounded sequence.

In the following, we prove that $\{b_m\}$ is an unbounded sequence by showing that for $m \geq 2$, it is always true that $b_{m+1} - b_m \geq b_m - b_{m-1}$ (this is because $b_{m+1} - b_2 \geq (m-1)(b_2 - b_1)$ from the summation method by splitting, leading to unboundedness of $\{b_m\}$).

In fact, suppose that $b_{m+1} = a_{n+2}$ for $m \geq 2$. Then $a_{n+2} > a_{n+1}$ by the definition of $\{b_m\}$. If $a_{n+1} > a_n$, then $b_m = a_{n+1}$, while $b_{m-1} = a_{n-1}$ or $a_n \geq a_{n-1}$. Hence,

$$b_{m+1} - b_m = a_{n+2} - a_{n+1} = a_n = a_{n+1} - a_{n-1}$$

$$= b_m - b_{m-1} \geq b_m - b_{m-1}.$$

If $a_{n+1} < a_n$, then $b_m = a_n$, while $b_{m-1} = a_{n-1}$ or $a_{n-2} \geq a_{n-1}$ (since $b_{m-1} = a_{n-2}$ requires $a_{n-2} > a_{n-1}$). Thus,

$$b_{m+1} - b_m = a_{n+2} - a_n = a_{n+1}$$

$$= a_n - a_{n-1} = b_m - a_{n-1}$$

$$\geq b_m - b_{m-1}.$$

Therefore, the inequality $b_{m+1} - b_m \geq b_m - b_{m-1}$ is always satisfied. Hence, $\{a_n\}$ is an unbounded sequence.

Solutions for Chapter 25

Periodic Sequences

1. 9. From the condition, $x_n + x_{n+1} + x_{n+2} = 20$ and $x_{n+1} + x_{n+2} + x_{n+3} = 20$. Subtracting them gives $x_{n+3} = x_n$. So, $\{x_n\}$ is a periodic sequence of period 3. Thus, $x_1 = x_4 = 9$ and $x_2 = x_{11} = 7$. Hence, $x_{2014} = x_1 = 9$.

2. x_0. Since $a_{n+3} = \frac{1+x_{n+2}}{x_{n+1}} = \frac{1+\frac{1+x_{n+1}}{x_n}}{x_{n+1}} = \frac{1+x_n+x_{n+1}}{x_n x_{n+1}} = \frac{x_{n+1}x_n-1+x_{n+1}}{x_n x_{n+1}} = \frac{1+x_{n-1}}{x_n} = \frac{x_n x_{n-2}}{x_n} = x_{n-2}$, we see that $\{x_n\}$ is a periodic sequence of period 5. Hence, $x_{2010} = x_0$.

3. -1. A direct calculation finds $x_1 = 1, x_2 = -2, x_3 = -1, x_4 = 2, x_5 = 1$, and $x_6 = -2$. From this and mathematical induction, we can show that $\{x_n\}$ is a periodic sequence of period 4. Hence,

$$\sum_{k=1}^{2014} x_k = x_1 + x_2 + 503(x_1 + x_2 + x_3 + x_4) = -1.$$

4. $-\frac{2015}{2013}$. Let $k = \sqrt{2}+1$ and $m = \sqrt{2}-1$. Then $mk = 1$ and $k-m = 2$. From the given condition, we see that $x_{n+1} = \frac{kx_n-1}{x_n+k} = \frac{x_n-m}{mx_n+1}$ and $x_{n+2} = \frac{x_n-1}{x_n+1}$. Similarly, $x_{n+4} = \frac{-1}{x_n}$ and $x_{n+8} = x_n$. Thus, $x_{n+8t} = x_n$ $(t \in \mathbf{Z}_+)$. Also $x_5 = \frac{-1}{2014}$, so

$$x_{2015} = x_7 = \frac{x_5 - 1}{x_5 + 1} = \frac{-1 - 2014}{-1 + 2014} = -\frac{2015}{2013}.$$

5. Suppose that there is a period T. If $T = 2^p \cdot q$, where $q = 4n+3$, then $a_{2^k} = a_{2^k+T} = a_{2^p(2^{k-p}+q)} = a_{2^{k-p}+q} = 0$ for $k \geq p+2$, a contradiction to $a_{2^k} = 1$. If $T = 2^p \cdot q$, where $q = 4n+1$, then $a_{2^k} = a_{2^k+3T} = a_{2^p(2^{k-p}+3q)} = a_{2^{k-p}+3q} = 0$ for $k \geq p+2$, also a contradiction. Hence, the sequence is not periodic.

6. Let $A_n = a_n + b_n + c_n + d_n$ and $B_n = a_n^2 + b_n^2 + c_n^2 + d_n^2$. Then $\{A_n\}$ and $\{B_n\}$ are both periodic and satisfy $A_{n+1} = 2A_n$ and $B_{n+2} = 2B_{n+1} + 2A_n^2$. Consequently, $A_{n+1} = 2^n \cdot A_1$. The periodicity of $\{A_n\}$ then implies that $A_n = 0$, from which $B_{n+2} = 2B_{n+1}$. The same argument implies that $B_n = 0$ for $n \geq 2$. Hence,

$$a_2 = b_2 = c_2 = d_2 = 0.$$

7. By the definition, we have $f_1(11) = (1+1)^2 = 4, f_2(11) = f_1(4) = 4^2 = 16, f_3(11) = f_1(16) = (1+6)^2 = 49, f_4(11) = f_1(49) = (4+9)^2 = 169, f_5(11) = f_1(169) = (1+6+9)^2 = 256, f_6(11) = f_1(256) = (2+5+6)^2 = 169$, and $f_7(11) = f_1(169) = (1+6+9)^2 = 256$. Thus, the values of $f_n(11)$ when $n \geq 4$ are alternatively 169 and 256, that is, $\{f_n(11)\}$ is a mixed periodic sequence of period 2. Therefore, $f_{100}(11) = 169$.

8. If $a_n > a$, then $a_{n+1} = a_n + a$ when a_n is an odd number, and so $a_{n+2} = \frac{a_n + a}{2}$; and $a_{n+1} = \frac{a_n}{2}$ when a_n is an even number. So, it is always true that $a_n > \min\{a_{n+1}, a_{n+2}\}$. In this way, we see that there exists $n \in \mathbf{N}_+$ such that $a_n \leq a$. Furthermore, $a_m \leq 2a$ and a_m is a positive integer for any $m \geq n$. Consequently by the drawer principle, there exist $k \geq n$ and $T \in \mathbf{N}_+$ such that $a_{k+T} = a_k$. This means that $\{a_n\}$ is a periodic sequence of period T.

9. Calculate directly the first several terms $0, 1, 8, 63, 496, 3905, 30744$ of U_n. Note that $U_3 = 63$ and $U_6 = 30744$ are both multiples of 3 and 7, but the other terms $U_1, U_2, U_4,$ and U_5 are neither multiples of 3, nor multiples of 7. We conjecture that

$$3 \mid U_n \Rightarrow 7 \mid U_n. \tag{25.1}$$

In the following, we show that (25.1) is true. First consider module 3. From the recursive formula, we know that $U_{n+3} - U_n = 9(U_{n+2} - U_{n+1})$, namely $U_{n+3} \equiv U_n \pmod{3}$. The sequence $\{U_n \pmod 3\}$ is a periodic one of period 3, and its first three terms are $0, 1$, and -1. So,

$$3 \mid U_n \Leftrightarrow 3 \mid n. \tag{25.2}$$

Then take module 7. By the same token, the sequence $\{U_n \pmod 7\}$ is a periodic one of period 6, and its first six terms are $0, 1, 1, 0, -1$, and -1. Thus,

$$7 \mid U_n \Leftrightarrow 3 \mid n. \tag{25.3}$$

From (25.2) and (25.3), we know that (25.1) is correct, and so there are no terms of the form $3^\alpha \cdot 5^\beta$ (α and β are positive integers). (Remark: The method for this problem is very clever. This requires an accumulation of experiences in solving this type of problems, and needs an ability of keen observations.)

10. We prove: For a positive integer k satisfying $2^k > \max S_0$, the choice of $N = 2^k$ meets the requirement.

    In fact, consider polynomials in the sense of mod 2. According to the structure of the set sequence, we have

$$\sum_{j \in S_n} x^j \equiv (1 + x) \sum_{j \in S_{n-1}} x^j \equiv \cdots \equiv (1 + x)^n \sum_{j \in S_0} x^j \ (\text{mod } 2),$$

while $(x + y)^2 = x^2 + 2xy + y^2 \equiv x^2 + y^2 \ (\text{mod } 2)$. Hence, for any $n \in \mathbf{N}_+$, we can show by using mathematical induction that $(x+y)^{2^n} \equiv x^{2^n} + y^{2^n} \ (\text{mod } 2)$.

Summarizing the above conclusions, when $N = 2^k > \max S_0$, we have

$$\sum_{j \in S_n} x^j \equiv (1 + x)^{2^k} \sum_{j \in S_{n-1}} x^j \equiv \left(1 + x^{2^k}\right) \sum_{j \in S_0} x^j \ (\text{mod } 2)$$

$$= \left(1 + x^N\right) \sum_{j \in S_0} x^j.$$

This means that $S_N = S_0 \cup \{N + a \mid a \in S_0\}$. And the proposition is thus proved.

11. Define a sequence $\{x_m\}$ as follows:

$$x_m = \begin{cases} 1, & \text{if } m \in X, \\ 0, & \text{if } m \notin X. \end{cases}$$

Note that the sets $X + a_1, \ldots, X + a_n$ constitute a partition of $\mathbf{Z} \Leftrightarrow$ for any $t \in \mathbf{Z}$, the sets $X + a_1 + t, \ldots, X + a_n + t$ constitute a partition of $\mathbf{Z}$. We may assume that $0 = a_1 < a_2 < \cdots < a_n$.

Using the partition of $\mathbf{Z}$ by $X, X + a_2, \ldots, X + a_n$, we know that $m \in X \Leftrightarrow$ for any $j \in \{2, \ldots, n\}$, there holds that $m \notin X + a_j \Leftrightarrow$ for any $j \in \{2, \ldots, n\}$, there holds that $m - a_j \notin X$.

Thus, $x_m = (1 - x_{m-a_2}) \cdots (1 - x_{m-a_n})$. This means that the value of each term in the sequence $\{x_m\}$ is determined by the value of a previous term a_n. Hence, there are at most 2^{a_n} different possibilities for the values of the numbers array $(x_m, x_{m+1}, \ldots, x_{m+a_n})$. By the drawer principle, there are positive integers m_1 and m_2 with $m_1 < m_2$ such that

$$(x_{m_1}, x_{m_1+1}, \ldots, x_{m_1+a_n}) = (x_{m_2}, x_{m_2+1}, \ldots, x_{m_2+a_n}).$$

Therefore, $\{x_m\}$ is a periodic sequence of period $N \, (= m_2 - m_1)$.

On the other hand, since $X, X + a_2, \ldots, X + a_n$ constitute a partition of $\mathbf{Z}$, the sets $X - a_n, X + a_2 - a_n, \ldots, X + a_{n-1} - a_n$ also form a partition of $\mathbf{Z}$. From the same analysis as above, $x_m = (1 - x_{m+a_n})(1 - x_{m+a_n-a_2}) \cdots (1 - x_{m+a_n-a_{n-1}})$. Using the recursive relation backward, we find that $\{x_m\}$ is a purely periodic sequence of period N. This indicates: For any $m \in \mathbf{N}_+$, there is $x_m = x_{m+N}$. In other words, $m \in X \Leftrightarrow m + N \in X$. Therefore, $X + N = X$.

12. (1) When i is an integer greater than 1, we first prove: With respect to any prime factor p of x_i, the sequence $\{x_n\}$ is periodic from the ith term on in the sense of mod p (that is, with $x_1, \ldots, x_{i-1}$ removed, the remaining sequence is periodic). In fact, note that in the sense of mod p, there are only finitely many cases for the number pairs (x_m, x_{m+1}) (each term has only p possibilities, so there are at most p^2 different number pairs). Thus, there are $r \geq i$ and $T \in \mathbf{N}_+$ such that $(x_{r+T}, x_{r+T+1}) \equiv (x_r, x_{r+1}) \pmod{p}$. If $r = i$, then from $x_{n+1} = x_n x_{n-1} + 1$, we know that $\{x_n\}$ is a periodic sequence from the ith term on in the sense of mod p. Consider the case $r > i$. If $x_r \equiv 0 \pmod{p}$, then $x_{r+1} \equiv 1 \pmod{p}$. While $(x_i, x_{i+1}) \equiv (0, 1) \pmod{p}$, it follows that $(x_i, x_{i+1}) \equiv (x_r, x_{r+1}) \pmod{p}$. (Note: Here we used $i \geq 2$, since only in this case there can be $x_{i+1} = x_i x_{i-1} + 1 \equiv \pmod{p}$.) We know $(x_i, x_{i+1}) \equiv (x_r, x_{r+1}) \pmod{p}$. Hence, $\{x_n\}$ is periodic sequence of period $r - i$ from the ith term on in the sense of mod p. If $x_r \not\equiv 0 \pmod{p}$, then $x_{r-1+T} x_{r+T} = x_{r+T-1} - 1 \equiv x_{r+1} - 1 = x_{r-1} x_r \pmod{p}$. Taking consideration of $x_{r+T} \equiv x_r \not\equiv 0 \pmod{p}$, we obtain $x_{r-1+T} \equiv x_{r-1} \pmod{p}$. That is,

$$(x_{r-1}, x_r) \equiv (x_{r-1+T}, x_{r+T}) \pmod{p}.$$

This indicates that r can be moved forward by one position. Continuing in this way with constant forwarding as above, we know that $\{x_n\}$ is a periodic sequence from the ith term on in the sense of mod p. Now, assume that $x_i = p_1^{\alpha_1} \cdots p_k^{\alpha_k}$, where $p_1 < \cdots < p_k$ are prime numbers and $\alpha_1, \ldots, \alpha_k \in \mathbf{N}_+$. For $p_l \in \{p_1, \ldots, p_k\}$, let $\{x_n\}$ be the periodic sequence of period T_l from the ith term on in the sense of mod p_l, and denote $T = [T_1, \ldots, T_k]$, which is the least common multiple of $T_1, \ldots, T_k$. Let $j = i + nT$, where

$$n = i \cdot \max\{\alpha_1, \ldots, \alpha_k\}.$$

Then for $1 \leq l \leq k$, since $x_j = x_{i+nT} \equiv x_i \equiv 0 \pmod{p}$, so $p_l^{i \cdot \alpha_l} \mid x_j^{i \cdot \alpha_l}$, and then $p_l^{i \cdot \alpha_l} \mid x_j^j$. By noting $x_i = p_1^{\alpha_1} \cdots p_k^{\alpha_k}$, we have $x_i^i \mid x_j^j$, proving the proposition.

(2) Choose $x_1 = 33$ and $x_2 = 23$. Then the terms of the sequence $\{x_n\}$ in the sense of mod 33 are $0, 23, 1, 24, 25, 7, 11, 12, 1, 13, 14, 18, 22,$ $1, 23, 24, 25, \ldots$, and the sequence is periodic of period 13 from the 4th term on in the sense of mod 33. When $j > 1$, we have $33 \nmid x_j$. Therefore, the conclusion of (2) is negative.

Solutions for Chapter 26

Polar Coordinates

1. 6.   When $m = n$, the curve represented by $\rho = \frac{1}{1 - C_m^n \cos\theta}$ is a parabola. When $m \neq n$, let $e = C_m^n > 1$. Note that $C_3^1 = C_3^2, C_4^1 = C_4^3, C_5^2 = C_5^3$, and $C_5^1 = C_5^4$. So, there are totally $10 - 4 = 6$ different values for the eccentricity $e > 1$.

2. Hyperbola. Since

$$\rho = \frac{5}{3 - 4\cos\theta + 4\sin\theta} = \frac{5}{3 - 4\sqrt{2}\cos\left(\theta + \frac{\pi}{4}\right)},$$

by the unified equation of conic sections, $e = \frac{4\sqrt{2}}{3} > 1$.

3. $\left[\frac{\pi}{4}, \frac{3\pi}{4}\right]$.   By the unified approach to conic sections in polar coordinates, we have the equation $\rho = \frac{4}{1 - \cos\theta}$ of the parabola. From the given condition, we know that $\theta \in [0, \pi]$ and $\frac{4}{1 - \cos\theta} + \frac{4}{1 + \cos\theta} \leq 16$, that is $\sin^2\theta \geq \frac{1}{2}$, namely $|\sin\theta| \geq \frac{\sqrt{2}}{2}$. Hence, $\theta \in \left[\frac{\pi}{4}, \frac{3\pi}{4}\right]$.

4. Set up a polar coordinates system with F as the pole and the symmetric axis of the parabola as the polar axis. Then the polar equation of the parabola is $\rho = \frac{p}{1 - \cos\theta}$. Let α be the angle between PQ and the polar axis. Then the polar radii of P and Q are

$$\rho_1 = |FP| = \frac{p}{1 - \cos\alpha}, \quad \rho_2 = |FQ|\frac{p}{1 + \cos\alpha}.$$

Since the polar radius of the middle point S of PQ is

$$\rho_0 = |FS| = \frac{1}{2}(\rho_1 - \rho_2) = \frac{1}{2}\left(\frac{p}{1 - \cos\alpha} - \frac{p}{1 + \cos\alpha}\right) = \frac{p\cos\alpha}{\sin^2\alpha},$$

$$|FR| = \frac{|FS|}{\cos\alpha} = \frac{p}{\sin^2\alpha} = \frac{1}{2}(\rho_1 + \rho_2) = \frac{1}{2}|PQ|.$$

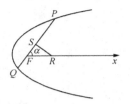

Figure 26.1 Figure for Exercise 4.

5. Denote $P_1(\rho_1, \theta_1)$ and $P_2(\rho_2, \theta_1 + \pi)$. By the unified equation $\rho = \frac{p}{1 - \cos\theta}$, we have $\rho_1 = \frac{p}{1 - \cos\theta_1}$ and $\rho_2 = \frac{p}{1 + \cos\theta_1}$. Hence,

$$\frac{1}{\rho_1} + \frac{1}{\rho_2} = \frac{1 - \cos\theta_1}{p} + \frac{1 + \cos\theta_1}{p} = \frac{2}{p}.$$

6. Set up a polar coordinates system with the left focus F as the pole and FX as the polar axis. Then the equation of the ellipse is $\rho = \frac{7}{4 - 3\cos\theta}$. Since $\angle AFB = \angle BFC = \angle CFA$, they are all $\frac{2\pi}{3}$. Without loss of generality, denote $A(\rho_1, \theta)$, and then we have $B\left(\rho_2, \theta + \frac{2\pi}{3}\right)$ and $C\left(\rho_3, \theta + \frac{4\pi}{3}\right)$. Thus

$$\frac{1}{|AF|} + \frac{1}{|BF|} + \frac{1}{|CF|} = \frac{1}{\rho_1} + \frac{1}{\rho_2} + \frac{1}{\rho_3}$$

$$= \frac{4 - 3\cos\theta}{7} + \frac{4 - 3\cos\left(\theta + \frac{2\pi}{3}\right)}{7}$$

$$+ \frac{4 - 3\cos\left(\theta + \frac{4\pi}{3}\right)}{7} = \frac{12}{7}.$$

7. Set up a polar coordinates system with the origin O as the pole and OX as the polar axis. Then the equation of the circle A is $\rho = 2\sin\theta$. Denote $P(\rho, \theta)$ and $Q(\rho_1, \theta)$. Then $\rho - \rho_1 = 2 - \rho\sin\theta$, namely $\rho - 2\sin\theta = 2 - \rho\sin\theta$. Simplifying it gives $\rho = 2$ or $\theta = -\frac{\pi}{2}$. Since P is on the ray OQ, so the desired equation of the trajectory is $\rho = 2$ $(0 < \theta < \pi)$ or $\theta = -\frac{\pi}{2}$. The corresponding equation in the rectangular coordinates is $x^2 + y^2 = 4$ $(y > 0)$ or $x = 0$ $(y > 0)$.

8. With F as the pole and the straight line perpendicular to l as the polar axis, set up a polar coordinates system. Denote $N(\rho_1, \theta)$ and $M(\rho_2, \theta + \pi)$ $\left(-\frac{\pi}{2} < \theta < \frac{\pi}{2}\right)$. Then $\rho_2 \cos(\theta + \pi) = -p$, that is, $\rho_2 = \frac{p}{\cos\theta}$. Also, since $\frac{|FN|}{|MN|} = \frac{1}{|MF|}$, namely $\rho_1\rho_2 = \rho_1 + \rho_2$, we obtain that

$\rho_1 = \frac{\rho_2}{\rho_2 - 1} = \frac{p}{p - \cos\theta}$. Consequently,

$$\rho_1 + \rho_2 = \frac{p}{p - \cos\theta} + \frac{p}{\cos\theta} = \frac{p^2}{\cos\theta(p - \cos\theta)} \geq \frac{p^2}{\frac{p^2}{4}} = 4,$$

and the equality is achieved when $\cos\theta = \frac{p}{2}$:

(1) If $0 < p \leq 2$, then the minimum value of $\rho_1 + \rho_2$ is 4;

(2) if $p > 2$, then $\rho_1 + \rho_2$ obtains the minimum value $\frac{p^2}{p-1}$ when $\cos\theta = 1$.

9. Set up a polar coordinates system with the left focus as the pole and the positive direction of the x-axis as the polar axis. Then the equation of the ellipse is $\rho = \frac{ep}{1 - e\cos\theta}$. Denote $A(\rho_1, \alpha)$ and $B(\rho_2, \alpha + \pi)$ the two end points of the side of the parallelogram passing through the pole. Then $|AB| = \rho_1 + \rho_2 = \frac{2ep}{1 - e^2\cos^2\theta}$. Thus, the distance of the two parallel lines is

$$h = |F_1 F_2| \sin\alpha = 2c\sin\alpha,$$

so $S = |AB|h = \frac{4ab^2c}{\frac{b^2}{\sin\alpha} + c^2\sin\alpha}$, where S is the area of the parallelogram.

(i) When $0 < \frac{b}{c} \leq 1$, we have $S \leq \frac{4ab^2c}{2bc} = 2ab$;

(ii) when $\frac{b}{c} > 1$, we see that $S \leq \frac{4ab^2c}{a^2} = \frac{4b^2}{a}$.

10. With F as the pole and the positive semi-axis of the x-axis as the polar axis, set up a polar coordinates system. Then the equation of the parabola is $\rho = \frac{p}{1 - \cos\alpha}$ (here α is the polar angle).

Draw $AM \perp$ the x-axis and $BN \perp$ the x-axis, where M and N are the feet of a perpendicular. Then $|FA| = \frac{p}{1 - \cos\theta}, |AM| = \frac{p\sin\theta}{1 - \cos\theta}$, and $|FM| = \frac{p\cos\theta}{1 - \cos\theta}$. Since $|OF| = \frac{p}{2}$, so $|OM| = |OF| + |FM| = \frac{p(1 + \cos\theta)}{2(1 - \cos\theta)}$, and $\tan\angle AOM = \frac{|AM|}{|OM|} = \frac{2\sin\theta}{1 + \cos\theta}$. Also

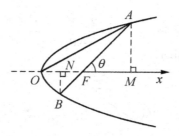

Figure 26.2 Figure 1 for Exercise 10.

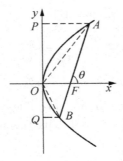

Figure 26.3   Figure 2 for Exercise 10.

$|FB| = \frac{p}{1+\cos\theta}, |BN| = \frac{p\sin\theta}{1+\cos\theta}, |FN| = \frac{p\cos\theta}{1+\cos\theta}, |ON| = \frac{p}{2} - \frac{p\cos\theta}{1+\cos\theta} = \frac{p(1-\cos\theta)}{2(1+\cos\theta)}$, and $\tan\angle BON = \frac{2\sin\theta}{1-\cos\theta}$. Consequently,

$$\tan\angle AOB = \tan(\angle AOM + \angle BON) = -\frac{4}{3\sin\theta}.$$

Therefore, $\angle AOB = \pi - \arctan\frac{4}{3\sin\theta}$.

11. (1) Without loss of generality, assume that F is the left focus. As in Figure 26.4, with F as the pole and the x-axis as the polar axis, set up a polar coordinates system. Then the polar coordinates equation of the ellipse $\frac{x^2}{a^2} + \frac{y^2}{b^2} = 1$ is $\rho = \frac{ep}{1-e\cos\theta}$ $(\rho > 0)$, where $e = \frac{c}{a}, p^2 = \frac{b^2}{c}$, and $c = \sqrt{a^2 - b^2}$.

Denote $A(\rho_1, \theta), B(\rho_2, \theta + \pi), C\left(\rho_3, \theta + \frac{\pi}{2}\right)$, and $D\left(\rho_4, \theta + \frac{3\pi}{2}\right)$ with $\theta \in \left[0, \frac{\pi}{2}\right]$. Then $\frac{1}{|FA|} + \frac{1}{|FB|} = \frac{1}{\rho_1} + \frac{1}{\rho_2} = \frac{1-e\cos\theta}{ep} + \frac{1+e\cos\theta}{ep} = \frac{2}{ep} = \frac{2a}{b^2}$ is a fixed number. Also

$$\frac{1}{|AB|} + \frac{1}{|CD|} = \frac{1}{\rho_1 + \rho_2} + \frac{1}{\rho_3 + \rho_4}$$

$$= \frac{1 - e^2\cos^2\theta}{2ep} + \frac{1 - e^2\sin^2\theta}{2ep}$$

$$= \frac{2 - e^2}{2ep} = \frac{a^2 + b^2}{2ab^2}$$

is a fixed number.

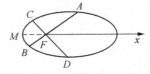

Figure 26.4   Figure for Exercise 11.

(2) We calculate out that

$$|AB| + |CD| = \rho_1 + \rho_2 + \rho_3 + \rho_4$$

$$= \frac{ep}{1 - e\cos\theta} + \frac{ep}{1 + e\cos\theta} + \frac{ep}{1 + e\sin\theta} + \frac{ep}{1 - e\sin\theta}$$

$$= 2ep \left(\frac{1}{1 - e^2\cos^2\theta} + \frac{1}{1 - e^2\sin^2\theta} \right)$$

$$= \frac{2ep(2 - e^2)}{1 - e^2 + e^4\sin^2\theta\cos^2\theta} \geq \frac{2ep(2 - e^2)}{1 - e^2 + \frac{1}{4}e^4} = \frac{8ab^2}{a^2 + b^2},$$

and if and only if $\theta = \frac{\pi}{4}$ it has the minimum value $\frac{8ab^2}{a^2 + b^2}$.
  Also

$$|AB||CD| = \frac{|AB| + |CD|}{\frac{1}{|AB|} + \frac{1}{|CD|}} \geq \frac{\frac{8ab^2}{a^2 + b^2}}{\frac{a^2 + b^2}{2ab^2}} = \frac{16a^2b^4}{(a^2 + b^2)^2},$$

and if and only if $\theta = \frac{\pi}{4}$ it has the minimum value $\frac{16a^2b^4}{(a^2 + b^2)^2}$ (all can be calculated directly).

(3) We know that the polar coordinates of M are $(a - c, \pi)$. Then

$$S_{\triangle ABM} = S_{\triangle AMF} + S_{\triangle BMF} = \frac{1}{2}|MF|(\rho_1 + \rho_2)\sin\theta$$

$$= \frac{1}{2}(a - c)\left(\frac{ep}{1 - e\cos\theta} + \frac{ep}{1 + e\cos\theta} \right)\sin\theta$$

$$= (a - c)\frac{ep}{1 - e^2\cos^2\theta}\sin\theta = (a - c)\frac{ab^2}{b^2 + c^2\sin^2\theta} \cdot \sin\theta$$

$$= \frac{ab^2(a - c)}{\frac{b^2}{\sin\theta} + c^2\sin\theta}.$$

Thus, $S_{\triangle ABM} \leq \frac{ab^2(a-c)}{2bc} = \frac{ab(a-c)}{2c}$ when $b \leq c$, and if and only if $\sin^2\theta = \frac{b^2}{c^2}$, namely $\theta = \arcsin\frac{b}{c}$, it reaches the maximum value.
  When $b > c$, from the monotonicity, $(S_{\triangle ABM})_{\max} = \frac{ab^2(a-c)}{b^2+c^2} = \frac{b^2(a-c)}{a}$, and if and only if $\theta = \frac{\pi}{2}$, it obtains the maximum value.
  In summary,

$$(S_{\triangle ABM})_{\max} = \begin{cases} \dfrac{ab(a - c)}{2c}, & b \leq c; \\[3mm] \dfrac{b^2(a - c)}{a}, & b > c. \end{cases}$$

12. (1) With the center O of the ellipse as the pole and the x-axis as the polar axis, set up a polar coordinates system. Then the polar coordinates equation of the ellipse is $\frac{1}{\rho^2} = \frac{\cos^2\theta}{a^2} + \frac{\sin^2\theta}{b^2}$. Denote $A(\rho_1, \theta)$ and $B\left(\rho_2, \theta + \frac{\pi}{2}\right)$ with $\theta \in \left[0, \frac{\pi}{2}\right)$. Then $\frac{1}{|OA|^2} + \frac{1}{|OB|^2} = \frac{\cos^2\theta}{a^2} + \frac{\sin^2\theta}{b^2} + \frac{\sin^2\theta}{a^2} + \frac{\cos^2\theta}{b^2} = \frac{1}{a^2} + \frac{1}{b^2}$ is a fixed number.

(2) A direct computation gives that

$$\frac{1}{|OA|} + \frac{1}{|OB|} = \sqrt{\frac{\cos^2\theta}{a^2} + \frac{\sin^2\theta}{b^2}} + \sqrt{\frac{\sin^2\theta}{a^2} + \frac{\cos^2\theta}{b^2}}$$

$$= \frac{1}{ab}\left(\sqrt{a^2\sin^2\theta + b^2\cos^2\theta} + \sqrt{a^2\cos^2\theta + b^2\sin^2\theta}\right)$$

$$= \frac{1}{ab}\sqrt{a^2 + b^2 + 2\sqrt{a^2b^2(\sin^4\theta + \cos^4\theta) + (a^4 + b^4)\sin^2\theta\cos^2\theta}}$$

$$= \frac{1}{ab}\sqrt{a^2 + b^2 + 2\sqrt{a^2b^2 + \frac{1}{4}(a^2 - b^2)^2\sin^2 2\theta}}.$$

So, when $\theta = 0$, the desired minimum value is $\frac{1}{ab}\sqrt{a^2 + b^2 + 2ab} = \frac{a+b}{ab}$; when $\theta = \frac{\pi}{4}$, the desired maximum value is

$$\frac{1}{ab}\sqrt{a^2 + b^2 + 2 - \frac{1}{2}(a^2 + b^2)} = \frac{\sqrt{2(a^2 + b^2)}}{ab}.$$

(3) The polar coordinates equation of the ellipse can be rewritten as $\rho^2 = \frac{a^2b^2}{a^2\sin^2\theta + b^2\cos^2\theta}$, so

$$S_{\triangle AOB} = \frac{1}{2}\rho_1\rho_2$$

$$= \frac{1}{2}a^2b^2\sqrt{\frac{1}{(a^2\sin^2\theta + b^2\cos^2\theta)(a^2\cos^2\theta + b^2\sin^2\theta)}}$$

$$= \frac{1}{2}a^2b^2\sqrt{\frac{1}{a^2b^2 + \frac{1}{4}(a^2 - b^2)\sin^2 2\theta}}.$$

Hence, when $\theta = \frac{\pi}{4}$, the desired minimum value of the area is $\frac{a^2b^2}{a^2+b^2}$; when $\theta = 0$, the maximum value of the area is $\frac{1}{2}ab$.

Solutions for Chapter 27

Analytic Method for Plane Geometry

1. Two thirds of a circle with radius 1. With O as the origin and the straight line containing the line segment OA as the x-axis, set up a rectangular coordinates system. Let $G(x, y)$ be the center of gravity of $\triangle ABC$. For the two points B and C, since $\angle BAC = \frac{\pi}{3}$, we have $\angle BOC = \frac{2}{3}\pi$. It is easy to see that the distance of the middle point $D(x_0, y_0)$ of the chord BC to the point O is $3\cos\frac{\pi}{3} = \frac{3}{2}$, so $x_0^2 + y_0^2 = \left(\frac{3}{2}\right)^2 \left(-\frac{3}{2} \le x_0 \le \frac{3}{4}\right)$. Also $\frac{|\overrightarrow{AD}|}{|\overrightarrow{DG}|} = -3$, thus $x_0 = \frac{3x-3}{2}$ and $y_0 = \frac{2y}{2}$. Substituting them into the equation that the point D satisfies, and assembling the resulting expression, we obtain

$$(x - 1)^2 + y^2 = 1 \left(0 \le x \le \frac{3}{2}\right).$$

2. 0. Establish a coordinates system as shown in Figure 27.1. Let $|\overrightarrow{OA}| = r, \overrightarrow{OB} = p, \overrightarrow{OC} = q, \angle AOB = \alpha$, and $\angle AOC = \beta$. Then $\overrightarrow{OA} = (r, 0), \overrightarrow{OB} = (p\cos\alpha, p\sin\alpha)$, and $\overrightarrow{OC} = (q\cos\beta, q\sin\beta)$. Thus,

$$S_{\triangle OAC}\overrightarrow{OB} + S_{\triangle OAB}\overrightarrow{OC} + S_{\triangle OBC}\overrightarrow{OA}$$

$$= \frac{1}{2}pr\sin\alpha(q\cos\beta, q\sin\beta) + \frac{1}{2}pq\sin(\beta - \alpha)(r, 0)$$

$$\quad - \frac{1}{2}qr\sin\beta(p\cos\alpha, p\sin\alpha)$$

$$= \frac{1}{2}pqr(\sin\alpha\cos\beta + \sin(\beta - \alpha) - \sin\beta\cos\alpha, \sin\beta\sin\alpha - \sin\beta\sin\alpha)$$

$$= \mathbf{0}.$$

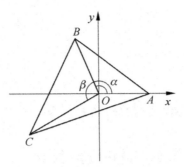

Figure 27.1 Figure for Exercise 2.

3. **A circle.** Establish a rectangular coordinates system, so that the base of the flag pole of height h is $(-a, 0)$ and the base of the flag pole of height k is $(a, 0)$. Let the elevations of the point $p(x, y)$ with respect to the top of the poles be equal. Then $\dfrac{h}{\sqrt{(x+a)^2+y^2}} = \dfrac{k}{\sqrt{(x-a)^2+y^2}}$, that is,

$$(k^2 - h^2)x^2 + 2a(h^2 + k^2)x + (k^2 - h^2)y^2 + a^2(k^2 - h^2) = 0.$$

Since $k \neq h$, the equation is reduced to $x^2 + y^2 + 2a\dfrac{k^2+h^2}{k^2-h^2}x + a^2 = 0$, and the corresponding point trajectory is a circle.

4. Draw $AO \perp BC$, intersecting BC at O. With BC and AO as the x-axis and the y-axis respectively, set up a coordinates system. Denote $A(0, c)$, $M(n, 0)$, and $C(n+a, 0)$. Then we have the coordinates $(n - a, 0)$ of B, and the equation of the straight line PQ is

$$kx - y + b = 0. \tag{27.1}$$

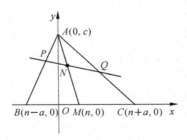

Figure 27.2 Figure for Exercise 4.

Let $\frac{|AP|}{|PB|} = \lambda$ and the coordinates of P be (x_1, y_1). Then

$$\begin{cases} x_1 = \dfrac{\lambda(n-a)}{1+\lambda}, \\ y_1 = \dfrac{c}{1+\lambda}. \end{cases}$$

Substituting the coordinates (x_1, y_1) of P into (27.1) gives $\lambda = \frac{c-b}{k(n-a)+b}$, so $\frac{|AB|}{|AP|} = \frac{k(n-a)+c}{c-b}$. By the same token, $\frac{|AM|}{|AN|} = \frac{kn+c}{c-b}$ and $\frac{|AC|}{|AQ|} = \frac{k(n+a)+c}{c-b}$. Hence, $2\frac{|AM|}{|AN|} = \frac{|AB|}{|AP|} + \frac{|AC|}{|AQ|}$.

5. As in Figure 27.3, set up a rectangular coordinates system with the center of the circle of diameter AB and radius R as the origin O and the straight line AB as the x-axis. Then the equation of the circle O is

$$x^2 + y^2 = R^2. \tag{27.2}$$

Denote $Q(x_0, y_0)$. Then $x_0^2 + y_0^2 = R^2$. The equation of the circle centered at Q with QH as radius is

$$(x - x_0)^2 + (y - y_0)^2 = y_0^2. \tag{27.3}$$

(27.2) $-$ (27.3) and after a simplification, we obtain the equation

$$2x_0 x + 2y_0 y = x_0^2 + R^2 \tag{27.4}$$

of the straight line CD, and the equation

$$x = x_0 \tag{27.5}$$

of the straight line QH. Substituting (27.5) into (27.4) gives $2y_0 y = y_0^2$. Since $Q \neq A$ and B, so $y_0 \neq 0$. Thus, $y = \frac{y_0}{2}$, that is, the coordinates of the intersection point of the straight lines CD and QH are $\left(x_0, \frac{y_0}{2}\right)$. Therefore, the straight line CD bisects the line segment QH.

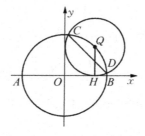

Figure 27.3   Figure for Exercise 5.

6. With the straight line containing BC as the x-axis and the middle point of BC as the origin, set up a rectangular coordinates system. Let the radius of the semicircle be r. Then we have $B(-r, 0)$ and $C(r, 0)$. Let the equations of the straight lines AB and AC be $y = \frac{b}{a+r}(x+r)$ and $y = \frac{b}{a-r}(x-r)$ respectively, and let the equation of the semicircle be $x^2 + y^2 = r^2$ ($y \geq 0$). We can find the coordinates of the intersection points D and E of the straight lines AB and AC with the semicircle, respectively:

$$(x_D, y_D) = \left(\frac{r[(a+r)^2 - b^2]}{(a+r)^2 + b^2}, \frac{2br(a+r)}{(a+r)^2 + b^2} \right),$$

$$(x_E, y_E) = \left(\frac{-r[(a-r)^2 - b^2]}{(a-r)^2 + b^2}, \frac{-2br(a-r)}{(a-r)^2 + b^2} \right).$$

Let $\frac{|GE|}{|FD|} = \lambda$. Then $\lambda = \frac{y_E}{y_D}$. From $x_M = \frac{x_G + \lambda x_D}{1 + \lambda}$, we find $x_M = a$. Hence, $AM \perp BC$.

7. By symmetry, we know that $BD \perp AC$ and $|OD| = |OB|$. With the straight line containing BD as the x-axis and O as the origin, set up a rectangular coordinates system. By the condition, we may denote $A(0, a), B(b, 0), C(0, c)$, and $D(d, 0)$. Then the straight line equation of AB is

$$\frac{x}{b} + \frac{y}{a} = 1 \tag{27.6}$$

and the straight line equation of BC is

$$\frac{x}{b} + \frac{y}{c} = 1.$$

Let the slopes of EF and GH be k and k', respectively. Then

$$EF : y = kx,$$

$$GH : y = k'x. \tag{27.7}$$

Solving (27.6) and (27.7) together to get the coordinates of G:

$$x_G = \frac{ab}{a + bk'}, \quad y_G = \frac{abk'}{a + bk'}.$$

By the same token, substituting c for a and k for k', we have the coordinates of F:

$$x_F = \frac{bc}{c + bk}, \quad y_F = \frac{bck}{c + bk}.$$

Then the equation of GF (not including the two points G and F) is

$$\frac{y - \frac{abk'}{a+bk'}}{x - \frac{ab}{a+bk}} = \frac{y - \frac{bck}{c+bk}}{x - \frac{bc}{c+bk}}.$$

Let $y = 0$. Then we get the horizontal coordinate of the intersection point I of GF and BD:

$$x_I = \frac{abc(k - k')}{(a + bk')ck' - (c + bk)ak'}.$$

Similarly, substituting $d = -b$ for b and interchanging k and k', we obtain the horizontal coordinate of the intersection point J of EH and BD:

$$x_J = \frac{abc(k' - k)}{(a + bk')ck' - (c + bk)ak'}.$$

After comparison we find that $x_I = -x_J$, so $|OI| = |OJ|$.

8. Choose the center of the circumcircle of $\triangle ABC$ as the origin to establish a coordinates system. Denote $A(x_1, y_1), B(x_2, y_2)$, and $C(x_3, y_3)$. Then the coordinates of the center of gravity are $G\left(\frac{x_1+x_2+x_3}{3}, \frac{y_1+y_2+y_3}{3}\right)$. Let the radius of the circumcircle be r. Then

$$r^2 = x_1^2 + y_1^2 = x_2^2 + y_2^2 = x_3^2 + y_3^2.$$

From the circular power theorem,

$$\frac{|AG|}{|GP|} = \frac{|AG|^2}{|AG||GP|} = \frac{|AG|^2}{r^2 - |OG|^2}.$$

By the same token,

$$\frac{|BG|}{|GM|} = \frac{|BG|^2}{r^2 - |OG|^2}, \quad \frac{|CG|}{|GN|} = \frac{|CG|^2}{r^2 - |OG|^2}.$$

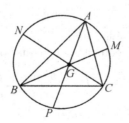

Figure 27.4   Figure for Exercise 8.

Also from

$$|AG|^2 + |BG|^2 + |CG|^2$$

$$= \sum_{i=1}^{3}\left[\left(x_i - \frac{x_1 + x_2 + x_3}{3}\right)^2 + \left(y_i - \frac{y_1 + y_2 + y_3}{3}\right)^2\right]$$

$$= x_1^2 + x_2^2 + x_3^2 + y_1^2 + y_2^2 + y_3^2 - \frac{2}{3}(x_1 + x_2 + x_3)\sum_{i=1}^{3} x_i$$

$$+ 3\left(\frac{x_1 + x_2 + x_3}{3}\right)^2 - \frac{2}{3}(y_1 + y_2 + y_3)\sum_{i=1}^{3} y_i$$

$$+ 3\left(\frac{y_1 + y_2 + y_3}{3}\right)^2$$

$$= (x_1^2 + y_1^2) + (x_2^2 + y_2^2) + (x_3^2 + y_3^2)$$

$$- 3\left(\frac{x_1 + x_2 + x_3}{3}\right)^2 - 3\left(\frac{y_1 + y_2 + y_3}{3}\right)^2$$

$$= 3r^2 - 3|OG|^2,$$

$$\frac{|AG|^2}{r^2 - |OG|^2} + \frac{|BG|^2}{r^2 - |OG|^2} + \frac{|CG|^2}{r^2 - |OG|^2} = 3.$$

In other words,

$$\frac{|AG|}{|GP|} + \frac{|BG|}{|GM|} + \frac{|CG|}{|GN|} = 3.$$

9. Establish a rectangular coordinates system with the straight line containing BC as its x-axis and the point A on its y-axis. Let $\angle ABC = 2\alpha, \angle ACB = 2\beta, x_B = -b$, and $x_C = c$. Since A is on the y-axis, so $b\tan 2\alpha = c\tan 2\beta$, from which the equation of BO_2 is $y = \tan\alpha(x+b)$ and the equation of CO_2 is $y = \cot\beta(x - c)$. The coordinates of their intersection point are $O_2\left(\frac{b\tan\alpha + c\cot\beta}{\cot\beta - \tan\alpha}, \frac{b+c}{\cot\beta - \tan\alpha} \cdot \frac{\tan\alpha}{\tan\beta}\right)$. Thus, we can write $F\left(\frac{b\tan\alpha + c\cot\beta}{\cot\beta - \tan\alpha}, 0\right)$. Also, since $FH \perp BO_2$, so $k_{HF} = -\frac{1}{k_{BO_2}} = -\cot\alpha$. Consequently, the equation of HF is

$$y = -\cot\alpha\left(x - \frac{b\tan\alpha + c\cot\beta}{\cot\beta - \tan\alpha}\right).$$

By the same token, the equation of EG is

$$y = \cot\beta\left(x - \frac{c\tan\beta + b\cot\alpha}{\tan\beta - \cot\alpha}\right).$$

Since

$$\frac{\cot\alpha \cdot \frac{b\tan\alpha + c\cot\beta}{\cot\beta - \tan\alpha}}{-\cot\beta \cdot \frac{c\tan\beta + b\cot\alpha}{\tan\beta - \cot\alpha}} = \frac{\tan 2\beta \tan\beta + \frac{\tan 2\alpha}{\tan\alpha}}{\tan 2\alpha \tan\alpha + \frac{\tan 2\beta}{\tan\beta}},$$

$$\tan\alpha \tan 2\alpha - \frac{\tan 2\alpha}{\tan\alpha} = \frac{\tan 2\alpha}{\tan\alpha}\left(-\frac{2\tan\alpha}{\tan 2\alpha}\right) = -2,$$

and

$$\tan\beta \tan 2\beta - \frac{\tan 2\beta}{\tan\beta} = -2,$$

we have

$$\cot\alpha \cdot \frac{b\tan\alpha + c\cot\beta}{\cot\beta - \tan\alpha} = -\cot\beta \cdot \frac{c\tan\beta + b\cot\alpha}{\tan\beta - \cot\alpha}.$$

Therefore, HF and EG both pass through the point $\left(0, \cot\alpha \cdot \frac{b\tan\alpha + c\cot\beta}{\cot\beta - \tan\alpha}\right)$.

10. With the straight line BC as the x-axis and the straight line containing the altitude DA as the y-axis, set up a rectangular coordinates system, as Figure 27.5 shows. Let the coordinates of A, B, and C be $(0, 2p), (2q, 0)$, and $(2r, 0)$, respectively, where $p > 0, q < 0$, and $r > 0$. Then the equation of the perpendicular bisector of the line segment BC is $x = q + r$ and the equation of the perpendicular bisector of the line segment CA is $y - p = \frac{r}{p}(x - r)$. By solving them we get the coordinates $\left(q + r, \frac{qr + p^2}{p}\right)$ of the circumcenter O. Also the equation of the altitude DA of the side BC is $x = 0$ and the equation of the altitude BE of the side AC is $y = \frac{r}{p}(x - 2q)$. So, solving them gets the coordinates

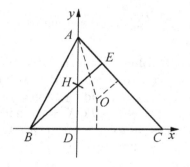

Figure 27.5 Figure for Exercise 10.

$\left(0, -\frac{2qr}{p}\right)$ of the orthocenter H. Hence, $9R^2 - d^2 = 9|OA|^2 - |OH|^2 = 9\left[(q+r)^2 + \left(\frac{qr+p^2}{p} - 2p\right)^2\right] - \left[(q+r)^2 + \left(\frac{qr+p^2}{p} + \frac{2qr}{p}\right)^2\right] = 8(p^2 + q^2 + r^2 - qr)$, and $a^2 + b^2 + c^2 + d^2 = |BC|^2 + |CA|^2 + |AB|^2 = (2q - 2r)^2 + [(2r)^2 + (2p)^2] + [(2q)^2 + (2p)^2] = 8(p^2 + q^2 + r^2 - qr)$. Therefore, $a^2 + b^2 + c^2 = R^2 - d^2$.

11. **Lemma.** Suppose that for an acute $\triangle A_1BC$ that is not an isosceles one, the altitude A_1F of the side BC is a fixed straight line. From F draw $FD_1 \perp A_1B$ at D_1 and $FE_1 \perp A_1C$ at E_1. Then all D_1E_1 pass through a fixed point. See Figure 27.6 for an illustration.

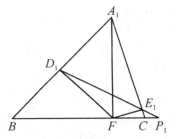

Figure 27.6   Figure 1 for Exercise 11.

Proof of Lemma. Since the four points A_1, D_1, F, and E_1 are on the same circle, P_1F is a tangent line to the circle, and $FE_1 \perp A_1C$, thus we have $\angle A_1CF = \angle A_1FE_1 = \angle A_1D_1E_1$. It follows that the four points B, C, E_1, and D_1 are on the same circle.

So, $|P_1F|^2 = |P_1E_1||P_1D_1| = |P_1C||P_1B| = (|P_1F| - |CF|)(|P_1F| + |FB|)$.

Hence, $|P_1F| = \frac{|BF||CF|}{|BF|-|CF|}$ is a fixed number, namely all D_1E_1 pass through a fixed point.

Proof of Original Problem. Set up a rectangular coordinates system as Figure 27.7 shows, and assume that the two tangent lines intersect the extended line of BC at B_1 and C_1, respectively. Let the equation of the circle O be $x^2 + y^2 = 1$, $\angle TOX = 2\alpha$, and $\angle SOX = 2\beta$. Then the equation of A_1C_1 is $\cos 2\alpha \cdot x + \sin 2\alpha \cdot y - 1 = 0$ and the equation of A_1B_1 is $\cos 2\beta \cdot x + \sin 2\beta \cdot y - 1 = 0$.

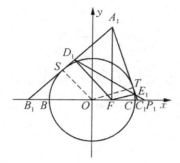

Figure 27.7   Figure 2 for Exercise 11.

So, $x_F = x_{A_1} = \frac{\cos(\alpha+\beta)}{\cos(\alpha-\beta)}$ is a fixed number, and $|A_1F| = y_{A_1} = \frac{\sin(\alpha+\beta)}{\sin(\alpha-\beta)}$.

From the lemma, we know that

$$|P_1F| = \frac{|B_1F||C_1F|}{|B_1F| - |C_1F|} = \frac{|A_1F|\tan 2\alpha \tan 2\beta}{\tan 2\alpha + \tan 2\beta}$$

$$= \frac{\sin 2\alpha \sin 2\beta}{2\cos(\alpha+\beta)\cos(\alpha-\beta)}$$

$$= \frac{\cos 2(\alpha-\beta) - \cos 2(\alpha+\beta)}{4\cos(\alpha+\beta)\cos(\alpha-\beta)}$$

$$= \frac{\cos^2(\alpha-\beta) - \cos^2(\alpha+\beta)}{2\cos(\alpha+\beta)\cos(\alpha-\beta)}$$

$$= \frac{\cos(\alpha-\beta)}{2\cos(\alpha+\beta)} - \frac{\cos(\alpha+\beta)}{2\cos(\alpha-\beta)}$$

$$= \frac{1}{2x_F} - \frac{x_F}{2}$$

is a fixed number, namely all D_1E_1 pass through a fixed point.

12. (1) As in Figure 27.8, set up a rectangular coordinates system with the center of the two circles as the origin and the point P on the negative half x-axis. Then the coordinates of the point P are $(-r, 0)$, and the equations of the two circles are $x^2 + y^2 = R^2$ and $x^2 + y^2 = r^2$, respectively. Let the coordinates of the three points A, B, and

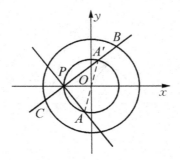

Figure 27.8   Figure for Exercise 12.

C be $A(x_1, y_1), B(x_2, y_2)$, and $C(x_3, y_3)$. Then $|BC|^2 + |CA|^2 + |AB|^2 = (x_2 - x_3)^2 + (y_2 - y_3)^2 + (x_3 - x_1)^2 + (y_3 - y_1)^2 + (x_1 - x_2)^2 + (y_1 - y_2)^2 = 4R^2 + 2r^2 - 2(x_1 x_2 + x_2 x_3 + x_3 x_1 + y_1 y_2 + y_2 y_3 + y_3 y_1)$. Since $PA \perp PB$, so $0 = \overrightarrow{PA} \cdot \overrightarrow{PB} = (x_1 + r)(x_2 + r) + y_1 y_2$, that is, $x_1 x_2 + y_1 y_2 = -r^2 - r(x_1 + x_2)$. Hence,

$$x_1 x_2 + x_2 x_3 + x_3 x_1 + y_1 y_2 + y_2 y_3 + y_3 y_1$$
$$= -r^2 - r(x_1 + x_2) + x_3(x_1 + x_2) + y_3(y_1 + y_2)$$
$$= -(x_1 + x_2)(r - x_3) + y_3(y_1 + y_2) - r^2. \tag{27.8}$$

Let the other intersection point of BC and the smaller circle be A'. By $PB \perp PA$ we get the coordinates $(-x_1, -y_1)$ of A'. Also from the property of concentric circles, we see that the line segments BC and $A'P$ have the same middle point. Thus, $x_2 + x_3 = -r - x_1$ and $y_2 + y_3 = -y_1$, namely

$$\begin{cases} x_1 + x_2 = -r - x_3, \\ y_1 + y_2 = -y_3. \end{cases} \tag{27.9}$$

Substituting the above into (27.8) gives that $x_1 x_2 + x_2 x_3 + x_3 x_1 + y_1 y_2 + y_2 y_3 + y_3 y_1 = -(-r - x_3)(r - x_3) + y_3(-y_3) - r^2 = r^2 - x_3^2 - y_3^2 - r^2 = -R^2$. Therefore,

$$|BC|^2 + |CA|^2 + |AB|^2 = 4R^2 + 2r^2 + 2R^2 = 6R^2 + 2r^2.$$

The set of all possible values is $\{6R^2 + 2r^2\}$.

(2) Let the coordinates of the middle point of AB be (x, y). From (27.9), $2x = x_1 + x_2 = -r - x_3$ and $2y = y_1 + y_2 = -y_3$, namely

$$\begin{cases} x_3 = -r - 2x, \\ y_3 = -2y. \end{cases}$$

Substituting them into the equation of the bigger circle, we obtain

$$(-r - 2x)^2 + (-2y)^2 = R^2,$$

that is, $\left(x + \frac{r}{2}\right)^2 + y^2 = \frac{R^2}{4}$. Therefore, the middle point trajectory of AB is the circle centered at $\left(-\frac{r}{2}, 0\right)$ of radius $\frac{R}{2}$.

Solutions for Chapter 28

Synthetic Problems for Complex Numbers

1. 0.   By the condition, $|z_1||z_3| = |z_2|^2$. Note that $|z_2| = 2|z_1|$. Then $|z_3| = 4|z_1|$, that is, $\sqrt{9a^2 + 16} = 4\sqrt{a^2 + 1}$. Solving it gives $a = 0$.

2. 12.   It is easy to see that $z_1 + z_2 + z_3 = 8 + 6i$, so

$$|z_1| + |z_2| + |z_3| \geq |z_1 + z_2 + z_3| = 10.$$

   If and only if $\frac{2-a}{1-b} = \frac{3+2a}{2+3b} = \frac{3-a}{3-2b} = \frac{8}{6}$, the expression $|z_1| + |z_2| + |z_3|$ achieves its minimum value, from which $a = \frac{7}{3}$ and $b = \frac{5}{4}$. Thus, $3a + 4b = 12$.

3. $160°$.   From $(z^3)^2 + z^3 + 1 = 0$, we get $z^3 = \omega$ or $\overline{\omega}$. Then z is a cubic root of ω or $\overline{\omega}$, and so $\arg z = k \cdot 120° \pm 40°$ $(k = 0, 1, 2)$. The result follows.

4. 2728 or 1999.   Since $z^3 = 27$, that is $(z - 3)(z^2 + 3z + 9) = 0$, if $z = 3$, then the original expression $= 2728$. If $z^2 + 3z + 9 = 0$, then the original expression $= z^3(z^2 + 3z) + 2242 = 27(-9) + 2242 = 1999$.

5. π.   Note that the opposite angles of the inscribed quadrilateral of a circle are supplementary to each other, and take care of the direction of the rotation angles. Then we can get the answer π.

6. -999.   The given condition implies $x_1 = \overline{x_2}$. Since $\frac{x_1^2}{x_2} \in \mathbf{R}$, we have $\frac{x_1^2}{x_2} = \frac{\overline{x_1}^2}{\overline{x_2}}$. So $x_1^3 = x_2^3$, from which $\left(\frac{x_1}{x_2}\right)^3 = 1$. Hence, $\frac{x_1}{x_2} = \omega$ or ω^2. Due to the fact that

$$\omega^{2^{2k}} + \omega^{2^{2k+1}} = \omega + \omega^2 = -1,$$

   the original expression $= -999$.

7. -2.   The original expression equals

$$\frac{\varepsilon^4}{1+\varepsilon} + \frac{\varepsilon}{1+\varepsilon^2} + \frac{\varepsilon^5}{1+\varepsilon^3} = \frac{2(1+\varepsilon+\varepsilon^2+\varepsilon^4+\varepsilon^5+\varepsilon^6)}{1+\varepsilon+\varepsilon^2+\varepsilon^3+\varepsilon^5+\varepsilon^4+\varepsilon^6} = -2.$$

8. $\left[\frac{\sqrt{2}}{2}, \frac{\sqrt{5}}{2}\right)$.   The condition implies that $z_2 = -i, A + C = 120°$, and $p + z_2 = \cos A + i \cos C$. So,

$$|p + z_2|^2 = \cos^2 A + \cos^2 C = 1 + \frac{1}{2}(\cos 2A + \cos 2C)$$

$$= 1 - \frac{1}{2}\cos(A - C).$$

Since $A - C = 120° - 2C$ and $0° < C < 120°$, we have $-120° < A - C < 120°$, namely $-\frac{1}{2} < \cos(A - C) \le 1$. In other words, $\frac{1}{2} \le |p + z_2|^2 < \frac{5}{4}$, from which $\frac{\sqrt{2}}{2} \le |p + z_2| < \frac{\sqrt{5}}{2}$.

9. (1) Since $a_0 = \alpha + \beta, a_1 = \alpha x_1 + \beta x_2$, and $a_2 = \alpha x_1^2 + \beta x_2^2$, so

$$a_0 x_1 - a_1 = \beta(x_1 - x_2), \quad a_1 x_1 - a_2 = \beta x_2(x_1 - x_2).$$

Then $a_1 x_1 - a_2 = x_2(a_0 x_1 - a_1)$, namely $a_2 - (x_1 + x_2)a_1 + a_0 = 0$. Since a_0, a_1, and a_2 are nonzero rational numbers, it follows that $x_1 + x_2$ is a rational number.

(2) Since x_1 and x_2 are not real numbers and $x_1 + x_2$ is a rational number, we may assume that $x_1 = b_1 + ci$ and $x_2 = b_2 - ci$, where b_1, b_2, and c are real numbers and $c \neq 0$. From

$$a_1 = \alpha(b_1 + ci) + \beta(b_2 - ci) = (\alpha b_1 + \beta b_2) + (\alpha - \beta)ci,$$

and the condition that α and β are real numbers, $(\alpha - \beta)c = 0$, from which $\alpha = \beta$.

10. Assume that the equation has an imaginary root $p + qi$ $(p, q \in \mathbf{R}$ and $q \neq 0)$. Then it has the imaginary root $p - qi$. Put them into the equation respectively to get two equalities. Subtracting them gives that

$$2qi\left[\frac{1}{(p + a_1)^2 + q^2} + \frac{1}{(p + a_2)^2 + q^2} + \cdots + \frac{1}{(p + a_n)^2 + q^2}\right] = 0,$$

but this equality cannot be satisfied.

11. Since α is a root of the equation, we know that

$$|\alpha|^n + |\alpha|^{n-1} + \cdots + |\alpha| + 1$$

$$\ge |\alpha^n \cos\theta_n + \alpha^{n-1}\cos\theta_{n-1} + \cdots + \alpha\cos\theta_1 + \cos\theta_0|$$

$$= 2,$$

that is,

$$|\alpha|^n + |\alpha|^{n-1} + \cdots + |\alpha| \geq 1. \tag{28.1}$$

If $|\alpha| \leq \frac{1}{2}$, then $|\alpha|^n + |\alpha|^{n-1} + \cdots + |\alpha| < 1$, a contradiction to (28.1).

12. Denote $f_\alpha(z) = (z + \alpha)^2 + \alpha\overline{z}$. Then

$$f_\alpha(z_1) - f_\alpha(z_2) = (z_1 + \alpha)^2 + \alpha\overline{z_1} - (z_2 + \alpha)^2 - \alpha\overline{z_2}$$

$$= (z_1 + z_2 + 2\alpha)(z_1 - z_2) + \alpha(\overline{z_1} - \overline{z_2}). \tag{28.2}$$

If there exist complex numbers z_1 and z_2 ($|z_1|, |z_2| < 1$, and $z_1 \neq z_2$) such that $f_\alpha(z_1) = f_\alpha(z_2)$, then from (28.2) we know that

$$|\alpha(\overline{z_1} - \overline{z_2})| = |-(z_1 + z_2 + 2\alpha)(z_1 - z_2)|.$$

Using $|\overline{z_1} - \overline{z_2}| = |\overline{z_1 - z_2}| = |z_1 - z_2| \neq 0$, we obtain

$$|\alpha| = |z_1 + z_2 + 2\alpha| \geq 2|\alpha| - |z_1| - |z_2| > 2|\alpha| - 2.$$

In other words, $|\alpha| < 2$.

On the other hand, for any complex number α satisfying $|\alpha| < 2$, let $z_1 = -\frac{\alpha}{2} + \beta i$ and $z_2 = -\frac{\alpha}{2} - \beta i$, where $0 < \beta < 1 - \frac{|\alpha|}{2}$. Then $z_1 \neq z_2$ and $\left|-\frac{\alpha}{2} \pm \beta i\right| \leq \left|-\frac{\alpha}{2}\right| + |\beta| < 1$, so $|z_1| < 1$ and $|z_2| < 1$. Now, substituting

$$z_1 + z_2 = -\alpha, \; z_1 - z_2 = 2\beta i, \; \overline{z_1} - \overline{z_2} = \overline{2\beta i} = -2\beta i$$

into (28.2), we obtain that $f_\alpha(z_1) - f_\alpha(z_2) = 2\alpha\beta i - 2\alpha\beta i = 0$, namely $f_\alpha(z_1) = f_\alpha(z_2)$.

In summary, the set of α that satisfy the requirement is

$$\{\alpha \mid \alpha \text{ is a complex number such that } |\alpha| \geq 2\}.$$

Solutions for Chapter 29

Mathematical Induction (II)

1. Since $a_n = \frac{1}{27}(a_{n+3} - a_{n+2} + 3a_{n+1})$, by inverse induction we can show that $a_n \in \mathbf{Q}$ for any $n \in \mathbf{N}$. In particular, $a, b, c \in \mathbf{Q}$.

2. We shall prove that there exist infinitely many $n \in \mathbf{N}_+$ $(n > 1)$ such that $n \mid 2^n + 2$ and $n - 1 \mid 2^n + 1$.

   Note that $n = 2$ has the above property. Now, assume that $n \geq 2$ possesses the above property. Let $m = 2^n + 2$. We show that m also has the same property. In fact, since $n - 1 \mid 2^n + 1$ and $2^n + 1$ is an odd number, we may assume that $2^n + 1 = (n - 1)q$ with q an odd number. Then $2^{m-1} + 1 = 2^{2^n+1} + 1 = (2^{n-1})^q + 1 = (2^{n-1} + 1)[(2^{n-1})^{q-1} - (2^{n-1})^{q-2} + \cdots + 1]$. Thus, $2^{n-1} + 1 \mid 2^{m-1} + 1$, and so $2^n + 2 \mid 2^m + 2$, that is, $m \mid 2^m + 2$. On the other hand, from $n - 1 \mid 2^n + 1$, we know that $n - 1$ is an odd number, from which n is an even number. Hence, from $n \mid 2^n + 2$, we may assume that $2^n + 2 = np$ with p an odd number, and $2^m + 1 = (2^n)^p + 1 = (2^n + 1)[(2^n)^{p-1} - (2^n)^{p-2} + \cdots + 1]$. In other words, $2^n + 1 \mid 2^m + 1$, namely $m - 1 \mid 2^m + 1$.

3. Choose $n_1 = 1$. Then $n_1 \mid 2^{n_1} + 1$. Now, assume that there exists n_k such that $n_k \mid 2^{n_k} + 1$. Let $n_{k+1} = 2^{n_k} + 1$. Then n_{k+1} is an odd number and $n_{k+1} = n_k q$, where q is an odd number. Thus, $2^{n_{k+1}} + 1 = (2^{n_k})^q + 1 = (2^{n_k} + 1)[(2^{n_k})^{q-1} - \cdots + 1]$. This indicates that $2^{n_k} + 1 \mid 2^{n_{k+1}} + 1$, that is, $n_{k+1} \mid 2^{n_{k+1}} + 1$. Also $n_k < n_{k+1}$ from the definition of n_k. In this way, we have defined an infinite sequence $n_1, n_2, \ldots, n_k, \ldots$ that satisfies the requirement.

4. Consider the two propositions

$$P(n) : S_{2n-1} = \frac{1}{2}n(4n^2 - 3n + 1),$$

$$Q(n) : S_{2n} = \frac{1}{2}n(4n^2 + 3n + 1).$$

When $n = 1$, $a_1 = 1$, we have $\frac{1}{2} \cdot 1 \cdot (4 \cdot 1^3 - 3 \cdot 1 + 1) = 1$, so $P(1)$ is satisfied.

Suppose that $P(k)$ is satisfied for $n = k$, namely, $S_{2k-1} = \frac{1}{2}k(4k^2 - 3k + 1)$. Then

$$S_{2k} = \frac{1}{2}k(4k^2 - 3k + 1) + 3k = \frac{1}{2}k(4k^2 + 3k + 1),$$

in other words, $Q(k)$ is satisfied. If $Q(k)$ is true, namely $S_{2k} = \frac{1}{2}k(4k^2 + 3k + 1)$, then

$$S_{2k+1} = \frac{1}{2}k(4k^2 + 3k + 1) + 3k(k + 1) + 1$$

$$= \frac{1}{2}(k + 1)\left[(4(k + 1)^2 - 3(k + 1) + 1\right],$$

namely $P(k + 1)$ is valid. Hence, $P(n)$ and $Q(n)$ are satisfied for all positive integers n.

5. First, we show that $f(1) = 1$ and $f(3) = 3$. From $0 < f(1) < f(2) = 2$, we see that $f(1) = 1$. Since

$$f(3)f(7) = f(21) < f(22) = f(2)f(11) = 2f(11)$$

$$< 2f(14) = 2f(2)f(7) = 4f(7),$$

so $f(3) < 4$ (but $f(3) > f(2) = 2$). Hence, $f(3) = 3$.

Then we can prove that $f(2^m + 1) = 2^m + 1$ for all positive integers m.

6. First $\cos\theta = \sqrt{1 - \sin^2\theta} = \frac{a^2 - b^2}{a^2 + b^2}$. Let $B_n = (a^2 + b^2)^n \cos n\theta$. Then $A_1 = 2ab$ and $B_1 = a^2 - b^2$ are both integers. Assume that A_k and B_k are both integers. Since $A_{k+1} = A_k B_1 + B_k A_1$ and $B_{k+1} = B_k B_1 + A_k A_1$, we see that A_{k+1} and B_{k+1} are both integers.

7. Let $f(m, n) = \frac{(2m)!(2n)!}{m!n!(m+n)!}$. Then $f(0, n) = \frac{(2n)!}{n!n!} = C_{2n}^n$ is an integer. Suppose that when $m = k$, no matter what nonnegative integer n is, $f(k, n)$ is always an integer. Since

$$f(k + 1, n) = 4f(k, n) - f(k, n + 1),$$

$f(k + 1, n)$ is also an integer.

8. After calculating the first several terms, we conjecture: $a_{5n+1} = 5n+1$, $a_{5n+2} = 5n+4, a_{5n+3} = 5n+2, a_{5n+4} = 5n+5$, and $a_{5n+5} = 5n+3$. We prove the above conjecture by mathematical induction. When $n = 0$, the conjecture is true by the recursive formula. Suppose that the above conjecture is true for all $n < m$. Consider the case $n = m$. By the inductive hypothesis, $(a_1, a_2, \ldots, a_{5m})$ is a permutation of $1, 2, \ldots, 5m$, so $a_{5m} - 2 = 5m - 4 \in \{a_1, a_2, \ldots, a_{5m}\}$. Hence, $a_{5m+1} = a_{5m} + 3 = 5m + 1$. Similarly, $a_{5m+2} = 5m + 4$, while $a_{5m+2} - 2 = 5m + 2 \notin \{a_1, \ldots, a_{5m+2}\}$, so $a_{5m+3} = 5m + 2$. Repeating the above discussion, we know that $a_{5m+4} = a_{5m+3} + 3 = 5m + 5$ and $a_{5m+5} = a_{5m+4} - 2 = 5m+3$. Thus, our conjecture is true. In summary, every positive integer greater than 1 appears exactly one time in the sequence $a_1, a_2, \ldots$. Hence, by combining the fact that each square k^2 is congruent to $0, 1$, or 4 with respect to the module 5, we know that for each of the three cases, there exists a corresponding term in the sequence $\{a_n\}$ with subscript $\equiv 4, 1$, or 2 (mod 5) that is equal to k^2. This means that there exists $n \in \mathbf{N}_+$ such that $a_n = a_{n-1} + 3 = k^2$.

9. We use mathematical induction with respect to n.

When $n = 3$, the proposition is obviously true.

When $n = 4$, for the square $ABCD$, there must be two vertices with the same color, say the red color for A and C, green color for B, and blue color for D. Then partitioning the square $ABCD$ with a diagonal line satisfies the requirement.

Assume that the proposition is true for $3, 4, \ldots, n - 1$.

Consider the case of n.

(1) When one color appears just one time, say the color of the vertex A is different from the colors of other $n - 1$ vertices, let B be a vertex that is not beside A. Connect the points A and B to form the line segment AB. Then the diagonal AB divides the n-polygon into two polygons Γ_1 and Γ_2 with the number of vertices less than n each. By the inductive hypothesis, the polygons Γ_1 and Γ_2 can be partitioned to satisfy the requirement, and hence there is a partition of the n-polygon that satisfies the property.

(2) When each color appears at least two times, denote $d(X, Y)$ the minimal number of the vertices passed from vertex X to vertex Y, and $S(X, Y)$ denotes a corresponding path.

Let $d_1(R_1, R_2) = \min\{d(X, Y) \mid X \neq Y \text{ have the same color}\}$, in which R_1 and R_2 are assumed to be both red color, without loss of generality.

Since the same color vertices are not adjacent to each other, so there is at least one vertex P of non-red color in $S(R_1, R_2)$, say the color of P is green.

Then the number of blue vertices in $S(R_1, R_2)$ cannot be more than 1. Otherwise, it will contradict the minimality of $d(R_1, R_2)$. Thus, there is at least one blue vertex Q that is not in $S(R_1, R_2)$.

Connect P and Q. Then PQ divides the n-polygon into two polygons Γ_1 and Γ_2. The three colors all appear in Γ_i ($i = 1$ and 2). By the inductive hypothesis, Γ_i can be partitioned into triangles with vertices of different colors. Hence, a partition of the n-polygon is obtained that satisfies the condition.

This completes the induction proof.

10. Define a sequence $\{y_n\}$ as follows: $y_1 = y_2 = 1$, and $y_{n+2} = [1 + 2\rho(n + 1)]y_{n+1} - y_n$ for $n = 1, 2, \ldots$.

We establish the following conclusions in succession.

(i) For any $n \in \mathbf{N}_+$, it always holds that $x_n = \frac{y_n}{y_{n+1}}$.

We prove it by induction with respect to n. The induction transition can be done in the following way:

$$\frac{1}{x_{n+1}} = 1 + 2\rho(n + 1) - x_n = 1 + 2\rho(n + 1) - \frac{y_n}{y_{n+1}}$$

$$= \frac{1}{y_{n+1}}\{[1 + 2\rho(n + 1)]y_{n+1} - y_n\} = \frac{y_{n+2}}{y_{n+1}},$$

so $x_{n+1} = \frac{y_{n+1}}{y_{n+2}}$.

(ii) For any $n \in \mathbf{N}_+$, it is true that $y_{2n+1} = y_{n+1} + y_n$ and $y_{2n} = y_n$.

We prove it by induction with respect to n. If (ii) is true for n, then

$$y_{2n+2} = [1 + 2\rho(2n + 1)]y_{2n+1} - y_{2n} = y_{2n+1} - y_{2n} = y_{n+1},$$

$$y_{2n+3} = [1 + 2\rho(2n + 2)]y_{2n+2} - y_{2n+1}$$

$$= \{1 + 2[1 + \rho(n + 1)]\}y_{2n+2} - y_{2n+1}$$

$$= 2y_{n+1} + [1 + \rho(n + 1)]y_{n+1} - (y_n + y_{n+1})$$

$$= y_{n+1} + [1 + \rho(n + 1)]y_{n+1} - y_n = y_{n+1} + y_{n+2}.$$

This and the initial step guarantee the validity of (ii).

(iii) In the following, we show that for any $p, q \in \mathbf{N}_+$ such that the greatest common divisor of p and q is 1, the number pair (p, q) appears exactly one time in the set $\{(y_n, y_{n+1}) \mid n = 1, 2, \ldots\}$.

We prove it by induction with respect to $p + q$. When $p + q = 2$, we have $p = q = 1$. Then $(p, q) = (y_1, y_2)$. By conclusion (ii), when $n \geq 2$, at least one of y_n and y_{n+1} is greater than 1, from which $(y_n, y_{n+1}) \neq (y_1, y_2)$. Thus, conclusion (iii) is true for $p + q = 2$.

Now, assume that the conclusion is satisfied for all positive integer pairs (p, q) such that $p + q < m$ ($m \geq 3$ and $m \in \mathbf{N}_+$), and the greatest common divisor of p and q is 1. Consider the case that $p + q = m$. Here $p \neq q$.

(I) When $p < q$, since the greatest common divisor of p and q is 1, the greatest common divisor of p and $q - p$ is 1, too. Also $(q - p) + p = q < m$, hence by the inductive hypothesis, there exists a unique $n \in \mathbf{N}_+$ such that $(p, q - p) = (y_n, y_{n+1})$. Then $(p, q) = (y_n, y_n + y_{n+1}) = (y_{2n}, y_{2n+1})$ (using (ii)).

On the other hand, if there exist $k < l$ with $k, l \in \mathbf{N}_+$ such that

$$(p, q) = (y_k, y_{k+1}) = (y_l, y_{l+1}),$$

then $y_k = y_l$ and $y_{k+1} = y_{l+1}$. If k and l are even numbers, then from (ii), $(p, q - p)$ has two different representations, contradictory to the inductive hypothesis. But if k is an odd number, then $y > y_{k+1}$, a contradiction to $p < q$. Thus, k can only be an even number. By the same token, l can only be an even number. Hence, there is only one $n \in \mathbf{N}_+$ such that

$$(p, q) = (y_n, y_{n+1}).$$

(II) When $p > q$, the discussion is similar to (I).

To summarize, the original proposition is proved to be true.

11. (1) One of the integers m and n equals 3 and the other is an even number; or (2) m and n are both greater than or equal to 4, and one of them is a multiple of 3.

First, we show the necessity of the above conditions. In fact, obviously $3 \mid mn$. When one of m and n is 3, say $m = 3$, if n is an odd number, according to the discussion about the coverage of the left upper corner by the "L"-shapes (the notation "L" here means the union of three small squares, forming a shape of L), one can only use the method shown in Figure 29.1 for the coverage (otherwise the left two checks of the third line cannot be covered). At this time the second "L" must be as drawn in Figure 29.1. Remove the first two columns and continue the discussion. When n is an odd number, the most right column cannot be divided into the "L"-shapes. Hence, the conditions in the answer are necessary.

We further show the sufficiency of the conditions in the answer. Note that two "L"-shapes can form a 2×3 rectangle, So, when one of m and n is a multiple of 3 and the other is an even number, one can obtain completely "L"-shapes. By the symmetry of m and n, in order to prove the sufficiency, it is enough to consider the case of $3 \mid n, n \geq 6, m \geq 5$, and m is an odd number, for which we show that the table can be completely partitioned by all "L"-shapes.

Using mathematical induction, it is sufficient to provide 5×6 or 5×9 tables with a complete partition by "L"-shapes (see Figures 29.2 and 29.3). For other $m \times n$ tables, after obtaining a 5×6 or 5×9 table from the left upper corner via a partition of the table, the remaining part can be covered by 2×3 rectangles that can be covered by "L"-shapes.

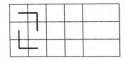

Figure 29.1   Figure 1 for Exercise 11.

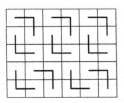

Figure 29.2   Figure 2 for Exercise 11.

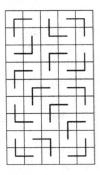

Figure 29.3   Figure 3 for Exercise 11.

12. Since $f(1,n) = 1 = C_{n+1-2}^0$ and $f(m,1) = 1 = C_{m+1-2}^{m+1}$ for all positive integers n and m, so $P(m,1)$ and $P(1,m)$ are true.

    Suppose that $P(m+1,n)$ and $P(m,n+1)$ are true, that is,

    $$f(m+1,n) \le C_{m+n-1}^m, \quad f(m,n+1) \le C_{m+n-1}^{m-1}.$$

    Then $f(m+1,n+1) \le f(m+1,n) + f(m,n+1) \le C_{m+n-1}^m + C_{m+n-1}^{m-1} = C_{m+n}^m$. In other words, the proposition $P(m+1,n+1)$ is satisfied. By the method of double induction, the original proposition is valid.

Solutions for Chapter 30

Proof by Contradiction

1. Use proof by contradiction. If there exists $n \in \mathbf{N}_+$ such that $2^m - 1 \mid 2^n + 1$, then $n > m$ (note that here we used $m > 2$).

   Since $2^n + 1 = 2^{n-m}(2^m - 1) + 2^{n-m} + 1$, from $2^m - 1 \mid 2^n + 1$, we see that $2^m - 1 \mid 2^{n-m} + 1$. Substituting $n - m$ for the above n to continue the discussion, we are lead to the conclusion that n is larger than any positive multiple of m, which is a contradiction.

2. Suppose that n is a composite number. Let $n = pq$ with $p \leq q$ and p a prime number.

   If $p < q$, then $n \mid q!$ while $n \mid n!$. This means that $q!$ and $n!$ have the same remainder when divided by n, a contradiction to the condition.

   If $p = q$ and $p > 2$, then $n \mid (2p)!$ and $2p < n$. This indicates that $(2p)!$ and $n!$ have the same remainder when divided by n, another contradiction to the condition. If $p = q$ and $p = 2$, then $n = 4$. Now, 2! and 3! have the same remainder when divided by 4, also a contradiction to the condition.

3. Use proof by contradiction. We may assume that there are at most two negative numbers among $a_1, a_2, \ldots, a_{12}$. Then there exists $1 \leq k \leq 9$ such that a_k, a_{k+1}, a_{k+2}, and a_{k+3} are all nonnegative numbers. By the given condition,

$$\begin{cases} a_{k+1}(a_k - a_{k+1} + a_{k+2}) < 0, \\ a_{k+2}(a_{k+1} - a_{k+2} + a_{k+3}) < 0. \end{cases}$$

Since $a_{k+1} \geq 0$ and $a_{k+2} \geq 0$, the above inequalities ensure that $a_{k+1} > 0$ and $a_{k+2} > 0$. Hence,

$$\begin{cases} a_k - a_{k+1} + a_{k+2} < 0, \\ a_{k+1} - a_{k+2} + a_{k+3} < 0. \end{cases}$$

Adding the two inequalities gives $a_k + a_{k+3} < 0$, which contradicts $a_k \geq 0$ and $a_{k+1} \geq 0$. Therefore, there are at least 3 positive numbers and 3 negative numbers among $a_1, a_2, \ldots, a_{12}$.

4. Without loss of generality, assume that $a_1 \leq a_2 \leq \cdots \leq a_n$. Suppose that for all $k \in \{1, 2, \ldots, n\}$, there hold that

$$\sum_{i=1}^{n} |a_i - a_k| > \sum_{i=1}^{n} |b_i - a_k|. \tag{30.1}$$

In particular, (30.1) is satisfied for $k = 1$ and n. So,

$$\sum_{i=1}^{n} |a_i - a_1| + \sum_{i=1}^{n} |a_i - a_n| > \sum_{i=1}^{n} (|b_i - a_1| + |b_i - a_n|) \tag{30.2}$$

Note that the left hand side of (30.2) equals

$$\sum_{i=1}^{n} (a_i - a_1) + \sum_{i=1}^{n} (a_n - a_i) = n(a_n - a_1)$$

and the right hand side of (30.2) is greater than or equal to

$$\sum_{i=1}^{n} |(b_i - a_1) - (b_i - a_n)| = n|a_n - a_1| = n(a_n - a_1).$$

This is a contradiction. Hence, the original proposition is valid.

5. Note that all prime numbers that are greater than 3 and less than 41 have the following representations:

$$3^2 - 2^2 = 5, \quad 3^2 - 2 = 7, \quad 3^3 - 2^4 = 11, \quad 2^8 - 3^5 = 13,$$
$$3^4 - 2^5 = 17, \quad 3^3 - 2^2 = 23, \quad 2^5 - 3 = 29, \quad 2^5 - 1 = 31,$$
$$2^6 - 3^3 = 37.$$

Below we prove: There do not exist $a, b \in \mathbf{N}_+$ such that $|3^a - 2^b| = 41$.

Case 1. Suppose that $3^a - 2^b = 41$. Taking module 3 both sides (note that $a > 1$) gives $(-1)^b \equiv 1 \pmod 3$, so b is an even number, and clearly $b > 0$. Taking module 4 both sides gives $(-1)^a \equiv 1 \pmod 4$, so a is also an even number. Let $a = 2m$ and $b = 2n$. Then

$$(3^m - 2^n)(3^m + 2^n) = 41,$$

from which $3^m - 2^n = 1$. Thus, $3^m + 2^n = 41$. Adding the two equalities, we obtain $3^m = 21$, a contradiction.

Case 2. If $2^b - 3^a = 41$, then $b > 3$. Taking module 8 both sides gives $3^b \equiv -1 \equiv 7 \pmod 8$. But $3^b \equiv 1$ or $3 \pmod 8$, a contradiction.

Therefore, the desired prime number is 41.

6. Since $C_4^3 = 4$, so there are four different triangles obtained via picking up three numbers from x_1, x_2, x_3, and x_4 as the side lengths. In other words, any three numbers as side lengths can make different triangles. In the following, we use proof by contradiction to show the proposition.

Suppose that there exist three numbers among x_1, x_2, x_3, and x_4 that cannot construct a triangle with them as the side lengths. By symmetry, we may assume that the three numbers are x_1, x_2, and x_3 that satisfy $x_1 \geq x_2 + x_3$. First note that

$$(x_1 + x_2 + x_3 + x_4)\left(\frac{1}{x_1} + \frac{1}{x_2} + \frac{1}{x_3} + \frac{1}{x_4}\right)$$

$$= 4 + \frac{x_1}{x_2} + \frac{x_1}{x_3} + \frac{x_2}{x_1} + \frac{x_3}{x_1} + \left[\left(\frac{x_1}{x_4} + \frac{x_4}{x_1}\right) + \left(\frac{x_2}{x_3} + \frac{x_3}{x_2}\right)\right.$$

$$\left. + \left(\frac{x_2}{x_4} + \frac{x_4}{x_2}\right) + \left(\frac{x_3}{x_4} + \frac{x_4}{x_3}\right)\right]$$

$$\geq 4 + x_1\left(\frac{1}{x_2} + \frac{1}{x_3}\right) + \frac{x_2 + x_3}{x_1} + 2 \cdot 4$$

$$= 12 + x_1\left(\frac{1}{x_2} + \frac{1}{x_3}\right) + \frac{x_2 + x_3}{x_1}.$$

From $(x_2 + x_3)\left(\frac{1}{x_2} + \frac{1}{x_3}\right) \geq 4$, we get

$$\frac{1}{x_2} + \frac{1}{x_3} \geq \frac{4}{x_2 + x_3}.$$

Denote $\frac{x_1}{x_2 + x_3} = t \ (t \geq 1)$. Then

$$(x_1 + x_2 + x_3 + x_4)\left(\frac{1}{x_1} + \frac{1}{x_2} + \frac{1}{x_3} + \frac{1}{x_4}\right)$$

$$\geq 12 + \frac{4x_1}{x_2 + x_3} + \frac{x_2 + x_3}{x_1}$$

$$= 12 + 4t + \frac{1}{t}.$$

By the condition, $12 + 4t + \frac{1}{t} < 17$, that is $4t + \frac{1}{t} - 5 < 0 \ (t \geq 1)$.

But in fact, when $t \geq 1$,

$$4t + \frac{1}{t} - 5 = \frac{4t^2 - 5t + 1}{t} = \frac{(4t-1)(t-1)}{t} \geq 0.$$

Thus, we reach a contradiction. Hence, the original proposition is true.

7. Since there are only finitely many different scores, there must be one case that has the maximal square sum of players' scores. Then the scores $p_1, p_2, \ldots, p_{10}$ of all players must be different, since if $p_i = p_j$, then changing the win-lose result of the two players, namely replacing p_i and p_j with $p_i - 1$ and $p_j + 1$ or $p_i + 1$ and $p_j - 1$ respectively, thus from

$$(p_i - 1)^2 + (p_j + 1)^2 - (p_i^2 + p_j^2) = 2 > 0$$

and the fact that other players' scores remain the same, the square sum will be strictly increased, which contradicts the maximality of the square sum.

Therefore, when $p_i = i - 1$ ($i = 1, 2, \ldots, 10$), the value of $\sum_{i=1}^{10} p_i^2$ is maximal, the maximum value being

$$\sum_{i=0}^{9} i^2 = 285.$$

So, the square sum of the scores of all participants does not exceed 285.

8. Use proof by contradiction. Assume that these n points are not all on the same straight line. Then there must be points among the n points that are off the straight line determined by any two of the n points. For each of such straight lines, calculate the distances of all points off the line to the line, and since there are only finitely many such distances, there must be a minimal one, denoted as d_0.

Suppose that d_0 is the distance of the point A to the straight line determined by points B and C. Draw $AH \perp BC$ with H the foot of the perpendicular. Then $d_0 = |AH|$ (see Figure 30.1). By the assumption, there is at least another point E among the n points on the straight line BC. Clearly, there are at least two points among the three points B, C, and E that are on the same side of H, say C and E are on the same side of H and $|HE| < |HC|$ (the point E may coincide with the point H).

Draw $EF \perp AC$ with F the foot of the perpendicular, and denote $d_1 = |EF|$. From the minimality of d_0, we know that $d_0 \leq d_1$. Since

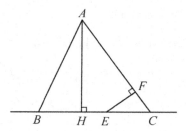

Figure 30.1 Figure for Exercise 8.

the right $\triangle CEF \sim$ the right $\triangle CAH$, so

$$\frac{d_1}{d_0} = \frac{|EF|}{|AH|} = \frac{|CE|}{|AC|} = \frac{|CP|}{|AC|} < 1,$$

that is, $d_1 < d_0$, which is a contradiction. Therefore, all the n points are on the same straight line.

9. Let the unordered pair $\{a_j, b_j\}$ represent the players of the jth competition, and denote

$$S = \{\{a_j, b_j\} \mid j = 1, 2, \ldots, 14\}.$$

Let M be a nonempty subset of S such that all players who appear in the pairs of M are different. Obviously, there are only finitely many such subsets, and assume that the subset with the most number of pairs is M_0. Let $|M_0| = r$. Clearly, it is enough to show that $r \geq 6$.

Suppose that $r \leq 5$. Since M_0 is the subset of S with most number of different players, those $20 - 2r$ players who do not appear in M_0 do not play each other; otherwise it would contradict the definition of M_0. This means that the competitions that these $20 - 2r$ players participate in must be performed with those $2r$ players in M_0. Since each player participates in at least one competition, so besides the r competitions from M_0, there are at least $20 - 2r$ competitions. In other words, the total number of competitions is at least

$$r + (20 - 2r) = 20 - r \geq 15,$$

which is a contradiction to the assumption that the total number of competitions is 14. This proves that $r \geq 6$.

10. Let $f(x) = ax^2 + bx + c$. Then

$$\begin{cases} f(1) = a + b + c, \\ f(-1) = a - b + c, \\ f(0) = c \end{cases} \Rightarrow \begin{cases} a = \dfrac{f(1) + f(-1) - 2f(0)}{2}, \\ b = \dfrac{f(1) - f(-1)}{2}, \\ c = f(0), \end{cases}$$

which implies that

$$f(x) = \frac{x^2 + x}{2} f(1) + \frac{x^2 - x}{2} f(-1) + (1 - x^2)f(0).$$

Let m be a real root of the equation $f(x) = 0$. Clearly, $m < 0$.
Suppose that $a, b,$ and c are all greater than $\frac{1}{4} f(1)$, namely

$$\begin{cases} \dfrac{f(1) + f(-1) - 2f(0)}{2} > \dfrac{1}{4} f(1), \\ \dfrac{f(1) - f(-1)}{2} > \dfrac{1}{4} f(1), \\ f(0) > \dfrac{1}{4} f(1). \end{cases}$$

Then

$$\begin{cases} 2f(0) - \dfrac{1}{2} f(1) < f(-1) < \dfrac{1}{2} f(1), \\ f(0) > \dfrac{1}{4} f(1) > 0. \end{cases}$$

It follows that

$$\begin{aligned}
f(m) &= \frac{m^2 + m}{2} f(1) + \frac{m^2 - m}{2} f(-1) + (1 - m^2)f(0) \\
&> \frac{m^2 + m}{2} f(1) + \left[2f(0) - \frac{1}{2} f(1) \right] \frac{m^2 - m}{2} + (1 - m^2)f(0) \\
&= \frac{m^2 + 3m}{4} f(1) + (1 - m)f(0) \\
&> \frac{m^2 + 3m}{4} f(1) + \frac{1 - m}{4} f(1) \\
&= \frac{(m + 1)^2}{4} f(1).
\end{aligned}$$

This contradicts the equality $f(m) = 0$.

Hence, not all a, b, and c are greater than $\frac{1}{4}f(1)$, that is, $\min\{a, b, c\} \leq \frac{1}{4}(a + b + c)$.

11. When $x < \frac{1}{2}$, we have $f(x) - x = \frac{1}{2} > 0$; when $x \geq \frac{1}{2}$,

$$f(x) - x = x^2 - x < 0.$$

Divide the interval $(0, 1)$ into two subintervals

$$I_1 = \left(0, \frac{1}{2}\right), \quad I_2 = \left[\frac{1}{2}, 1\right).$$

Then $(a_n - a_{n-1})(b_n - b_{n-1}) = [f(a_{n-1}) - a_{n-1}][f(b_{n-1}) - b_{n-1}] < 0$ if and only if a_{n-1} and b_{n-1} are in the two different subintervals.

Suppose that a_k and b_k are in the same subintervals for all $k = 1, 2, \ldots$.

Let $d_k = |a_k - b_k|$. If $a_k, b_k \in I_1$, then

$$d_{k+1} = |a_{k+1} - b_{k+1}| = \left|a_k + \frac{1}{2} - \left(b_k + \frac{1}{2}\right)\right| = d_k,$$

and if $a_k, b_k \in I_2$, then $\min\{a_k, b_k\} \geq \frac{1}{2}$ and

$$\max\{a_k, b_k\} = \min\{a_k, b_k\} + d_k \geq \frac{1}{2} + d_k.$$

This means that

$$d_{k+1} = |a_{k+1} - b_{k+1}| = |a_k^2 - b_k^2| = |(a_k - b_k)(a_k + b_k)|$$

$$\geq |a_k - b_k| \left(\frac{1}{2} + \frac{1}{2} + d_k\right) = d_k(1 + d_k) \geq d_k.$$

Hence, the sequence $\{d_k\}$ is monotonically increasing. In particular, $d_k \geq d_0 > 0$ for any positive integer k.

Furthermore, if $a_k, b_k \in I_2$, then

$$d_{k+2} \geq d_{k+1} \geq d_k(1 + d_k) \geq d_k(1 + d_0);$$

if $a_k, b_k \in I_1$, then $a_{k+1}, b_{k+1} \in I_2$, so

$$d_{k+2} \geq d_{k+1}(1 + d_{k+1}) \geq d_{k+1}(1 + d_0) \geq d_k(1 + d_0).$$

Thus, in the both cases, we always have

$$d_{k+2} \geq d_k(1 + d_0).$$

By mathematical induction, for any positive integer m,

$$d_{2m} \geq d_0(1 + d_0)^m. \qquad (30.3)$$

When m is large enough, the right hand side of (30.3) is greater than 1, but $b_{2m} < 1$ since $a_{2m}, b_{2m} \in (0, 1)$. This is a contradiction.

Therefore, there exists a positive integer n such that a_{n-1} and b_{n-1} are in the two different subintervals.

12. Let $b_i = \frac{1}{n-1+a_i}$ with $i = 1, 2, \ldots, n$. Then $b_i < \frac{1}{n-1}$ and

$$a_i = \frac{1 - (n-1)b_i}{b_1}, \; i = 1, 2, \ldots, n.$$

Thus, the condition of the problem is transformed to

$$\sum_{i=1}^{n} \frac{1 - (n-1)b_i}{b_1} = \sum_{i=1}^{n} \frac{b_1}{1 - (n-1)b_i}.$$

In the following, we use proof by contradiction. Suppose that

$$b_1 + b_2 + \cdots + b_n < 1. \qquad (30.4)$$

The Cauchy inequality implies that for $i = 1, 2, \ldots, n$,

$$\sum_{j \neq i}[1 - (n-1)b_j] \sum_{j \neq i} \frac{1}{1 - (n-1)b_j} \geq (n-1)^2.$$

By (30.4), we have

$$\sum_{j \neq i}[1 - (n-1)b_j] < (n-1)b_i,$$

so

$$\sum_{j \neq i} \frac{1}{1 - (n-1)b_j} > \frac{n-1}{b_i}.$$

It follows that

$$\sum_{j \neq i} \frac{1 - (n-1)b_i}{1 - (n-1)b_j} > (n-1)\frac{1 - (n-1)b_i}{b_i}.$$

Summing up the above inequalities for $i = 1, 2, \ldots, n$, we obtain that

$$\sum_{i=1}^{n} \sum_{j \neq i} \frac{1 - (n-1)b_i}{1 - (n-1)b_j} > (n-1)\sum_{i=1}^{n} \frac{1 - (n-1)b_i}{b_i},$$

namely

$$\sum_{j=1}^{n}\sum_{j\neq i}\frac{1-(n-1)b_i}{1-(n-1)b_j} > (n-1)\sum_{i=1}^{n}\frac{1-(n-1)b_i}{b_i}. \qquad (30.5)$$

But $\sum_{i\neq j}[1-(n-1)b_i] < b_j(n-1)$ from (30.4), thus by using (30.5), we have

$$(n-1)\sum_{j=1}^{n}\frac{b_j}{1-(n-1)b_j} > (n-1)\sum_{i=1}^{n}\frac{1-(n-1)b_i}{b_i},$$

which gives a contradiction.

Solutions for Chapter 31

Construction Method

1. $\left(3, \frac{7}{2}\right)$.   According to the condition, we have $ac = b^2$ and $a + c = b(b-1)$.

   Construct the equation $x^2 - b(b-1)x + b^2 = 0$ about x.

   The question is then transformed to that the two roots of the equation are inside the intervals $(1, 3)$ and $(3, 7)$, respectively, which is equivalent to

   $$\begin{cases} 1 - b(b-1) + b^2 > 0, \\ 3^3 - 3b(b-1) + b^2 < 0, \\ 7^2 - 7b(b-1) + b^2 > 0 \end{cases} \Rightarrow b \in \left(3, \frac{7}{2}\right).$$

2. 12.   Construct line segments $OA, OB,$ and OC such that $|OA| = x, |OB| = \frac{y}{\sqrt{3}}, |OC| = z, \angle AOB = 90°,$ and $\angle BOC = 150°.$ Then $\angle COA = 120°,$ and $|AB| = |BC| = |CA| = 2.$

   From $S_{\triangle AOB} + S_{\triangle BOC} + S_{\triangle COA} = S_{\triangle ABC}$, we have

   $$\frac{1}{2}\frac{y}{\sqrt{3}}x + \frac{1}{2}\frac{y}{\sqrt{3}}z \sin 150° + \frac{1}{2}xz \sin 120°$$

   $$= \frac{1}{2} \cdot 2 \cdot 2 \sin 60°,$$

   from which $2xy + yz + 3xz = 12.$

3. 6.   Denote $A(0, 3), B(-\sqrt{3}, 0), C(\sqrt{3}, 0),$ and $P(x, y).$ Then

   $$f(x, y) = |PA| + |PB| + |PC|.$$

Note that $\angle ABO = \angle BAC = \angle ACO = 60°$, so $D(0,1)$ is a Fermat point of $\triangle ABC$. Thus, the minimum value of $f(x,y)$ is obtained when $P = D$, and the minimum value is

$$|DA| + |DB| + |DC| = 6.$$

4. We use Pythagorean numbers to make a construction. Pick any array of Pythagorean numbers (x, y, z) (not necessarily primitive). Let

$$a = x|4y^2 - z^2|, \ b = y|4x^2 - z^2|, \ c = 4xyz.$$

Then

$$
\begin{aligned}
a^2 + b^2 &= x^2(3y^2 - x^2)^2 + y^2(3x^2 - y^2)^2 \\
&= (x^2 + y^2)^3 = (z^3)^2, \\
a^2 + c^2 &= x^2(4x^2 + z^2)^2, \\
b^2 + c^2 &= y^2(4x^2 + z^2)^2.
\end{aligned}
$$

Since there are infinitely many arrays of Pythagorean numbers, so there are infinitely many arrays of three numbers that satisfy the condition.

In particular, when $x = 3, y = 4$, and $z = 5$, we have $a = 117, b = 44$, and $c = 240$, and furthermore,

$$117^2 + 44^2 = 125^2, \ 117^2 + 240^2 = 267^2, \ 44^2 + 240^2 = 241^2.$$

5. We use combinatorial numbers to construct a polynomial satisfying the condition. Let

$$P(x,y) = x + \frac{(x+y)(x+y+1)}{2}.$$

For any natural number n, since the sequence $\{C_r^2\}$ $(r \in \mathbf{N}_+)$ increases to infinity as r increases to infinity, there exists a unique natural number h such that

$$C_h^2 \le n < C_{h+1}^2. \tag{31.1}$$

Let $k = n - C_h^2$ and $m = h - k - 1$. Then h and k are natural numbers, $k < C_{h+1}^2 - C_h^2 = h$, and

$$P(k,m) = k + C_h^2 = n. \tag{31.2}$$

Conversely, if there are natural numbers k and m that satisfy (31.2), then (31.1) is satisfied ($h = k + m + 1$), hence h is uniquely determined. Therefore, $k = n - C_h^2$ and $m = h - k - 1$ are uniquely determined in succession.

In summary, there exists a polynomial satisfying the given condition.

6. Consider the following four sets:

$$A_i = \{\text{all positive integers of } S \text{ that can be divided by } i\},$$

where $i = 2, 3, 5,$ and 7. Denote $A = A_2 \cup A_3 \cup A_5 \cup A_7$. By the principle of inclusion and exclusion, it is easy to calculate that the number of elements in A is 216. Since any five numbers picked from A must have two in the same A_i, so they are not relatively prime. Thus, $n \geq 217$.

We show below that any subset containing 217 elements must have five numbers that are relatively prime. For this purpose, consider the following sets:

$$B_1 = \{1 \text{ and all prime numbers of } S\},$$
$$B_2 = \{2^2, 3^2, 5^2, 7^2, 11^2, 13^2\},$$
$$B_3 = \{2 \cdot 131, 3 \cdot 89, 5 \cdot 53, 7 \cdot 37, 11 \cdot 23, 13 \cdot 19\},$$
$$B_4 = \{2 \cdot 127, 3 \cdot 83, 5 \cdot 47, 7 \cdot 31, 11 \cdot 19, 13 \cdot 17\},$$
$$B_5 = \{2 \cdot 113, 3 \cdot 79, 5 \cdot 43, 7 \cdot 29, 11 \cdot 17\},$$
$$B_6 = \{2 \cdot 109, 3 \cdot 73, 5 \cdot 41, 7 \cdot 23, 11 \cdot 13\}.$$

It is easy to see that the number of elements of B_1 is 60. Let

$$B = B_1 \cup B_2 \cup B_3 \cup B_4 \cup B_5 \cup B_6.$$

Then the number of elements in B is 88, so the number of the elements in S that are not in B is $280 - 88 = 192$. Thus, among any 217 numbers from S, there are $217 - 192 = 25$ numbers that belong to B. Hence, there exists some i with $1 \leq i \leq 6$ such that the 217 numbers contain five numbers in B_i, and the five numbers are relatively prime.

To summarize, the natural number n satisfying the requirement is 217.

7. Define an auxiliary function

$$f(x) = [x] + \left[x + \frac{1}{n}\right] + \cdots + \left[x + \frac{n-1}{n}\right] - [nx].$$

Then for any $x \in \mathbf{R}$,

$$f\left(x + \frac{1}{n}\right) = \left[x + \frac{1}{n}\right] + \left[x + \frac{2}{n}\right] + \cdots + [x + 1] - [nx + 1]$$

$$= [x] + \left[x + \frac{1}{n}\right] + \cdots + \left[x + \frac{n-1}{n}\right] - [nx]$$

$$= f(x).$$

Thus, $f(x)$ is a periodic function of period $\frac{1}{n}$. To show that $f(x) \equiv 0$ for all real numbers x, it is enough to prove that $f(x) \equiv 0$ when $x \in \left[0, \frac{1}{n}\right)$, but this is obvious. hence the proposition is proved.

8. Assume that m is a positive odd number. Let $(\sqrt{2}+1)^m = \sqrt{2}x_m + y_m$, where $x_m, y_m \in \mathbf{N}_+$. By the binomial theorem, $(\sqrt{2}-1)^m = \sqrt{2}x_m - y_m$. Thus, $2x_m^2 - y_m^2 = 1$, namely $2x_m^2 = y_m^2 + 1$. It follows that $y_m^4 < 2x_m^2 y_m^2 = y_m^2(y_m^2 + 1) < (y_m^2 + 1)^2$. Hence,

$$y_m^2 < \sqrt{2}x_m y_m < y_m^2 + 1.$$

Let $n = x_m y_m$. Then $a_n = [\sqrt{2n}] = [\sqrt{2x_m y_m}] = y_m^2$.

9. When $x_1^2 + x_2^2 + x_3^2 = 1$, the original inequality is obviously true. When $x_1^2 + x_2^2 + x_3^2 < 1$, construct a quadratic function

$$
\begin{aligned}
f(t) &= (x_1^2 + x_2^2 + x_3^2 - 1)t^2 - 2(x_1 y_1 + x_2 y_2 + x_3 y_3 - 1)t \\
&\quad + (y_1^2 + y_2^2 + y_3^2 - 1) \\
&= (x_1 t - y_1)^2 + (x_2 t - y_2)^2 + (x_3 t - y_3)^2 - (t-1)^2.
\end{aligned}
$$

The graph of the function is a parabola that is open downward. Since

$$f(1) = (x_1 - y_1)^2 + (x_2 - y_2)^2 + (x_3 - y_3)^2 \geq 0,$$

so the parabola must have an intersection point with the x-axis, from which the discriminant

$$
\begin{aligned}
\Delta = 4(x_1 y_1 + x_2 y_2 + x_3 y_3 - 1)^2 - 4(x_1^2 + x_2^2 \\
+ x_3^2 - 1)(y_1^2 + y_2^2 + y_3^2 - 1) \geq 0.
\end{aligned}
$$

Therefore,

$$(x_1 y_1 + x_2 y_2 + x_3 y_3 - 1)^2 \geq (x_1^2 + x_2^2 + x_3^2 - 1)(y_1^2 + y_2^2 + y_3^2 - 1).$$

10. Construct a sequence as follows:

$$u_1 = a + b, \quad u_{k+1} = b + a(u_1 u_2 \cdots u_k) \ (k \in \mathbf{N}_+).$$

In the following, we show that any two terms of the sequence $\{u_n\}$ are relatively prime and each term is relatively prime with b.

We use mathematical induction. Obviously, $(u_1, b) = 1$. Suppose that u_k and u_i are relatively prime and u_i is relatively prime with

b $(1 \leq i \leq k)$. Then

$$(u_{k+1}, u_i) = (b + au_1u_2 \cdots u_k, u_i) = (b, u_i) = 1,$$

$$(u_{k+1}, b) = (au_1u_2 \cdots u_k, b) = 1.$$

Hence, the terms of the sequence $\{u_n\}$ are pairwise relatively prime, and each term of $\{u_n\}$ is relatively prime with b.

11. Construct a polynomial

$$f(x) = (x + a_1)(x + a_2) \cdots (x + a_{100}) - 1.$$

From the condition of the problem, $f(b_i) = 0$ for $i = 1, 2, \ldots, 100$. So, by the factorization theorem (since the degree of $f(x)$ is 100 and the leading term coefficient is 1), we see that

$$f(x) = (x - b_1)(x - b_2) \cdots (x - b_{100}).$$

Thus, we have the identity

$$(x + a_1)(x + a_2) \cdots (x + a_{100}) - 1 = (x - b_1)(x - b_2) \cdots (x - b_{100}).$$

Let $x = a_i$ $(i = 1, 2, \ldots, 100)$ and substitute it into the above. We obtain

$$-1 = (-1)^{100}(a_i + b_1)(a_i + b_2) \cdots (a_i + b_{100}),$$

that is, the product of all the numbers in the ith row is -1. Hence, the proposition is proved.

12. If A contains infinitely many positive integers that are relatively prime with respect to each other, then all such numbers constitute a subset B, which makes the conclusion valid.

If there exists a prime number p_1 that is a factor of infinitely many numbers of A, then the set

$$A_1 = \left\{ \frac{a}{p_1} \;\middle|\; \frac{a}{p_1} \in A \right\}$$

is an infinite one. Continuing in this way (by using A_1 in place of A, and so on), since the number of factors of every number of A is not greater than 1990, there must be an infinite set

$$A_k = \left\{ \frac{a}{p_1 p_2 \cdots p_k} \;\middle|\; \frac{a}{p_1 p_2 \cdots p_k} \in A \right\},$$

in which each prime number p_i is a factor of only finitely many numbers.

Pick any $a_1 \in A_k$. After picking $a_1, a_2, \ldots, a_n$ that are pairwise mutually prime, since each prime number is a factor of only finitely many numbers in A_k, there exists a_{n+1} in A_k, which is relatively prime with each of $a_1, a_2, \ldots, a_n$. Therefore, an infinite subset B_k of A_k is obtained such that elements of B_k are relatively prime with respect to each other.

Multiplying $p_1 p_2 \cdots p_k$ to each number of B_k, we obtain an infinite subset of A, in which the greatest common factor of every pair of numbers is always $p_1 p_2 \cdots p_k$.

Printed in the United States
by Baker & Taylor Publisher Services